World Geography

SOUTH & CENTRAL AMERICA

Regions | Physical Geography | Biogeography and Natural Resources
Human Geography | Economic Geography | Gazetteer

Second Edition

FLAGS OF THE WORLD

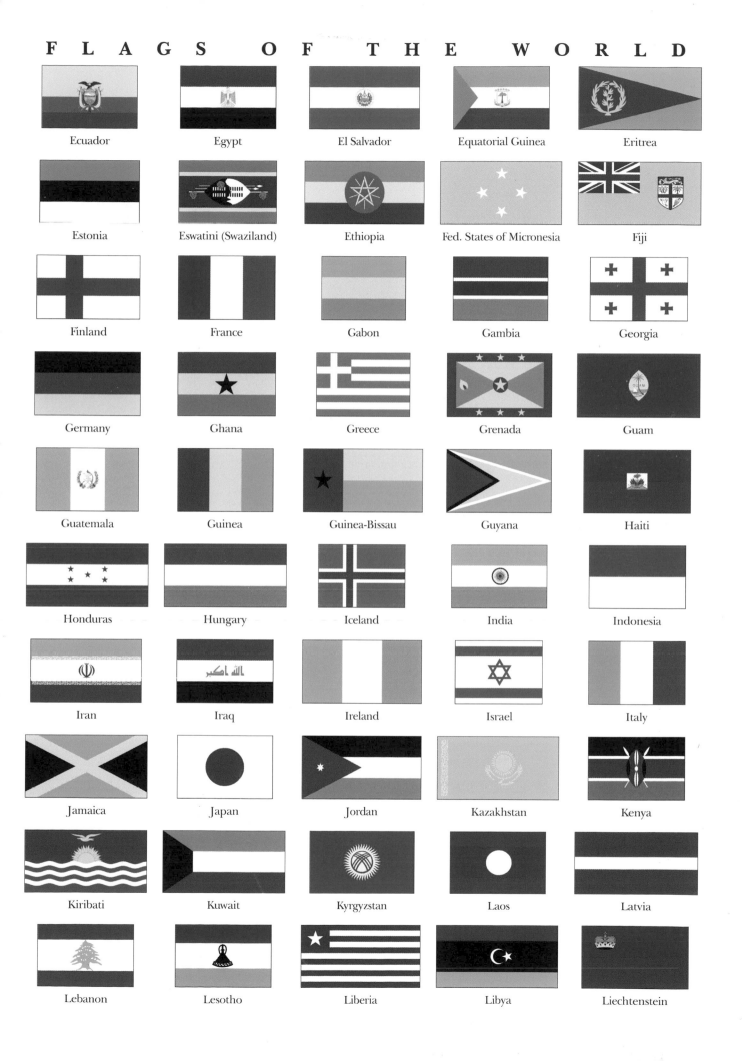

Ecuador

Egypt

El Salvador

Equatorial Guinea

Eritrea

Estonia

Eswatini (Swaziland)

Ethiopia

Fed. States of Micronesia

Fiji

Finland

France

Gabon

Gambia

Georgia

Germany

Ghana

Greece

Grenada

Guam

Guatemala

Guinea

Guinea-Bissau

Guyana

Haiti

Honduras

Hungary

Iceland

India

Indonesia

Iran

Iraq

Ireland

Israel

Italy

Jamaica

Japan

Jordan

Kazakhstan

Kenya

Kiribati

Kuwait

Kyrgyzstan

Laos

Latvia

Lebanon

Lesotho

Liberia

Libya

Liechtenstein

World Geography

SOUTH & CENTRAL AMERICA

Regions | Physical Geography | Biogeography and Natural Resources
Human Geography | Economic Geography | Gazetteer

Second Edition

Volume 1

Editor
Joseph M. Castagno
Educational Reference Publishing, LLC

SALEM PRESS
A Division of EBSCO Information Services, Inc.
Ipswich, Massachusetts
GREY HOUSE PUBLISHING

Cover photo: South America from outer space. Image by 1xpert

Publisher's Cataloging-In-Publication Data
(Prepared by The Donohue Group, Inc.)

Names: Castagno, Joseph M., editor.
Title: World geography / editor, Joseph M. Castagno, Educational Reference Publishing, LLC.
Description: Second edition. | Ipswich, Massachusetts : Salem Press, a division of EBSCO Information Services, Inc. ; Amenia, NY : Grey House Publishing, [2020] | Interest grade level: High school. | Includes bibliographical references and index. | Summary: A six-volume geographic encyclopedia of the world, continents and countries of each continent. In addition to physical geography, the set also addresses human geography including population distribution, physiography and hydrology, biogeography and natural resources, economic geography, and political geography. | Contents: Volume 1. South & Central America — Volume 2. Asia — Volume 3. Europe — Volume 4. Africa — Volume 5. North America & the Caribbean — Volume 6. Australia, Oceania & the Antarctic.
Identifiers: ISBN 9781642654257 (set) | ISBN 9781642654288 (v. 1) | ISBN 9781642654318 (v. 2) | ISBN 9781642654301 (v. 3) | ISBN 9781642654295 (v. 4) | ISBN 9781642654271 (v. 5) | ISBN 9781642654325 (v. 6)
Subjects: LCSH: Geography—Encyclopedias, Juvenile. | CYAC: Geography—Encyclopedias. | LCGFT: Encyclopedias.
Classification: LCC G133 .W88 2020 | DDC 910/.3—dc23

First Printing
PRINTED IN CANADA

CONTENTS

Publisher's Note

North Americans have long thought of the field of geography as little more than the study of the names and locations of places. This notion is not without a basis in fact: Through much of the twentieth century, geography courses forced students to memorize names of states, capitals, rivers, seas, mountains, and countries. Both students and educators eventually rebelled against that approach, geography courses gradually fell out of favor, and the future of geography as a discipline looked doubtful. Happily, however, the field has undergone a remarkable transformation, starting in the 1990s. Geography now has a bright and pivotal significance at all levels of education.

While learning the locations of places remains an important part of geography studies, educators recognize that place-name recognition is merely the beginning of geographic understanding. Geography now places much greater emphasis on understanding the characteristics of, and interconnections among, places. Modern students address such questions as how the weather in Brazil can affect the price of coffee in the United States, why global warming threatens island nations, and how preserving endangered plant and animal species can conflict with the economic development of poor nation.

World Geography, Second Edition, addresses these and many other questions. Designed and written to meet the needs of high school students, while being accessible to both middle school and undergraduate college students, these six volumes take an integrated approach to the study of geography, emphasizing the connections among world regions and peoples. The set's six volumes concentrate on major world regions: South and Central America; Asia; Europe; Africa, North America; and Australia, Oceania, and the Antarctic. Each volume begins with common overview information related to the geography, maps and mapmaking. The core essays in the volumes begin with an overview section to provide global context and then goes on to examine important geographic aspects of the regions in that area of the world: its physical geography; biogeography and natural resources; human geography (including its political geography); and economic geography. These essays range in length from three to ten pages. A gazetteer for the region indicates major political, geographic, and man-made features throughout the region.

A robust appendix found in each volume provides further information:

- The Earth in Space (The Solar System, Earth's Moon, The Sun and the Earth, The Seasons);
- Earth's Interior (Earths Internal Structure, Plate Tectonics, Volcanoes, Geologic Time Scale);
- Earth's Surface (Internal Geological Processes, External Processes, Fluvial and Karst Processes, Glaciation, Desert Landforms, Ocean Margins);
- Earth's Climates (The Atmosphere, Global Climates; Cloud Formation, Storms);
- Earth's Biological Systems (Biomes);
- Natural Resources (Soils, Water);
- Exploration and Transportation (Exploration and Historical Trade Routes, Road Transportation, Railways, Air Transportation);
- Energy and Engineering (Energy Sources, Alternative Energies, Engineering Projects);
- Industry and Trade (Manufacturing, Globalization of Manufacturing and Trade, Modern World Trade Patterns);
- Political Geography (Forms of Government, Political Geography, Geopolitics, National Park Systems);
- Boundaries and Time Zones (International Boundaries, Global Time and Time Zones);
- Global Education (Themes and Standards in Geography Education);
- Global Data (The World Gazetteer of Oceans and Continents, The World's Oceans and Seas, Major Land Areas of the World, Major Islands of the World, Countries of the World (including population and pollution density), Past and Projected World Population Growth, 1950-2050, The World's Largest Countries by Area, The World's Smallest Countries by Area, The World's Largest Countries by Population,

The World's Smallest Countries by Population, The World's Most Densely Populated Countries, The World's Least Densely Populated Countries, The World's Most Populous Cities, Major Lakes of the World, Major Rivers of the World, The Highest Peaks in Each Continent, Major Deserts of the World, Highest Waterfalls of the World).

- A Glossary, General Bibliography, and Index completes the backmatter.

The regional divisions in the set make it possible to study specific countries or parts of the world. Pairing the specific regional information, organized by regions, physical geography, biogeography and natural resources, human geography, economic geography, and a gazetteer, with global information makes it possible for students to see the connections not only between countries and places within the region, but also between the regions and the entire global system, all within a single volume.

To make this set as easy as possible to use, all of its volumes are organized in a similar fashion, with six major divisions—Regions (organized into subregions by volume), Physical Geography, Biogeography and Natural Resources, Human Geography, and Economic Geography. The number of subregions in each volume varies, depending upon the major world division being examined—Asia, for example, includes the following regions: China; Japan, Korea, and Taiwan; Southeast Asia; South Asia; Mongolia and Asian Russia; the Transcaucus; and the Middle East.

Physical geography considers a world region's physiography, hydrology, and climatology. Biogeography and natural resources explores renewable and nonrenewable resources, flora, and fauna. Human geography addresses the people, population distribution, culture regions, urbanization, and political geography of the area. Economic geography considers the regions agriculture, industries, engineering projects, transportation, trade, and communications.

Gazetteers include descriptive entries on hundreds of important places, especially those mentioned in the volume's essays. A typical entry gives the place name and location, indicating the category into which the place falls (mountain, river, city, country, lake, etc). The entries also include statistics relevant to the categories of place (height of mountains, length of rivers, population of cities and countries).

A feature new to this edition is the discussion questions included throughout the volume. These questions are meant to foster discussion and further research into the topics related to the history, current issues, and future concerns related to physical, human, economic, and political geography.

Both a physical and a social science, geography is unique among social sciences in the demands it makes for visual support. For this reason, *World Geography* contains more than 100 maps, more than 1300 photographs, and scores of other graphical elements. In addition, essays are punctuated with more than 500 textual sidebars and tables, which amplify information in the essays and call attention to especially important or interesting points.

Both English and metric measures are used throughout this set. In most instances, English measures are given first, followed by their metric equivalents in parentheses. It should be noted that in cases of measures that are only estimates, such as the areas of deserts or average heights of mountain ranges, the metric figures are often rounded off to estimates that may not be exact equivalents of the English-measure estimates. In order to enhance clarity, units of measure are not abbreviated in the text, with these exceptions: kilometers are abbreviated as km. and square kilometers as sq. km. This exception has been made because of the frequency with which these measures appear.

Reference works such as this would be impossible without the expertise of a large team of contributing scholars. This project is no exception. Salem Press would like to thank the more than 175 people who wrote the signed essays and contributed entries to the gazetteers. A full list of contributors follows this note. We recognize the efforts of Dr. Ray Sumner, of California's Long Beach City College, for the expertise and insights that she brought to the previous editions of this book, and which have formed the strong foundation for this new edition. We also acknowledge the work of the editor of this current volume, Joe Castagno, Educational Reference Publishing, LLC.

INTRODUCTION

When Henry Morton Stanley of the *New York Herald* shook the hand of David Livingstone on the shore of Central Africa's Lake Tanganyika in 1871, the moment represented the high point of geography to many people throughout the world. A Scottish missionary and explorer, Livingstone had been out of contact with the outside world for nearly two years, and European and American newspapers had buzzed with speculation about his disappearance. At that time, so little was known about the geography of the interior of Africa that Stanley's finding Livingstone was acclaimed as a brilliant triumph of explorations.

The field of geography in Stanley and Livingstone's time was—and to a large extent still is—synonymous with explorations. Stories of epic journeys, both historic and contemporary, continue to exert a powerful attraction on readers. Mountains, deserts, forest, caves, and glaciers still draw intrepid explorers, while even more armchair travelers are thrilled by accounts and pictures of these exploits and discoveries. We all love to travel—to the beach, into the mountains, to our great national parks, and to foreign countries. In the need and desire to explore our surroundings, we are all geographers.

Numerous geographical societies welcome both professional geographers and the general public into their membership, as they promote a greater knowledge and understanding of the earth. The National Geographic Society, founded in 1888 "for the increase and diffusion of geographical knowledge," has awarded more than 11,000 grants for scientific exploration and research. Each year, the society invests millions of dollars in expeditions and fieldwork related to environmental concerns and global geographic issues. The findings are recorded in the pages of the familiar yellow-bordered *National Geographic* magazine, now produced in 40 local-language editions in many countries around the world, publishing around 6.8 million copies monthly, with some 60 million readers. The magazine, along with the National Geographic International television network, reaches more than 135 million readers and viewers worldwide and has more than 85 million subscribers.

An even older geographical association is Great Britain's' Royal Geographical Society, which grew out of the Geographical Society of London, founded in 1830 with the "sole object" of promoting "that most important and entertaining branch of knowledge—geography." Over the century that followed, the Royal Geographical Society focused on exploration of the continents of Africa and Antarctica. In the society's London headquarters adjacent to the Albert Hall, visitors can still view such historic artifacts as David Livingstone's cap and chair, as well as diaries, sketches, and maps covering the great period of the British Empire and beyond. Today the society assists more than five hundred field expeditions every year.

With the aid of satellites and remote-sensing instruments, we can now obtain images and data from almost anywhere on Earth. However, remote and inaccessible places still invite the intrepid to visit and explore them in person. Although the outlines of the continents have now been completed, and their interiors filled in with details of mountains and rivers, cities and political boundaries, remote places still exert a fascination on modern urbanites.

The enchantment of tales about strange sights and courageous journeys has been with us since the ancient voyages of Homer's *Ulysses*, Marco Polo's travels to China, and the nautical expeditions of Christopher Columbus, Ferdinand Magellan, and James Cook. While those great travelers are from the remote past, the age of exploration is far from over—a fact repeatedly demonstrated by the modern Norwegian navigator Thor Heyerdahl. Moreover, new journeys of discovery are still taking place. In 1993, after dragging a sled wearily across the frigid wastes of Antarctica for more than three months, Sir Ranulph Twisleton-Wykeham-Fiennes announced that the age of exploration is not dead. Six years later, in 1999, the long-missing body of British mountain climber George Mallory was found on the slopes of Mount Everest, near whose top he had mysteriously vanished in 1924. That discovery sparked a new wave of admiration and respect for explorers of such courage and endurance.

How many people have been enthralled by the bravery of Antarctic explorer Robert Falcon Scott and the noble sacrifice his injured colleague Lawrence Oates made in 1912, when he gave up his life in order not to slow down the rest of the expedition? There can be no doubt that the thrills and the dangers of exploring find resonance among many modern readers.

The struggle to survive in environments hostile to human beings reminds us of the power of our planet Earth. Significant books on this theme have included Jon Krakauer's *Into Thin Air* (1998), an account of a disastrous expedition climbing Mount Everest, and Sebastian Junger's *The Perfect Storm* (1997), the story of the worst gale of the twentieth century and its effect on a fishing fleet off the East Coast of North America. *Endurance* (1998), the epic of Sir Ernest Shackleton's survival and leadership for two years on the frozen Arctic, attracts the same people who avidly read *Undaunted Courage* (1996) the story of Meriwether Lewis and William Clark's epic exploration of the Louisiana Purchase territories in the early nineteenth century. In 1997 *Seven Years in Tibet* premiered, a popular film about the Austrian Heinrich Harrer, who lived in Tibet in the mid-twentieth century. The more urban people become, the greater their desire for adventurous, remote places, a least vicariously, to raise the human spirit.

There are, of course, also scientific achievements associated with modern exploration. In November 1999, the elevation of Mount Everest, the world tallest peak was raised by 7 feet (2.1 meters) to a new height of 29,035 feet (8,850 meters) above sea level; the previously accepted height had been based on surveys made during the 1950s. This new value was the result of Global Positioning System (GPS) technology enabling a more accurate measurement than had been possible with land-based earthbound surveying equipment. A team of climbers supported by the National Geographic Society and the Boston Museum of Science was equipped with GPS equipment, which enabled a fifty-minute recording of data based on satellite signals. At the same time, the expedition was able to ascertain that Mount Everest is moving northeast, atop the Indo-Australian Plate, at a rate of approximately 2.4 inches (10 centimeters) per year.

In 2000, the International Hydrographic Organization named a "new" ocean, the Southern Ocean, which encompasses all the water surrounding Antarctica up to 60° south latitude. With an area of approximately 7.8 million square miles (20.3 million sq. km.), the Southern Ocean is about twice the size of the entire United States and ranks as the world's fourth largest ocean, after the Pacific, Atlantic, and Indian Oceans, but just ahead of the Arctic Ocean.

Despite the humanistic and scientific advantage of geographic knowledge, to many people today, geography is a subject where one merely memorized longs lists of facts dealing with "where" questions. (Where is Andorra? Where is Prince Edward Island? Where is Kalamazoo?) or "what" questions (What is the highest mountain in South America? What is the capital of Costa Rica?) This approach to the study of geography has been perpetuated by the annual National Geographic Bee, conducted in the United States each year for students in grades four through eight. Participants in the competition display an astonishing recall of facts but do not have the opportunity of showing any real geographic thought. To a geographer, such factual knowledge is simply a foundation for investigating and explaining the much more important questions dealing with "why"—"Why is the Sahara a desert?"

Geographers aim to understand why environments and societies occur where and as they do, and how they change. Geography must be seen as an integrative science; the collection of factual data and evidence, as in exploration, is the empirical foundation for deductive reasoning. This leads to the creation of a range of geographical methods, models, theories, and analytical approaches that serve to unify a very broad area of knowledge—the interaction between natural and human environments. Although geography as an academic discipline became established in nineteenth century Germany, there have always been geographers, in the sense of people curious about their world. Humans have always wanted to know about day and night, the shape of the earth, the nature of climates, differences in plants and animals, as well as what lies beyond the horizon. Today, as we hear about and actually experience the sweeping effects of globalization, we need more than ever to develop our geographic skills. Not only are we connected by economic ties to the countries of the world, but we must also appreciate the consequences of North America's high standard of living.

Political boundaries are artificial human inventions, but the natural world is one biosphere. As concern over global warming escalates, national leaders meet to seek a solution to the emission of greenhouse gases, rising ocean levels, and mass extinctions. Are we connected to our environment? At a time when the rate of species extinction is a hundred times above normal, and the human population is crowding in increasing numbers into huge urban centers, we have, nevertheless, taken time each year in April to celebrate Earth Day since 1970. We need now to realize that every day is Earth Day.

Geography languished in the United States in the 1960s, as social studies was taught with a history emphasis in schools. American students became alarmingly disadvantaged in geographic knowledge, compared with most other countries. Fortunately, members of the profession acted to restore geography to the curriculum. In 1984, the National Geographic Society undertook the challenge of restoring geography in the United States. The society turned to two organizations active in geographic education: The Association of American Geographers, the professional geographers" group with more than 10,000 members, mostly in higher education in the United States; and the National Council for Geographic Education that supports geography teaching at all levels—from kindergarten through university, with members that include U.S. and international teachers, professors, students, businesses, and others who support geography education. The council administers the Geographic Alliances, found in every state of the United Sates, with a national membership of about 120,000 schoolteachers. Together, they produced the "Guidelines in Geographic Education," which introduced the Five

Themes of Geography, to enhance the teaching of geography in schools. Using the themes of Location, Place, Human/Environment Interaction, Movement and Regions, teachers were able to plan and conduct lessons in which students encountered interesting real-world examples of the relevance and importance of geography. Continued research into geographic education led to the inclusion of geography in 1990 as one of the core subjects of the National Education Goals, along with English, mathematics, science, and history. .

Another milestone was the publication in 1994 of "Geography for Life," the national Geography Standards. The earlier Five Themes were subsumed under the new Six Essential Elements: The World in Spatial Terms; Places and Regions; Physical Systems; Human Systems; Environment Systems; Environment and Society; and The Uses of Geography. Eighteen geography standards are included, describing what a geographically informed person knows and understands. States, schools, and individual teachers have welcomed the new prominence of geography, and enthusiastically adopted new approaches to introduce the geography standards to new learners. The rapid spread of computer technology, especially in the field of Geographical Information Science, has also meant a new importance for spatial analysis, a traditional area of geographical expertise. No longer is geography seen as an outdated mass of useless or arcane facts; instead, geography is now seen, again, to be an innovative an integrative science, which can contribute to solving complex problems associated with the human-environmental relationship in the twenty-first century.

Geographers may no longer travel across uncharted realms, but there is still much we long to explore, to learn, and seek to understand, even if it is only as "armchair" geographers. This reference work, *World Geography*, will help carry readers on their own journeys of exploration.

Ray Sumner
Long Beach City College

Joseph M. Castagno
Educational Reference Publishing, LLC

CONTRIBUTORS

Emily Alward
Henderson, Nevada Public Library

Earl P. Andresen
University of Texas at Arlington

Debra D. Andrist
St. Thomas University

Charles F. Bahmueller
Center for Civic Education

Timothy J. Bailey
Pittsburg State University

Irina Balakina
Writer/Editor, Educational Reference Publishing LLC

David Barratt
Nottingham, England

Maryanne Barsotti
Warren, Michigan

Thomas F. Baucom
Jacksonville State University

Michelle Behr
Western New Mexico University

Alvin K. Benson
Brigham Young University

Cynthia Breslin Beres
Glendale, California

Nicholas Birns
New School University

Olwyn Mary Blouet
Virginia State University

Margaret F. Boorstein
C.W. Post College of Long Island University

Fred Buchstein
John Carroll University

Joseph P. Byrne
Belmont University

Laura M. Calkins
Palm Beach Gardens, Florida

Gary A. Campbell
Michigan Technological University

Byron D. Cannon
University of Utah

Steven D. Carey
University of Mobile

Roger V. Carlson
Jet Propulsion Laboratory

Robert S. Carmichael
University of Iowa

Joseph M. Castagno
Principal, Educational Reference Publishing LLC

Habte Giorgis Churnet
University of Tennessee at Chattanooga

Richard A. Crooker
Kutztown University

William A. Dando
Indiana State University

Larry E. Davis
College of St. Benedict

Ronald W. Davis
Western Michigan University

Cyrus B. Dawsey
Auburn University

Frank Day
Clemson University

M. Casey Diana
University of Illinois at Urbana-Champaign

Stephen B. Dobrow
Farleigh Dickinson University

Steven L. Driever
University of Missouri, Kansas City

Sherry L. Eaton
San Diego City College

Femi Ferreira
Hutchinson Community College

Helen Finken
Iowa City High School

Eric J. Fournier
Samford University

Anne Galantowicz
El Camino College

Hari P. Garbharran
Middle Tennessee State University

Keith Garebian
Ontario, Canada

Laurie A. B. Garo
University of North Carolina, Charlotte

Jay D. Gatrell
Indiana State University

Carol Ann Gillespie
Grove City College

Nancy M. Gordon
Amherst, Massachusetts

Noreen A. Grice
Boston Museum of Science

Johnpeter Horst Grill
Mississippi State University

Charles F. Gritzner
South Dakota State University

C. James Haug
Mississippi State University

Douglas Heffington
Middle Tennessee State University

Thomas E. Hemmerly
Middle Tennessee State University

Jane F. Hill
Bethesda, Maryland

Carl W. Hoagstrom
Ohio Northern University

Catherine A. Hooey
Pittsburg State University

Robert M. Hordon
Rutgers University

Kelly Howard
La Jolla, California

Paul F. Hudson
University of Texas at Austin

Huia Richard Hutton
University of Hawaii/Kapiolani Community College

Raymond Pierre Hylton
Virginia Union University

Solomon A. Isiorho
Indiana University/Purdue University at Fort Wayne

Ronald A. Janke
Valparaiso University

Albert C. Jensen
Central Florida Community College

Jeffry Jensen
Altadena, California

Bruce E. Johansen
University of Nebraska at Omaha

Kenneth A. Johnson
State University of New York, Oneonta

Walter B. Jung
University of Central Oklahoma

James R. Keese
California Polytechnic State University, San Luis Obispo

Leigh Husband Kimmel
Indianapolis, Indiana

Denise Knotwell
Wayne, Nebraska

James Knotwell
Wayne State College

Grove Koger
Boise Idaho Public Library

Alvin S. Konigsberg
State University of New York at New Paltz

Doris Lechner
Principal, Educational Reference Publishing LLC

Steven Lehman
John Abbott College

Denyse Lemaire
Rowan University

Dale R. Lightfoot
Oklahoma State University

Jose Javier Lopez
Minnesota State University

James D. Lowry, Jr.
East Central University

Jinshuang Ma
Arnold Arboretum of Harvard University Herbaria

Dana P. McDermott
Chicago, Illinois

Thomas R. MacDonald
University of San Francisco

Robert R. McKay
Clarion University of Pennsylvania

Nancy Farm Männikkö
L'Anse, Michigan

Carl Henry Marcoux
University of California, Riverside

Christopher Marshall
Unity College

Rubén A. Mazariegos-Alfaro
University of Texas/Pan American

Christopher D. Merrett
Western Illinois University

John A. Milbauer
Northeastern State University

Randall L. Milstein
Oregon State University

Judith Mimbs
Loftis Middle School

Karen A. Mulcahy
East Carolina University

B. Keith Murphy
Fort Valley State University

M. Mustoe
Omak, Washington

Bryan Ness
Pacific Union College

Kikombo Ilunga Ngoy
Vassar College

Joseph R. Oppong
University of North Texas

Richard L. Orndorff
University of Nevada, Las Vegas

Bimal K. Paul
Kansas State University

Nis Petersen
New Jersey City University

Mark Anthony Phelps
Ozarks Technical Community College

John R. Phillips
Purdue University, Calumet

Alison Philpotts
Shippensburg University

Julio César Pino
Kent State University

Timothy C. Pitts
Morehead State University

Carolyn V. Prorok
Slippery Rock University

P. S. Ramsey
Highland Michigan

Robert M. Rauber
University of Illinois at Urbana-Champaign

Ronald J. Raven
State University of New York at Buffalo

Neil Reid
University of Toledo

Susan Pommering Reynolds
Southern Oregon University

Nathaniel Richmond
Utica College

Edward A. Riedinger
Ohio State University Libraries

Mika Roinila
West Virginia University

Thomas E. Rotnem
Brenau University

Joyce Sakkal-Gastinel
Marseille, France

Helen Salmon
University of Guelph

Elizabeth D. Schafer
Loachapoka, Alabama

Kathleen Valimont Schreiber
Millersville University of Pennsylvania

Ralph C. Scott
Towson University

Guofan Shao
Purdue University

Wendy Shaw
Southern Illinois University, Edwardsville

R. Baird Shuman
University of Illinois, Champaign-Urbana

Sherman E. Silverman
Prince George's Community College

Roger Smith
Portland, Oregon

Robert J. Stewart
California Maritime Academy

Toby R. Stewart
Alamosa, Colorado

Ray Sumner
Long Beach City College

Paul Charles Sutton
University of Denver

Glenn L. Swygart
Tennessee Temple University

Sue Tarjan
Santa Cruz, California

Robert J. Tata
Florida Atlantic University

John M. Theilmann
Converse College

Virginia Thompson
Towson University

Norman J. W. Thrower
University of California, Los Angeles

Paul B. Trescott
Southern Illinois University

Robert D. Ubriaco, Jr.
Illinois Wesleyan University

Mark M. Van Steeter
Western Oregon University

Johan C. Varekamp
Wesleyan University

Anthony J. Vega
Clarion University

William T. Walker
Chestnut Hill College

William D. Walters, Jr.
Illinois State University

Linda Qingling Wang
University of South Carolina, Aiken

Annita Marie Ward
Salem-Teikyo University

Kristopher D. White
University of Connecticut

P. Gary White
Western Carolina University

Thomas A. Wikle
Oklahoma State University

Rowena Wildin
Pasadena, California

Donald Andrew Wiley
Anne Arundel Community College

Kay R. S. Williams
Shippensburg University

Lisa A. Wroble
Redford Township District Library

Bin Zhou
Southern Illinois University, Edwardsville

REGIONS

OVERVIEW

THE HISTORY OF GEOGRAPHY

The moment that early humans first looked around their world with inquiring minds was the moment that geography was born. The history of geography is the history of human effort to understand the nature of the world. Through the centuries, people have asked of geography three basic questions: What is Earth like? Where are things located? How can one explain these observations?

Geography in the Ancient World
In the Western world, the Greeks and the Romans were among the first to write about and study geography. Eratosthenes, a Greek scholar who lived in the third century BCE, is often called the "father of geography and is credited with first using the word geography (from the Greek words *ge*, which means "earth," and *graphe*, which means "to describe"). The ancient Greeks had contact with many older civilizations and began to gather together information about the known world. Some, such as Hecataeus, described the multitude of places and peoples with which the Greeks had contact and wrote of the adventures of mythical characters in strange and exotic lands. However, the ancient Greek scholars went beyond just describing the world. They used their knowledge of mathematics to measure and locate. The Greek scholars also used their philosophical nature to theorize about Earth's place in the universe.

One Greek scholar who used mathematics in the study of geography was Anaximander, who lived from 610 to 547 BCE. Anaximander is credited with being the first person to draw a map of the world to scale. He also invented a sundial that could be used to calculate time and direction and to distinguish the seasons. Eratosthenes is also famous for his mathematical calculations, in particular of the circumference of Earth, using observations of the Sun. Hipparchus, who lived around 140 BCE, used his mathematical skills to solve geographic problems and was the first person to introduce the idea of a latitude and longitude grid system to locate places.

Such early Greek philosophers as Plato and Aristotle were also concerned with geography. They discussed such issues as whether Earth was flat or spherical and if it was the center of the universe, and debated the nature of Earth as the home of humankind.

Whereas the Greeks were great thinkers and introduced many new ideas into geography, the Roman contribution was to compile and gather available knowledge. Although this did not add much that was new to geography, it meant that the knowledge of the ancient world was available as a base to work from and was passed down across the centuries. Geogra-

---

CURIOSITY: THE ROOT OF GEOGRAPHY

The earliest human beings, as they hunted and gathered food and used primitive tools in order to survive, must have had detailed knowledge of the geography of their part of the world. The environment could be a hostile place, and knowledge of the world meant the difference between life and death. Human curiosity took them one step further. As they lived in an ancient world of ice and fire, human beings looked to the horizon for new worlds, crossing continents and spreading out to all areas of the globe. They learned not only to live as a part of their environment, but also to understand it, predict it, and adapt it to their needs.

phy in the ancient world is often said to have ended with the great work of Ptolemy (Claudius Ptolemaeus), who lived from 90 to 168 CE. Ptolemy is best known for his eight-volume *Guide to Geography*, which included a gazetteer of places located by latitude and longitude, and his world map.

Geography in China

The study of geography also was important in ancient China. Chinese scholars described their resources, climate, transportation routes, and travels, and were mapping their known world at the same time as were the great Western civilizations. The study of geography in China begins in the Warring States period (fifth century BCE). It expands its scope beyond the Chinese homeland with the growth of the Chinese Empire under the Han dynasty. It enters its golden age with the invention of the compass in the eleventh century CE (Song dynasty) and peaks with fifteenth century CE (Ming dynasty) Chinese exploration of the Pacific under admiral Zheng He during the treasure voyages.

Geography in the Middle Ages

With the collapse of the Roman Empire in the fifth century CE, Europe entered into what is commonly known as the Early Middle Ages. During this time, which lasted until the fifteenth century, the geographic knowledge of the ancient world was either lost or challenged as being counter to Christian teachings. For example, the early Greeks had theorized that Earth was a sphere, but this was rejected during the Middle Ages. Scholars of the Middle Ages believed that the world was a flat disk, with the holy city of Jerusalem at its center.

The knowledge and ideas of the ancient world might have been lost if they had not been preserved by Muslim scholars. In the Islamic countries of North Africa and the Middle East, some of the scholarship of the ancient world was sheltered in libraries and universities. This knowledge was extensively added to as Muslims traveled and traded across the known world, gathering their own information.

Among the most famous Muslim geographers were Ibn Battutah, al-Idrisi, and Ibn Khaldun. Ibn Battutah traveled east to India and China in the fourteenth century. Al-Idrisi, at the command of King Roger II of Sicily, wrote *Roger's Book*, which systematically described the world. Information from *Roger's Book* was engraved on a huge planisphere (disk), crafted in silver; this once was considered a wonder of the world, but it is thought to have been destroyed. Ibn Khaldun (1332-1406) is best known for his written world history, but he also was a pioneer in focusing on the relationship of human beings to their environment.

The Age of European Exploration

Beginning in the fifteenth century, the isolation of Europe came to an end, and Europeans turned their attention to exploration. The two major goals of this sudden surge in exploration were to spread the Christian faith and to obtain needed resources. In 1418 Prince Henry the Navigator established a school for navigators and began to gather the tools and knowledge needed for exploration. He was the first of many Europeans to travel beyond the limits of the known world, mapping, describing, and cataloging all that they saw.

The great wave of European exploration brought new interest in geography, and the monumental works of the Greeks and Romans—so carefully preserved by Muslim scholars—were rediscovered and translated into Latin. The maps produced in the Middle Ages were of little use to the explorers who were traveling to, and beyond, the limits of the known world. Christopher Columbus, for example, relied on Ptolemy's work during his voyages to the Americas, but soon newer, more accurate maps were drawn and, for the first time, globes were made. A particularly famous map, which is still used as a base map, is the Mercator projection. On the world map produced by Gerardus Mercator (born Geert de Kremer) in 1569, compass directions appear as straight lines, which was a great benefit on navigational charts.

When the age of European exploration began, even the best world maps crudely depicted only a few limited areas of the world. Explorers quickly began to gather huge quantities of information, making detailed charts of coastlines, discovering new continents, and eventually filling in the maps of those continents with information about both the natural and human features they encountered. This age of exploration is often said to have ended when Roald Amundsen planted the Norwegian flag at the South Pole in 1911. At that time, the world map became complete, and human beings had mapped and explored every part of the globe. However, the beginning of modern geography is usually associated with the work of two nineteenth century German geographers: Alexander von Humboldt and Carl Ritter.

The Beginning of Modern Geography

The writings of Alexander von Humboldt and Carl Ritter mark a leap into modern geography, because these writers took an important step beyond the work of previous scholars. The explorers of the previous centuries had focused on gathering information, describing the world, and filling in the world map with as much detail as possible. Humboldt and Ritter took a more scientific and systematic approach to geography. They began not only to compile descriptive information, but also to ask why: Humboldt spent his lifetime looking for relationships among such things as climate and topography (landscape), while Ritter was intrigued by the multitude of connections and relationships he observed within human geographic patterns. Both Humboldt and Ritter died in 1859, ending a period when information-gathering had been paramount. They brought geography into a new age in which synthesis, analysis, and theory-building became central.

European Geography

After the work of Humboldt and Ritter, geography became an accepted academic discipline in Europe, particularly in Germany, France, and Great Britain. Each of these countries emphasized different aspects of geographic study. German geographers continued the tradition of the scientific view, using observable data to answer geographic questions. They also introduced the concept that geography could take a chorological view, studying all aspects, physical and human, of a region and of the interrelationships involved.

The chorological view came to dominate French geography. Paul Vidal de la Blache (1845-1918) was

> ### NATIONAL GEOGRAPHIC SOCIETY AND GEOGRAPHIC RESEARCH
>
> In 1888 the National Geographic Society was founded to support the "increase and diffusion of geographic knowledge" of the world. In its first 110 years, the society funded more than five thousand expeditions and research projects with more than 6,500 grants. By the 1990s it was the largest such foundation in the world, and the results of its funded projects are found on television programs, video discs, video cassettes, and books, as well as in the *National Geographic* magazine, established in 1888. Its productions are cutting-edge resources for information about archaeology, ethnology, biology, and both cultural and physical geography.

the most prominent French geographer. He advocated the study of small, distinct areas, and French geographers set about identifying the many regions of France. They described and analyzed the unique physical and human geographic complex that was to be found in each region. An important concept that emerged from French geography was "possibilism." German geographers had introduced the notion of environmental determinism—that human beings were largely shaped and controlled by their environments. Possibilism rejected the concept of environmental determinism, asserting that the relationship between human beings and the environment works in two directions: The environment creates both limits and opportunities for people, but people can react in different ways to a given environment, so they are not controlled by it.

British geographers, influenced by the French approach, conducted regional surveys. British regional studies were unique in their emphasis on planning and geography as an applied science. From this work came the concept of a functional region—an area that works together as a unit based on interaction and interdependence.

American Geography

Prior to World War II, only a small group of people in the United States called themselves geographers. They were mostly influenced by German ideas, but the nature of geography was hotly debated. Two schools of geographers were philosophical adversaries. The Midwestern School, led by Richard Hartshorne, believed that description of unique regions was the central task of geography.

The Western (or Berkeley) School of geography, led by Carl Sauer, agreed that regional study was important, but believed it was crucial to go beyond description. Sauer and his followers included genesis and process as important elements in any study. To understand a region and to know where it is going, they argued, one must look at its past and how it got to its present state.

In the 1930s, environmental determinism was introduced to U.S. geography but ultimately was rejected. Although geography in both Europe and the United States was essentially an all-male discipline, the United States produced the first famous woman geographer, Ellen Churchill Semple (1863-1932).

World War II illustrated the importance of geographic knowledge, and after the war came to an end in 1945, geographers began to come into their own in the United States. From the end of World War II to the early 1960s, U.S. geographers produced many descriptive regional studies.

In the early 1960s, what is often called the quantitative revolution occurred. The development of computers allowed complex mathematical analysis to be performed on all kinds of geographic data, and geographers began to analyze a wide range of problems using statistics. There was great enthusiasm for this new approach to geography at first, but beginning in the 1970s, many people considered a purely mathematical approach to be somewhat sterile and thought it left out a valuable human element.

In the 1980s and 1990s, many new ways to look at geographic issues and problems were developed, including humanism, behaviorism, Marxism, feminism, realism, structuration, phenomenology, and postmodernism, all of which bring human beings back into focus within geographical studies.

Geography in the Twenty-first Century

Geography increasingly uses technology to analyze global space and answer a wide range of questions related to a host of concerns including issues related to the environment, climate change, population, rising sea levels, and pollution. The Geographic Information System (GIS), in particular, provides a powerful way for people trained in geography to understand geographic issues, solve geographic problems, and display geographic information. Geographers continue to adopt a wide variety of philosophies, approaches, and methods in their quest to answer questions concerning all things spatial.

Wendy Shaw

MAPMAKING IN HISTORY

Cartography is the science or art of making maps. Although workers in many fields have a concern with cartography and its history, it is most often associated with geography.

Maps of Preliterate Peoples

The history of cartography predates the written record, and most cultures show evidence of mapping skills. The earliest surviving maps are those carved in stone or painted on the walls of caves, but modern preliterate peoples still use a variety of materials to express themselves cartographically. For example, the Marshall Islanders use palm fronds, fiber from coconut husks (coir), and shells to make sea charts for their inter-island navigation. The Inuit use animal skins and driftwood, sometimes painted, in mapping. There is a growing interest in the cartography of early and preliterate peoples, but some of their maps do not fit readily into a more traditional concept of cartography.

Mapping in Antiquity

Early literate peoples, such as those of Egypt and Mesopotamia, displayed considerable variety in their maps and charts, as shown by the few maps from these civilizations that still exist. The early Egyptians painted maps on wooden coffin bases to assist the departed in finding their way in the afterlife; they also made practical route maps for their mining operations. It is thought that geometry developed from the Egyptians' riverine surveys. The Babylonians made maps of different scales, using clay tablets with cuneiform characters and stylized symbols, to create city plans, regional maps, and "world" maps. They also divided the circle in the sexigesimal system, an idea they may have obtained from India and that is commonly used in cartography to this day.

The Greeks inherited ideas from both the Egyptians and the Mesopotamians and made signal contributions to cartography themselves. No direct evidence of early Greek maps exists, but indirect evidence in texts provides information about their cosmological ideas, culminating in the concept of a perfectly spherical earth. This they attempted to measure and divide mathematically. The idea of climatic zones was proposed and possibly mapped, and the large known landmasses were divided into first two continents, then three.

Perhaps the greatest accomplishment of the early Greeks was the remarkably accurate measurement of the circumference of Earth by Eratosthenes (276-196 BCE). Serious study of map projections began at about this time. The gnomonic, orthographic, and stereographic projections were invented before the Christian era, but their use was confined to astronomy in this period. With the possible single exception of Aristarchus of Samos, the Greeks believed in a geocentric universe. They made globes (now lost) and regional maps on metal; a few map coins from this era have survived.

Later Greeks carried on these traditions and expanded upon them. Claudius Ptolemy invented two projections for his world maps in the second century CE. These were enormously important in the European Renaissance as they were modified in the light of new overseas discoveries. Ptolemy's work is known mainly through later translations and reconstructions, but he compiled maps from Greek and Phoenician travel accounts and proposed sectional maps of different scales in his *Geographia*. Ptolemy's prime meridian (0 degrees longitude) in the Canary Islands was generally accepted for a millennium and a half after his death.

Roman cartography was greatly influenced by later Greeks such as Ptolemy, but the Romans themselves improved upon route mapping and surveying. Much of the Roman Empire was subdivided by instruments into hundredths, of which there is a cartographic record in the form of marble tablets. In Rome, a small-scale map of the world known to the Romans was made on metal by Marcus Vipsanius Agrippa, the son-in-law of Augustus Caesar, and displayed publicly. This map no longer exists, however.

Cartography in Early East Asia

As these developments were taking place in the West, a rich cartographic tradition developed in Asia, particularly China. The earliest survey of China (Yu Kung) is approximately contemporaneous with the oldest reported mapmaking activity of the Greeks. Later, maps, charts, and plans accompanied Chinese texts on various geographical themes. Early rulers of China had a high regard for cartography—the science of princes. A rectangular grid was introduced by Chang Heng, a contemporary of Ptolemy, and the south-pointing needle was used for mapmaking in China from an early date.

These traditions culminated in Chinese cartographic primacy in several areas: the earliest printed maps (about 1155 CE), early printed atlases, and terrestrial globes (now lost). Chinese cartography greatly influenced that in other parts of Asia, particularly Korea and Japan, which fostered innovations of their own. It was only after the introduction of ideas from the West, in the Renaissance and later, that Asian cartographic advances were superseded.

Islamic Cartography

A link between China and the West was provided by the Arabs, particularly after the establishment of Islam. It was probably the Arabs who brought the magnetized needle to the Mediterranean, where it was developed into the magnetic compass.

Some scholars have argued that the Arabs were better astronomers than cartographers, but the Arabs did make several clear advances in mapmaking. Both fields of study were important in Muslim science, and the astrolabe, invented by the Greeks in antiquity but developed by the Arabs, was used in both their astronomical and terrestrial surveys. They made and used many maps, as indicated by the output of their most famous cartographer, al-Idrisi (who lived about 1100–1165). Some of his work still exists, including a zonal world map and detailed charts of the Mediterranean islands.

At about the same time, the magnetic compass was invented in the coastal cities of Italy, which gave rise to advanced navigational charts, including information on ports. These remarkably accurate charts were used for navigating in the Mediterranean Sea. They were superior to the European maps of the Middle Ages, which often were concerned with religious iconography, pilgrimage, and crusade. The scene was now set for the great overseas discoveries of the Europeans, which were initiated in Portugal and Spain in the fifteenth century.

In the next four centuries, most of the coasts of the world were visited and mapped. The early, projectionless navigational charts were no longer adequate, so new projections were invented to map the enlarged world as revealed by the European overseas explorations. The culmination of this activity was the development of the projection, in 1569, of Gerardus Mercator, which bears his name and is of special value in navigation.

Early Modern Mapmaking
Europeans began mapping their own countries with greater accuracy. New surveying instruments were invented for this purpose, and a great land-mapping activity was undertaken to match the worldwide coastal surveys. For about a century, the Low Countries of Belgium, Luxembourg, and the Netherlands dominated the map and chart trades, producing beautiful hand-colored engraved sheet wall maps and atlases.

France and England established new national observatories, and by the middle of the seventeenth century, the Low Countries had been eclipsed by France in surveying and making maps and charts. The French adopted the method of triangulation of Mercator's teacher, Gemma Frisius. Under four generations of the Cassini family, a topographic survey of France more comprehensive than any previous survey was completed. Rigorous coastal surveys were undertaken, as well as the precise measurement of latitude (parallels).

The invention of the marine chronometer by John Harrison made it possible for ships at sea to determine longitude. This led to the production of charts of all the oceans, with England's Greenwich eventually being adopted as the international prime meridian.

Quantitative, thematic mapping was advanced by astronomer Edmond Halley (1656–1742) who produced a map of the trade winds; the first published magnetic variation chart, using isolines; tidal charts; and the earliest map of an eclipse. The Venetian Vincenzo Coronelli made globes of greater beauty and accuracy than any previous ones. In the German lands, the study of map projections was vigorously pursued. Johann H. Lambert and others invented a number of equal-area projections that were still in use in the twentieth century.

Ideas developed in Europe were transmitted to colonial areas, and to countries such as China and Russia, where they were grafted onto existing cartographic traditions and methods. The oceanographic explorations of the British and the French built on the earlier charting of the Pacific Ocean and its islands by native navigators and the Iberians.

Nineteenth Century Cartography
Cartography was greatly diversified and developed in the nineteenth century. Quantitative, thematic mapping was expanded to include the social as well as the physical sciences. Alexander von Humboldt used isolines to show mean air temperature, a method that later was applied to other phenomena. Contour lines gradually replaced less quantitative methods of representing terrain on topographic maps. Such maps were made of many areas, for ex-

ample India, which previously had been poorly mapped.

Extraterrestrial (especially lunar) mapping, had begun seriously in the preceding two centuries with the invention of the telescope. It was expanded in the nineteenth century. In the same period, regular national censuses provided a large body of data that could be mapped. Ingenious methods were created to express the distribution of population, diseases, social problems, and other data quantitatively, using uniform symbols.

Geological mapping began in the nineteenth century with the work of William Smith in England, but soon was adopted worldwide and systematized, notably in the United States. The same is true of transportation maps, as the steamship and the railway increased mobility for many people. Faster land travel in an east-west direction, as in the United States, led to the official adoption of Greenwich as the international prime meridian at a conference held in Washington, D.C., in 1884. Time zone maps were soon published and became a feature of the many world atlases then being published for use in schools, offices, and homes.

A remarkable development in cartography in the nineteenth century was the surveying of areas newly occupied by Europeans. This occurred in such places as the South American republics, Australia, and Canada, but was most evident in the United States. The U.S. Public Land Survey covered all areas not previously subdivided for settlement. Property maps arising from surveys were widely available, and in many cases, the information was contained in county and township atlases and maps.

Modern Mapping and Imaging

Cartography was revolutionized in the twentieth century by aerial photography, sonic sounding, satellite imaging, and the computer. Before those developments, however, Albrecht Penck proposed an ambitious undertaking—an International Map of the World (IMW). Cartography historically had been a nationalistic enterprise, but Penck suggested a map of the world in multiple sheets produced cooperatively by all nations at the scale of 1:1,000,000 with uniform symbols. This was started in the first half of the twentieth century but was not completed, and was superseded by the World Aeronautical Chart (WAC) project, at the same scale, during and after World War II.

The WAC project owed its existence to flight information made available following the invention of the airplane. Both photography and balloons were developed before the twentieth century, but the new, heavier-than-air craft permitted overlapping aerial photographs to be taken, which greatly facilitated the mapping process. Aerial photography revolutionized land surveys—maps could be made at less cost, in less time, and with greater accuracy than by previous methods. Similarly, marine surveying was revolutionized by the advent of sonic sounding in the second half of the twentieth century. This enabled mapping of the floor of the oceans, essentially unknown before this time.

Satellite imaging, especially continuous surveillance by Landsat since 1972, allows temporal monitoring of Earth. The computer, through Geographical Information Systems (GIS) and other technologies, has greatly simplified and speeded up the mapping process. During the twentieth century, the most widely available cartographic product was the road map for travel by automobile.

Spatial information is typically accessed through apps on computers and mobile devices; traditional maps are becoming less common. The new media also facilitate animated presentations of geographical and extraterrestrial distributions. Cartographers remain responsive to the opportunities provided by new technologies, materials, and ideas.

Norman J. W. Thrower

Mapmaking and New Technologies

The field of geography is concerned primarily with the study of the curved surface of Earth. Earth is huge, however, with an equatorial radius of 3,963 miles (6,378 km.). How can one examine anything more than the small patch of earth that can be experienced at one time? Geographers do what scientists do all of the time: create models. The most common model of Earth is a globe—a spherical map that is usually about the size of a basketball.

A globe can show physical features such as rivers, oceans, the continents, and even the ocean floor. Political globes show the division of Earth into countries and states. Globes can even present views of the distant past of Earth, when the continents and oceans were very different than they are today. Globes are excellent for learning about the distributions, shapes, sizes, and relationships of features of Earth. However, there are limits to the use of globes.

How can the distribution of people over the entire world be described at one glance? On a globe, the human eye can see only half of Earth at one time. What if a city planner needs to map every street, building, fire hydrant, and streetlight in a town? To fit this much detail on a globe, the globe might have to be bigger than the town being mapped. Globes like these would be impossible to create and to carry around. Instead of having to hire a fleet of flatbed trucks to haul oversized globes, the curved surface of the globe can be transformed to a flat plane.

The method used to change from a curved globe surface to a flat map surface is called a map projection. There are hundreds of projections, from simple to extremely complex and dating from about two thousand years ago to projections being invented today. One of the oldest is the gnomonic projection. Imagine a clear globe with a light inside. Now imagine holding a piece of paper against the surface of the globe. The coastlines and parallels of latitude and meridians of longitude would show through the globe and be visible on the paper. Computers can do the same thing because there are mathematical formulas for nearly all map projections.

Geometric Models for Map Projections

One way to organize map projections is to imagine what kind of geometric shape might be used to create a map. Like the paper (a plane surface) against the globe described above, other useful geometric shapes include a cone and a cylinder. When the rounded surface of any object, including Earth, is flattened there must be some stretching, or tearing. Map projections help to control the amount and kinds of distortion in maps. There are always a few exceptions that cannot be described in this way, but using geometric shapes helps to classify projections into groups and to organize the hundreds of projections.

Another way to describe a map projection is to consider what it might be good for. Some map projections show all of the continents and oceans at their proper sizes relative to one another. Another type of projection can show correct distances between certain points.

Map Projection Properties

When areas are retained in the proper size relationships to one another, the map is called an equal-area map, and the map projection is called an equal-area projection. Equal-area (also called equivalent or homolographic) maps are used to measure areas or view densities such as a population density.

If true angles are retained, the shapes of islands, continents, and oceans look more correct. Maps made in this way are called conformal maps or conformal map projections. They are used for navigation, topographic mapping, or in other cases when it is important to view features with a good representation of shape. It is impossible for a map to be both equal-area and conformal at the same time. One or the other must be selected based on the needs of the map user or mapmaker.

One special property—distance—can only be true on a few parts of a map at one time. To see how far it is between places hundreds or thousands of miles apart, an equidistant projection should be used. There will be several lines along which distance is true. The azimuthal equidistant projection shows true distances from the center of the map outward. Some map projections do not retain any of these properties but are useful for showing compromise views of the world.

Modern Mapmaking

Modern mapmaking is assisted from beginning to end by digital technologies. In the past, the paper map was both the primary means for communicating information about the world and the database used to store information. Today, the database is a digital database stored in computers, and cartographic visualizations have taken the place of the paper map. Visualizations may still take the form of paper maps, but they also can appear as flashes on computer screens, animations on local television news programs, and even on screens within vehicles to help drivers navigate. Communication of information is one of the primary purposes of making maps. Mapping helps people to explore and analyze the world.

Making maps has become much easier and the capability available to many people. Desktop mapping software and Internet mapping sites can make anyone with a computer an instant cartographer. The maps, or cartographic visualizations, might be quite basic but they are easy to make. The procedures that trained cartographers use to make map products vary in the choice of data, software, and hardware, but several basic design steps should always take place.

First, the purpose and audience for whom the map is being made must be clear. Is this to be a general reference map or a thematic map? What image should be created in the mind of the map reader? Who will use the map? Will it be used to teach young children the shapes of the continents and oceans, or to show scientists the results of advanced research? What form will the cartographic visualization take?

SLIDING ROCKS GET DIGITAL TREATMENT

Dr. Paula Messina studied the trails of rocks that slide across the surface of a flat playa in Death Valley, California. The sliding rocks have been studied in the past, but no one had been able to say for certain how or when the rocks moved. It was unclear whether the rocks were caught in ice floes during the winter, were blown by strong winds coming through the nearby mountains, or were moved by some other method.

Messina gave the mystery a totally digital treatment. She mapped the locations of the rocks and the rock trails using the global positioning system (GPS) and entered her rock trail data into a geographic information system (GIS) for analysis. She was able to determine that ice was not the moving agent by studying the pattern of the trails. She also used digital elevation models (DEM) and remotely sensed imagery to model the environment of the playa. She reported her results in the form of maps using GIS' cartographic output capabilities. While she did not solve completely the mystery of the sliding rocks, she was able to disprove that winter ice caused the rocks to slide along together in rafts and that there are wind gusts strong enough to move the biggest rock on the playa.

Will it be a paper map, a graphic file posted to the Internet, or a video?

The answers to these questions will guide the cartographer in the design process. The design process can be broken down into stages. In the first stage of map design, imagination rules. What map type, size and shape, basic layout, and data will be used? The second stage is more practical and consists of making a specific plan. Based on the decisions made in the first stage, the symbols, line weights, colors, and text for the map are chosen. By the end of this stage, there should be a fairly clear plan for the map. During the third stage, details and specifications are finalized to account for the production method to be used. The actual software, hardware, and methods to be used must all be taken into consideration.

What makes a good map? Working in a digital environment, a mapmaker can change and test various designs easily. The map is a good one when it communicates the intended information, is pleasing to look at, and encourages map readers to ask thoughtful questions.

New Technologies

Mapping technology has gone from manual to magnetic, then to mechanical, optical, photochemical, and electronic methods. All of these methods have overlapped one another and each may still be used in some map-making processes. There have been recent advances in magnetic, optical, and most of all, electronic technologies.

All components of mapping systems—data collection, hardware, software, data storage, analysis, and graphical output tools—have been changing rapidly. Collecting location data, like mapping in general, has been more accessible to more people. The development of the Global Positioning System (GPS), an array of satellites orbiting Earth, gives anyone with a GPS receiver access to location information, day or night, anywhere in the world. GPS receivers are also found in planes, passenger cars, and even in the backpacks of hikers.

Satellites also have helped people to collect data about the world from space. Orbiting satellites collect images using visible light, infrared energy, and other parts of the electromagnetic spectrum. Active sensing systems send out radar signals and create images based on the return of the signal. The entire world can be seen easily with weather satellites, and other specialized satellite imagery can be used to count the trees in a yard.

These great resources of data are all stored and maintained as binary, computer-readable information. Developments in laser technology provide large amounts of storage space on media such as optical disks and compact disks. Advances in magnetic technology also provide massive storage capability in the form of tape storage, hard drives, and cloud storage. This is especially important for saving the large databases used for mapping.

Computer hardware and software continue to become more powerful and less expensive. Software continues to be developed to serve the specialized needs that mapping requires. Just as word processing software can format a paper, check spelling and grammar, draw pictures and shapes, import tables and graphics, and perform dozens of other functions, specialized software executes maps. The most common software used for mapping is called Geographic Information System (GIS) software. These systems provide tools for data input and for analysis and modeling of real-world spatial data, and provide cartographic tools for designing and producing maps.

Karen A. Mulcahy

SOUTH AMERICA

South America is a region of great diversity, contrast, and change. Students in the United States often mistakenly perceive South America in stereotypes, depicting it as a tropical land inhabited by poor peasant farmers, bound to Roman Catholic traditionalism, and shaken by revolution, military dictators, and drug lords. The physical and human diversity of South America makes it possibly the most heterogeneous of all the continents; the least understood; and thanks in part to its historical and cultural links to Europe and North America, one of the most interesting world regions to study and visit.

Physical Dimensions

South America covers an area of 6.9 million square miles (17.9 million sq. km.), which is about three-quarters the size of North America. The continent's northernmost point is Punta de Gallinas, on the Caribbean Sea at 12 degrees north latitude. South America extends southward 4,500 miles (7,200 km.) to the frigid Antarctic waters off Cape Horn, located at 56 degrees south latitude. Apart from Antarctica, South America extends farther south than any other continent. The widest east-west distance is 3,300 miles (5,300 km.) between João Pessoa in Brazil and Punta Pariñas in Peru. The Tropic of Capricorn bisects South America, placing about 55 percent of the continent within the tropics, but leaving a significant portion in the middle latitudes. The equator passes through Ecuador, Colombia, and Brazil.

Physical Geography

South America has virtually every physiographic, climatic, and vegetation type found on Earth. Among its notable landforms are the Andes Mountains, which span the entire western edge of the continent, and three of the world's top five river systems in terms of volume of water. The mighty Amazon River flows through Brazil but has tributaries originating in eight other countries. The Orinoco River drains northern South America. The Paraná River system drains southern Brazil, Paraguay, and Argentina, and includes the Gran Pantanal, which is the largest wetland area in the world, five times the size of the Florida Everglades. South America's great plains include the Llanos of Venezuela and Colombia, the Gran Chaco of Paraguay, and the fertile Pampas region of Argentina.

Its vast size, variety of landforms, and large latitudinal span lend the continent a truly remarkable variation in climate and vegetation. A large portion of the continent lies within the tropics. The humid areas receive up to 160 inches (4,100 millimeters) of rainfall a year and are blanketed in a thick rainforest canopy. But perhaps surprisingly, 45 percent of the continent lies outside the tropics.

Central Chile enjoys a Mediterranean-type climate similar to that of southern California. Southern Chile has a marine west coast climate not unlike that of the coasts of Oregon and Washington. Southern Brazil and the Buenos Aires area of Argentina have a humid subtropical climate similar to the mid-Atlantic states in the United States.

Arid regions include the steppes of the Argentine Patagonia and the bone-dry Atacama Desert of southern Peru and northern Chile, where decades can pass without measurable rainfall. In the Andes Mountains, elevation greatly influences temperature, precipitation, and vegetation. On the equator, areas at sea level are tropical, the mountain valleys have pleasant springlike weather, and the high peaks are covered in tundra and frozen snowcaps.

Amazon Basin

The Amazon Basin contains the world's largest rain forest and about one-fifth of the world's freshwater.

PANAMA

VENEZUELA

GUYANA

SURINAME

FRENCH
GUIANA

North

Atlantic

Ocean

Medellin •

• Bogotá
★

Cali •

**Galápagos
Islands**

Quito ★

COLOMBIA

Amazon   Basin

Belem •

ECUADOR

PERU

BRAZIL

A N D E S

Lima ★

BOLIVIA
★
La Paz

Brasilia
★

Arica •

South

Pacific

Ocean

PARAGUAY

Sao Paulo •

Rio de Janeiro •

Asuncion ★

M O U N T A I N S

CHILE

ARGENTINA

Santiago ★

URUGUAY
★
★ Montevideo

Buenos
Aires

South

Atlantic

Ocean

Falkland Islands

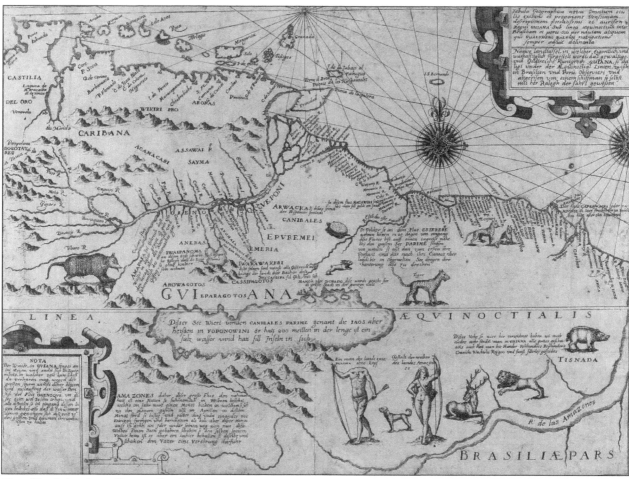

Map of South America as illustrated by Theodor de Bry, an engraver, goldsmith, editor and publisher, famous for his depictions of early European expeditions to the Americas. (Herzog August Bibliothek, Wolfenbüttel)

The Amazon River carries the greatest volume of water of any river on Earth, emptying 46 million gallons (175 million liters) per second at its mouth. At 4,000 miles (6,400 km.) in length, the Amazon River is the world's second-longest river—after Africa's Nile—stretching from the slopes of the Andes Mountains eastward to the Atlantic Ocean.

The rain forests of the Amazon contain more species of plants and animals than any other place on Earth. A single patch of rain forest about the size of a typical American shopping mall might contain 12 million varieties of insects, 500 varieties of birds, 100 varieties of other animals, and 250 varieties of trees. By contrast, all of North America has about 1,000 species of trees. This vast forested area also plays a vital role as one of Earth's "lungs," absorbing carbon dioxide and releasing oxygen into the air. To some, the Amazon Basin is perceived as a

frontier with untapped potential. Due to development pressures from logging, agriculture, cattle ranching, mining, road building, and poorly planned settlement schemes, about 17 percent of the Amazonian rain forest has been destroyed over the past fifty years, and the loss rate is on the rise. The World Wildlife Fund (WWF) estimates that 27 percent—more than a quarter—of the Amazon biome will be without trees by 2030 if the current rate of deforestation continues.

Population and Culture

South America contains thirteen countries: Argentina, Bolivia, Brazil, Chile, Colombia, Ecuador, French Guiana, Guyana, Paraguay, Peru, Suriname, Uruguay, and Venezuela. The size of each country varies widely. Brazil is the largest, with a land area of 3.3 million square miles (8.5 million sq. km.). This

Nicaraguan women wearing the Mestizaje costume, which is a traditional costume worn to dance the Mestizaje dance. The costume demonstrates the Spanish influence upon Nicaraguan clothing. (Image by Omar Caldera)

makes it slightly smaller than the United States, and the fifth-largest country in the world. French Guiana is the smallest, about the size of Maine. The total population of South America is 414 million. Brazil has the largest population with 208 million, while the Guianas (French Guiana, Guyana, and Suriname) have fewer than 2 million people combined.

The image of a typical South American is a person of mixed indigenous and European ancestry, or mestizo. While mestizos are the majority in the region as a whole, a detailed analysis reveals a complex mosaic of peoples. The European conquest in

the sixteenth century led to a dramatic decline of the indigenous population. Nevertheless, indigenous peoples still make up 20 to 30 percent of the populations of Bolivia, Peru, and Ecuador. Other countries have notable Amerindian minorities.

Perhaps 10 million African slaves were brought to work on the plantations of Latin America. Every country in South America has a significant population of African descent, especially in the coastal zones, the Guianas, and northeast Brazil. In Argentina, Uruguay, Chile, and southern Brazil, descendants of European immigrants, especially from Spain, Portugal, Italy, and Germany, make up the majority.

Immigrants have come to South America from many other countries, including the United Kingdom, France, Netherlands, India, China, Indonesia, and Lebanon. Brazil has the largest Japanese population outside of Japan. Extensive intermarriage has created practically every racial combination imaginable. However, this mixing has not created an egalitarian society free of prejudice.

A Japanese-Brazilian Xintoist priest participating in a Japanese festival in Praça do Japão, in Curitiba, Paraná. (Image by Jean Colemonts)

The Atacama Dry lake, in Chile. At the horizon, the Tumisa, Lejía and Miñiques volcanoes. (Image by Francesco Mocellin)

Government

As the third decade of the twenty-first century begins, South America is still dominated by people of European ancestry, or those who have adopted a European lifestyle. These groups still control the economies and governments of the region, and large numbers of people remain poor and marginalized. While every country in South America has been democratically electing its government since the 1990s, even now, well into the twenty-first century, democratic institutions remain weak in many countries, and corruption and inequitable access to political power are pervasive. Patterns of language and religion also reveal much complexity. Spanish is the predominant language on the continent, but Portuguese is the official language of Brazil. Many native languages remain in use. Quechua, the language of the Inca Empire, is spoken by as many as 10 million people in the Andean countries. The Amazon Basin still has many indigenous groups with their own languages. English, Dutch, and French are spoken in the Guianas.

The Roman Catholic Church has been the dominant religious faith in South America since the conquest by the Spanish and Portuguese, although Evangelical Protestantism has emerged as the region's fastest-growing religious movement. More than 19 percent of the people in South America are Protestant, dominated by sects such as the Assembly of God, Mormons, and Seventh-day Adventists. Fifteen percent of Chileans and 22 percent of Brazilians are Protestant. Even the Roman Catholic Church has undergone division and reform efforts. For example, in the 1950s and 1960s, a progressive movement within the church, known as liberation theology, gained strength in the region. The movement sought to connect with the poor and address issues of social inequity, justice, and political alienation. Various points of disagreement between church leaders and liberation theologists have since been largely resolved, especially since Jorge Mario Bergoglio, an Argentinian, was elected in 2013 to the papacy as Pope Francis. He is the first South American to lead the Roman Catholic Church.

Las Lajas Sanctuary is a basilica church located in the southern Department of Nariño, municipality of Ipiales, Colombia. The place is a popular pilgrimage location since the apparition of the Virgin Mary in 1754. The first shrine was built by 1750 and was replaced by a bigger one in 1802 including a bridge over the canyon of the Guáitara River. The present temple, of Gothic Revival style, was built between 1916 and 1949. (Image by Diego Delso)

Economic Patterns

The economies of South America are characterized by great contrasts, change, and potential. The region has some of the most richly endowed countries on Earth. At the beginning of the twenty-first century, the average annual income per person in South America was US$8,511 (compared to US$59,500 in the United States); but when compared to many places in Africa and Asia where people earn less than US$1,000 per year, South Americans fare better. However, there are significant differences among countries within the region. With an annual per-capita income of US$24,600, Chile is the most affluent South American country. Uruguay is next with US$22,400, followed by Argentina with US$20,900. The poorest countries are Bolivia, with an average income of just US$7,600, followed by Guyana with US$8,100.

Agricultural patterns vary from rain-forest dwellers, who practice slash-and-burn cultivation; to subsistence farmers in the Andes, who raise potatoes,; to modern commercial plantations producing bananas and coffee for export; to massive cattle ranches, or *estancias*, on the Pampas of Argentina. Brazil and Colombia are the world's largest and third-largest exporters of coffee, respectively (Vietnam is the second-largest). Ecuador is the world's fifth-largest producer of bananas, and Chile exports large quantities of high-value fruits. Peru and Bolivia are the largest producers of coca for cocaine production, while Colombia leads in the production and export of refined cocaine. Despite media images, however, illegal drugs make up only a small part of the total economy of South America.

Mining is important to many South American countries. Brazil is the third-largest producer of iron ore in the world. Chile leads the world in copper production, and tin and aluminum are also important exports from the region. During the last fifty years, manufacturing and service industries have grown to dominate the economies of virtually every South American country.

In recent decades, South America made significant progress in the development of global marketing and regional trade through the efforts of the Andean Community and Mercosur (a trading bloc comprising Brazil, Argentina, Uruguay, and Paraguay). Nevertheless, development patterns remain uneven. Brazil and Argentina have sophisticated manufacturing sectors that produce aircraft, automobiles, and high-technology equipment; Chile developed high-income agricultural exports; while Bolivia and Guyana still depend on the export of lower-value minerals and agricultural products.

Urbanization

Eighty percent of the people of South America reside in cities—approximately the same percentage as in the United States. South America has five cities with populations of over 5 million—Buenos Aires,

Argentina (13 million), São Paulo, Brazil (10 million); Lima, Peru (7 million); Bogotá, Colombia (7 million); and Rio de Janeiro, Brazil (6 million).

South American cities continue to see a flood of migrants from rural areas in search of jobs and a better life. This trend has been driven by industrialization policies, agricultural modernization emphasizing large farms and mechanization, and high birth rates in rural areas. All of South America's major cities are surrounded by shantytowns, where the poor struggle to erect makeshift housing and find basic urban services. These areas stand in stark contrast to historic city centers and affluent residential neighborhoods and shopping districts.

Urbanization rates vary widely among the individual countries. Uruguay is the most urban country, with 95 percent of the people living in cities. Other highly urban countries are Argentina (92 percent), Chile (90 percent), and Venezuela (89 percent). The least urban countries are Guyana (30 percent), Paraguay (65 percent), and Bolivia (72 percent).

James R. Keese

CENTRAL AMERICA

North America and South America are connected by a long isthmus made up of seven small and mostly Hispanic countries: Belize, Costa Rica, El Salvador, Guatemala, Honduras, Nicaragua, and Panama. Together, these countries have a population of roughly 44 million people living on a landmass about a quarter the size of nearby Mexico. It is in the land that the wealth of Central America lies. With only 28 percent of the region's gross domestic product coming from industry, the bulk of the nations' economies, as well as their people, depend on the area's small private farms, large plantations, and stock ranches.

Physical Geography

Most of Central America is dominated by three physical zones that are, at times, spatially disconnected: the interior highlands, the Pacific coastal region, and the Caribbean coastal lowlands.

The highland interior of countries such as Guatemala and Nicaragua was first occupied by Amerindians—the native peoples of the Americas—who farmed small subsistence plots. Hispanic culture has spread into the area, displacing the Indigenous cultures and introducing larger-scale agriculture and cattle-based industries. All of the Central American nations, except Belize, share this physiographic zone. The interior highlands in many ways serve as the core of the region.

Five capital cities are located in the highland interior zone: San Salvador, El Salvador; Tegucigalpa, Honduras; Guatemala City, Guatemala; San José, Costa Rica; and Managua, Nicaragua. This zone is the most densely settled physical area within Central America. It is also where the most intensive coffee production occurs.

To the west of the highlands lies the Pacific region, with large-scale plantations focusing on such crops as bananas and, in the drier portions, cattle ranching. On the east side of the isthmus, the Caribbean coastal region has seen an upsurge of agriculture by people displaced through resettlement programs from the interior's heavily populated cities. The area has also become popular as a tourist destination, including vacations focused on ecological themes.

The region as a whole experiences similar weather patterns. From the northern tip of Belize to the southern portion of Panama, only 9 degrees of latitude are spanned in the Central American landmass. Therefore, the area is well within a temperate climate range that is affected only by altitude, which causes the climate to vary from tropical to subtropical.

A cordillera, or chain of mountains, runs the length of the isthmus, connecting with the Sierra Madre del Sur in Mexico in the north and the northern portions of the Andes in Colombia. Mountains in Central America rise to a maximum of almost 14,000 feet (more than 4,300 meters) at Tajumulco, a volcano in western Guatemala.

With such a range in altitudes, Central America sustains a wide variety of species. These plants and animals inhabit rain and cloud forest in the higher elevations and mangrove swamps in the coastal lowlands. Such variation in elevation in a small geographic area also reduces the amount of available arable land—from a low of 3 percent in Belize to a high of 36 percent in El Salvador. Bananas are the main crop. The heavy reliance on this tropical fruit and the political power of fruit companies gave rise to the pejorative term "banana republic" for countries in this region.

History

Indigenous civilizations rose and fell in various parts of Central America before the arrival of the Spanish. Most notable of these were the Maya, who

constructed temples and cities, such as Tikal and Xunantunich, in Guatemala, Honduras, El Salvador, and Belize.

Christopher Columbus first visited Central America in 1502, and the explorer Vasco Nuñez de Balboa crossed the Isthmus of Panama in 1513. In the 1520s, the Spanish began colonization in earnest. Guatemala, known as the Captaincy General of Guatemala, was the Spanish colonial capital for Central America. Soon after the independence of Mexico from Spain in 1821, the Central America region declared its independence. In 1823 the newly

independent region formed the United Provinces of Central America. Isolation, poor communication, and personal and regional rivalries soon led to the demise of this federation, and by the late 1830s, the United Provinces of Central America no longer existed. All the countries of Central America declared their independence except Belize, formerly known as British Honduras, which did not gain independence from Great Britain until 1981.

The nineteenth century was dominated by internal conflicts among the region's liberal and conservative factions and by United States and British ef-

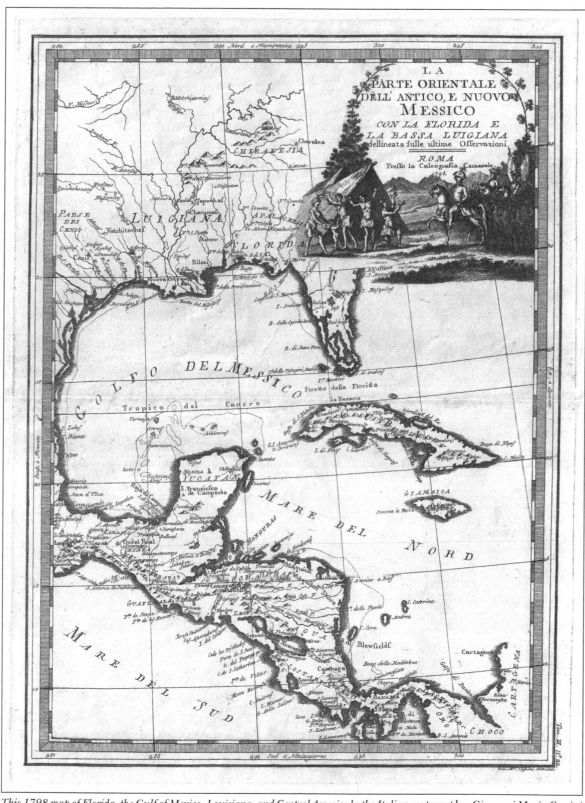

This 1798 map of Florida, the Gulf of Mexico, Louisiana, and Central America by the Italian cartographer Giovanni Maria Cassini is cartographically primitive. Florida is represented as an archipelago. Cassini also fails to attach the Mississippi River to the Missouri. In the upper right quadrant there is a cartouche depicting a Spanish conquistador army entering an American Indian village. The naked villagers, including children, are clearly distraught and fearful, and one can read Cassini's criticism of Spain's brutal colonial policy in the Americas. This map was published in Cassini's 1798 Nuovo Atlante Geografico Universale.

forts to expand their political influence. During this time, coffee was introduced into the region and quickly became important to the area's economy. By the end of the nineteenth century, the production of coffee had created a class of elite landowners, furthering the gap between the wealthy and the poor. Military regimes provided some stability for areas in the region from the late nineteenth century into the twentieth century, but often at the cost of human rights. By the beginning of the twentieth century, the United States was the dominant foreign power within the region.

During the early twentieth century, the United States took a more visible role in Central America's development, especially by way of the Panama Canal. The United States dug the canal, which was opened in 1914. The United States remained in control of the Canal Zone until January 1, 2000,

when Panama assumed control. Since this time, the role of the United States within the region has diminished somewhat, while the role of China (which has become South America's top trading partner) has been growing.

Economies

Central American economies traditionally have had a limited base, which has resulted in slow economic growth and human development. Even by 2020, the major difference in the region's economies persists between the population centers of the interior highlands and the plantation agricultural zones of the lowlands. Most Central Americans live in the highlands. The coastal areas, especially those of the eastern portions of the region and the Isthmus of Panama, are the least densely populated.

The five rowed volcanos in the coat of arms of Central America was inspired by the Cordillera de Apaneca volcanic range of El Salvador, visible from the city of Sonsonate, which became the capital of the Federal Republic of Central America in 1834. (Image by Pablo Nuñez)

Laughing Bird Caye in the Sapodilla Cayes Marine Reserve, Belize. (Image by Victoria Reay)

Demography

The population of Central America grew greatly during the 1960s and 1970s, primarily because of the region's high fertility rates. (At the time, they were around six children per woman in most Central American countries.) The growth rates began dropping in the late 1990s and have steadily declined ever since due to urbanization and advances in literacy and family planning. By the 2020s, in countries such as Nicaragua, Costa Rica, Belize, Panama, and El Salvador, fertility rates were around or lower than the population-replacement rate of 2.1 births per woman. Declining birth rates and increased life expectancy are creating aging populations and associated issues—the trend long experienced by the more developed nations of North America and Europe. These Central American countries are now facing a challenge of meeting the rising demand for social services and healthcare for their aging citizens.

In poverty-stricken Guatemala and Honduras, which still have low levels of literacy and family planning, fertility rates hover at three or four children per woman. They are even higher among their rural and indigenous populations. This demographic trend may not drop significantly in the near future, since the bulk of the population in Guatemala and Honduras remains young and in their prime childbearing years. Such growth would place continuing demands on the economic, social, and environmental systems of these countries.

Social Divisions

Ethnic and racial differences traditionally have been the basis of Central America's social class distinctions. For example, Amerindians, who still make up a large portion of the population, particularly in Guatemala, have been relegated to the lower rungs of the socioeconomic ladder. They have maintained a pattern of subsistence farming tied to local market towns, often outside the dynamic realm of a greater world economic system.

In contrast, people of European descent and mestizos (persons of mixed indigenous and European background) often hold the social, political, and economic power—a relic of the past, when power meant owning large tracts of land for raising either crops or livestock. This power base is seen not only in land ownership but in control of manufacturing and agribusinesses.

Between the upper class of European and mestizo heritage and those of indigenous ancestry, there are smaller ethnic groups. These include persons of African heritage brought to the area as slave labor or freedmen to work in agriculture or construction. Asians have migrated to the region to fill a host of middle-level jobs, most in the service sector, adding another dynamic piece to the Central American cultural quilt.

Central America is a region both physically and culturally. It comprises some of the poorest nations in Latin America, Costa Rica and Panama being the exceptions. Its agricultural products are subject to world economic fluctuations and volatile markets. This, along with periods of political and military turmoil, has slowed socioeconomic development for the region as a whole. Many Central American countries are looking to tourism, especially ecotourism, to diversify and more adequately compete in the global economic theater.

Douglas Heffington
Judith Mimbs

DISCUSSION QUESTIONS: REGIONS

Q1. How did the European settlement patterns in South America differ from those in Central America? List three factors that led to these differences. Are these factors still significant today?

Q2. How are the societies in South and Central America structured? What roles did ethnic and racial backgrounds play in the evolution of the region's social divisions?

Q3. Why are South and Central America collectively referred to as "Latin America"? Why is Mexico excluded from discussions of Central America?

PHYSICAL GEOGRAPHY

OVERVIEW

CLIMATE AND HUMAN SETTLEMENT

"Everyone talks about the weather," goes an old saying, "but nobody does anything about it." If everyone talks about the weather, it is because it is important to them—to how they feel and to how their bodies and minds function. There is plenty they can do about it, from going to a different location to creating an artificial indoor environment.

Climate

The term "climate" refers to average weather conditions over a long period of time and to the variations around that average from day to day or month to month. Temperature, air pressure, humidity, wind conditions, sunshine, and rainfall—all are important elements of climate and differ systematically with location. Temperatures tend to be higher near the equator and are so low in the polar regions that very few people live there. In any given region, temperatures are lower at higher altitudes. Areas close to large bodies of water have more stable temperatures. Rainfall depends on topography: The Pacific Coast of the United States receives a great deal of rain, but the nearby mountains prevent it from moving very far inland. Seasonal variations in temperature are larger in temperate zones.

Throughout human history, climate has affected where and how people live. People in technologically primitive cultures, lacking much protective clothing or housing, needed to live in mild climates, in environments favorable to hunting and gathering. As agricultural cultivation developed, populations located where soil fertility, topography, and climate were favorable to growing crops and raising livestock. Areas in the Middle East and near the Mediterranean Sea flourished before 1000 BCE.

Many equatorial areas were too hot and humid for human and animal health and comfort, and too infested with insect pests and diseases.

Improvements in technology allowed settlement to range more widely north and south. Sturdy houses and stables, internal heating, and warm clothing enabled people to survive and be active in long cold winters. Some peoples developed nomadic patterns, moving with herds of animals to adapt to seasonal variations.

A major challenge in the evolution of settled agriculture was to adapt production to climate and soil conditions. In North America, such crops as cotton, tobacco, rice, and sugarcane have relatively restricted areas of cultivation. Wheat, corn, and soybeans are more widely grown, but usually further north. Winter wheat is an ingenious adaptation to climate. It is sown and germinates in autumn, then matures and is harvested the following spring. Rice, which generally grows in standing water, requires special environmental conditions.

Tropical Problems

Some scholars argue that tropical climates encourage life to flourish but do not promote quality of life. In hot climates, people do not need much caloric intake to maintain body heat. Clothing and housing do not need to protect people from the cold. Where temperatures never fall below freezing, crops can be grown all year round. Large numbers of people can survive even where productivity is not high. However, hot, humid conditions are not favorable to human exertion nor (it is claimed) to mental, spiritual, and artistic creativity. Some tropical areas, such as South India, Bangladesh, Indonesia, and Central Africa, have developed large populations living at relatively low levels of income.

Slavery

Efforts to develop tropical regions played an important part in the rise of the slave trade after 1500 CE. Black Africans were kidnapped and forceably transported to work in hot, humid regions. The West Indian islands became an important location for slave labor, particularly in sugar production. On the North American continent, slave labor was important for producing rice, indigo, and tobacco in colonial times. All these were eclipsed by the enormous growth of cotton production in the early years of U.S. independence. It has been estimated that the forced migration of Africans to the Americas involved about 1,800 Africans per year from 1450 to 1600, 13,400 per year in the seventeenth century, and 55,000 per year from 1701 to 1810. Estimates vary wildly, but at least 7.7 million Africans were forced to migrate in this process.

European Migration

Migration of European peoples also accelerated after the discovery of the New World. They settled mainly in temperate-zone regions, particularly North America. Although Great Britain gained colonial dominion over India, the Netherlands over present-day Indonesia, and Belgium over a vast part of central Africa, few Europeans went to those places to live. However, many Chinese migrated throughout the Nanyang (South Sea) region, becoming commercial leaders in present-day Malaysia, Thailand, Indonesia, and the Philippines, despite the heat and humidity. British emigrants settled in Australia and New Zealand, Spanish and Italians in Argentina, Dutch (Boers) in South Africa—all temperate regions.

Climate and Economics

Most of the economic progress of the world between 1492 and 2000 occurred in the temperate zones, primarily in Europe and North America. Climatic conditions favored agricultural productivity. Some scholars believe that these areas had climatic conditions that were stimulating to intellectual and tech-

IRELAND'S POTATO FAMINE AND EUROPEAN EMIGRATION

Mass migration from Europe to North America began in the 1840s after a serious blight destroyed a large part of the potato crop in Ireland and other parts of Northern Europe. The weather played a part in the famine; during the autumns of 1845 and 1846 climatic conditions were ideal for spreading the potato blight. The major cause, however, was the blight itself, and the impact was severe on low-income farmers for whom the potato was the major food.

The famine and related political disturbances led to mass emigration from Ireland and from Germany. By 1850 there were nearly a million Irish and more than half a million Germans in the United States. Combined, these two groups made up more than two-thirds of the foreign-born U.S. population of 1850. The settlement patterns of each group were very different. Most Irish were so poor they had to work for wages in cities or in construction of canals and railroads. Many Germans took up farming in areas similar in climate and soil conditions to their homelands, moving to Wisconsin, Minnesota, and the Dakotas.

nological development. They argue that people are invigorated by seasonal variation in temperature, sunshine, rain, and snow. Storms—particularly thunderstorms—can be especially stimulating, as many parents of young children have observed for themselves.

Climate has contributed to the great economic productivity of the United States. This productivity has attracted a flow of immigrants, which averaged about 1 million a year from 1905 to 1914. Immigration approached that level again in the 1990s, as large numbers of Mexicans crossed the southern border of the United States, often coming for jobs as agricultural laborers in the hot conditions of the Southwest—a climate that made such work unattractive to many others.

Unpredictable climate variability was important in the peopling of North America. During the 1870s and 1880s, unusually favorable weather encouraged a large flow of migration into the grain-producing areas just west of the one-hundredth meridian. Then came severe drought and much agrarian distress. Between 1880 and 1890, the combined population of Kansas and Nebraska increased by about a million, an increase of 72 percent. During the 1890s, however, their combined population was virtually constant, indicating that a large out-migration was offsetting the natural increase. Much of the area reverted to pasture, as climate and soil conditions could not sustain the grain production that had attracted so many earlier settlers.

Climate variability can be a serious hazard. Freezing temperatures for more than a few hours during spring can seriously damage fruits and vegetables. A few days of heavy rain can produce serious flooding.

Recreation and Retirement

Whenever people have been able to separate decisions about where to live from decisions about where to work, they have gravitated toward pleasant climatic conditions. Vacationers head for Caribbean islands, Hawaii, the Crimea, the Mediterranean Coast, even the Baltic coast. "The mountains" and "the seashore" are attractive the world over. Paradoxically, some of these areas (the Caribbean, for instance) have monotonous weather year-round and thus have not attracted large inflows of permanent residents. Winter sports have created popular resorts such as Vail and Aspen in Colorado, and numerous older counterparts in New England. Large numbers of Americans have retired to the warm climates in Florida, California, and Arizona. These areas then attract working-age adults who earn a living serving vacationers and retirees. Since these locations are uncomfortably hot in summer, their attractiveness for residence had to await the coming of air conditioning in the latter half of the twentieth century.

Human Impact on Climate

Climate interacts with pollution. Bad-smelling factories and refineries have long relied on the wind to disperse atmospheric pollutants. The city of Los Angeles, California, is uniquely vulnerable to atmospheric pollution because of its topography and wind currents. Government regulations of automobile emissions have had to be much more stringent there than in other areas to keep pollution under control.

Human activities have sometimes altered the climate. Development of a large city substitutes buildings and pavements for grass and trees, raising summer temperatures and changing patterns of water evaporation. Atmospheric pollutants have contributed to acid rain, which damages vegetation and pollutes water resources. Many observers have also blamed human activities for a trend toward global warming. Much of this has been blamed on carbon dioxide generated by combustion, particularly of fossil fuels. A widespread and continuing rise in temperatures is expected to raise water levels in the oceans as polar icecaps melt and change the relative attractiveness of many locations.

Paul B. Trescott

FLOOD CONTROL

Flood control presents one of the most daunting challenges humanity faces. The regions that human communities have generally found most desirable, for both agriculture and industry, have also been the lands at greatest risk of experiencing devastating floods. Early civilization developed along river valleys and in coastal floodplains because those lands contained the most fertile, most easily irrigated soils for agriculture, combined with the convenience of water transportation.

The Nile River in North Africa, the Ganges River on the Indian subcontinent, and the Yangtze River in China all witnessed the emergence of civilizations that relied on those rivers for their growth. People learned quickly that residing in such areas meant living with the regular occurrence of life-threatening floods.

Knowledge that floods would come did not lead immediately to attempts to prevent them. For thousands of years, attempts at flood control were rare. The people living along river valleys and in floodplains often developed elaborate systems of irrigation canals to take advantage of the available water for agriculture and became adept at using rivers for transportation, but they did not try to control the river itself. For millennia, people viewed periodic flooding as inevitable, a force of nature over which they had no control. In Egypt, for example, early people learned how far out over the riverbanks the annual flooding of the Nile River would spread and accommodated their society to the river's seasonal patterns. Villagers built their homes on the edge of the desert, beyond the reach of the flood waters, while the land between the towns and the river became the area where farmers planted crops or grazed livestock.

In other regions of the world, buildings were placed on high foundations or built with two stories on the assumption that the local rivers would regularly overflow their banks. In Southeast Asian countries such as Thailand and Vietnam, it is common to see houses constructed on high wooden posts above the rivers' edge. The inhabitants have learned to allow for the water levels' seasonal changes.

Flood Control Structures
Eventually, societies began to try to control floods rather than merely attempting to survive them. Levees and dikes—earthen embankments constructed to prevent water from flowing into low-lying areas—were built to force river waters to remain within their channels rather than spilling out over a floodplain. Flood channels or canals that fill with water only during times of flooding, diverting water

away from populated areas, are also a common component of flood control systems. Areas that are particularly susceptible to flash floods have constructed numerous flood channels to prevent flooding in the city. For example, for much of the year, Southern California's Los Angeles River is a small stream flowing down the middle of an enormous, 20- to 30-foot-deep (6–9 meters) concrete-lined channel, but winter rains can fill its bed from bank to bank. Flood channels prevent the river from washing out neighborhoods and freeways.

Engineers designed dams with reservoirs to prevent annual rains or snowmelt entering the river upstream from running into populated areas. By the end of the twentieth century, extremely complex flood control systems of dams, dikes, levees, and flood channels were common. Patterns of flooding that had existed for thousands of years ended as civil engineers attempted to dominate natural forces.

The annual inundation of the Egyptian delta by the flood waters of the Nile River ceased in 1968 following construction of the 365-foot-high (111 meters) Aswan High Dam. The reservoir behind the 3,280-foot-long (1,000-meter) dam forms a lake almost ten miles (16 km.) wide and almost 300 miles (480 km.) long. Flood waters are now trapped behind the dam and released gradually over a year's time.

Environmental Concerns
Such high dams are increasingly being questioned as a viable solution for flood control. As human understanding of both hydrology and ecology have improved, the disruptive effects of flood control projects such as high dams, levees, and other engineering projects are being examined more closely.

Hydrologists and other scientists who study the behavior of water in rivers and soils have long known that vegetation and soil types in watersheds can have a profound effect on downstream flooding. The removal of forest cover through logging or clearing for agriculture can lead to severe flooding in the future. Often that flooding will occur many miles downstream from the logging activity. Devastating floods in the South Asian country of Bangladesh, for example, have been blamed in part on clear-cutting of forested hillsides in the Himalaya Mountains in India and Nepal. Monsoon rains that once were absorbed or slowed by forests now run quickly off mountainsides, causing rivers to reach unprecedented flood levels. Concerns about cause-and-effect relationships between logging and flood control in the mountains of the United States were one reason for the creation of the U.S. Forest Service in the nineteenth century.

In populated areas, even seemingly trivial events such as the construction of a shopping center parking lot can affect flood runoff. When thousands of square feet of land are paved, all the water from rain runs into storm drains rather than being absorbed slowly into the soil and then filtered through the watertable. Engineers have learned to include catch basins, either hidden underground or openly visible but disguised as landscaping features such as ponds, when planning a large paving project.

Wetlands and Flooding

Less well known than the influence of watersheds on flooding is the impact of wetlands along rivers. Many river systems are bordered by long stretches of marsh and bog. In the past, flood control agencies often allowed farmers to drain these areas for use in agriculture and then built levees and dikes to hold the river within a narrow channel. Scientists now know that these wetlands actually serve as giant sponges in the flood cycle. Flood waters coming down a river would spread out into wetlands and be held there, much like water is trapped in a sponge.

Draining wetlands not only removes these natural flood control areas but worsens flooding problems by allowing floodwater to precede downstream faster. Even if life-threatening or property-damaging floods do not occur, faster-flowing water significantly changes the ecology of the river system. Waterborne silt and debris will be carried farther. Trying to control floods on the Mississippi River has

had the unintended consequence of causing waterborne silt to be carried farther out into the Gulf of Mexico by the river, rather than its being deposited in the delta region. This, in turn, has led to the loss of shore land as ocean wave actions washes soil away, but no new alluvial deposits arrive to replace it.

In any river system, some species of aquatic life will disappear and others replace them as the speed of flow of the water affects water temperature and the amount of dissolved oxygen available for fish. Warm-water fish such as bass will be replaced by cold-water fish such as trout, or vice versa. Biologists estimate that more than twenty species of freshwater mussels have vanished from the Tennessee River since construction of a series of flood control and hydroelectric power generation dams have turned a fast-moving river into a series of slow-moving reservoirs.

Future of Flood Control

By the end of the twentieth century, engineers increasingly recognized the limitations of human interventions in flood control. Following devastating floods in the early 1990s in the Mississippi River drainage, the U.S. Army Corps of Engineers recommended that many towns that had stood right at the river's edge be moved to higher ground. That is, rather than trying to prevent a future flood, the Corps advised citizens to recognize that one would inevitably occur, and that they should remove themselves from its path. In the United States and a number of other countries, land that has been zoned as floodplains can no longer be developed for residential use. While there are many things humanity can do to help prevent floods, such as maintaining well-forested watersheds and preserving wetlands, true flood control is probably impossible. Dams, levees, and dikes can slow the water down, but eventually, the water always wins.

Nancy Farm Männikkö

ATMOSPHERIC POLLUTION

Pollution of the Earth's atmosphere comes from many sources. Some forces are natural, such as volcanoes and lightning-caused forest fires, but most sources of pollution are byproducts of industrial society. Atmospheric pollution cannot be confined by national boundaries; pollution generated in one country often spills over into another country, as is the case for acid deposition, or acid rain, generated in the midwestern states of the United States that affects lakes in Canada.

Major Air Pollutants

Each of eight major forms of air pollution has an impact on the atmosphere. Often two or more forms of pollution have a combined impact that exceeds the impact of the two acting separately. These eight forms are:

1. Suspended particulate matter: This is a mixture of solid particles and aerosols suspended in the air. These particles can have a harmful impact on human respiratory functions.

2. Carbon monoxide (CO): An invisible, colorless gas that is highly poisonous to air-breathing animals.

3. Nitrogen oxides: These include several forms of nitrogen-oxygen compounds that are converted to nitric acid in the atmosphere and are a major source of acid deposition.

4. Sulfur oxides, mainly sulfur dioxide: This sulfur-oxygen compound is converted to sulfuric acid in the atmosphere and is another source of acid deposition.

5. Volatile organic compounds: These include such materials as gasoline and organic cleaning solvents, which evaporate and enter the air in a vapor state. VOCs are a major source of ozone formation in the lower atmosphere.

6. Ozone and other petrochemical oxidants: Ground-level ozone is highly toxic to animals and plants. Ozone in the upper atmosphere, however, helps to shield living creatures from ultraviolet radiation.

7. Lead and other heavy metals: Generated by various industrial processes, lead is harmful to human health even at very low concentrations.

8. Air toxics and radon: Examples include cancer-causing agents, radioactive materials, or asbestos. Radon is a radioactive gas produced by natural processes in the earth.

All eight forms of pollution can have adverse effects on human, animal, and plant life. Some, such as lead, can have a very harmful effect over a small range. Others, such as sulfur and nitrogen oxides, can cross national boundaries as they enter the atmosphere and are carried many miles by prevailing wind currents. For example, the radioactive discharge from the explosion of the Chernobyl nuclear plant in the former Soviet Union in 1986 had harmful impacts in many countries. Atmospheric radiation generated by the explosion rapidly spread over much of the Northern Hemisphere, especially the countries of northern Europe.

Impacts of Atmospheric Pollution

Atmospheric pollution not only has a direct impact on the health of humans, animals, and plants but also affects life in more subtle, often long-term, ways. It also affects the economic well-being of people and nations and complicates political life.

Atmospheric pollution can kill quickly, as was the case with the killer smog, brought about by a temperature inversion, that struck London in 1952 and led to more than 4,000 pollution-related deaths. In the late 1990s, the atmosphere of Mexico City was so polluted from automobile exhausts and industrial pollution that sidewalk stands selling pure oxygen to people with breathing problems became thriving businesses. Many of the heavy metals and organic constituents of air pollution can cause cancer when people are exposed to large doses or for long periods of time. Exposure to radioactivity in the atmosphere can also increase the likelihood of cancer.

In some parts of Germany and Scandinavia in the 1990s, as well as places in southern Canada and the southern Appalachians in the United States, certain types of trees began dying. There are several possible reasons for this die-off of forests, but one potential culprit is acid deposition. As noted above, one byproduct of burning fossil fuels (for example, in coal-fired electric power plants) is the sulfur and nitrous oxides emitted from the smokestacks. Once in the atmosphere, these gases can be carried for many miles and produce sulfuric and nitric acids.

These acids combine with rain and snow to produce acidic precipitation. Acid deposition harms crops and forests and can make a lake so acidic that aquatic life cannot exist in it. Forests stressed by contact with acid deposition can become more susceptible to damage by insects and other pathogens. Ozone generated from automobile emissions also kills many plants and causes human respiratory problems in urban areas.

Air pollution also has an impact on the quality of life. Acid pollutants have damaged many monuments and building facades in urban areas in Europe and the United States. By the late 1990s, the distance that people could see in some regions, such as the Appalachians, was reduced drastically because of air pollution.

The economic impact of air pollution may not be as readily apparent as dying trees or someone with a respiratory ailment, but it is just as real. Crop damage reduces agricultural yield and helps to drive up the cost of food. The costs of repairing buildings or monuments damaged by acid rain are substantial. Increased health-care claims resulting from exposure to air pollution are hard to measure but are a cost to society nevertheless.

It is impossible to predict the potential for harm from rapid global warming arising from greenhouse gases and the destruction of the ozone layer by chlorofluorocarbons (CFCs), but it could be cata-

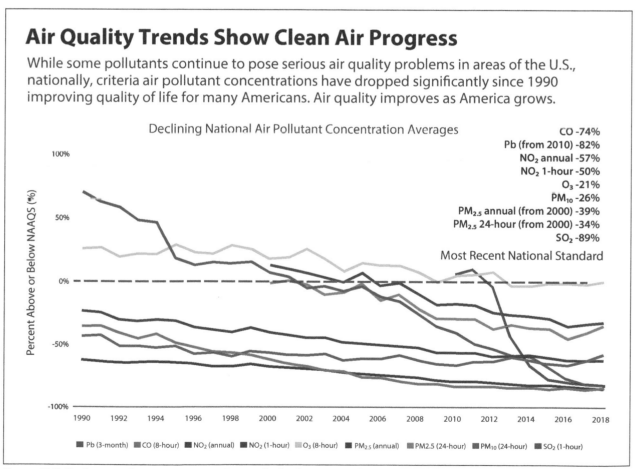

Air Quality Trends Show Clean Air Progress

While some pollutants continue to pose serious air quality problems in areas of the U.S., nationally, criteria air pollutant concentrations have dropped significantly since 1990 improving quality of life for many Americans. Air quality improves as America grows.

Declining National Air Pollutant Concentration Averages

CO -74%
Pb (from 2010) -82%
NO_2 annual -57%
NO_2 1-hour -50%
O_3 -21%
PM_{10} -26%
$PM_{2.5}$ annual (from 2000) -39%
$PM_{2.5}$ 24-hour (from 2000) -34%
SO_2 -89%

Most Recent National Standard

Percent Above or Below NAAQS (%)

Pb (3-month)   CO (8-hour) NO_2 (annual) NO_2 (1-hour) O_3 (8-hour) $PM_{2.5}$ (annual)   PM2.5 (24-hour) PM_{10} (24-hour) SO_2 (1-hour)

Source: U.S. Environmental Protection Agency, Our Nation's Air, Status and Trends Through 2018.

strophic. Rapid global warming would cause the sea level to rise because of the melting of the polar ice caps. Low-lying coastal areas would be flooded, or, in the case of Bangladesh, much of the country. Global warming would also change crop patterns for much of the world.

Solutions for Atmospheric Pollution

Although there is still some debate, especially among political leaders, most scientists recognize that air pollution is a problem that affects both the industrialized and less-industrialized world. In their rush to industrialize, many nations begin generating substantial amounts of air pollution; China's extensive use of coal-fired power plants is just one example.

The major industrial nations are the primary contributors to atmospheric pollution. North America, Europe, and East Asia produce 60 percent of the world's air pollution and 60 percent of its food supply. Because of their role in supplying food for many other nations, anything that damages their ability to grow crops hurts the rest of the world. In 2018, about 76 million tons of pollution were emitted into the atmosphere in the United States. These emissions mostly contribute to the formation

of ozone and particles, the deposition of acids, and visibility impairment.

Many industrialized nations are making efforts to control air pollution, for example, the Clean Air Act of 1970 in the United States or the international Montreal Accord to curtail CFC production. Progress is slow and the costs of reducing air pollution are often high. Worldwide, bad outdoor air caused an estimated 4.2 million premature deaths in 2016, about 90 percent of them in low- and middle-income countries, according to the World Health Organization. Indoor smoke is an ongoing health threat to the 3 billion people who cook and heat their homes by burning biomass, kerosene, and coal.

In the year 2019 the record of the nations of the world in dealing with air pollution was a mixed one. There were some signs of progress, such as reduced automobile emissions and sulfur and nitrous oxides in industrialized nations, but acid deposition remains a problem in some areas. CFC production has been halted, but the impact of CFCs on the ozone layer will continue for many years. However, more nations are becoming aware of the health and economic impact of air pollution and are working to keep the problem from getting worse.

John M. Theilmann

DISEASE AND CLIMATE

Climate influences the spread and persistence of many diseases, such as tuberculosis and influenza, which thrive in cold climates, and malaria and encephalitis, which are limited by the warmth and humidity that sustains the mosquitoes carrying them. Because the earth is warming as a result of the generation of carbon dioxide and other "greenhouse gases" from the burning of fossil fuels, there is intensified scientific concern that warm-weather dis-

eases will reemerge as a major health threat in the near future.

Scientific Findings

The question of whether the earth is warming as a result of human activity was settled in scientific circles in 1995, when the Second Assessment Report of the Intergovernmental Panel on Climate Change, a worldwide group of about 2,500 experts, was issued. The panel concluded that the earth's temperature had increased between 0.5 to 1.1 degrees Farenheit (0.3 to 0.6 degrees Celsius) since reliable worldwide records first became available in the late nineteenth century. Furthermore, the intensity of warming had increased over time. By the 1990s, the temperature was rising at the most rapid rate in at least 10,000 years.

The Intergovernmental Panel concluded that human activity—the increased generation of carbon dioxide and other "greenhouse gases"—is responsible for the accelerating rise in global temperatures. The amount of carbon dioxide in the atmosphere has risen nearly every year because of increased use of fossil fuels by ever-larger human populations experiencing higher living standards.

In 1998, Paul Epstein of the Harvard School of Public Health described the spread of malaria and dengue fever to higher altitudes in tropical areas of the earth as a result of warmer temperatures. Rising winter temperatures have allowed disease-bearing insects to survive in areas that could not support them previously. According to Epstein, frequent flooding, which is associated with warmer temperatures, also promotes the growth of fungus and provides excellent breeding grounds for large numbers of mosquitoes. Some experts cite the flooding caused by Hurricane Floyd and other storms in North Carolina during 1999 as an example of how global warming promotes conditions ideal for the spread of diseases imported from the Tropics.

Heat, Humidity, and Disease

During the middle 1990s, an explosion of termites, mosquitoes, and cockroaches hit New Orleans, following an unprecedented five years without frost. At the same time, dengue fever spread from Mexico across the border into Texas for the first time since records have been kept. Dengue fever, like malaria, is carried by a mosquito that is limited by temperature and humidity. Colombia was experiencing plagues of mosquitoes and outbreaks of the diseases they carry, including dengue fever and encephalitis, triggered by a record heat wave followed by heavy rains. In 1997 Italy also had an outbreak of malaria. An outbreak of zika in 2015–16, related to a virus spread by mosquitoes, raised concerns regarding the safety of athletes and spectators at the 2016 Summer Olympics in Rio de Janeiro and led to travel warnings and recommendations to delay getting pregnant for those living or traveling in areas where the mosquitoes are active.

The global temperature is undeniably rising. According to the National Oceanic and Atmospheric Administration, July 2019, was the hottest month since reliable worldwide records have been kept, or about 150 years. The previous record had been set in July 2017.

The rising incidence of some respiratory diseases may be related to a warmer, more humid environment. The American Lung Association reported that more than 5,600 people died of asthma in the United States during 1995, a 45.3 percent increase in mortality over ten years, and a 75 percent increase since 1980. Roughly a third of those cases occurred in children under the age of eighteen. Asthma is now one of the leading diseases among the young. Since 1980, there has been a 160 percent increase in asthma in children under the age of five.

Heat Waves and Health

A study by the Sierra Club found that air pollution, which will be enhanced by global warming, could be responsible for many human health problems, including respiratory diseases such as asthma, bronchitis, and pneumonia.

According to Joel Schwartz, an epidemiologist at Harvard University, air pollution concentrations in the late 1990s were responsible for 70,000 early deaths per year and more than 100,000 excess hospitalizations for heart and lung disease in the United States. Global warming could cause these numbers to increase 10 to 20 percent in the United States, with significantly greater increases in countries that are more polluted to begin with, according to Schwartz.

Studies indicate that global warming will directly kill hundreds of Americans from exposure to extreme heat during summer months. The U.S. Centers for Disease Control and Prevention have found that between 1979 and 2014, the death rate as a direct result of exposure to heat (underlying cause of death) generally hovered around 0.5 to 1 deaths per million people, with spikes in certain years). Overall, a total of more than 9,000 Americans have died from heat-related causes since 1979, according to death certificates. Heat waves can double or triple the overall death rates in large cities. The death toll in the United States from a heat wave during July 1999 surpassed 200 people. As many as 600 people died in Chicago alone during the 1990s due to heat waves. The elderly and very young have been most at risk.

Respiratory illness is only part of the picture. The Sierra Club study indicated that rising heat and humidity would broaden the range of tropical diseases, resulting in increasing illness and death from diseases such as malaria, cholera, and dengue fever, whose range will spread as mosquitoes and other disease vectors migrate.

The effects of El Niño in the 1990s indicate how sensitive diseases can be to changes in climate. A study conducted by Harvard University showed that warming waters in the Pacific Ocean likely contrib-

uted to the severe outbreak of cholera that led to thousands of deaths in Latin American countries. Since 1981, the number of cases of dengue fever has risen significantly in South America and has begun to spread into the United States. According to health experts cited by the Sierra Club study, the outbreak of dengue near Texas shows the risks that a warming climate might pose. Epstein and the Sierra Club study concur that if tropical weather expands, tropical diseases will expand.

In many regions of the world, malaria is already resistant to the least expensive, most widely distributed drugs. According to the World Health Organization (WHO), there were 219 million cases of malaria globally in 2017 and 435,000 malaria deaths, representing a decrease in malaria cases and deaths rates of 18 percent and 28 percent since 2010. Of those deaths, 403,000 (approximately 93 percent) were in the WHO African Region.

Bruce E. Johansen

PHYSIOGRAPHY AND HYDROLOGY

PHYSIOGRAPHY OF SOUTH AMERICA

South America has a spectacular physical environment. Its Andes range is the second-highest mountain chain in the world. The Amazon River carries more water than any other river on Earth. Its Amazon Basin is the world's largest drainage basin, comparable in size to the whole United States. Angel Falls in Venezuela is the world's highest waterfall—ten times higher than North America's Niagara Falls. Lake Titicaca in Bolivia is the world's highest navigable lake; Mount Guallatiri in Chile is the world's tallest active volcano. Quito, Ecuador, is the

world's largest city on the equator. The Amazon rain forest is the world's largest. When British naturalist Charles Darwin visited the continent during the 1830s, he was amazed at the diversity and grandeur of South America's natural environment.

The Andes Mountains

South America's great mountain chain extends the entire west coast of South America, stretching 4,500 miles (7,200 km.) from Venezuela to Tierra del Fuego. Although their crests occur only an average

View of Cuernos del Paine in Torres del Paine National Park

PHYSICAL GEOGRAPHY OF SOUTH AMERICA

of about 110 miles (180 km.) from the continent's Pacific coastline, the Andes are nevertheless the continental divide for South America. Rivers on the western slopes of the mountains are generally short, swift streams that enter the Pacific Ocean without creating large fertile valleys or navigable waterways. Rivers on the eastern slopes are long, sometimes mighty rivers that carry sediments from the mountains into the Atlantic Ocean.

The mountains are high and continuous. Many peaks exceed 17,000 feet (5,200 meters); some soar above 20,000 feet (6,100 meters). The Western Hemisphere's highest mountain, Mount Aconcagua, rises 22,834 feet (6,960 meters) between Chile and Argentina. Most of the passes through the Andes are at an elevation of 10,000 feet (3,050 meters). Given these extreme heights, the Andes have presented a formidable barrier to east-west travel from prehistoric times to the present.

Geologists estimate that the Andes were formed by a series of crustal uplifts and volcanic activity from 65 million to 70 million years ago. The region that is now the Andes was then a long, narrow trough that gradually filled with sediments washed from surrounding highlands. Continuous pressure from movement of the great crustal plates of the Pacific region in an east-northeast direction folded, bent, warped, and broke the rocks that eventually became the Andes. All this was accompanied by vulcanism, the movement of molten rock both under the crust and through volcanoes and cracks in the crust, onto Earth's surface. This mountain-building process proceeded sporadically over the millennia, creating a complex system of mountain ridges, highland and lowland basins, plateaus, and fault scarps. Today, many earthquakes and an occasional volcanic eruption are proof that the mountain-building process is still going on.

Earth scientists do not agree on any systematic way to divide the Andes into logical regions, so they should be examined according to their general physical characteristics.

Northern Andes

Starting near the border between Ecuador and Colombia, the Andes divide into three distinct parallel mountain chains separated by deep valleys. The sierras (mountains) are named Sierra Occidental (west), Central, and Oriental (east). The western

MAJOR ACTIVE ANDEAN VOLCANOES

Volcano	Country	Height		Last Eruption
		Meters	Feet	
Guallatiri	Chile	6,071	19,913	1985
Cotopaxi	Ecuador	5,911	19,388	2016
El Misti	Peru	5,823	19,101	1985
Tupungatito	Chile	5,640	18,499	1987
Lascar	Chile	5,592	18,342	2015
Nevado del Ruiz	Colombia	5,321	17,458	2018
Nevado del Sangay	Ecuador	5,320	17,454	1934-ongoing
Tolima	Colombia	5,215	17,105	1943
Tungurahua	Ecuador	5,023	16,475	2014
Puracé	Colombia	4,800	15,744	1977
Guagua Pinchicah	Ecuador	4,784	15,692	2002
Lautaro	Chile	3,380	11,151	1979
Llaima	Chile	3,125	10,250	1984
Villarica	Chile	2,847	9,341	2019
Osorno	Chile	2,660	8,727	1869

range has the highest average elevation—12,800 feet (3,900 meters)—but the central range has several peaks exceeding 18,000 feet (5,500 meters). Bogotá, Colombia's capital city, is located at 8,660 feet (2,680 meters) in a high basin of the Sierra Oriental. The western and central Sierras end in the swampy Caribbean coastal plains, while the eastern branch forms two arms around Lake Maracaibo, extending eastward beyond Caracas, Venezuela.

Central Andes

From the northern border of Ecuador to the northern parts of Chile and Argentina, the Andes are a complex mixture of high ridges, transverse valleys, and high platforms topped by scattered volcanoes. The mountain mass is narrowest in Ecuador and widest in southern Peru and Bolivia, where it is made up of two high ridges separated by high basins called *altiplanos*. Quito, Ecuador's capital, sits in one at 9,350 feet (2,850 meters), and La Paz, Bolivia's capital, is located at 11,900 feet (3,625 meters).

The Peruvian/Bolivian *altiplano* is one of the world's largest interior basins to occur at such a high altitude. With a surface elevation from 11,900 to 12,800 feet (3,625 to 3,900 meters), the basin measures 500 by 80 miles (800 by 130 km.). Several mountain peaks around the basin exceed 19,700 feet (6,000 meters). Lake Titicaca, the world's highest navigable lake, at 12,500 feet (3,810 meters) altitude, is more than 110 miles (180 km.) long and extremely deep. The lake contains freshwater, even though it has no outlet to the sea, because a short river empties its waters into a salt lake farther south. Near-freezing temperatures occur almost nightly, and no trees grow under these conditions. Potatoes, which survive by growing underground, were first domesticated here and were the Amerindians' staple food.

Southern Andes

From the borders of Argentina, Bolivia, and Chile to Tierra del Fuego, the Andes form a single chain of massive mountains. Opposite Chile's Atacama Desert, the mountains are high, steep, and quite dry. The runoff of winter snows sends small rivers eastward into Argentina, but many of these never have enough water to reach the sea. Most passes through the mountains require a climb of more than 10,000 feet (3,000 meters).

Between Mendoza, Argentina, and Santiago, Chile, soars Mount Aconcagua, where the Andes reach their highest point in Mount Aconcagua. This extinct volcano has perpetual ice and snow at its summit and is the tallest mountain in the Western Hemisphere. From here to Tierra del Fuego, the Andes show more distorted peaks and valleys as a result of glaciers eroding the mountain masses. Ice and snow also produced the lakes and ski slopes of Chile and Argentina. Both countries have created national parks in this region to preserve the area's natural beauty.

While the Andes decline to about 6,000 feet (1,800 meters) at South America's southern tip, the action of ice in sculpting the terrain has carved the land into rugged and picturesque formations. Southern Chile is a land of volcanoes, fjords, lakes, hanging valleys, remnant glaciers, innumerable islands, and glacial deposits; only Alaska has similar features on so grand a scale. In middle Chile, the line of permanent ice and snow is about 12,000 feet (3,650 meters); but progressing toward Antarctica, it falls to 5,000 feet (1,500 meters) on Mount Osorno, and to 2,300 feet (700 meters) at Tierra del Fuego. Puerto Montt (41 degrees south latitude, 73 degrees west longitude) is the start of the Chilean fjord region.

During the Ice Age, mountain glaciers covered this area and most of Argentine Patagonia. Glaciers ground mountains into barren rocky surfaces with weird shapes, gouged out deep, rounded valleys, left perpendicular valleys hanging up to several hundred feet above their main river, and built up large deposits of loose rock and sand.

The Eastern Highlands

Comprising two separate areas—the Guiana Highlands and the Brazilian Plateau—the Eastern Highlands are quite different from the Andes Mountains. Their maximum elevation is only 9,075 feet (2,770 meters) in Venezuela's Mount Roraima; a mountain northeast of Rio de Janeiro, Brazil, reaches a similar altitude. The average elevation of both highlands is less than 2,000 feet (610 meters). The Eastern Highlands are mostly made up of highly eroded plateaus of sedimentary rocks over ancient crystalline rock and lava flows. These crustal features are much older than the Andes. Lack of earthquakes and volcanic activity in the area today show that the landforms are no longer in the mountain-building phase. After eroding for so many millennia, the highlands have many flat surfaces and chains of low hills. Resistant granite rocks stand above the general plateau and form such features as Sugarloaf Mountain in Rio de Janeiro's harbor and Corcovado Peak, up on which the famous statue of Christ the Redeemer stands, overlooking Rio de Janeiro.

Guiana Highlands

This area covers most of southern Venezuela, Suriname, French Guiana, and a small part of southern Brazil. Most of the mountainous parts of this plateau are on its western and southern fringes. Sloping generally toward the Caribbean and Atlantic coastlines, it has major rivers rising in the south, crossing the high tablelands, and dropping over waterfalls before entering the coastal plains. A tributary of Venezuela's Caroní River bursts from an aquifer near the top of a high tableland and drops 3,212 feet (980 meters) in Angel Falls, the world's highest waterfall. Forests and grasslands form a complex pattern on the highlands, the type and density of tropical vegetation depending on rainfall and soil types. Rough terrain, infertile soils, and thick forests make the Guiana Highlands a remote and underdeveloped region.

Mineral deposits have been worked for many years in the highlands, and the discovery of more deposits is likely. Gold and diamonds have attracted prospectors since the nineteenth century. Iron ore and bauxite have been mined extensively; their presence indicates that soils are poor for agriculture. Near Venezuela's Orinoco River is Cerro Bolívar, a mountain of high-grade iron ore that has been mined since the 1940s.

Heliamphora chimantensis, is a species of marsh pitcher plant endemic to the Chimantá Massif (a Venezuelan part of the Guiana Shield). It has been recorded from Apacará and Chimantá Tepuis. (Image by Andreas Eils)

The Brazilian Highlands

This region is located on the southern side of the Amazon Basin, extending all the way to Brazil's eastern seaboard, from Fortaleza to Pôrto Alegre—an area almost the size of Alaska. Averaging about 2,000 feet (610 meters), the surface of the highlands is broken by rounded crystalline mountains in the south and scattered masses trending parallel to the east coast. The São Francisco River rises in southern Brazil and runs in a deep canyon for 2,000 miles (3,200 km.) northeast and parallel to the coast before it descends in a series of waterfalls to the Atlantic Ocean just south of Recife.

Many tributaries of the Amazon have cut deep canyons into the highlands, adding to the roughness of the general terrain. The southern end of the highlands, the Paraná Plateau, is made up of thick layers of diabase lava. Reaching about 6,000 feet (1,800 meters), this lava plateau is among the world's largest. Rich soils for coffee and other crops form from the lava. The mild climate of the midlatitudes and the rich soils make this one of Brazil's most productive farm regions.

The Brazilian Highlands end in steep cliffs along the eastern seaboard, but their northern side grades gently into the Amazon plain. The sharp edge of the highlands, called an escarpment, lies within sight of the coast. Rising 2,000 feet (610 meters) or higher in a series of steps, the escarpment has made transportation from the coast to the interior difficult and costly. The Brazilian capital, Brasília, was built on the highlands to encourage people to move into the nation's interior.

The Amazon Basin

There is great concern about destruction of the Amazon rain forest, the world's largest rain forest. Hundreds of species of plants have been counted in half a square mile, and the number of other life-forms in the same area is unknown. Many plants are used to make medicines, and loss of the rain forest could mean loss of potential new medicines and much of the world's biological diversity. The Amazon Basin is one of the least populated places on Earth; continued population growth there will increase pressure to develop more of the region. New mineral discoveries in the basin also encourage development. The basin is thought to contain large petroleum and natural gas fields.

The Amazon Basin covers most of Brazil and parts of Venezuela, Colombia, Ecuador, Peru, and Bolivia, an area about the same size as the United States. This is the largest area of lowland in South America. A vast inland sea 5 million years ago, the structural basin that became the Amazon was filled with sediments from the Andes and Eastern Highlands. More than 1,000 rivers flow into the basin and form branches of the mighty Amazon. These rivers deposit sediments in some places and cut valleys into the surface at others.

Although the basin is only about 300 feet (91 meters) above sea level at the Peruvian border, the plain has numerous hills and rolling terrain. Low

spots become vast swamps; the huge floodplains along the rivers become lakes during the rainy season from November through June, but the hilly land remains permanently above water. Widest in the west, the plain narrows to about 250 miles (400 km.) in the east where the Guiana Highlands and Brazilian Highlands almost meet. The Amazon has not created an impressive delta at its mouth, because the huge amount of silt brought across the continent by the rivers is carried north by ocean currents. Hourly, 170 trillion gallons (640 trillion liters) of freshwater are flushed into the sea, pushing sea water about 100 miles (160 km.) back from the Amazon's mouth.

The equator passes through South America just north of the Amazon River. Rain comes to the basin in all seasons, but November through June bring floods when most of the rivers overflow. Average annual rainfall in the basin is 100 inches (1,500 millimeters). Flooding brings new soil to the river lowlands, but also covers good farmland with water. Warm temperatures of 80 degrees Fahrenheit (27 degrees Celsius) and plentiful rain provide the right combination for a tropical rain forest.

In many places, there is a complete covering of interlaced leaves high up in the giant trees. The forest floor is dark, and little vegetation grows on the ground; where sunlight does reach the floor, masses of vegetation flourish in dense tangles, making it difficult to walk among the trees. Patches of grassland and scrub forest occupy especially well-drained land, or land where trees have been cut and burned. Soils away from the rivers are not fertile for crops, and disturbance of the forest-soil-water ecology can create a sterile wasteland until nature has time to restore the environmental balance. Rubber, Brazil nuts, and hardwood lumber are valuable Amazonian products.

Orinoco Lowlands

Located in Colombia and Venezuela, the Orinoco Lowlands have some of the same characteristics of the Amazon Basin. Relatively flat land slopes from the Andes Mountains and Guiana Highlands. The Orinoco River and its tributaries drain the *llanos* (plains) and form a large swampy delta opposite the island of Trinidad. The major difference between the Amazon Basin and the Orinoco Lowlands is that the Orinoco has a six-month-long dry season and a six-month-long wet season. Another difference is that the natural vegetation of the *llanos* is grass, not trees; this is a land of tropical savanna. Soils are poor for crops on both plains.

From June through October, so much rain falls that large areas of the *llanos* are flooded. Dry winters cause many rivers to disappear and grass becomes hard, brittle, and parched. Cattle roaming on open pastures have been the traditional use of this almost unpopulated land. Even for cattle, both rainy and dry seasons make life difficult.

The Paraná-Plata Plains

This region starts at the edge of the Andes foothills and Brazilian Plateau at about 16 degrees south latitude, 60 degrees west longitude. Patagonia borders the southern part of this lowland, which runs into the sea at Buenos Aires Province, Argentina. These lowlands are made up of a huge basin that was filled by sediments washed down from the Andes and the Brazilian Highlands over thousands of years. The northern and western parts of the plains have sediments that are 10,000 feet (3,050 meters) deep before bedrock is reached.

At Buenos Aires, the sediments are 965 feet (300 meters) deep. Few stone fragments or pebbles can be found on the plains, and tall buildings cannot be constructed on this loose subsoil. In the south, the plains have many characteristics similar to the U.S. plains from Denver to Chicago. The Paraná-Plata Plains comprise the Pantanal, Chaco, and Pampas.

The Pantanal

This tropical swamp, located in southwest Brazil, is made up of the drainage basin of the upper Paraguay River. About the size of South Dakota, it is one of the world's largest wetlands. The Paraguay River

The name Pantanal comes from the Portuguese word, pântano, *meaning wetland, bog, swamp, quagmire or marsh. Floodplain ecosystems such as the Pantanal are defined by their seasonal inundation and desiccation. They shift between phases of standing water and phases of dry soil, when the water table can be well below the root region.*

and its tributaries overflow in the rainy season (February) and the region becomes a maze of swamps, bays, canals, lakes, and marshes. Even in the dry season, the water table is never far below the surface. Water birds, fish, reptiles, and many species of mammals thrive in this environment. World environmentalists have made this a prime target for conservation efforts. The Pantanal Conservation Area is now a UNESCO World Heritage Centre.

The Chaco

This vast alluvial plain stretches from tropical latitudes just south of the Pantanal to the middle latitudes of northern Argentina. Larger than Texas, it extends 700 miles (1,125 km.) in a north-south direction. Receiving much less rain than the Pantanal, it is a dry tropical/semitropical region crossed by rivers that dry up in the winter dry season and flood into swamps during the wet summers. As a result of salt deposits building up, many low basins have cactus and other dry-land plants. The region is poor for agriculture. *Chaco* means hunting land in the local Amerindian language, and the area teems with wildlife of all types. Some *quebracho* forests in wetter areas support lumbering and tannin industries; livestock graze and cotton grows in favored places.

The Pampas

One of the world's greatest food-producing regions, the Pampas are Argentina and Uruguay's mid-latitude plains, similar to the U.S. Midwest. Mendoza is in the dry foothills of the Andes, as Denver is in the Rockies; Buenos Aires is at the end of the humid prairies, as Chicago is. Mendoza's elevation is 2,500 feet (760 meters), and the Pampas grade gently to Buenos Aires at sea level. Tall grass is the natural vegetation of the humid east; short grass and scrub plants cover the arid west. Rivers from the mountains water the oases at the Pampa's western margins, making them garden spots for fruits and vegetables. Only in the far southeast corner of Buenos Aires Province do low mountains break the flat surface of the Pampas.

Rich soils develop under midlatitude grasslands that receive adequate rain, allowing such excellent animal feed as alfalfa, corn, and soybeans to grow. Beef is a major source of income for farms in this part of the Pampas, just as it is in Iowa and Illinois.

The Pampas (from the Quechua: pampa, meaning "plain") are fertile South American lowlands that cover more than 750,000 km² (289,577 sq. mi.) and include the Argentine provinces of Buenos Aires, La Pampa, Santa Fe, Entre Ríos and Córdoba; all of Uruguay; and Brazil's southernmost state, Rio Grande do Sul. (Image by Alex Pereira)

A glacier cave located on the Perito Moreno Glacier in Argentina. (Image by Martin St-Amant)

In the drier west, wheat and sheep find ideal conditions, like their counterparts in eastern Colorado and western Nebraska.

Patagonia

The high, cold, windswept tablelands of southern Argentina are called Patagonia. Rocks that form the tablelands are either flat-lying sedimentary rocks or old lava flows. Seaside cliffs range from only a few feet to 300 feet (90 meters) in the north. The tablelands rise in steplike fashion toward the Andes, where they reach 5,000 feet (1,500 meters) in some places. Low mountains of resistant crystalline rocks stick up above the general surface in a few places.

At the Andean foothills, there are many lakes and shallow depressions where glacial troughs and cirques are partly filled with meltwater from the mountains' snow cover. This is good sheep country.

The Colorado, Chubut, and Negro are among the few rivers with enough water to have cut steep-sided and flat-bottomed valleys into the tablelands all the way to the sea. Farmers have settled in some of these valleys, taking advantage of irrigation waters, good soils, and protection from the strong, cold southwest winds.

Most of Patagonia is classified as a cold desert. Permanent settlements have succeeded only in the sheltered valleys and seaports, and near mineral deposits. Off the coast of Patagonia is a broad continental shelf, and marine life is plentiful in this part of the south Atlantic. Petroleum around Comodoro Rivadavia and coal in the Río Gallegos district support small centers of urban development. Patagonia's huge empty spaces, about the size of Texas, are likely to remain empty.

Robert J. Tata

HYDROLOGY OF SOUTH AMERICA

Most areas of South America are well supplied with water. The only region of the continent classified as a desert is the Atacama of northern Chile, where some locations have registered no significant precipitation for more than 100 years. Other areas, such as the interior of northeast Brazil, coastal Peru, the highlands of southeast Peru and western Bolivia, and the eastern slope of the Andes Mountains in Argentina, receive only slight and intermittent precipitation.

Snowfall in South America is restricted to the higher mountains and extreme south, but heavy rainfall occurs in many regions. In the southwest portion of the continent, the continuous westerly winds off the Pacific Ocean are forced to rise along the western side of the Andes. The uplift causes water vapor to condense on the windward side of the mountains (the orographic effect), resulting in rainfall. Southern Chile, therefore, is one of the wettest areas of the continent.

Similar areas of heavy rainfall exist along the southeast coast of Brazil and the Pacific shores of Ecuador and Colombia in northwest South America. The greatest amount of precipitation occurs in the middle of the continent, in the equatorial regions of northwest Brazil and the eastern provinces of Bolivia, Peru, Ecuador, and Colombia. Throughout the year, sunlight beats down on the surface, causing the moist air to rise and condense into massive thunderstorms that produce torrential downpours in the late afternoon.

The highest mountains in South America are the Andes, which extend from the Caribbean Sea in the north to Tierra del Fuego in the far south. Formed along the western edge of the continent where the South American and Pacific plates grind together and force crustal rock dramatically upward, the mountains divert most of the rainfall runoff toward the east. Only a few short streams, such as the Esmeraldas and Guayas Rivers of Ecuador, the Santa River of Peru, and the Bío-Bío River in Chile, flow down the mountains into the Pacific Ocean.

Most of the rain in South America falls on the large, relatively flat interior surface, where the runoff merges with streams flowing toward the east from the Andes. The water collects into ever-broader rivers, eventually making its way to the Atlantic Ocean after a journey that can exceed several thousand miles.

The heavy precipitation and warm temperatures in the north-central region of the continent help produce the tropical rain forest, the most diverse, yet one of the most fragile, ecological systems in the world. The climate is also partially responsible for the low agricultural productivity in the region, because the heavy rainfall causes important plant nutrients to be dissolved and washed out of the upper layers of the soil, in a process called leaching. The characteristics of the climate, vegetation, and soil have an important impact on the population and lifestyle along the rivers in central South America.

The water moving from the west and central regions toward the east collects in two dominant river basins: the Amazon system in the north, and the Paraná/Paraguay/Uruguay system in the south. Although not as large, the Orinoco in Venezuela and the São Francisco in Brazil are also major rivers that help drain water toward the east, while the Essequibo in Guyana and the Colorado in Argentina are locally significant. A few streams flow directly north from the Andes into the Caribbean Sea, the most important being the Magdalena and Cauca Rivers of Colombia.

Amazon River System

The largest rivers in South America are located in the heart of the continent. With a total length of nearly 4,000 miles (more than 6,400 km.), more than 1,000 tributaries, a capture basin of approximately 2,007,200 square miles (5.2 million sq. km.), and a discharge of more than 6 million cubic feet (170,000 cubic meters) per second at the mouth, the Amazon, in many ways, is the largest river in the world. Named after a mythological Greek tribe of

MAJOR WATERSHEDS OF SOUTH AMERICA

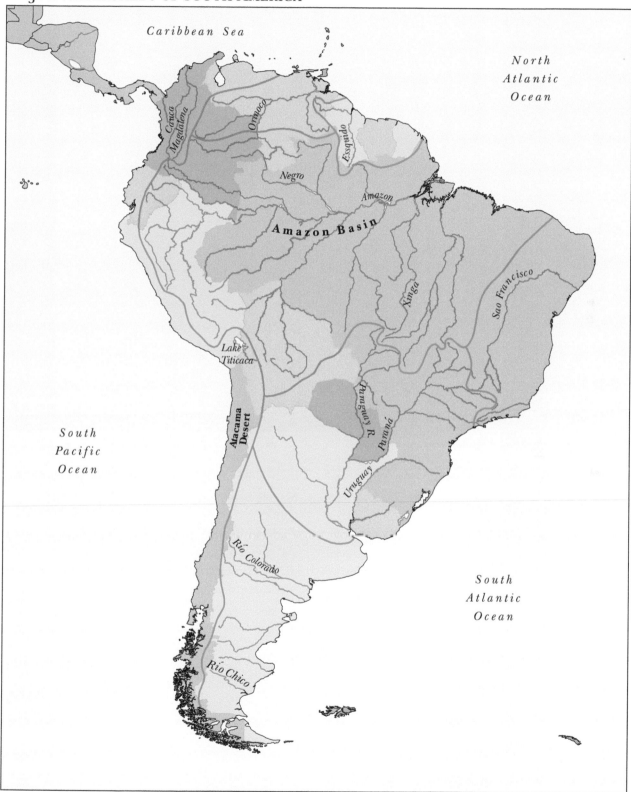

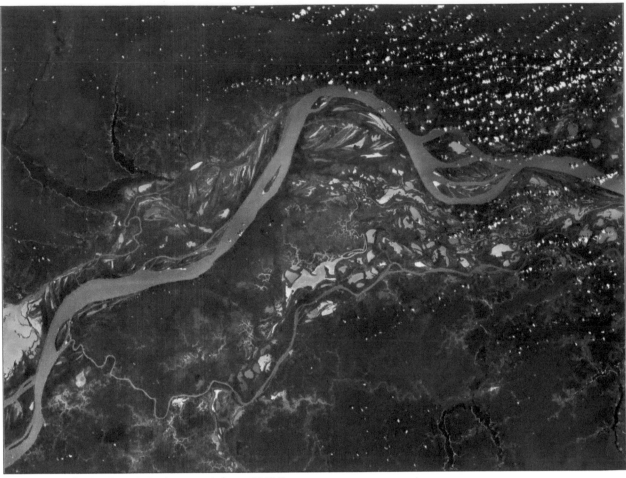

Amazon River flowing through the Amazon rainforest. (NASA)

warrior women, the stream begins high in the Andes of southwestern Peru. Water collects in the Ucayali and Maranon, which flow north along the eastern side of the mountains and eventually come together south of the port of Iquitos. As the merged flow moves downstream from the city, it takes on the name Solimões in Peru. Only when the river reaches Brazil is it called the Amazon.

Along its journey to the mouth (the place where a river empties into the sea), the volume of water is increased from several large tributaries that join along the southern flank. The largest of these are the Madeira, Tapajós, and Xingú. Sometimes the Tocantíns and its large tributary, the Araguáia, are considered to be part of the Amazon system, for they empty into the Atlantic within the complex Amazon delta region near the city of Belém and just south of the large island of Marajó. From the north, the only major tributary is the Negro. which passes through the large and historically important city of Manaus. The river's dark color is caused by the tannic acid that it transports.

Most tributaries feed the Amazon from the south, and they drain a large interior area of the continent that is dominated by the tropical savanna climate. The most important feature of the weather in these areas is the seasonality of the rainfall. During the summer months, which here and in other areas south of the equator are December, January, and February, the savanna climate produces heavy rainfall. This occurs because during this season, the Sun's position directly overhead causes intense convectional uplift and afternoon thunderstorms. Six months later, the Sun will have moved north across the equator, so the areas experience descending and drying air, high surface barometric pressure, and little precipitation.

The Itaipu Dam is a hydroelectric dam on the Paraná River located on the border between Brazil and Paraguay. The name Itaipu *was taken from an isle that existed near the construction site. In the Guarani language, Itaipu means "the sounding stone." The Itaipu Dam's hydroelectric power plant produced the most energy of any in the world as of 2016, setting a new world record of 103,098,366 megawatt hours (MWh); it surpassed the Three Gorges Dam plant in energy production in 2015 and 2016. (Image by Jonas de Carvalho)*

The amount of water transported by the south-joining Amazon tributaries reflects this climate variability, and because most of the water in the Amazon comes from these rivers, the volume in the Amazon itself varies greatly during the year. The Amazon experiences low water levels during June, July, and August; six months later, high water marks are reached. This also has a major effect on the tropical-forest wildlife and the inhabitants along the rivers.

Beyond the natural levees at the water's edge exist large lowland areas, known in Brazil as *várzeas*, that are subject to regular flooding caused by the variability in the flow. The swampy nature of these areas causes difficulties for people attempting to farm the land and build roads and bridges. The *várzeas* are also favorite homes for many species of tropical birds, insects, and reptiles, including huge anaconda snakes.

Because of the depth and large volume of water that flows down the Amazon River, it can be used by oceangoing ships for much of its course. These vessels travel from the mouth to the interior ports of Santarém, Manaus, and Iquitos with little difficulty. For some types of bulky cargo, transportation between Lima, on the western side of the mountains of Peru, and Iquitos, in the Peruvian eastern lowlands, is less expensive by ship than over land. Despite the greater distance of passage through the Panama Canal, around northern South America, and up the entire length of the lower Amazon, ship transport is less costly than overland travel through the Andes Mountains.

The Amazon and its tributaries also are used by a wide variety of smaller boats, ranging from the dugout canoes of the indigenous peoples to the large riverboats that provide supplies to the residents along the water's edge. Because the population in the interior of South America is sparse and building roads is difficult, the rivers provide the only connection to the outside world for many people.

Settlements tend to cluster along the streams, the places where people meet to purchase or trade for items such as pots and pans, textile fabric, fuel, and other necessities of life. The river edge is also where people are able to sell some of the things that are produced in the area, such as agricultural products, rubber extract, and the fish caught in the river or game hunted in the forest. The river trading boats, many of which are operated by families who live aboard, are important to the people of the Amazon region.

Paraná River System

In many ways the Paraná system, which also includes the Paraguay and Uruguay Rivers, is the most important in South America. The Paraná River originates in central Brazil; flows through the southeast region of that country, along the international borders between Brazil and Paraguay, and between Argentina and Paraguay; goes through northern Argentina; and finally enters the Atlantic Ocean at the wide estuary known as the Río de la Plata.

The approximate total length of the Paraná is 2,485 miles (4,000 km.). Along its course, it is joined by the Paraguay River, which also originates

in Brazil, bounds the territories of Brazil and Bolivia, and then bisects the landlocked country of Paraguay. From their beginnings in Brazil, the waters of the Uruguay River form the border between Uruguay and Argentina before merging into the Paraná/Paraguay at the estuary. The combined rivers discharge about 2.8 million cubic feet (80,000 cubic meters) of water per second into the Atlantic Ocean and drain a capture basin of almost 1.7 million square miles (4.4 million sq. km.).

In comparison to the Amazon system in the north, the Paraná system is not as long, discharges less water, and drains a smaller area. The greater importance of the rivers in the south comes from their location. They form boundaries among South America's most economically vibrant countries and flow through the industrial and urban areas of the continent. Waterborne trade on the Paraná has always been important, even though the riverbeds, unlike those of the Amazon, need to be constantly dredged, levied, and otherwise maintained to prevent ships from running aground.

Since the colonial period, the waters of these rivers have been fought over frequently. They played a significant role in the Paraguayan War (sometimes called the War of the Triple Alliance) of the 1860s, which pitted Argentina, Brazil, and Uruguay against the small nation of Paraguay. The gun-

SOUTH AMERICA'S MIGHTIEST RIVERS

River	Flows Into	Length Kilometers	Length Miles	World Rank
Amazon	Atlantic Ocean	7,000	4,300	2
Paraná	Rio de la Plata	4,880	3,030	11
Purús	Amazon River	2,960	1,840	15
Madeira	Amazon River	3,250	2,020	16
São Francisco	Atlantic Ocean	2,914	1,811	17
Japurá	Amazon River	2,820	1,750	22
Tocantins	Amazon River	2,450	1,520	25
Orinoco	Atlantic Ocean	2,140	1,330	26
Paraguay	Paraná River	2,695	1,675	27
Negro	Amazon River	2,250	1,400	33
Xingu	Amazon River	1,640	1,020	37

Source: Migiro, Geoffrey. "The Longest Rivers in South America." WorldAtlas, Mar. 8, 2018, worldatlas.com/articles/the-longest-rivers-in-south-america.html.

boat battles and land conflicts destroyed most of the population of Paraguay, although they left its territory more or less intact. More recently, the rivers have caused international disputes over water rights. The nations along the lower course of the Paraná have claimed that the vital water necessary to support farms, cities, and boat traffic has been diverted for other uses upstream in Brazil. Some disputes have arisen because of the construction of large hydroelectric projects in the region.

In some areas of Brazil, the Paraná and its tributaries flow from areas that are underlain by hard volcanic rocks (basalt) onto regions of softer sedimentary formations. When this happens, waterfalls develop because the harder rocks upstream are more resistant to weathering and erosion than the softer surfaces of the lower course.

Waterfalls

The most famous of the waterfalls is located on the Iguaçu, attracting many tourists to the area where that river joins the Paraná near the borders among Brazil, Argentina, and Paraguay. Other waterfalls have disappeared, however, because dams have been constructed across the major rivers and flooded the valleys. This has been done to supply the electricity needed by the factories and homes of this area of South America, especially in Brazil's industrial southeast.

One of the largest hydroelectric projects in the world was completed to the north of the still exis tent Iguaçu Falls. A similar series of falls on the Paraná, once known as the Guaíra Falls, was completely submerged in 1982 by the artificial lake created by the construction of the Itaipú dam. Most of the generated electricity is transmitted via high-tension lines back to the factories near the city of São Paulo, Brazil.

A major project affecting the Paraná-Paraguay river system has caused much debate since the end of the twentieth century. It involves plans for dredging and channeling the riverbed along the Paraguay River, in the region north of the city of Corumbá on the Bolivia-Brazil border, in order to create an industrial waterway (the Paraguay-Parana Hidrovia). In this region, the Paraguay passes through the Pantanal, a wide area where a substantial percentage of the land remains submerged through the rainy season. The hidrovia is supposed to greatly reduce the costs of transporting goods from the area, in particular soybeans—Brazil's top export.

The Pantanal is the world's largest tropical wetlands ecosystem; it attract thousands of species of birds and animals and is considered to be one of the richest wildlife repositories in the world. Many fear that modification of the drainage caused by disturbances to the Paraguay River will destroy the complex ecology of this region. Environmentalists in Brazil and from around the world have appealed to the government to halt the proposed effort and were successful in delaying the project. The debate

Angel Falls (Spanish: Salto Ángel; *Pemon language:* Kerepakupai Meru *meaning "waterfall of the deepest place," or* Parakupá Vená, *meaning "the fall from the highest point") is the world's highest uninterrupted waterfall. The waterfall drops over the edge of the Auyán-tepui mountain in the Canaima National Park, a UNESCO World Heritage site in the Gran Sabana region of Bolívar State in Venezuela. (Image by Paulo Capiotti)*

over its potential ecological consequences versus economic benefits continues.

In the 1990s Brazil undertook a large project involving the Paraná and its major Brazilian tributary, the Tietê. The many dams on these rivers were fitted with locks with the capacity to lift large vessels around the barriers. The Paraná-Tietê waterway is now Brazil's key shipping route between its southern and southwestern regions, transporting soybeans, corn, cellulose, and other agricultural products to the industrial heart of Brazil, São Paulo.

Many major cities are located along the Paraná and its tributaries. In Brazil, these include São Paulo, Campinas, Piracicaba, Sorocaba, Ribeirão Preto, and São José do Rio Prêto. Corumbá in western Brazil and Paraguay's capital city of Asunción are located on the Paraguay, as well as Uruguay's and Argentina's largest cities and capitals, Montevideo and Buenos Aires. Other important cities on the lower course of the Paraná and Río de la Plata include Santa Fe, Paraná, Rosario, and La Plata.

No other region of the world comes close to generating as much of its energy from rivers as South America. In Brazil, for instance, hydroelectric power accounts for 65 percent of electricity supply. The country boasts several of the world's largest hydroelectric facilities. Venezuela's Simón Bolívar Hydroelectric Plant (Guri Dam) and its reservoir are also among the world's largest. All other South American countries have built or are building large dams, which have long been viewed as symbols of national pride.

Other Rivers

Several other rivers in South America are important to the countries through which they flow. In Venezuela, the Orinoco has played a major role in the country's history. Approximately 1,600 miles (2,574 km.) long and draining an area of 364,900 square miles (945,000 sq. km.), it flows toward the east through a grassy plains region known as the *llanos*. The cowboylike animal herders from this region were an important element of Simón Bolívar's army when he fought for the independence of South America from Spain in the early nineteenth century.

Today, the Orinoco is used to transport raw materials such as iron ore to the east coast, from there, they are exported to the industrial centers of Western Europe and the United States. Important cities on the Orinoco are Ciudad Bolívar and Ciudad Guayana. The latter is a planned city established as a manufacturing center to take advantage of locally available raw materials and hydroelectricity. The city has grown dramatically, and now supports a population nearly three times its planned maximum.

In Colombia, the Magdalena and Cauca Rivers were important to the settlement of the country. Both provided links between the interior Andean highlands and the communities of the northern coast. During the colonial period, the Magdalena was used to transport people into the interior and bring gold and emeralds down to the ports of Barranquilla and Cartagena. The Cauca flows through Colombia's most productive areas, and near its banks can be found the large cities of Medellin, Cali, and Manizales.

Brazil's São Francisco River flows more than 1,800 miles (2,900 km.) through the dry interior regions of the eastern states of Minas Gerais, Bahia, Pernambuco, Sergipe, and Alagoas. The region is used primarily for raising cattle, and the river provides a means for moving animals to markets near the coast. As with the rivers of the Paraná system farther south, the hydroelectric potential of the São Francisco has also been tapped, most notably at the Paulo Afonso Falls on the northern border of the state of Bahia.

Lakes

South America includes relatively few significant lakes. There are no massive systems of lakes such as the Great Lakes of North America or the large inland seas of central Asia. Most lakes in South America are produced by the dams along the rivers. The largest human-made lake in the world is the Dr. W. J. van Blommestein Meer reservoir in Suriname.

In the north, Lake Maracaibo near Venezuela's coast is the most important body of freshwater. Technically it is not a lake, for there is an outlet to the sea, but the passage is shallow and its waters are fresh. The rocks under the lake include extensive petroleum deposits, and although reserves are no longer as extensive as they were at one time, the area remains a major oil producer. Refineries, pipelines, and population growth in places like Maracaibo City provide evidence of this resource.

In the highland border between Peru and Bolivia, Lake Titicaca has nurtured human settlement for several thousand years. Indigenous cultures have made the lakeshore their home, and the population is dominated by the Aymara and Quechua-speaking descendants of the Incas and previous cultures. The lake is deep, and the large volume of water has a moderating effect on the harsh, cold climate of the highlands. The people live off the fish and other wildlife in the water and grow meager crops in the hard soil. They travel across the waters of the lake on boats made from the fiber of reeds that grow along its shore.

Melting snows on the surrounding mountains flow into Lake Titicaca, and excess water in the lake flows out to the southeast in the Desaguadero River. The landscape is quite dry, but small communities exist along the river. Eventually the river, by this time a trickle, empties into Lake Poopó—an extensive area on a map but of little significance because of its high saline content and shallow depth.

Among the other lakes in South America, few have any great significance. Along the coast of extreme south Brazil, the Patos and da Mangueira Lakes are large but not very important. The southern Brazilian city of Porto Alegre fronts onto the Lagoa dos Patos. This freshwater lagoon connects Porto Alegre to the Atlantic Ocean at Rio Grande. Glacially formed lakes in the Andes of Chile and Argentina have great natural beauty and draw tourists from the urban centers of the east to the nearby ski slopes.

Lake Atitlán in the Guatemalan Highlands of the Sierra Madre mountain range is in the Sololá Department of southwestern Guatemala. It is the deepest lake in Central America. (Image by Francisco Anzola)

A reed boat on Lake Titicaca. (Image by Thomas Quine)

In summary, the hydrology of South America reflects the climate and terrain. Although impressive, the Amazon system has had limited economic impact because it drains regions that are sparsely populated and not very productive. The Paraná system in the south, however, has been economically and politically important. This area of South America is developing rapidly, and the southern rivers continue to play a vital role in the economies of Brazil, Argentina, Uruguay, and Paraguay.

Cyrus B. Dawsey

PHYSIOGRAPHY AND HYDROLOGY OF CENTRAL AMERICA

The seven countries between Colombia and Mexico—Guatemala, Belize, Honduras, El Salvador, Nicaragua, Costa Rica, and Panama—make up Central America. El Salvador is the smallest at 8,100 square miles (21,000 sq. km.), close to the size of Massachusetts. Nicaragua is the largest, with 50,300 square miles (130,000 sq. km.), about the size of New York State. The combined area of all seven countries is 203,100 square miles (526,000 sq. km.).

The region is narrowest in Panama, about 35 miles (56 km.) wide, and widest, at 350 miles (560 km.), across Honduras. North to south, Central America measures 1,200 miles (1,930 km.). If Central America were to be placed over the eastern United States, it would reach from the tip of Florida to Washington, D.C.

PHYSICAL GEOGRAPHY OF CENTRAL AMERICA

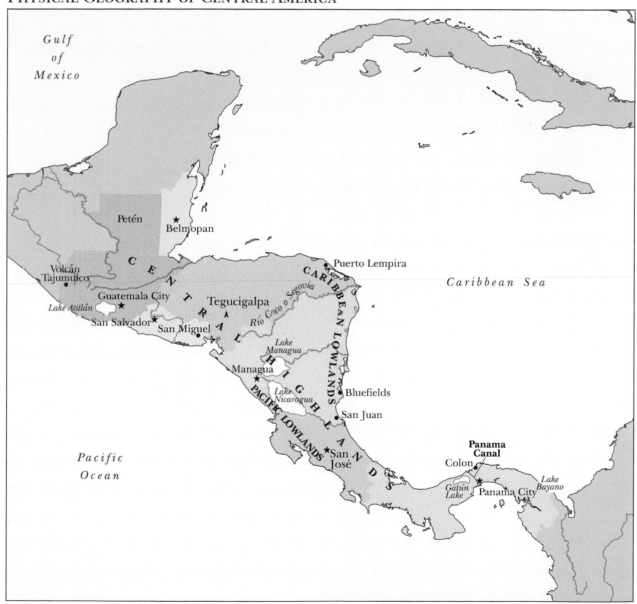

Seismic Activity

When Pangaea, the supercontinent first described by Alfred Wegener in 1912 as a single continental landmass, broke apart about 200 million years ago, North America and South America were not connected. As tectonic activity intensified, a series of islands was formed that ultimately joined. This created the isthmus link, about 50 million years ago.

Seismic activity is induced by the convergence of the Cocos, Caribbean, North American, South American, and Nazca Plates. These plate movements cause the Cocos Plate to dive under the Caribbean Plate. Along the western edge of Nicaragua, a rift has occurred. A fault between the North American Plate and the Caribbean Plate, similar to the San Andreas Fault in California, exists across southern Guatemala. These tectonic movements generate the earthquakes and volcanoes that persist in the region.

Seismic activity is a major concern to inhabitants of the area. Active volcanoes extend in the west from Panama to Guatemala and beyond. Only Belize is without an active volcano. Small-magnitude earthquakes occur frequently and are widespread across the western portion of the region. Most earthquakes cause little or no damage, but violent quakes do occur. The last major earthquake struck near Guatemala's capital in 2018. It caused the deaths of more than 165 people.

Physiographic Regions

Central America has four physiographic regions. Granitic rock underlies most of the region, except in the Petén area of Guatemala, where limestone is dominant.

The Caribbean Lowlands, with elevations below 1,000 feet (300 meters), are irregular; their width varies from a few miles in places to as much as 75 miles (120 km.) in others. These lowlands are relatively flat and crossed by many rivers. It is a region of deltas, shallow lagoons, and salt marshes. Soils range from those that are heavily leached and have low fertility to highly fertile, alluvial soils. Most of the native vegetation has been removed. The east-

Inner crater of Irazu Volcano, seen from rim of outer crater, during eruption of 1963, showing rising ash cloud. (Image by MGLeonard)

ern peninsula shared by Nicaragua and Honduras is by far the largest of these lowlands.

The Central Highlands—above 1,000 feet (300 meters)—were formed by tectonic uplift with a volcanic band along the western edge. High, rugged relief and interior valleys are typical. Elevations range from about 1,000 feet to 6,500 feet (2,000 meters), but several peaks exceed 6,500 feet. The highest peak is Volcán Tajumulco in Guatemala, at 13,767 feet (4,196 meters). These highlands extend from Panama to Guatemala.

The Pacific Lowlands are the least extensive of the four regions. The largest of these lowlands is the combined Guanacaste region of Costa Rica and lake region of Nicaragua. A unique feature off the coast of Panama is the great tidal range. On the Caribbean side, the tidal range is about 28 inches (700 millimeters); on the Pacific side, the range is nearly

Guatemala Earthquake 1976. Rails bent in Gualán. (Figure 42-A, U.S. Geological Survey Professional paper 1002.)

275 inches (7,000 millimeters)—almost 23 feet (7 meters).

The Petén, in northern Guatemala, has different characteristics from the rest of Central America. Flat to moderately rolling topography dominates the area. Elevation ranges from about 490 feet (150 meters) to about 740 feet (225 meters). Geologically, the Petén is an extension of Mexico's Yucatán peninsula and was not formed by the tectonic activity that created most of Central America. Soils are of low fertility and very porous.

Hydrology

In the south, from Panama to Honduras and El Salvador, there are hundreds of watercourses. Those rivers flowing west from the continental divide into the Pacific Ocean are generally short but swift, and some are intermittent. Those draining east are longer and slower moving. Because these rivers originate in the Central Highlands, many have been harnessed by dams for inexpensive hydroelectric power. In the north of Guatemala, most rivers flow north to Mexico or east through Belize. Since the soil in this area is porous, some rivers drain underground.

Gatun Lake, created by damming the Chagres River, supplies the water to operate the locks of the Panama Canal. Lake Nicaragua is the largest lake in the region. This lake, together with Lake Managua, was once a bay connected to the Pacific Ocean. When Lake Nicaragua became isolated from the ocean, many marine species became trapped. Over time, the lake became fresh, and saltwater species such as sharks and tarpon adapted to the new conditions. Lake Atitlán in Guatemala is more than 985 feet (300 meters) deep in spots.

Donald Andrew Wiley

The Great Blue Hole is a giant marine sinkhole off the coast of Belize. It lies near the center of Lighthouse Reef, a small atoll 70 km (43 mi) from the mainland and Belize City. The hole was formed during several episodes of quaternary glaciation when sea levels were much lower. The Great Blue Hole is a part of the larger Belize Barrier Reef Reserve System, a World Heritage Site of the United Nations Educational, Scientific and Cultural Organization (UNESCO). (Image by U.S. Geological Survey)

DISCUSSION QUESTIONS:
PHYSIOGRAPHY AND HYDROLOGY

Q1. How did the Andes Mountains come into existence? What is meant when the Andes are referred to as South America's continental divide?

Q2. Where in South America is the Amazon Basin? Why is it considered so important? What concerns regarding the Amazon Basin are raised by environmentalists?

Q3. Why do geographers refer to Central America as an isthmus? How did the Central American landmass come into existence?

CLIMATOLOGY

SOUTH AMERICA

South America covers 13 percent of Earth's total landmass—about 7 million square miles (18 million sq. km.)—and extends over 67 degrees of latitude. The continent's diversity of climates is a product of six factors: latitude, the equator, the varying width of the continent, elevation, prevailing winds, and ocean currents.

Latitude

South America's position straddles the equator, making it primarily a tropical continent. Its landmass is widest between the Tropic of Cancer and the Tropic of Capricorn. Only Uruguay and the southern parts of Brazil, Argentina, and Chile extend south of the Tropic of Capricorn. The rest of the continent is entirely within the tropics, and tropical climates in South America extend over 62 percent of its landmass.

The Equator

South America is divided into unequal portions by the equator, which runs through the northern part of the continent at the mouth of the Amazon River. Only 12 degrees of latitude out of the 67 spanned by South America are in the Northern Hemisphere. The climatic belts of South America are not symmetrical in relation to the equator as they are in Africa.

Continental Width

The greater east-to-west extent of the northern portion of the continent, along the equator, does not create a continentality effect as it does in Africa because insolation varies little during the year. The southern part of the continent, which is much narrower, allows the maritime influence to extend farther inland.

Elevation

The presence of high mountain ranges acts as a climatic barrier. The Andes of South America deeply modify the climate of South America. These mountains prevent winds from the South Pacific anticyclone (high-pressure area) from reaching the continent's interior. South America's high mountains and plateau modify the low-elevation climates by lowering temperature, increasing precipitation on the windward side of the mountains, and creating a rain shadow effect on the lee side.

Prevailing Winds

South America is dominated by two wind belts: the trade winds and the westerlies. The trade winds or tropical easterlies blow toward the equator out of the subtropical anticyclones centered near 30 degrees north and 30 degrees south latitude. Because of the rotation of Earth, winds flowing from these high-pressure areas are deflected to their right in the Southern Hemisphere and to their left in the Northern Hemisphere. Therefore, the air moving toward the equator flows from east to west (the tropical easterlies or trade winds), while the air moving toward the poles flows from west to east (the westerlies).

The subtropical highs are located over the Pacific and Atlantic Oceans. The strongest high-pressure region is located off the east coast of South America, so that strong westerly winds blow toward the continent. Two low-pressure centers affect wind patterns inland: one located at about 60 degrees south latitude, north of the South Polar region, and the other extending from northeastern Brazil to the Gran Chaco region of Argentina. The trade winds flow toward the low pressure that is centered just below the equator in January and on the equator in July. The low pressure is associated with uplift, producing intense convective rains and attracting moisture-laden winds from the Atlantic.

The circulation of air over South America is influenced by winds related to these anticyclones. The trade winds coming from the subtropical high located over the south Atlantic Ocean do not blow over the Andes, which also prevent winds from the South Pacific anticyclone from reaching the interior of the continent.

The North Pacific anticyclone has little influence on the climate of northwestern South America; its effects are felt only as far south as Costa Rica. It seldom affects Colombia and Venezuela. The thermal continental and equatorial low-pressure belts are of seasonal importance. The thermal low-pressure center is maintained by the warming of the continent during the summer season (December to February) when air pressure is lowest in the center of the continent, near 20 degrees south latitude and 60 degrees west longitude.

Ocean Currents

Three ocean currents have a strong influence on South American climates. The Peru (or Humboldt) Current in the southwest and the Falkland Current in the southeast both carry cold water from the temperate zone into the tropics. The Brazil Current brings warm water from the equatorial region of the Atlantic Ocean toward the east coast of South America, south of the equator, along the coast of Brazil. Cold currents chill the winds that blow over them and reduce the amount of moisture that reaches the neighboring shores; warm currents increase rainfall.

On occasion, during an El Niño event, the South Equatorial Current flows eastward instead of westward and brings warm water to the west coast of South America. The name "El Niño" (the child) was coined in Peru because this phenomenon occurs around Christmas (a reference to the Christ Child).

Higher pressure than normal on the western Pacific Ocean weakens the trade winds that normally flow westward. The South Equatorial Current flows eastward across the Pacific Ocean. The warm surface water slows the normal upwelling of the Peru Current. Surface water, depleted in nutrients, replaces the nutrient-rich water of the Peru Current. The whole food chain—phytoplankton, fish, marine mammals, and birds—suffers from the lack of nourishment, and massive fish kills occur, leaving local fishing-based economies in turmoil. Coastal Ecuador and northern Peru receive abundant rains and experience devastating floods. El Niño events happen every two to seven years. The strongest El Niño event of the twenty-first century occurred in 2015–2016.

On the west coast of South America, trade winds blowing parallel to, and slightly away from, the coast skim off warm surface water and contribute to the upwelling of cooler subsurface water; this cool water chills the air and further inhibits convection in the desert climate of coastal Peru and northern Chile. The Peru Current carries cold water to the coast of southern Ecuador, where the flow turns westward. Cold currents chill the winds that blow over them and reduce the amount of moisture that reaches the neighboring shores.

The east coast of South America is bathed by the cold Falkland Current in the south. Rio Gallegos, Argentina, has a maximum monthly temperature during the summer of 57 degrees Fahrenheit (13.1 degrees Celsius), which is much colder than would be expected at this latitude.

The coastal area of Brazil, south of the equator, is washed by the warm Brazil Current. Warm ocean currents heat the adjacent air, providing abundant water vapor and energy for convective storms. In the absence of an El Niño event, only the far northwestern coast of South America lies within reach of a warm current, the Equatorial Countercurrent.

Elevation

Elevation helps to explain the division of South America into climatic belts. The highest mountain, the Aconcagua, reaches about 22,800 feet (6,960 meters) above sea level. Many peaks are higher than 19,000 feet (5,800 meters). All climates are modified by elevation. The temperature range remains identical to the sea-level climate type, but each monthly temperature is decreased by about 11.7 degrees Fahrenheit (6.5 degrees Celsius) for every 3,280 feet (1,000 meters) rise in elevation.

Climate Regions

South America can be divided into several well-defined climatic regions, resulting from the interplay of the six major factors. These climates are tropical rain forest (tropical wet); tropical savanna (tropical wet-dry); monsoon; semiarid or steppe (called Sahel in Africa); subtropical dry (desert); humid subtropical (similar to the climate on the East Coast of North America from the Mid-Atlantic states to the northern part of Florida); Mediterranean; and mountain.

Tropical Rain Forest

The tropical rain forest (tropical wet) climate extends over 1.5 million square miles (4 million sq. km.) across the equator. In the east, it covers the Amazon Basin and part of the coast of Brazil. It extends farther to the north along the coast of the Guyanas to the Orinoco River delta. On the west coast, it covers the coastal region of Colombia from about 3 to 11 degrees north latitude. Heavy rainfall, evenly distributed all year, averages about 98 inches (2,500 millimeters) in Para, Brazil; about 112 inches (2,800 millimeters) in Iquitos, Peru; and 89 inches (2,300 millimeters) in Manaus, Brazil. In Buenaventura, the main harbor of Colombia and one of the wettest areas in the world, it rains more than 300 days a year, producing more than 354 inches (9,000 millimeters) of rain per year, the highest precipitation of the continent.

Laguna de Sonso tropical dry forest in Northern Andes. (Image by Mateo.gable)

The average temperature for a tropical wet climate at sea level is around 86 degrees Fahrenheit (30 degrees Celsius). The temperature range is low, usually less than 5 degrees Fahrenheit (3 degrees Celsius), because there are no great variations of solar radiation at noon and minimal changes in length of day throughout the year.

In the Amazon region, marked rainy seasons occur; the southern part receives the heaviest rain during the southern summer, while the northern part has its rainy season during the northern summer. There is no dry season and humidity is always high. The natural vegetation in the tropical rain forest is called *selva* in South America.

Tropical Savanna

The second type of tropical climate is the savanna: grassland with scattered trees, also known as tropical wet-dry. It is characterized by high temperatures, with monthly minimum temperatures above 64 degrees Fahrenheit (18 degrees Celsius), but with less precipitation than the tropical rain forest climate and a definite dry season. This climate covers 2.6 million square miles (6.8 million sq. km.), with 73 percent of this climate in Brazil only. It is found around the rain-forest belt in the Orinoco Basin, on the Brazilian Highlands, and in part of western Ecuador between Guayaquil and Quito. There is a prolonged dry season centered on the local winter. Natural vegetation has adapted to the two seasons of precipitation.

Monsoon

The monsoon is a seasonal wind that blows from the ocean during the summer. Monsoon climates are limited in South America. They only affect the coast of Colombia and, in Brazil, a very narrow area south of the Amazon River's mouth. Laden with moisture, these winds generate intense rainfall.

Monsoon phenomena occur because the equatorial lows over the oceans act like a magnet for the trade winds, giving rise to the intertropical convergence zone (ITCZ).

Within the ITCZ, the converging trade winds are forced upward by equatorial warming, creating active cloud bands and precipitation. In satellite images, the ITCZ can appear as two parallel cloud bands separated by a clear sky. Usually the ITCZ is located over the oceans north of the equator. In the summer, it reaches as far south as the mouth of the Amazon River on the Atlantic coast; on the Pacific coast, it is kept north by the cold waters of the Peru Current and the strong southeasterly trade winds. It is this migration of the ITCZ across the equator that creates monsoon phenomena.

Humid Subtropical

Humid subtropical climates characterized by lower winter temperatures are found south of the Tropic of Capricorn (in Paraguay, Uruguay, parts of Bolivia, Brazil, Argentina, and Chile) and in the Andes, where, in their rainfall pattern, they resemble tropical climates. On the Atlantic side, temperatures in the warmest month average 77 degrees Fahrenheit (25 degrees Celsius), but cold month averages vary from 63 degrees Fahrenheit (17 degrees Celsius) in the north (Asunción, Paraguay) to 50 degrees Fahrenheit (10 degrees Celsius) in Buenos Aires. Rainfall is greater than 1.5 inches (38 millimeters) each month in the east, decreasing to the west. Rainfall reaches its maximum during the local summer. Some frontal systems are strong enough to cross the Andes and induce winter rain and snow on the Argentine plains, when the encounter of subtropical and subpolar air masses causes depressions (midlatitude cyclones) with associated severe storms.

In southern Chile, south of 40 degrees south latitude, the climate is dominated by the westerlies blowing across the Pacific Ocean. Moisture picked up from the ocean is released as the winds rise over the west flank of the Andes, making this one of the wettest spots on the continent. Rainfall averages 102 inches (2,600 millimeters) at Valdivia, Chile, and probably twice that amount on the western slopes of the mountains. On the leeward side of the southern Andes, the westerlies are still strong, but having deposited their moisture on the western slopes of the mountains, they only exacerbate the dryness of Argentine Patagonia. The continuous flow of maritime air keeps the temperature at lower annual levels than in the corresponding North

American and European latitudes, discouraging settlement and farming. In winter, cold fronts from the south can penetrate north as far as the southern Brazilian Highlands and seriously damage the coffee and cocoa plantations.

Mediterranean Climates

A Mediterranean climate is present in South America only in central Chile, between 32 and 38 degrees south latitude. This climate is characterized by a mild, rainy winter and dry summer. Temperatures in South America are cooler than those at the same latitude in the Mediterranean Sea region.

Mountain Climates

The Andes, about 5,000 miles (8,000 km.) long, with many summits at elevations above 22,000 feet (6,700 meters), have a formidable impact on rainfall distribution. As humid air masses are forced up and over the windward side of the Andes, cooling causes condensation, often creating persistent rains associated with cloud forests. The opposite occurs on the leeward side of the mountains; as the descending air warms up, it absorbs water vapor and fosters arid conditions. This is called the rain shadow effect.

At the latitude of Peru, the Andes form two parallel cordilleras surrounding high plateaus called the *altiplano*. Near Lake Titicaca, the average annual temperature is only 34 to 36 degrees Fahrenheit (1 to 2 degrees Celsius). At altitudes above 3,000 to 4,000 feet (1,000 to 1,200 meters) the Andes experience frost, even near the equator.

More important than latitudinal differences in temperatures are temperature variations with elevation, since these influence vegetation adaptation in the tropical Andes. In general, there is an increase in rainfall with elevation, which peaks around 5,000 feet (1,500 meters). Different names have been given to the levels of elevation in the tropical Andes: *tierra caliente, tierra templada, tierra fria, tierra helada,* and *tierra Nevada*.

The *tierra caliente* is located between the foothills and the 71 to 75 degrees Fahrenheit (22 to 24 degrees Celsius) annual isotherm around 2,500 to 3,200 feet (800 to 975 meters). The *tierra caliente* ex-

hibits year-round, fairly high temperatures that are characteristic of tropical lowlands. Plant growth is limited only by lack of moisture. The *tierra templada,* stretching up to the 61 to 64 degrees Fahrenheit (16 to 18 degrees Celsius) isotherm at about 6,500 feet (2,000 meters), has tropical crops such as coffee and cocoa.

The upper limit for coffee cultivation is near 6,000 feet (2,000 meters). However, sometimes the *tierra templada* penetrates farther upward in narrow valleys, along which warm, humid air from the lowlands rises almost to the *tierra fria* boundary. In Ecuador, coffee and cotton are cultivated at up to 6,900 feet (2,100 meters) of elevation in the Mira Valley. In the eastern Andes of Bolivia, similar valleys are virtual corridors for subtropical vegetation and crops from the lowland, up to approximately 8,200 feet (2,500 meters).

The *tierra fria* (also called the montane zone), between 6,500 and 13,000 feet (2,000 and 4,000 meters) with a minimum annual temperature of between 43 and 64 degrees Fahrenheit (6 and 18 degrees Celsius), is marked by the occurrence of frost. The natural vegetation seems barely changed; height and diversity decrease with elevation. Together, *tierra templada* and *tierra fria* cover about 309,000 square miles (800,000 sq. km.).

The *tierra helada* is between the lowest boundary of the snow line and the frost line, which usually coincides with the 36 to 39 degrees Fahrenheit (2 to 4 degrees Celsius) isotherm between 15,000 and 20,000 feet (4,500 and 6,100 meters), depending on the humidity. An Alpine meadow covers the ground. The *tierra Nevada* is the area of permanent snow and ice.

Arid Climates

Arid climates cover more than 400,000 square miles (1 million sq. km.) in four areas of South America. Patagonia east of the Andes is in the rain shadow zone, and rainfall is low (about 4 inches/100 millimeters in San Juan, Argentina). The annual range in temperature is more than 36 degrees Fahrenheit (20 degrees Celsius), the highest in South America. Another arid zone is found in a narrow coastal strip along the Pacific coast between 5 and 30 degrees south latitude. This strip covers 150,000 square miles (400,000 sq. km.) and is 1,875 miles (3,000 km.) long.

The cold seas characterizing the Peru (Humboldt) Current and the proximity of the high Andes produce an inversion of normal atmospheric temperature, because air in contact with the water cools more rapidly than the upper strata of air. The result is a cloud layer about 1,200 feet (365 meters) thick, present at altitudes between 1,000 and 3,000 feet (300 and 1,000 meters). This cloud layer prevents the warming of the air near the ground. Temperatures are consequently low: Lima has an average temperature of 64 degrees Fahrenheit (18 degrees Celsius). The coast of Peru is the cloudiest desert in the world, with no sunshine for at least six months of the year. It almost never rains, but fog condensation provides some moisture. The yearly precipitation is low: only 0.04 inch (1 millimeter) in Arica and Iquique and 0.15 inch (4 millimeters) in Antofagasta (all in Chile).

Another desert extends from northeastern Colombia to Venezuela, covering a zone where rains are scarce and droughts prolonged. A final arid zone occurs in northeastern Brazil, between the Parnaiba and São Francisco Rivers. The interior highlands act as a wedge separating the sea winds from the northeast, all of which carry their moisture beyond the region. Average annual rainfall is less than 4 inches (100 millimeters) and the dry season can last seven months. Rainfall is irregular, resulting in the severe droughts that plague the region.

The seasonal movements of the ITCZ have great impact in tropical South America. When the ITCZ moves south along the coast of the Guianas toward northern Brazil during the summer, abundant rains can be expected. In years when the South Atlantic trades are quite strong, the ITCZ is prevented from moving southward along the coast of the Guianas. Summer rainfall will be scarce or fail altogether. Northeast Brazil then experiences serious droughts. These droughts, resulting in loss of life and mass migration, occur frequently. Notable drought years include 1970, 1983, 1991–1992, 1997, 2008–2009, and 2015–2017.

The air masses of temperate South America are determined by fronts, westerly winds, and air mass invasions from subantarctic latitudes. South of 40 degrees south latitude, particularly on the Pacific side of the continent, the climate is dominated by the westerlies blowing across the Pacific Ocean. Moisture gathered over the ocean is released over the west flank of the Andes, making this one of the wettest spots on the continent.

Denys Lemaire

CENTRAL AMERICA

The complex physical geography of Central America produces a variety of climates, including virtually every type of tropical and subtropical climate as well as a few midlatitude types. This diversity relates to several climatic controls, including latitude, elevation, maritime and continental influences, and the dominant wind systems. Some of these characteristics are shared by Central America's northern neighbor, Mexico, which is an extension of Central America.

Climate Controls

All seven of Central America's nations lie entirely within the tropics, and variations in their climates are functions primarily of altitude. By contrast, the chief control of the climate of their northern, subtropical neighbor, Mexico, is latitude. The greatest impact of latitude is in temperature seasonality. Because Central America is within the tropics, temperatures lack seasonality, with annual temperature ranges of only 15 degrees Fahrenheit (8 degrees

Celsius) or less. Daily temperature ranges are usually higher than annual ranges. By contrast, northern Mexico experiences hot summers and cool winters.

Central America is a mountainous region, and elevation exerts a control on climates. Average temperatures drop 3.5 degrees Fahrenheit per 1,000 feet (6 degrees Celsius per 1,000 meters), so that highlands lack tropical heat. The Spanish colonial settlers named the various altitude zones they found in Central America (and Mexico) after their temperature characteristics: *tierra caliente* (hot land), *tierra templada* (temperate land), *tierra fria* (cold land), and *tierra helada* (frozen land).

The *tierra caliente*, lying between sea level and about 3,000 feet (900 meters), is the true tropical lowlands, typical of the half of tropical Central America near sea level. It was an uncomfortable and unhealthy climate for Europeans, whose activities there were confined primarily to tropical plantation agriculture. Spaniards preferred the cooler *tierra templada*, between elevations of 3,000 and 6,000 feet (900 and 1,800 meters), where they built numerous cities and grew coffee or raised livestock. Most of the indigenous population also occupied that zone.

The *tierra fria*, between 6,000 and 12,000 feet (1,800 and 3,600 meters) experiences frost and cool temperatures, which limited European activity, especially in the upper reaches of this zone, although many indigenous people lived there. The highest zone, the *tierra helada*, is found only in Mexico.

Maritime and continental influences relate mostly to the warm ocean currents that bring warm air and moisture to most coastal locales of Central America. The east coasts are bathed by the Caribbean and Gulf of Mexico currents, having waters ranging between 73 and 84 degrees Fahrenheit (23 and 29 degrees Celsius). The Equatorial Countercurrent along the southern half of Central America's Pacific coast brings in waters of about 81 degrees Fahrenheit (27 degrees Celsius) all year.

These currents produce warm, moist air masses that dominate the climate of coastal locales, and some interior locales as well. Areas distant from the ocean experience considerably hotter summers and cooler winters, and have drier climates.

CENTRAL AMERICAN CLIMATIC HAZARDS

Central America an unusually large variety of climatic hazards because of their tropical location and its proximity to the North American continent. Much of the region north of the Tropic of Cancer suffers periodic or permanent drought, which limits population densities. Torrential rains and flooding, with accompanying landslides, can occur in any part of Central America, even in the driest regions, because of cold fronts, easterly waves, and hurricanes.

Northern Mexico and the tropical highlands are threatened by winter frost, which can damage or destroy crops. The highlands Guatemala receive severe summer thunderstorms and crop-damaging hailstorms. Cool northerly winds from North America sometimes extend into Nicaragua, and even occasionally to northern South America. These winds can bring frost or snow in the north and, at higher elevations, dust storms. Hurricanes, named after the Mayan god of rain and thunder, Huracan, are intense tropical cyclones that can strike both coasts. Between June and October, up to twelve hurricanes may hit the east coast annually, often seriously damaging human structures.

Northeasterly trade winds are the main wind systems influencing Central America. These blow across warm seas, bringing warmth and moisture to the windward (eastern) coasts and upland slopes. Such locales experience average annual precipitation of 80 to 120 inches (2,000 to 3,000 millimeters). Lee locations, including interior rain-shadow basins, have less precipitation, typically ranging between 20 and 40 inches (500 and 1,000 millimeters).

From December to April, North American cold fronts and the northerly winds behind them often intrude into Central America, triggering showers, even snow in higher elevations. Once the fronts pass, cold air arrives in what are called *nortes*, strong north winds. This influence is strongest in northern Mexico, decreasing southward, as the winds enter Central America proper.

Arenal Volcano, Costa Rica. (Image by BonniePics)

Climate Types

At least four climate types can be identified in Central America: tropical wet, tropical wet-dry, tropical dry, and highland. Because of the variety of climatic influences, there are variations within each of these four types.

The tropical wet climate is the climate of most of the *tierra caliente* in tropical windward locales. It stretches in a narrow band from the eastern escarpment of central Mexico into Central America as far south as Panama. A band of tropical wet climates also occurs along the southern extremes of the Pacific coast from Nicaragua to Panama, resulting from warm offshore currents. Temperature seasonality is lacking—it is always hot.

Annual precipitation usually exceeds 80 inches (2,000 millimeters), and there is no significant dry season. This climate is associated with native rain forests of broadleaf evergreen trees in great variety. However, humans living there have caused a serious decrease in the extent of this vegetation. Plantation

and subsistence crops typical of this climate include temperature-sensitive sugarcane, banana, coconut, mango, papaya, citrus, pineapple, cacao, manioc, and yam. In this climate zone, there are various health hazards, such as malaria, yellow fever, and dengue fever.

The tropical wet-dry climate occurs primarily in the lee of the trade winds, also within the *tierra caliente*. It is found in most of the Pacific coastal lowlands, interior lowlands, and the neighboring Yucatán Peninsula. Temperatures are tropical, always hot and lacking seasons. Precipitation is strongly seasonal, with less than 80 inches (2,000 millimeters) of rain falling between May and October. The four- to six-month dry season centers on December through March. Native vegetation consists of low, dry-season deciduous forests or scrub and cactus. Heavy alteration of this vegetation results from higher population densities than are found in the wet tropics.

The highland climates are cooler and wetter than the lowland climates around them. They coincide with the *tierra templada*, *tierra fria*, and *tierra helada* zones from northern Mexico through Panama. Despite little temperature seasonality in the tropical highlands, at least one month averages below 64 degrees Fahrenheit (18 degrees Celsius).

Precipitation in the highland climates is mostly seasonal, coming chiefly as summer thunderstorms. Annual precipitation varies from 30 inches (760 millimeters) upward, reaching extremes in windward locales, where seasonality is not a factor. Within the *tierra helada*, snowfalls are common all year, even well into the tropics, and the highest volcanoes support glaciers. Native vegetation is typically an oak-coniferous forest. Midlatitude commercial crops raised here to take advantage of cooler temperatures include maize, wheat, apple, pear, potato, cabbage, onion, and carrot. Much of the forest has been cleared for cattle pastures.

The dry climates, in which annual water losses through evaporation and transpiration exceed precipitation, occur in Central America, mostly in the interior plateau and in small lowland rain shadows. Dry climates include desert (arid) and steppe (semiarid) climates, whose annual precipitation is usually under 20 inches (500 millimeters). Temperatures are subtropical, since these regions lie mostly outside the tropics. Summers are even hotter than in the lowland tropics, and winters are cool. Native vegetation ranges from cactus and shrub in the deserts to oak-juniper woodlands in the wetter steppes. Commercial crop agriculture is limited to irrigated areas, where productivity is high.

P. Gary White

DISCUSSION QUESTIONS: CLIMATOLOGY

Q1. How does the climate of South America's west coast differ from that of its east coast? What factors contribute to the differences?

Q2. What is the intertropical convergence zone (ITCZ)? How and in what areas do the ITCZ's movements influence weather?

Q3. How does elevation influence the climates of Central America? How did the Spanish colonial settlers differentiate among the various altitude climate zones? In which zone did the Spaniards prefer to live?

Biogeography and Natural Resources

OVERVIEW

MINERALS

Mineral resources make up all the nonliving matter found in the earth, its atmosphere, and its waters that are useful to humankind. The great ages of history are classified by the resources that were exploited. First came the Stone Age, when flint was used to make tools and weapons. The Bronze Age followed; it was a time when metals such as copper and tin began to be extracted and used. Finally came the Iron Age, the time of steel and other ferrous alloys that required higher temperatures and more sophisticated metallurgy.

Metals, however, are not the whole story—economic progress also requires fossil fuels such as coal, oil, natural gas, tar sands, or oil shale as energy sources. Beyond metals and fuels, there are a host of mineral resources that make modern life possible: building stone, salt, atmospheric gases (oxygen, nitrogen), fertilizer minerals (phosphates, nitrates, and potash), sulfur, quartz, clay, asbestos, and diamonds are some examples.

Mining and Prospecting

Exploitation of mineral resources begins with the discovery and recognition of the value of the deposits. To be economically viable, the mineral must be salable at a price greater than the cost of its extraction, and great care is taken to determine the probable size of a deposit and the labor involved in isolating it before operations begin. Iron, aluminum, copper, lead, and zinc occur as mineral ores that are mined, then subjected to chemical processes to separate the metal from the other elements (usually oxygen or sulfur) that are bonded to the metal in the ore.

Some deposits of gold or platinum are found in elemental (native) form as nuggets or powder and may be isolated by alluvial mining—using running water to wash away low-density impurities, leaving the dense metal behind. Most metal ores, however, are obtained only after extensive digging and blasting and the use of large-scale earthmoving equipment. Surface mining or strip mining is far simpler and safer than underground mining.

Safety and Environmental Considerations

Underground mines can extend as far as a mile into the earth and are subject to cave-ins, water leakage, and dangerous gases that can explode or suffocate miners. Safety is an overriding issue in deep mines, and there is legislation in many countries designed to regulate mine safety and to enforce practices that reduce hazards to the miners from breathing dust or gases.

In the past, mining often was conducted without regard to the effects on the environment. In economically advanced countries such as the United States, this is now seen as unacceptable. Mines are expected to be filled in, not just abandoned after they are worked out, and care must be taken that rivers and streams are not contaminated with mine wastes.

Iron, Steel, and Coal

Iron ore and coal are essential for the manufacture of steel, the most important structural metal. Both raw materials occur in many geographic regions. Before the mid-nineteenth century, iron was smelted in the eastern United States—New Jersey, New York, and Massachusetts—but then huge hematite deposits were discovered near Duluth, Minnesota, on Lake Superior. The ore traveled by ship to steel mills in northwest Indiana and northeast Illinois, and coal came from Illinois or Ohio. Steel also was made in Pittsburgh and Bethlehem in Pennsylvania, and in Birmingham, Alabama.

After World War II, the U.S. steel industry was slow to modernize its facilities, and after 1970 it had great difficulty producing steel at a price that could compete with imports from countries such as Japan, Korea, and Brazil. In Europe, the German steel industry centered in the Ruhr River valley in cities such as Essen and Düsseldorf. In Russia, iron ore is mined in the Urals, in the Crimea, and at Krivoi Rog in Ukraine. Elsewhere in Europe, the French "minette" ores of Alsace-Lorraine, the Swedish magnetite deposits near Kiruna, and the British hematite deposits in Lancashire are all significant. Hematite is also found in Labrador, Canada, near the Quebec border.

Coal is widely distributed on earth. In the United States, Kentucky, West Virginia, and Pennsylvania are known for their coal mines, but coal is also found in Illinois, Indiana, Ohio, Montana, and other states. Much of the anthracite (hard coal) is taken from underground mines, where networks of tunnels are dug through the coal seam, and the coal is loosened by blasting, use of digging machines, or human labor. A huge deposit of brown coal is mined at the Yallourn open pit mine west of Melbourne, Australia. In Germany, the mines are near Garsdorf in Nord-Rhein/Westfalen, and in the United Kingdom, coal is mined in Wales. South Africa has coal and is a leader in manufacture of liquid fuels from coal. There is coal in Antarctica, but it cannot yet be mined profitably. China and Japan both have coal mines, as does Russia.

Aluminum

Aluminum is the most important structural metal after iron. It is extremely abundant in the earth's crust, but the only readily extractable ore is bauxite, a hydrated oxide usually contaminated with iron and silica. Bauxite was originally found in France but also exists in many other places in Europe, as well as in Australia, India, China, the former Soviet Union, Indonesia, Malaysia, Suriname, and Jamaica.

Much of the bauxite in the United States comes from Arkansas. After purification, the bauxite is combined with the mineral cryolite at high temperature and subjected to electrolysis between carbon electrodes (the Hall-Héroult process), yielding pure aluminum. Because of the enormous electrical energy requirements of the Hall-Héroult method, aluminum can be made economically only where cheap power (preferably hydroelectric) is available. This means that the bauxite often must be shipped long distances—Jamaican bauxite comes to the United States for electrolysis, for example.

Copper, Silver, and Gold

These coinage metals have been known and used since antiquity. Copper came from Cyprus and takes its name from the name of the island. Copper ores include oxides or sulfides (cuprite, bornite, covellite, and others). Not enough native copper occurs to be commercially significant. Mines in Bingham, Utah, and Ely, Nevada, are major sources in the United States. The El Teniente mine in Chile is the world's largest underground copper mine, and major amounts of copper also come from Canada, the former Soviet Union, and the Katanga region mines in Congo-Kinshasa and Zambia.

Silver often occurs native, as well as in combination with other metals, including lead, copper, and gold. Famous silver mines in the United States include those near Virginia City (the Comstock lode) and Tonopah, Nevada, and Coeur d'Alene, Idaho. Silver has been mined in the past in Bolivia (Potosi mines), Peru (Cerro de Pasco mines), Mexico, and Ontario and British Columbia in Canada.

Gold occurs native as gold dust or nuggets, sometimes with silver as a natural alloy called electrum. Other gold minerals include selenides and tellurides. Small amounts of gold are present in sea water, but attempts to isolate gold economically from this source have so far failed. Famous gold rushes occurred in California and Colorado in the United States, Canada's Yukon, and Alaska's Klondike region. Major gold-producing countries include South Africa, Siberia, Ghana (once called the Gold Coast), the Philippines, Australia, and Canada.

THE EXXON VALDEZ OIL SPILL

On March 24, 1989, the tanker Exxon Valdez, with a cargo of 53 million gallons of crude oil, ran aground on Bligh Reef in Prince William Sound, Alaska. Approximately 11 million gallons of oil were released into the water, in the worst environmental disaster of this type recorded to date. Despite immediate and lengthy efforts to contain and clean up the spill, there was extensive damage to wildlife, including aquatic birds, seals, and fish. Lawsuits and calls for new regulatory legislation on tankers continued a decade later. Such regrettable incidents as these are the almost inevitable result of attempting to transport the huge oil supplies demanded in the industrialized world.

Petroleum and Natural Gas

Petroleum has been found on every continent except Antarctica, with 600,000 producing wells in 100 different countries. In the United States, petroleum was originally discovered in Pennsylvania, with more important discoveries being made later in west Texas, Oklahoma, California, and Alaska. New wells are often drilled offshore, for example in the Gulf of Mexico or the North Sea. The United States depends heavily on oil imported from Mexico, South America, Saudi Arabia and the Persian Gulf states, and Canada.

Over the years, the price of oil has varied dramatically, particularly due to the attempts of the Organization of Petroleum Exporting Countries (OPEC) to limit production and drive up prices. In Europe, oil is produced in Azerbaijan near the Caspian Sea, where a pipeline is planned to carry the crude to the Mediterranean port of Ceyhan, in Turkey. In Africa, there are oil wells in Gabon, Libya, and Nigeria; in the Persian Gulf region, oil is found in Kuwait, Qatar, Iran, and Iraq. Much crude oil travels in huge tankers to Europe, Japan, and the United States, but some supplies refineries in Saudi Arabia at Abadan. Tankers must exit the Persian Gulf through the narrow Gulf of Hormuz, which thus assumes great strategic importance.

After oil was discovered on the shores of the Beaufort Sea in northern Alaska (the so-called North Slope) in the 1960s, a pipeline was built across Alaska, ending at the port of Valdez. The pipeline is heated to keep the oil liquid in cold weather and elevated to prevent its melting through the permanently frozen ground (permafrost) that supports it. From Valdez, tankers reach Japan or California.

Drilling activities occasionally result in discovery of natural gas, which is valued as a low-pollution fuel. Vast fields of gas exist in Siberia, and gas is piped to Western Europe through a pipeline. Algerian gas is shipped in the liquid state in ships equipped with refrigeration equipment to maintain the low temperatures needed. Britain and Northern Europe benefit from gas produced in the North Sea, between Norway and Scotland.

Shale oil, a plentiful but difficult-to-exploit fossil fuel, exists in enormous amounts near Rifle, Colorado. A form of oil-bearing rock, the shale must be crushed and heated to recover the oil, a more expensive proposition than drilling conventional oil wells. In spite of ingenious schemes such as burning the shale oil in place, this resource is likely to remain largely unused until conventional petroleum is used up. A similar resource exists in Alberta, Canada, where the Athabasca tar sands are exploited for heavy oils.

John R. Phillips

RENEWABLE RESOURCES

Most renewable resources are living resources, such as plants, animals, and their products. With careful management, human societies can harvest such resources for their own use without imperiling future supplies. However, human history has seen many instances of resource mismanagement that has led to the virtual destruction of valuable resources.

Forests

Forests are large tracts of land supporting growths of trees and perhaps some underbrush or shrubs. Trees constitute probably the earth s most valuable, versatile, and easily grown renewable resource. When they are harvested intelligently, their natural environments continue to replace them. However, if a harvest is beyond the environment's ability to restore the resource that had been present, new and different plants and animals will take over the area. This phenomenon has been demonstrated many times in overused forests and grasslands that reverted to scrubby brushlands. In the worst cases, the abused lands degenerated into barren deserts.

The forest resources of the earth range from the tropical rain forests with their huge trees and broad diversity of species to the dry savannas featuring scattered trees separated by broad grasslands. Cold, subarctic lands support dense growths of spruces and firs, while moderate temperature regimes produce a variety of pines and hardwoods such as oak and ash. The forests of the world cover about 30 percent of the land surface, as compared with the oceans, which cover about 70 percent of the global surface.

Harvested wood, cut in the forest and hauled away to be processed, is termed roundwood. Globally, the cut of roundwood for all uses amounts to about 130.6 billion cubic feet (3.7 billion cubic meters). Slightly more than half of the harvested wood is used for fuel, including charcoal.

Roundwood that is not used for fuel is described as industrial wood and used to produce lumber, veneer for fine furniture, and pulp for paper prod-

ucts. Some industrial wood is chipped to produce such products as subflooring and sheathing board for home and other building construction. Most roundwood harvested in Africa, South America, and Asia is used for fuel. In contrast, roundwood harvested in North America, Europe, and the former Soviet Union generally is produced for industrial use.

It is easy to consider forests only in the sense of the useful wood they produce. However, many forests also yield valuable resources such as rubber, edible nuts, and what the U.S. Forest Service calls special forest products. These include ferns, mosses, and lichens for the florist trade, wild edible mushrooms such as morels and matsutakes for domestic markets and for export, and mistletoe and pine cones for Christmas decorations.

There is growing interest among the industrialized nations of the world in a unique group of forest products for use in the treatment of human disease. Most of them grow in the tropical rain forests. These medicinal plants have long been known and used by shamans (traditional healers). Hundreds of pharmaceutical drugs, first used by shamans, have been derived from plants, many gathered in tropical rain forests. The drugs include quinine, from the bark of the cinchona tree, long used to combat malaria, and the alkaloid drug reserpine. Reserpine, derived from the roots of a group of tropical trees and shrubs, is used to treat high blood pressure (hypertension) and as a mild tranquilizer. It has been estimated that 25 percent of all prescriptions dispensed in the United States contain ingredients derived from tropical rain forest plants. The value of the finished pharmaceuticals is estimated at US$6.25 billion per year.

Scientists screening tropical rain forest plants for additional useful medical compounds have drawn on the knowledge and experience of the shamans. In this way, the scientists seek to reduce the search time and costs involved in screening potentially useful plants. Researchers hope that somewhere in the dense tropical foliage are plant products that could treat, or perhaps cure, diseases such as cancer or AIDS.

Many as-yet undiscovered medicinal plants may be lost forever as a consequence of deforestation of large tracts of equatorial land. The trees are cut down or burned in place and the forest converted to grassland for raising cattle. The tropical soils cannot support grasses without the input of large amounts of fertilizer. The destruction of the forests also causes flooding, leaving standing pools of water and breeding areas for mosquitoes, which can spread disease.

Marine Resources

When renewable marine resources such as fish and shellfish are harvested or used, they continue to reproduce in their environment, as happens in forests and with other living natural resources. However, like overharvested forests, if the marine resource is overfished—that is, harvested beyond its ability to reproduce—new, perhaps undesirable, kinds of marine organisms will occupy the area. This has happened to a number of marine fishes, particularly the Atlantic cod.

When the first Europeans reached the shores of what is now New England in the early seventeenth century, they encountered vast schools of cod in the local ocean waters. The cod were so plentiful they could be caught in baskets lowered into the water from a boat.

At the height of the New England cod fishery, in the 1970s, efficient, motor-driven trawlers were able to catch about 32,000 tons. The catch began to decline that year, mostly as a result of the impact of fifteen different nations fishing on the cod stocks. As a result of overfishing, rough species such as dogfish and skates constitute 70 percent of the fish in the local waters. Experts on fisheries management decided that fishing for cod had to be stopped.

The decline of the cod was attributed to two causes: a worldwide demand for more fish as food and great changes in the technology of fishing. The technique of fishing progressed from a lone fisher with a baited hook and line, to small steam-powered boats towing large nets, to huge diesel-powered trawlers towing monster nets that could cover a football field. Some of the largest trawlers were floating factories. The cod could be skinned, the edible parts cut and quick-frozen for market ashore, and the skin, scales, and bones cooked and ground for animal feed and oil. A lone fisher was lucky to be able to catch 1,000 pounds (455 kilograms) in one day. In contrast, the largest trawlers were capable of catching and processing 200 tons per day.

In the 1990s, the world ocean population of swordfish had declined dramatically. With a worldwide distribution, these large members of the billfish family have been eagerly sought after as a food fish. Because swordfish have a habit of basking at the surface, fishermen learned to sneak up on the swordfish and harpoon them. Fishermen began to catch swordfish with fishing lines 25 to 40 miles (40 to 65 kilometers) long. Baited hooks hung at intervals on the main line successfully caught many swordfish, as well as tuna and large sharks. Whereas the harpoon fisher took only the largest (thus most valuable) swordfish, the longline gear was indiscriminate, catching and killing many swordfish too small for the market, as well as sea turtles and dolphins

As a result of the catching and killing of both sexually mature and immature swordfish, the reproductive capacity of the species was greatly reduced. Harpoons killed mostly the large, mature adults that had spawned several times. Longlines took all sizes of swordfish, including the small ones that had not yet reached sexual maturity and spawned. The decline of the swordfish population was quickly obvious in the reduced landings. But things have changed remarkably, thanks to a 1999 international plan that rebuilt this stock several years ahead of schedule. Today, North Atlantic swordfish is one of the most sustainable seafood choices.

Albert C. Jensen

NONRENEWABLE RESOURCES

Nonrenewable resources are useful raw materials that exist in fixed quantities in nature and cannot be replaced. They differ from renewable resources, such as trees and fish, which can be replaced if managed correctly. Most nonrenewable resources are minerals—inorganic and organic substances that exhibit consistent chemical composition and properties. Minerals are found naturally in the earth's crust or dissolved in seawater. Of roughly 2,000 different minerals, about 100 are sources of raw materials that are needed for human activities. Where useful minerals are found in sufficiently high concentrations—that is, as ores—they can be mined as profitable commercial products.

Economic nonrenewable resources can be divided into four general categories: metallic (hardrock) minerals, which are the source of metals such as iron, gold, and copper; fuel minerals, which include petroleum (oil), natural gas, coal, and uranium; industrial (soft rock) minerals, which provide materials like sulfur, talc, and potassium; and construction materials, such as sand and gravel.

Nonrenewable resources are required as direct or indirect parts of all the products that humans use. For example, metals are necessary in industrial sectors such as construction, transportation equipment, electrical equipment and electronics, and consumer durable goods—long-lasting products such as refrigerators and stoves. Fuel minerals provide energy for transportation, heating, and electrical power. Industrial minerals provide ingredients needed in products ranging from baby powder to fertilizer to the space shuttle. Construction materials are used in roads and buildings.

Location

When minerals have naturally combined together (aggregated) they are called rocks. The three general rock categories are igneous, sedimentary, and metamorphic. Igneous rocks are created by the cooling of molten material (magma). Sedimentary rocks are caused when weathering, erosion, trans-portation, and compaction or cementation act on existing rocks.

Metamorphic rocks are created when the other two types of rock are changed by heat and pressure. The availability of nonrenewable resources from these rocks varies greatly, because it depends not only on the natural distribution of the rocks but also on people's ability to discover and process them. It is difficult to find rock formations that are covered by the ocean, material left by glaciers, or a rain forest. As a result, nonrenewable resources are distributed unevenly throughout the world.

Some nonrenewable resources, such as construction materials, are found easily around the world and are available almost everywhere. Other nonrenewable resources can only be exploited profitably when the useful minerals have an unusually high concentration compared with their average concentration in the earth's crust. These high concentrations are caused by rare geological events and are difficult to find. For example, an exceptionally rare nonrenewable resource like platinum is produced in only a few limited areas.

No one country or region is self-sufficient in providing all the nonrenewable resources it needs, but some regions have many more nonrenewable resources than others. Minerals can be found in all types of rocks, but some types of rocks are more likely to have economic concentrations than others. Metallic minerals often are associated with shields (blocks) of old igneous (Precambrian) rocks. Important shield areas near the earth's surface are found in Canada, Siberia, Scandinavia, and Eastern Europe. Another important shield was split by the movement of the continents, and pieces of it can be found in Brazil, Africa, and Australia.

Similar rock types are in the mountain formations in Western Europe, Central Asia, the Pacific coast of the Americas, and Southeast Asia. Minerals for construction and industry are found in all three types of rocks and are widely and randomly distributed among the regions of the world.

The fuel minerals—petroleum and natural gas—are unique in that they occur in liquid and gaseous states in the rocks. These resources must be captured and collected within a rock site. Such a site needs source rock to provide the resource, a rock type that allows the resource to collect, and another surrounding rock type that traps the resource. Sedimentary rock basins are particularly good sites for fuel collection. Important fuel-producing regions are the Middle East, the Americas, and Asia.

Impact on Human Settlement

Nonrenewable resources have always provided raw materials for human economic development, from the flint used in early stone tools to the silicon used in the sophisticated chips in personal computers. Whole eras of human history and development have been linked with the nonrenewable resources that were key to the period and its events. For example, early human culture eras were called the Stone, Bronze, and Iron Ages.

Political conflicts and wars have occurred over who owns and controls nonrenewable resources and their trade. One example is the Persian Gulf War of 1991. Many nations, including the United States, fought against Iraq over control of petroleum production and reserves in the Middle East.

Since the actual production sites often are not attractive places for human settlement and the output is transportable, these sites are seldom important population centers. There are some exceptions, such as Johannesburg, South Africa, which grew up almost solely because of the gold found there. However, because it is necessary to protect and work the production sites, towns always spring up near the sites. Examples of such towns can be found near the quarries used to provide the material for the great monuments of ancient Egypt and in the Rocky Mountains of North America near gold and silver mines. These towns existed because of the nonrenewable resources nearby and the needs of the people exploiting them; once the resource was gone, the towns often were abandoned, creating "ghost towns," or had to find new purposes, such as tourism.

More important to human settlement is the control of the trade routes for nonrenewable resources. Such controlling sites often became regions of great wealth and political power as the residents taxed the products that passed through their community and provided the necessary services and protection for the traveling traders. Just one example of this type of development is the great cities of wealth and culture that arose along the trade routes of the Sahara Desert and West Africa like Timbuktu (in present-day Mali) and Kumasi (in present-day Ghana) based on the trade of resources like gold and salt.

Even with modern transportation systems, ownership of nonrenewable resources and control of their trade is still an important factor in generating national wealth and economic development. Modern examples include Saudi Arabia's oil resources, Egypt's control of the Suez Canal, South Africa's gold, Chile's copper, Turkey's control over the Bosporus Strait, Indonesia's metals and oil, and China's rare earth element.

Gary A. Campbell

Natural Resources

South America

South America is the world's fourth-largest continent, so it comes as no surprise that it has a diversity of climates, terrains, altitudes, and latitudes. The continent is shaped somewhat like a large ice cream cone, with about 75 percent of the continent within tropical latitudes, and with the remainder—the point of the cone—in the middle latitudes. Tierra del Fuego extends to about 56 degrees south latitude, only 700 miles (1,126 km.) from the Antarctic Circle. Because of the way South America tapers to a pointed cone, sea-level temperatures at the far south never get as cold as corresponding latitudes in Canada or Russia.

South America has an abundance of natural resources, both renewable—fisheries, soils, forests, climate, and water—and nonrenewable— minerals and energy. Many of these resources are far from where they are needed and are costly to develop, limiting their ability to meet human needs and wants. Location and land quality can also be considered nonrenewable resources. For example, the amount of land available for growing populations to use is an important resource. The quality of land, including location, surface roughness, climate, soil fertility, and attractive landscapes, is part of the resource base. As population grows in a place, pressure to use more resources grows; conservation programs attempt to ensure that resources will be available when needed.

Renewable Resources: Fisheries

Chile and Peru were two of the first nations in the world to claim control of the oceans up to 200 miles (320 km.) from their coasts. The Peru (Humboldt) Current flows from the Antarctic and then along the coasts of Chile and Peru. This cold water contains great amounts of plankton, and countless numbers of fish that live in the current depend on the plankton as their principal food source. Peru and Chile usually rank second and third, respectively, in the world in total fish catch (after China). Most of the fish are ground up to make fishmeal, a source of protein for animal food and fertilizers. Most South American nations catch at least some fresh and saltwater fish for food. Most do this in an unorganized way. To keep fisheries a renewable resource, people have to avoid overfishing, which does not allow fish enough time to reproduce. Overfished fisheries soon lose such large quantities of fish that fishing becomes unproductive.

Soils

Three-fourths of South America is tropical, and tropical soils are not fertile for enough crops. They lack humus, potash, phosphates, and nitrogen, and they are acidic. High iron and aluminum content, drought, and too much water in some places are all problems that require expensive solutions to fix for farmers. Faced with these handicaps, South American farmers work hard but do not produce as much food as they would if soils and other environmental factors were better.

Brazil is a major exporter of corn (maize) and a major producer of rice but must import wheat and rice to feed its people. Most countries produce enough basic foods, but there are still food deficits in many highland indigenous communities and in poverty-stricken urban areas. Many tropical South American countries produce large quantities of so-called luxury foods: coffee from Brazil, Colombia,

SELECTED RESOURCES OF SOUTH AMERICA

Mahogany loggers in Belize, squaring logs for export around 1936. (Image by Paul Carpenter Standley)

and Venezuela; cacao (chocolate) from Ecuador, Colombia, and Brazil; sugar from Brazil, Venezuela, and Colombia; and bananas from Brazil, Ecuador, Colombia, and Venezuela. Illegal drugs are grown in Bolivia, Peru, and Colombia.

The southern cone middle-latitude countries of Chile, Argentina, and Uruguay produce surplus foods. The Pampas of Argentina and Uruguay are one of the world's great breadbaskets, producing wheat, corn, and soybeans, and beef, pork, and lamb. Argentina is the world's fifth-largest producer of wine, and Chile ranks sixth. Soils are excellent in these countries, and farms are highly productive.

Conservation programs to keep soils a renewable resource have started in South America. Land destruction is a big threat in mountainous terrain where erosion can wash away the topsoil, and in tropical environments where a disturbance of the natural ecology can ruin land. The United Nations and regional agencies operate demonstration projects for conservation in all countries. More than 300 national parks preserve fragile and unique physical environments.

Forests

Fully half of South America is forested. The largest forest areas are the tropical hardwood forests of Brazil, Colombia, Peru, and Venezuela, and the softwood pine forests in the Andes Mountain regions of Chile and Argentina. South American forest products are used throughout the world. Wood for construction and furniture come from South America's tropical and middle-latitude forests. Rosewood, Amazon cedar, mahogany, and balsa are harvested from the rain forests of Brazil, Peru, Colombia, and Venezuela. Rubber trees are native to Amazonia. In fact, Brazil had a monopoly on this valuable industrial product until trees were transplanted to Southeast Asia and Africa.

There are so many species of trees in the Amazon that rubber trees and nut trees are scattered among many miles of rain forest. This makes gathering rubber and nuts an expensive job. In Brazil's drier tropical forests, one species of palm tree gives carnauba wax, another forest product in demand throughout the world. Another species of palm growing on the semiarid Brazilian Plateau yields vegetable cooking oil.

Yet another dry tropical forest product is tannin from Paraguay's quebracho trees, used for tanning animal hides. Cinchona trees yield quinine for the treatment of malaria, and chicle from the sapodilla tree is the base for chewing gum. Pine wood is used in construction and for making pulp, paper, plastics, plywood, tar, pitch, and turpentine. The great distances from the forests to the consuming centers and the rough terrain that must be crossed make these products expensive to gather and ship

Climate

Climate and water resources are closely related. A region's climate is determined both by the amount of heat provided by the Sun and the amount of precipitation it receives. Because three-fourths of South America is within the tropics, it receives an abundance of energy from the Sun. Varied tropical climates can be found in the mountains. For every 1,000 feet (305 meters) that one travels up the side of a mountain, the temperature decreases about 3.3 degrees Fahrenheit (1.8 degrees Celsius). On a 15,000-foot (4,572-meter) mountain in the tropics, there would be five bands of climate types based on average temperature: warm land, temperate land, cold land, cold grassland (too cold for trees), and permanent ice and snow.

Different crops are found in each climate band, because plants have different temperature requirements. For example, rice grows in the warmest areas; coffee grows in the temperate regions; wheat grows in cold areas; pasture grasses grow in still colder regions; and nothing grows in permanent ice and snow. All these climates and crops are affected by differences in rainfall. Some tropical places get 100 inches (2,540 millimeters) of rain, while others get only 20 inches (510 millimeters). Rainforest vegetation would grow in the first region, but only cactuslike plants would grow in the latter.

This great ecological diversity means that many types of food and other plants can grow in a small area. However, in the warm and wet areas especially, many plant, animal, and human disease vectors also flourish. Much food is lost to disease, and expensive chemicals are needed to control diseases. Medicines can control most human and animal dis-

eases, but they, too, are expensive. All in all, for many South American countries, it is cheaper to import food from other countries than to invest large sums of money in their own land.

Excess energy from the Sun can have good and bad effects. If the technology were more widely developed, people probably could derive most of their electricity directly from the Sun. Tropical regions would benefit most when this happens. Tropical warmth means that cold-weather clothing is not necessary, and buildings do not need to be heated. Although air conditioning is often desirable, it is not as essential as heating systems are in cold climates. However, if the current global warming trend continues unchecked, it can have a potentially devastating effect for tropical areas, making temperatures there too warm for human habitation. Expanded production of solar energy by sunlight-rich South America can play an important role in stopping this trend. While still abundant, the continent's resources of fossil fuels—petroleum, coal, and natural gas—are finite. Solar energy is a renewable resource; it is inexhaustible. And unlike the production of fossil fuels, its production does not involve toxic fumes and greenhouse gases that cause global warming and climate change.

Vast amounts of plant biomass are produced under the wet tropical heat. The commercial exploitation of this potentially significant energy resource is in its early stages. Clean and efficient technologies are still to be developed. Wood-burning stoves, widespread in South America, cause air pollution; they emit carbon dioxide, one of the so-called greenhouse gases that trap heat in the atmosphere and lead to global warming. Even more adverse are the consequences of wide-scale tree burning, a common technique of clearing forested areas for agricultural and industrial uses. This deforestation practice not only emits more carbon dioxide into the atmosphere, it destroys the very trees that consume carbon dioxide and reduce its greenhouse effect.

Water

Great water environments exist in Amazonia, the high Andes, the Pantanal, and the other major river valleys. Fish, birds, reptiles, and water plants grow in large quantities in those places. The usefulness of these resources for human and animal food and for industrial materials has yet to be determined. One possible use is fish farming. Tropical ponds can produce large amounts of protein for human and animal food. Many sites for hydroelectric power lie in South America's mountains and plateaus. Brazil, Venezuela, Paraguay, Argentina, and Chile have developed river basins for power, irrigation, recreation, and navigation. More sites are available, but the cost to develop them is high. Once developed, given normal climate conditions, water resources are largely self-sustaining. A serious problem in every country is the need to develop more sources of clean drinking water.

Nonrenewable Resources: Minerals

Spanish conquerors frantically searched for gold and silver from the beginning of South America's colonial period. They found some gold in the Caribbean, a great deal of gold and silver in Central America, and large quantities of silver and gold in Colombia, Peru, and Bolivia. The Portuguese found gold later in Brazil. The fabulous silver reserves found in 1545 at Potosí, Bolivia, produced half the world's silver for the next fifty years. Trade monopolies were established between Spain and Portugal and the colonies. In many cases, these monopolies disallowed the exploitation of various South American minerals, such as mercury, which the mother countries also produced. After independence, especially in the late nineteenth and twentieth centuries, mineral exploitation accelerated. Some of this was done with the help of foreign capital and technology.

South America produces about 20 percent of the world's iron ore. Venezuela, Brazil, and Bolivia have especially large deposits; Brazil is the world's second leading producer after Australia. These deposits are found in remote places where few people live: the Guiana Highlands, the Amazon Basin, and southeastern Bolivia. Cheap strip-mining is

often carried out without environmental considerations.

Nearly half the world's copper is produced in Chile and Peru. Mines in the northern desert of Chile are especially rich, with 2.5 percent copper in the ore. Various obstacles have kept Peru from fully exploiting its copper resources. For example, a vast copper mine project to be developed near the city of Arequipa, Peru, was cancelled in 2011 due to environmental concerns. Bauxite (ore for aluminum) is another world-class resource, coming mostly from Guyana, Suriname, Venezuela, and Brazil. Around the year 1900, nitrates from northern Chile were the only source in the world for making gunpowder and fertilizers. Now nitrogen can be collected directly from the air, and so Chile's nitrates are used for iodine, a much less valuable product.

Peru, Bolivia, and Brazil are world leaders in tin exports (after China and Indonesia). All of Peru's tin is produced at the San Rafael mine in Puno. Bolivia's tin is extracted from solid rock ores located on the *altiplano* at elevations of up to 15,000 feet (4,570 meters), making it costly to work. Brazil's tin is found in gravel deposits in the northern Amazon Basin, where production costs are low.

MAJOR MINERALS OF SOUTH AMERICA

Mineral	*Country and Percentage of Global Production*
Bauxite	Brazil (14%); Suriname (1%)
Chromite	Brazil (1.1%)
Copper	Chile (33%); Peru (12%)
Gold	Peru (4.5%); Brazil (2.8%)
Iron ore	Brazil (4%)
Lead	Peru (7.8%)
Manganese	Brazil (5.5%)
Petroleum	Venezuela (3.8%)
Silver	Peru (16.9%); Chile (4.9%); Bolivia (4.6%)
Tin	Peru (8%), Bolivia (6%), Brazil (4%)
Tungsten	Bolivia (1.2%)
Zinc	Peru (13.4%); Bolivia (3.7%), Brazil (1.5%)

Source: Central Intelligence Agency, The World Factbook, 2019

Chuquicamata, the largest open pit copper mine in the world. (Image by Reinhard Mannheim)

Peru is a world leader in lead and zinc production. Industrial diamonds used in cutting tools are found in Brazil's Amazon region. Venezuela's small production of gem diamonds has been worked for many years. Gold, platinum, and emeralds are all mined in Colombia, and gold is another major product from Peru.

Energy Resources

South America has no large, high-quality coal resources. Coal is a valuable fuel for the smelting of metals, and its lack hurts such important industries as iron and steel production. Brazil, Colombia, Chile, Peru, and Argentina have coal deposits, but they are located in rough terrain and far from consuming centers.

Argentina, Bolivia, Brazil, Chile, Colombia, Ecuador, Peru, and Venezuela produce petroleum and natural gas, but only Venezuela is a world-class producer. Venezuela's oil is located in the Maracaibo Basin and near the Orinoco Delta. The Orinoco fields produce a heavy crude with much tar and sul-

fur content. These deposits are not as valuable as the lighter crudes. Still, Venezuela's proven oil reserves are the world's largest, and the country's economy overwhelmingly depends on petroleum exports. In recent years, Venezuela's oil production and exports have been dropping steadily. This decline is attributed to the policies of socialist president Hugo Chávez, who died in 2013, and his successor Nicolás Maduro. Their expulsion of international oil companies and the subsequent mismanagement of the industry were compounded by widespread corruption, falling oil prices, general political unrest, and international sanctions. Nevertheless, Venezuela has remained one of the most important suppliers of oil to the United States.

Uranium is mined in Brazil. Deposits have also been discovered in other South American countries, but their exploration and exploitation are in early stages.

Nuclear power plants operate only in Argentina (two plants are producing and one is under construction) and Brazil (one producing and one under construction). Geothermal energy resources are plentiful in the Andes, but to date, there is only one geothermal plant in South America, located in Chile. Solar energy is also underutilized in South America. Chile, however, is making strides, with about 7 percent of its energy derived from the Sun. Wind energy holds great potential in South America. In 2016, Brazil ranked eighth in the world among wind-power-producing countries. There is a strong tidal bore (incoming wave) from the Atlantic into the mouth of the Amazon that has energy-generating possibilities, but this has yet to be exploited. Tapping the energy potential of the great biomass of South America's forestsis in its early stages.

Location and Land Quality

Many people do not understand how location and land quality can be natural resources. The location of a person, place, or resource on Earth's surface determines how easy or difficult it is to reach other people, places, or resources elsewhere on the planet. Movement of anything from one place to

another takes time, costs money, and requires energy. If one place is close to people and resources, it is cheaper, faster, and requires less energy to bring all factors together to satisfy peoples' needs and wants. Unfortunately for Argentina, the factors required for steel production are located in Buenos Aires, Tucumán, and Río Gallegos, places separated by 2,000 miles (3,218 km.) of fairly rough terrain. This combination makes steel production and distribution costly and inefficient in Argentina.

In Colombia, the distances among resources is not great, but the mountainous terrain is so rugged that the time, costs, and energy needed to combine resources are excessive. Similar handicaps exist in many resource situations for most South American countries.

Global Location

South America is located mostly in the Southern Hemisphere, far from the great population centers in Asia, Europe, and North America. Despite jet aircraft and fast commercial shipping, the distances remain great: It is 8,000 miles (13,000 km.)

The Port of Callao in Peru has one of South America's few good natural harbors on the Pacific Ocean. (Image by Alex E. Proimos)

A cloud forest, also called a water forest, primas forest, or tropical montane cloud forest (TMCF), is a generally tropical or subtropical, evergreen, montane, moist forest characterized by a persistent, frequent or seasonal low-level cloud cover, usually at the canopy level, formally described in the International Cloud Atlas *(2017) as* silvagenitus. *The Parque Internacional la Amistad, Panama, is shown here. (Image by Dirk van der Made)*

from Colombia to Japan; 4,800 miles (7,725 km.) from Chile to the United States; and 7,335 miles (11,800 km.) from Uruguay to Spain. Improved transportation and communications systems have lowered the time, costs, and energy spent to send people, goods, or messages from South America to the rest of the world. Nevertheless, relative to locations in Europe, North America, or Asia, South America is still at a disadvantage. This results in lower production, less human contact, and fewer sales to world markets. Competition with other world resource centers brings reduced profits to South American producers.

Regional Variation in Resources

Brazil has the best-balanced resource base in South America. Amazon and Paraná Plateau hardwood and softwood forests, plentiful water resources in the Amazon Basin, fertile farmland in the southern states, and abundant minerals and energy resources throughout the country mean that Brazil's people should have comfortable living conditions. The fact that many Brazilians remain poor is partly caused by the country's late start in exploiting its resources and some inefficiencies in using them.

Argentina has the next best-balanced resource base, with the great food production of the Pampas, minerals in the Andes, and oil and natural gas in Patagonia. Beautiful mountain scenery also brings in substantial amounts of tourist income.

The Guianas, Uruguay, and Paraguay are resource-deficient, having only one or two major natural resources. Their small sizes partly account for this situation. Venezuela, Colombia, Ecuador, Peru,

Montecristo National Park, El Salvador. (Image by Elmer Guevara)

Bolivia, and Chile all have outstanding mineral and energy resources, but all are handicapped by rough terrain, making it inefficient to develop their resources. If all the South American countries were united in one common market, their resource bases could be rich enough to secure a comfortable future for their peoples.

Robert J. Tata

CENTRAL AMERICA

Spain once ruled most of Central America, but the region's mineral resources and sparse Amerindian population caused it to be largely ignored. After independence came to the region's countries in the mid-nineteenth century, their natural resources remained largely underdeveloped for several reasons. First, the Central American countries are small and typically lack significant resources. Second, the transportation infrastructure is poorly integrated, due in large part to rugged terrain. Third, these countries often lack the capacity to develop their own resources. Most of the development that has taken place often relies on foreign investment. As a result, only two resources have been widely exploited: forests and soils. The development of other resources, although accelerating somewhat, is frequently stymied by unrest and corruption.

Soils

All of the Central American countries depend heavily on agriculture in their domestic and export economies, and are thus reliant on productive soils. However, due to steep terrain and wide expanses of humid tropical climates, good agricultural soils are limited. There are highly fertile pockets, especially in the floodplains of larger stream valleys and in areas of weathered volcanic ash. These areas are mainly dominated by large-scale commercial agriculture, primarily for the production of sugarcane, bananas, coconuts, coffee, citrus, cotton, and tobacco.

Most of the farmland is found in areas less suitable for agriculture for a variety of reasons. The upland interior suffers from steep slopes that are prone to rapid erosion. Lowlands along the Pacific

SELECTED RESOURCES OF CENTRAL AMERICA

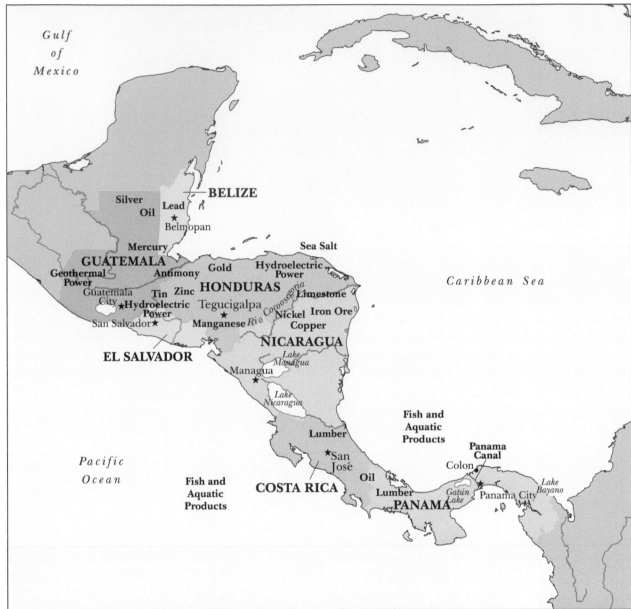

coast occupy the lee side of the uplands and are plagued by drought. Rain forests typically contain soils that are badly leached of nutrients, making farming unproductive. These less-suitable lands are usually owned by small-scale farmers. The holdings are tiny and fragmented. Farming in these areas tends to degrade the already marginal soils.

Small farms are largely devoted to food crops such as maize, beans, squash, plantains, manioc, and rice. Slash-and-burn agriculture, practiced by Amerindians since precolonial times, is still employed in some remote upland areas. In this system,

forest land is cleared by cutting and the slash is dried and burned. In the ashes are planted various crops, the most important of which is maize. Such plots can be used for only two or three years before they are abandoned and the cycle is started anew, at the expense of forests.

Grazing has expanded in both the natural grasslands (savannas) and within forest clearings. In some cases, African grass varieties have been introduced. Expansion of grazing pastures is in response to foreign demand for beef, especially by the United States. and, more recently, from China.

Minerals

Central America lacks rich deposits of minerals. Northern Central America has large areas of older igneous and metamorphic rocks, containing some silver, lead, mercury, antimony, and tin. More recent volcanic rocks may contain gold, silver, lead, zinc, and manganese. Certain granites have copper and molybdenum ores.

During the Spanish occupation, mining, primarily for gold and silver, was the chief economic activity. A variety of minerals are still mined, but the operations are generally small and marginal. Among the metallic minerals still produced are modest quantities of gold, silver, lead, zinc, manganese, nickel, iron ore, and copper. Other minerals include limestone (for cement) and sea salt, produced along the coast in evaporation ponds. The low value of limestone and salt preclude their export, and they supply domestic markets.

Oil has been found in several areas, but no large quantities have yet been produced. Notable is the Chapyal-Petén Basin of Guatemala. Costa Rica's Limón Basin shows potential. Exploration for oil is underway in various other areas, but results have been disappointing. To meet the demands of their industries and transportation sectors, Central American countries rely on imported oil. Since the region is poor in fossil fuels but rich in wood and other biomass, these are widely used in the residential sector as energy sources: most households in Guatemala and Nicaragua still use fuelwood for cooking, heating, and hot water. The use of biomass for fuel is also high in Honduras and El Salvador. Costa Rica is developing the production of energy-efficient and environmentally friendly biodiesel. Several other countries in Central America have initiated national programs aimed at transitioning from inefficient and polluting wood burning to liquid biofuels.

Forests

Every country in Central America once had abundant forests, most notably evergreen rain forests, but also coniferous softwoods in the uplands and deciduous forests in drier areas. The rain forests are extraordinarily rich in species diversity. One bio-

logical station of only 3,250 acres (1,300 hectares) in Costa Rica contains more species of plants than in all of California. In fact, Costa Rica alone has more than 9,000 species of flowering plants and 800 species of ferns.

In colonial times forest destruction was extensive, and only a fraction of the native forests remains. Wood was used primarily to shore underground mines, for fuel, for building construction, and for commercial purposes. Deforestation is occurring as a result of continued commercial wood harvesting, firewood gathering, and slash-and-burn agriculture.

Of the remaining forests, the rain forests are economically the most important. Among the many valuable tree species they contain are mahogany, lignum vitae, Spanish cedar, balsa, rosewood, ceiba, ironwood, sapodilla and chicosapote (for chewing gum), and rubber. The complexity of species of trees in the rain forests, plus the lack of road access, has hampered production. There have been attempts to establish plantations of certain commercial species. Mahogany trees require up to 100 years to reach economic size.

Most Central American countries have instituted reforestation and conservation programs in order to save their remaining rain forests. The leaders in this effort are Costa Rica, with 28 percent of its territory in preserve, and Panama, with 25 percent.

Marine Resources

Every Central American country has at least one coastline, providing access to marine resources. Fisheries are vast and varied, due to the variety of environments, which include mangrove swamps, shallow mudflats, coral reefs and islands, and adjacent ocean areas. Various marine animals have economic value, including crustaceans, scale fish, and sea turtles. The people of Central America generally do not consume marine products in significant quantities.

The most important crustaceans harvested here are shrimp and lobster. Shrimp favor muddy, shallow waters, whereas lobster are more common in coral beds. El Salvador and Honduras lead in crustacean harvesting. Scale fish include anchovies, herring, grouper, snappers, and sharks. Sea turtles are often raised on farms because their numbers have dwindled due to destruction of nesting grounds. Overall production of fish and other marine resources has declined in recent years due to overexploitation.

Inland Water Resources

All countries in Central America have numerous rivers. Rivers of the Caribbean watersheds are longer and have relatively even flow year-round due to humid climates. Those of the Pacific watersheds tend to be more seasonal and are prone to flash flooding. Most streams have steep gradients and frequent rapids and waterfalls, making them unusable for commercial boat navigation. Because of deforestation and erosion, water quality is typically low.

The chief use of streams, other than for domestic water and irrigation, is for hydroelectric power generation. The potential power production is large, and in the absence of significant fossil fuels the region is continuing to expand its use of hydropower. The Reventazón hydroelectric plant in Costa Rica is the largest in the region. Other leading power-generating dams include El Cajón (Honduras), the Fortuna (Panama), and Arenal (Costa Rica). A threat to reservoirs in this region is high rates of sedimentation due to erosion from the uplands. Sedimentation reduces the life span of reservoirs as well as the economic return of hydroelectric plants. The building of new hydroelectric dams continues, despite this problem.

Geothermal Power

Much of Central America lies within an active volcanic region; thus, geothermal energy is abundant. Several countries have begun to develop this energy resource, including El Salvador, Guatemala, Costa Rica, and Honduras. This form of energy is perhaps most valuable because it is abundant, nearly infinite, and clean and cheap. A serious impediment to its development is lack of local capital.

Other Resources

Central America has an abundance of beautiful locales that attract foreign tourists, especially ecotourists, including spectacular volcanoes, unspoiled tropical beaches, streams, rich rain forests, and a variety of fauna, along with rich cultural resources. Ecotourists come chiefly to enjoy the natural environment. Ecotourism is growing rapidly in this region, with the potential for even greater development. In addition to the foreign exchange gained, ecotourism carries with it the incentive to protect what remains of natural areas. Today, tourist attractions and ecotourism programs are being developed in many parts of the region.

P. Gary White

DISCUSSION QUESTIONS: NATURAL RESOURCES

Q1. Discuss how the discovery of gold and silver drew Spanish explorers to Central and South America. What nonprecious metals remain important to the local economies?

Q2. What role does hydroelectric power play in South and Central America? In what ways does the topography of this region lend itself to developing water-power facilities?

Q3. Why is the Brazilian tropical rain forest such a valuable resource? Discuss its role in climate, both local and worldwide. Why is the soil in a tropical rain forest considered unsuitable for farming?

FLORA AND FAUNA

FLORA OF SOUTH AMERICA

South America has the most diverse flora of any continent, primarily because of its unique location and geography. South America is relatively narrow, especially in the southern part, and long, extending from a little more than 10 degrees of latitude north of the equator to a little below 50 degrees of latitude south. An additional factor that increases its diversity of plant life is the high mountains, especially the Andes Mountains, which extend from north to south along the western part of the continent for much of its length. As a result, South America has such diverse biomes as tropical rain forests, tropical savannas, extremely dry deserts, temperate forests, and alpine tundra. The largest of these biomes are deserts, savannas, and tropical forests. The flora of South America is less well studied than those of the other continents. With the rapid rate of deforestation in places like the Amazon Basin, some plants may become extinct before being catalogued, let alone studied.

The subtropical desert biome is the driest biome in South America. It is restricted primarily to the west coast of the continent, from fewer than 10 degrees of latitude south of the equator to approximately 30 degrees of latitude south. This region is in fact considered the driest desert in the world, with an average annual precipitation of less than 0.25 inch (4 millimeters). These arid conditions prevail from the coast to relatively high elevations in the Andes. The Atacama Desert, in northern Chile, and the Patagonian Desert, in central Chile, are the most notable South American deserts. Smaller desert regions also occur in the rain-shadow portions of the Andes.

Next on the low-moisture scale are the savanna biomes. Annual rainfall averages from just under 27.6 inches (700 millimeters) to more than 55 inches (1,400 millimeters). Savanna occurs in two distinctly different areas of South America. The largest savanna area includes three distinctive regions: the cerrado, a region of gently rolling plateaus in east-central Brazil; the Pantanal, a lower-elevation region with extensive wetlands; and farther south, in southern Brazil, Uruguay, and northern Argentina, the famous grassland called the Pampas. The other savanna region, the *llanos*, is found in lower-elevation areas of Venezuela and Colombia. Portions of the *llanos* can be quite wet during part of the year and extremely dry the rest of the year.

Although a few of the forests in South America are dry, most are rain forests, receiving annual precipitation from more than 79 inches (2,000 millimeters) to almost 118 inches (3,000 millimeters). The Amazon rain forest, the world's largest, accounts for more than three-fourths of the rain forest

THE BAYMEN AND BELIZE

As piracy in the Caribbean declined, many British seamen, called "baymen," turned to the cutting of logwood along the Caribbean coast of contemporary Belize, Honduras, and Nicaragua. Logwood contains a core from which red and brown dyes were produced for use in English woolen mills. Once logwood was depleted, the baymen began cutting mahogany. Despite attempts by the Spanish to remove them, the baymen became well established, leading to the creation of the colony of British Honduras, renamed Belize at independence in 1981.

*A staff member tends the Victoria Regia, a giant water lily from Guiana, in the tropical house at Kew Gardens. According to the original caption: "Grown from a seed the size of a pea in February, it develops leaves up to seven foot six in diameter by July. Underside of leaf is ribbed to help it float, and covered with prickles to keep off fish. Flowers appear in July and August, changing from white to pink on their second day." (*The Work of Kew Gardens in Wartime, *Surrey, England, UK, 1943)*

area in South America. During the rainy season, the Amazon region receives so much rain that the entire river system floods extensively. The Amazon rain forest is one of the most species-rich areas of the world, but it is being rapidly destroyed by logging, ranching, and other human activities. Smaller rain forests are located along the southeastern coast of Brazil and in the northern part of Venezuela.

Covering much smaller areas are the Mediterranean and temperate forest biomes. A small Mediterranean region in central Chile is characterized by cool, wet winters and warm, dry summers with annual precipitation of 28 inches (700 millimeters). In the far south of Chile and Argentina is a small area of temperate forest, transitioning to alpine tundra in the far south. Temperatures are relatively cool and mild year-round (except in the far south, where it can be extremely cold in the winter), and annual precipitation is approximately 18 inches (450 millimeters). This area includes Tierra del Fuego, near notorious Cape Horn, the southern tip of South America where many ships have been lost.

Plants of the Subtropical Desert

In the Atacama Desert, one of the world's driest, some moisture is available, but it is limited to certain zones. Coastal regions below 3,280 feet (1,000 meters) receive regular fog (called *camanchacas*); most plants in this zone rely on fog as their primary source of water. At midelevation areas there is no regular fog and rainfall is almost nonexistent; thus, there is virtually no plant cover. At higher elevations, the rising air has cooled sufficiently to produce moderate amounts of rainfall, although the vegetation is still desertlike.

As in most deserts, the plants have adapted by either growing near water, storing water, or germinating and growing only when rains have soaked the ground adequately. Shrubs fall into the first category and typically grow near streambeds where their roots can reach a permanent source of water. Rainfall is so low in the Atacama Desert that even cacti (which normally store water) can hardly acquire enough water from rainfall alone, so many plants absorb a portion of their water from fog. In addition to the cacti, bromeliads (members of the pineapple family) also survive on fog for moisture. Some bromeliads are popular houseplants. They commonly are called air plants because their moisture needs are met solely by what they absorb from the air through special hairlike structures on their leaves.

Although the Atacama Desert often looks barren, when substantial moisture becomes available, plants called ephemerals change that appearance, seemingly overnight. Ephemerals are typically annuals that remain dormant as seeds in the dry soil. When moisture increases, they quickly germinate,

Illustration showing watercress, midwife toad and shells—the aquatic plants and wildlife of Suriname, South America. (Wellcome Collection)

grow, flower, and set seed before dry conditions prevail again. In the days and weeks following a good rain, many grasses appear and provide a backdrop for endless varieties of showy flowers. Some of the species will appear familiar to those acquainted with desert flowers from the Sonora or Mojave Deserts of the U.S. Southwest, but many are endemic to the Atacama Desert. Among the more showy flowers are species of *Alstroemeria* (commonly called irises, although they are actually in the lily family), *Nolana* (called pansies, although they are members of a family native only to Chile and Peru), *Leucocoryne* (spring onions), and *Calandrinia*.

A few familiar plants from North American deserts also occur here. *Ephedra* (called Mormon tea in North America), an unusual shrub somewhat related to pines, is abundant in places. In deserts like this, evaporation rates are high, which increases the salt content of some soils. Pigweed and saltbush often occur in these areas.

Conditions in the Patagonian desert are less harsh. The vegetation ranges from tussock grasslands near the Andes to more of a shrub-steppe community farther east. Needlegrass is especially abundant throughout the Patagonian region, and cacti are a common sight. In the shrub-steppe community in the eastern Patagonian desert, the shrubs quilembai and the cushionlike colapiche are common. Where the soil is salty, saltbush and other salt-tolerant shrubs grow.

Plants of the Tropical Savanna Biome

The cerrado region of east central Brazil and southward is not only the largest savanna biome of South America, it is also one of the most romanticized of the world's savannas. Like the Old West of North America, the grasslands of Brazil have cowboys who traditionally have used the cerrado for farming and cattle ranching. With ever-increasing pressure from agriculture, the cerrado is now under attack in various ways. Extensive use of fertilizers for agriculture, planting of trees for timber production, and introduction of foreign species, especially African grasses, are changing the cerrado. Frequent fires also have taken their toll. The cerrado contains more than 10,000 species of plants with about one-half of them being endemic (found only in the cerrado). In 2019, only 20 percent of the cerrado's original flora remained. Today, less than 3 percent of the cerrado is protected. A number of conservation groups are working frantically to save as much as possible of what is left.

The terms "savanna" and "cerrado" give the misleading impression that this area is one vast grassland. Some areas truly are grassland, such as the *campo limpo*, but other plant assemblages include *campo cerrado* (grassland with sparse shrubs), cerrado "proper" (grassland with scattered shrubs and trees), and *cerradão* (a more continuous cover of trees and shrubs). The last of these assemblages is closer to a forest than a savanna and is a transition type of habitat. Scattered randomly across the cerrado are forest gaps called *veradas*, which are spring-fed and have a high density of buriti or mauritia palms, a species of palm endemic to Brazil.

Two other savanna regions farther south are the Pantanal and the Pampas. Although the Pantanal is a savanna, during the rainy season it becomes a wetland and is a haven for aquatic plants, including water lilies so large and sturdy that a grown human can lie on one of the floating leaves without breaking through. Also common is water hyacinth, which has become a serious foreign invader in places like Florida. When the Pantanal dries out, grasslands appear in place of the water. Near the many rivers of the Pantanal and in waterlogged areas, acuri, carandá palms, and the beautiful paratudo tree, which produces a profusion of yellow flowers at the end of the rainy season, occur. This unique area is threatened by a variety of human activities, including navigation and artificial drainage projects, mining, agriculture, and urban waste.

The Pampas, like the great prairies that once covered central North America, is composed almost solely of grass. Trees and shrubs grow near bodies of water, but everywhere else grass predominates. Cattle ranching and wheat and corn farming are the primary occupations of the area and are thus the primary threat. Instead of a wet and dry season, as on the Pantanal, rain is distributed throughout the year, and because the area is farther south, it has a more temperate climate. Pampas grass has been exported as an ornamental plant. These lovely grasses reproduce asexually and create large numbers of seeds, making them a fairly invasive weed in some parts of the world.

The last major savanna region is the *llanos*, located at lower elevations in the drainage area of the Orinoco River in Venezuela and Colombia. Much like the Pantanal, this area has pronounced wet and dry seasons; the wet season is from May to October and the dry season is from November to April. At the lowest elevations, treeless grasslands persist after the water from the rainy season subsides. On the higher plains is a scattering of smaller trees, which on steeper slopes becomes more densely wooded. The mauritia palm can also be found here in poorly drained areas.

Plants of the Tropical Forest Biome

The Amazon rain forest is the largest contiguous rain forest in the world. It is so large and so lush with tree growth that it is actually responsible, in part, for the wet climate of the region. Its plant diversity is so great that it overwhelms even the most experienced botanists. Currently, no comprehensive plant guide exists for many parts of the Amazon rain forest. This should come as no surprise, considering that there are tens of thousands of plant species, with a large number having never been described.

This one-of-a-kind botanical and ecological treasure is being destroyed at a rapid and increasing pace of between 5,000 and 10,000 square miles (13,000 and 26,000 sq. km.) per year. Over 68,000 square miles (176,000 sq. km.) of tropical rain forest is already lost, mostly in Brazil, Peru, and Bolivia. If the current trend continues, by the year 2030 the deforested area will double in size, meaning that more than a quarter of the Amazon biome will be without trees.

The causes for this rapid destruction are primarily logging, agriculture, and cattle ranching. A common practice for preparing an area for cattle ranching or farming has been to simply burn the forest, not even necessarily logging it first, and then letting grass and other vegetation or crops grow in its place. The soils of the rain forest are so poor, however, that this practice usually depletes the soil within a few years and the land becomes a useless wasteland fit only for weeds. Forest burning also causes unintended wildfires, which have increased in recent years, especially in Brazil and Bolivia; they accelerate the deforestation even further. Mining, oil drilling, and dam construction also take a heavy toll. Even when trees are not burned, they are often simply cut down and left to rot in place. South American and international environmental organizations have waged anti-deforestation campaigns for decades, but with little success.

The Amazon rain forest is an extremely complex biome. The main plant biomass is composed of trees, which form a characteristically closed canopy that prevents most of the sunlight from reaching the forest floor. Consequently, the forest floor has little herbaceous growth, and most smaller plants tend to grow as epiphytes on the branches and trunks of trees. The trees that form the canopy are stratified into three fairly discrete levels. The lowest two levels are the most crowded and comprise smaller to medium-sized trees and relatively tall trees. These two levels alone block out much of the light from above. The highest level comprises extremely tall trees, often referred to as emergent trees because they stand out above the fairly continuous lower two layers.

Emergent trees are randomly scattered about and typically do not form a closed canopy like the lower two levels. Many of the tallest trees are buttressed at the base, an adaptation that seems to provide greater stability. Beneath the canopy, there are some smaller palms, shrubs, and ferns, but these never occur in high density, except where there is a break in the canopy that allows in greater light.

Although many of the trees of the rain forest are unfamiliar to everyone but an experienced botanist or logger, some species are well known, primarily because of their economic value. A favorite tree for making furniture is the mahogany. The "classic" mahogany trees are species of *Swietenia*, but many species from the *Meliaceae* (mahogany family) are also called mahogany and have fine wood as well. Because they are highly prized, many species of mahogany have drawn the attention of conservationists.

The South American rain forests are also the original source of rubber. Brazil had a monopoly on rubber until seeds were smuggled out and planted in Malaysia. Synthetic rubber has now replaced natural rubber for many applications. Another popular tree is the Brazil nut tree, an abundant food source that has been exploited by suppliers of mixed nuts.

The cacao tree, which is the source of chocolate, produces flowers and fruits that emerge right out of the trunk rather than from the branches, a type of fruiting called cauliflory. The fruits are football-shaped, although smaller, and produce up to about sixty cacao beans, which when removed can be processed to make chocolate.

Many of the smaller plants of the rain forest are epiphytes (or parasites, in some cases) on the trunks and branches of the canopy trees. Lianas, or vines, are a prominent component as well. The strangler fig begins as a liana and eventually grows to tree proportions by an unusual method. The fruits are eaten by monkeys, birds, or bats; their seed-filled droppings fall onto branches high in the canopy. The seeds germinate and send down a vine-like root all the way to the ground. Once this nutrient line is established, more roots grow down. Then branches, with leaves, begin growing up the tree, eventually surrounding it completely. Sometimes the host tree dies and rots, leaving just the tubelike strangler fig holding the shape of the dead tree.

The most common epiphytes in the Amazon rain forest include orchids, bromeliads, and even some cacti. The orchid flora is particularly diverse, with some odd species. Some orchids have flowers that mimic the females of certain species of wasps. Their similarity is so striking that male wasps readily attempt to mate with the flowers, thus transferring pollen from one orchid to another.

There is a large diversity of bromeliads, ranging from small, inconspicuous species to larger species, such as tank bromeliads, that can collect significant amounts of water in their central whorl of leaves. The water in these plants can contain a whole miniature ecosystem, complete with mosquito larvae, aquatic insects, and frogs. Ferns are another significant member of the epiphyte community. Some larger species of ferns, often called tree ferns, also grow in the understory.

Every year during the rainy season, the lowest elevation areas of the Amazon rain forest are flooded with several feet of water, which recedes after a few months. The trees, well adapted to this flooding cycle, flourish. A few even have unique adaptations; for example, some trees produce fruits that are eaten by fish, thus assuring the spread of their seeds. The flooding can be so extensive in some areas that the water reaches the lower parts of the canopy.

Coastal tropical rain forests also occur in northwestern and southeastern South America. These forests have many of the same characteristics and a similar flora to those found in the Amazon rain forest. Despite their similarities, there are many endemic species in each of these forests. In fact, endemism is extremely high throughout the rain forests of South America. Some tree species are so rare that they may be found in only a few square mile areas and nowhere else. Other rare species may be widespread in a particular rain forest, but occur only as single trees separated by many miles. One unique feature of the coastal rain forests is the mangroves. Where the tropical rain forest meets the ocean, certain species of these trees have become adapted to the tidal environment. Many mangrove trees have prop-roots, which make the trees look like they are growing on stilts. They also frequently have special root structures that extend above the water at high tide and allow the roots to breathe. Mangrove trees are also extremely salt tolerant.

Plants of the Mediterranean and Temperate Forest Biomes

One of the world's five Mediterranean climate regions is found in central Chile. This climate is characterized by warm dry summers and cool wet winters. The vegetation, called matorral, is composed primarily of leathery-leaved evergreen shrubs that are well adapted to the long summer drought. Overall, the vegetation is similar to the chaparrall found in the Mediterranean climate areas of California, although some of the genera and all of the species are different.

The only trees found in both California and the matorral are willows and mesquite. Oaks, dominant in Mediterranean California, are not found in the matorral, but southern beech is. Some shrubs, such as creosote bush and coyote brush, are found in both places. The matorral is the only Mediterranean area that has bromeliads. At lower elevation areas, somewhat inland, many of the shrubs are drought-deciduous, that is, they drop their leaves in the summer. In more inland parts of this biome, the espino tree is common. Historically, this native species was probably confined to disturbed habitats; with the many disturbances caused by humans, it has become more common.

Because South America extends so far south, it actually has a small region containing temperate forests. These forests range from temperate rain forest to drier temperate forests, and in all cases are typically dominated by southern beeches. The undergrowth is dominated by small evergreen trees and shrubs. Fuchsias, which are valued the world over for their showy flowers, are common in the undergrowth.

The monkey puzzle tree, an unusual conifer, is also abundant in some areas and has been grown as an ornamental in other parts of the world. Although not rich in species, the temperate rain forests of southern South America can be lush. In the far south, before the extreme climate restricts the vegetation to alpine tundra, a region of elfin woodlands predominates. These woodlands can be nearly impenetrable, with the densest growth often associated with patches of tall bamboo.

Bryan Ness

FAUNA OF SOUTH AMERICA

A wide range of animals, both ordinary and exotic, inhabit the continent of South America. The types of animals found in any geographical area are determined by the climate and the terrain. In the Andes, animals such as the llama have adapted to the terrain and climate of the high, steep mountains.

In the Galápagos Islands, off the coast of Ecuador, animals exist that are found nowhere else in the world. The English naturalist Charles Darwin studied the animals of the Galápagos and developed many of his ideas related to the theory of evolution from this study. Much farther south, not far from Antarctica, is the archipelago known as Tierra del Fuego. In this area live many penguins, whose layers of feathers help them to survive the frigid sea waters in that area.

Animals of the Andes

Four members of the camel family live in the Andes Mountains of Peru, Bolivia, Ecuador, and Chile: alpacas, vicunas, guanacos, and llamas. These camelids all have commercial value in the Andes as pack animals and for their meat and fur. Many people of the Andes also raise sheep on the mountains for both wool and meat.

More than 4,000 years ago, alpacas, which are raised for meat and for their fine cashmere-like fur, were reserved for the exclusive use of the Incas, who prized their coats. Alpacas live from fifteen to twenty-five years. The average adult is about 3 feet (1 meter) high at the shoulder and weighs up to 180 pounds (82 kilograms). In 2019 there were approximately 3.5 million alpacas in South America, mostly in Bolivia, Chile, and Peru.

Llamas

Closely related to the alpaca is the llama, domesticated as a work animal more than 3,000 years ago. Llamas, used primarily in Peru and Bolivia, have historically been the beasts of burden in the Andes Mountains. A single animal can carry about 200 pounds (90 kilograms) for twelve hours a day. However, they cannot be ridden and when they tire, they often simply lie down and refuse to move. They

Alpacas are kept in herds that graze on the level heights of the Andes of southern Peru, western Bolivia, Ecuador, and northern Chile at an altitude of 3,500 m (11,500 ft) to 5,000 m (16,000 ft) above sea level. Alpacas are not bred to be working animals but were bred specifically for their fiber. Alpaca fiber is used for making knitted and woven items, similar to sheep's wool. (Image by Philippe Lavoie)

HABITATS AND SELECTED VERTEBRATES OF SOUTH AMERICA

North
Atlantic
Ocean

Coatimundi

Giant Anteater

Anaconda

Marine Iguana

**GALÁPAGOS
ISLANDS**

Vampire Bat

Boa Constrictor

Sloth

Giant Land
Tortoise

Puma

Toucan

Giant Otter

Peba

Piranha

AMAZON RAIN FOREST

Vicuña

Capybara

Peccary

Tapir

Jaguar

Quetzal

Ocelot

Wooly Tree
Porcupine

Llama

Spectacled
Bear

Crab-Eating Zorro

EASTERN HIGHLANDS

Coypu Rat

Gray Fox

Andean
Condor

**COASTAL
REGION**

Plains
Viscacha

Bushmaster

Alpaca

James
Flamingo

Condor

GRASSLANDS

Agouti

Chinchilla

South
Pacific
Ocean

Rhea

Guanaco

Armadillo

Maned Wolf

Great
Anteater

**COASTAL
REGION**

South
Atlantic
Ocean

Patagonian
Weasel

Cavy

Cormorant

Sea Lion

Penguin

even spit at their drivers when they no longer want to work. Reaching heights of nearly 4.5 feet (1.4 meters) llamas are generally larger than alpacas. A llama's fur is usually white with patches of black and brown, but some are pure white and others pure black. Although llamas are used for work, they are also kept as pets in many homes in South America.

Vicunas

Vicuna fur has historical importance in South America. It was used by Incas to make cloth, and only members of Inca royalty could wear clothing made from this cloth. Anyone else found with such clothing was executed. By 1979 only 4,000 vicunas were left in South America. They had been hunted for their fleeces by poachers who killed the animals. Representatives of the governments of Bolivia, Chile, Ecuador, and Peru signed a treaty for protection of the vicuna. Twenty years later, there had been a resurgence of vicunas in Peru, Chile, and Argentina: in 1999, there were 103,000 in Peru, 30,000 in Argentina, 16,000 in Chile, and a small transplanted herd in Ecuador. By 2019, the vicuna population had grown to about 350,000. Even though *el chacu*, the communal hunting of vicuna, continued after 1979, laws limited these hunts to local people. Those local hunters sold the fiber from the animals, generating an important source of income for their families.

Vicunas weigh 90 pounds (11 kilograms) and are a little less than 3 feet (1 meter) in height at the shoulder. They have long necks, slender legs, padded cloven feet, large round eyes, and a fine, dense, tawny coat. Aside from their economic value, the vicuna is valuable for scientific study. They are highly communicative animals, signaling each other with body postures and ear and tail positions. They emit soft humming sounds to communicate bonding and greeting.

Guanacos

The fourth member of the camel family found in South America is the guanaco.

The guanaco is more adaptable than the other three camelids. It is found throughout the Andes, in the dry Atacama Desert of Chile, and in Tierra del Fuego, where it rains year-round. This animal, from which the llama was domesticated, began life in the semiarid desert and has developed physiological mechanisms for coping with both heat and dehydration. It is similar in structure to other camels, and holds the distinction of being the largest member of the camel family living in South America.

Large Cats

Throughout the Andes, from Argentina to Colombia, and on into Central America, pumas roam. This reddish-brown feline can reach lengths of about 6.5 feet (2 meters), not including its long tail. In some areas of South America, the puma is endangered. It is a carnivore whose natural prey are elk, deer, and small wild animals. It also preys on sheep and cattle, leading ranchers to hunt and kill the predators to protect their flocks and herds. Another member of the cat family is the elusive Andean mountain cat, rarely sighted by humans. This species is considered sacred by the native people of the *altiplano* of Bolivia, the Aymara. No more than

South American jaguar in the area of Miranda near the Amazon Basin. (Image by Bart van Dorp)

The spectacled bear (Tremarctos ornatus) *is the last remaining short-faced bear (subfamily Tremarctinae). Its closest relatives are the extinct Florida spectacled bear and the giant short-faced bears of the Middle to Late Pleistocene age (Arctodus and Arctotherium). They are the only surviving species of bear native to South America. The species is classified as Vulnerable by the IUCN because of habitat loss. (Image by Cburnett)*

2,500 Andean mountain cats live in the wild, and the species is classified as endangered.

In the forests of the Andes lives the spectacled bear. Its range extends as far north as Ecuador. This bear, which is vulnerable because of overhunting and destruction of its habitat, has a shaggy brown coat with yellow facial markings and a cream-colored muzzle, throat, and chest. It is the only bear native to South America.

Rodents

Several rodents are native to the Andes. Chinchillas were found living in crevices in the mountains when early Spanish explorers first arrived there. Living off berries and fruits in Peru, Chile, and Argentina, these rodents belonged to Inca royalty, who used their fur to make chinchilla stoles. In the latter part of the twentieth century, they were nearly extinct in the wild but existed in captivity. Some recovery in numbers has occurred since then, and in 2016, their status was upgraded from "critically endangered" to "endangered." Related to the chinchilla is the viscacha, which is prey for such animals as the Andean mountain cat. Mountain viscachas have long rabbit-like ears, and long, squirrel-like tails. East of the mountains lives the plains viscacha, which has shorter ears and a blunter head. The cavy, the South American guinea pig, lives in the crevices of the Andes.

Birds

Various exotic birds also live in the Andes, many of which also are found in the Amazon Basin to the east. Among these birds are the Andean cock-of-the-rock, the scarlet macaw, the quetzal, the Andean condor, and the James flamingo. The cock-of-the-rock is a huge dancing bird found in the mountain forests. The scarlet macaw, a brilliantly plumed member of the parrot family, is an endangered species. The quetzal had religious significance to early Andeans; even today, it is regarded as a symbol of the Andes Mountains. The Andean condor is found in the high plain area of Bolivia and Chile. With a wingspan of 12 feet (4 meters), it is the largest flying bird in the world. It can soar to 26,000 feet (7,925 meters) above sea level!

In the southern Andes lives the rhea, a flightless bird related to the ostrich that is often called the South American ostrich. The rhea is much smaller than an ostrich and has three toes on each foot,

The rheas are large ratites (flightless birds without a keel on their sternum bone) in the order Rheiformes, native to South America, distantly related to the ostrich and emu. (Image by Michael Palmer)

whereas an ostrich has two toes on each. Rheas live in flocks of twenty to thirty birds. They live primarily in Brazil's southern plains and in Argentina, Paraguay, and Uruguay.

Animals of the Amazon Basin

Representatives of almost one-half of all known varieties of animals live in South America, mostly in the Amazon Basin. The basin includes the rain forests, plateaus, rivers, and swamps southeast of the Andes Mountains.

The tapir, found in the Andes and in the forests east of the Andes, is South America's largest animal. With its short, hairy body, it resembles a pig, but is actually related to the horse and the rhinoceros. Tapirs are classified as either vulnerable or endangered. Their numbers have diminished as a result of hunting and habitat destruction.

The tapir's natural enemy, the jaguar, also is found in the eastern Andes and in the forests east of the high mountains. This feline, which was worshipped by pre-Columbian civilizations as a god, lives in the area between the southern United States and northern Argentina and is especially prevalent in Brazil. Jaguars are strong swimmers that live near rivers and other bodies of water. They are classified by conservation groups as threatened. Their numbers have declined

Andean condor, in Chilean national park Torres del Paine. (Image by Hugo Pedel)

Yellow-headed caracara on a capybara. The yellow-headed caracara (Milvago chimachima) *is found in tropical and subtropical South America and the southern portion of Central America. It is not a fast-flying aerial hunter, but is rather sluggish and often obtains food by scavenging. The capybara* (Hydrochoerus hydrochaeris) *is a mammal native to South America. It is the largest living rodent in the world. (Image by Charles J.*

in large part because farmers have worked to eliminate the jaguars from their lands, claiming that the cats preyed on their cattle and sheep.

Also living within the Amazon Basin are the giant anteater, the sloth, and the peba. The giant anteater has value to the environment and to farmers because it consumes up to 30,000 insects per day. The peba, a nine-banded armadillo found widely throughout South America, also contributes to agriculture by consuming insects and worms. This nocturnal animal is protected from its predators by a horn and bony plates covering its body. The sloth is the world's slowest-moving large mammal.

Within the Amazon Basin lives one of the world's most interesting rodents, the huge capybara. It generally is as large as a big dog, but can reach 4 feet (1.2 meters) in length and can weigh as much as 100 pounds (45 kilograms). Humans have little to fear from the amphibious capybara, however, since it is a vegetarian.

Both the land iguana and the lava lizard live in the Amazon Basin. Land iguanas can have a life span of up to twenty-five years and weigh up to 15 pounds (6.8 kilograms). They eat low-growing plants, shrubs, fallen fruits, and cactus tree pads. The lava lizard, about 1 foot (30.5 centimeters) long, is smaller than the land iguana. Lava lizards are beneficial to agriculture because they eat beetles, spiders, and ants.

The coatimundi inhabits areas from Arizona to northern Argentina. A member of the raccoon family, it is brown or rust-colored. It eats snails, fish, berries, insects, spiders, lizards, birds, eggs, and mice, and is often kept as a pet by South Americans.

Many types of birds live in the Amazon Basin, some deep within the rain forest, others closer to the mountains. Among the birds in the basin are hummingbirds, parrots, ospreys, macaws, boat-billed herons, great egrets, white-necked herons, least bitterns, and blue and yellow macaws. Toucans, which also live at high elevations up to 10,000 feet (3,050 meters), can be found deep within the Amazon rain forest.

The rain forest is also inhabited by many bats, squirrels, and parrots, which eat the fruits and nuts of the upper and lower canopies. In the lower canopy live lemurs, flying squirrels, and marmosets, small monkeys found mostly in eastern Brazil. Marmosets use their sweat glands for communication. Animals of the lower canopy eat the fruits, nuts, and insects that are found there. Within the rain forests also live vampire bats; these fascinating animals need to consume two tablespoons of blood per day in order to survive.

Other residents of the Amazon Basin include the yapok, a member of the opossum family that has webbed feet for swamp travel; the sapajou monkey, a small New World primate; and the octodont, an eight-toothed rodent also known as a spiny rat or a spiny hedgehog because of the sharp spines embedded in its fur.

The boa constrictor also lives in the jungles of this basin. This snake, which is usually 6 to 9 feet (2 to 3 meters) in length, but can reach 13 feet (4 meters), kills its prey by squeezing it to death, using its body coils to suffocate its victims. After killing its meal, the boa constrictor stretches its jaws wide apart and pulls the entire victim into its mouth. Us-

The scarlet macaw (Ara macao) *is a large red, yellow, and blue Central and South American parrot, a member of a large group of neotropical parrots called macaws. It is native to humid evergreen forests of tropical Central and South America. The scarlet macaw is the national bird of Honduras. (Image by Ben Lunsford)*

ing this method of killing, a boa is able to eat animals that are much larger than its head.

Another large snake native to the Amazon Basin is the anaconda. Most anacondas weigh several hundred pounds, reaching weights of 550 pounds (250 kilograms) and lengths of 36 feet (11 meters). The anaconda is found in the Guyanas and throughout tropical South America, east of the Andes. With eyes high on its head, the anaconda can submerge its body in water and watch for the approach of unsuspecting prey.

Several types of foxes are indigenous to the Amazon Basin and roam through lower mountain areas to the east and the west of the Amazon. Among them are the gray fox and the crab-eating zorro. The gray fox roams over the plains, the Pampas, the desert, and the low mountains, but its numbers are decreasing because farmers are cultivating lands

Anaconda (Eunectes murinus) *at the ophiological center, Leticia, Colombia. (Image by Dick Culbert)*

THE ANACONDA

Another name for the anaconda is the water boa, an appropriate name for a snake that is almost always found near water. Anacondas live in the Amazon and Orinoco basins of tropical South America, and their habitat extends to Trinidad. Like the crocodile, the anaconda has nostrils high on its snout so that it can swim with its head above water to breathe. The anaconda lies near the shore, waiting for its prey. When a deer, bird, or other prey comes to the water to drink, the anaconda quickly strikes, dragging its victim underwater to drown it. The anaconda then eats the unfortunate animal whole. A good meal can last an anaconda for several weeks, during which it will lie in the water, digesting its food.

that previously were its habitat. In Argentina, where it has been hunted for its skin, the gray fox has been placed on the endangered species list. In Chile it is protected by law, but enforcement of the law has been lax. The crab-eating zorro is found in Colombia, Venezuela, Suriname, eastern Peru, Bolivia, Paraguay, Uruguay, Brazil, and northern Argentina. An omnivore, it eats not only crabs but also insects, rodents, fruit, reptiles, and birds.

River Animals

Many animals spend all or part of their lives within the Amazon. The black caiman, an alligator, is one such animal. It can weigh as much as a ton. It will eat all vertebrates, including humans, if that is the only food available. Black caimans, once critically endangered, have seen a dramatic recovery in their populations largely as a result of conservation efforts.

The semi-aquatic brown water lizard also is found in the jungle area around the Amazon. Within the Amazon are manatees as well as the endangered Amazon river dolphin (also known as the boto or the pink dolphin). This is the only dolphin to have a neck. Giant otters live in the waters of the Amazon as do many types of fish, including piranhas, which, with their sharp teeth, can quickly strip the flesh from their prey.

Animals of the Eastern Highlands

The Eastern Highlands of South America host many unique animals, including the bush dog, woolly tree porcupine, maned wolf, peccary, bushmaster, and coypu rat, and many birds, including flamingos.

The bush dog is a wild dog, but, with its webbed feet, it resembles an otter more than a dog. Bush dogs live in packs and hunt small deer and rodents. They can be found from the rain forests into the grasslands, in Colombia, Venezuela, the Guyanas, Brazil, Paraguay, northeastern Argentina, eastern Bolivia, and eastern Peru.

The maned wolf is one of South America's most beautiful and revered animals. Weighing on average 100 pounds (45 kilograms), it is South America's largest canid. It is found mostly in Argentina, Brazil, and Paraguay, and has no natural enemies. An omnivore, it will eat nearly anything, including fruits, insects, and small vertebrates. It is estimated that 23,600 maned wolves exist in the wild. Because of its beautiful red and gold fur, the maned wolf is something of a tourist attraction. Many South Americans regard it as an important part of their cultural heritage. The rural people of the Serra da Canastra of Brazil even believe that the maned wolf has medicinal and supernatural powers.

The peccary and the bushmaster are also found in the Eastern Highlands. The peccary resembles the tapir, but is much smaller. It has a big head, sharp teeth, and prickly fur. It eats smaller animals and plants such as cactus flowers. The bushmaster, the largest venomous snake in the Americas, is a type of pit viper related to the rattlesnake. Like the rattlesnake, it shakes its tail before striking, but it has no rattles. Gray and brown with a diamond pattern, it averages 8 to 12 feet (2.4 to 3.7 meters) in length.

The coypu rat is also known as the swamp beaver. This relative of the muskrat is found in southern Brazil, Bolivia, and Colombia. The agouti, a rodent nearly 2 feet (0.6 meter) long, also lives in the Eastern Highlands. Farmers loathe the agouti because it eats sugar and banana plants. The ocelot, a cousin of the jaguar, also makes its home in this area. This slender cat's fur provides camouflage in the forests and deserts of the highlands.

ENDANGERED MAMMALS OF SOUTH AMERICA

Common Name	Scientific Name	Range	Status
Armadillo, giant	*Priodontes maximus*	Venezuela and Guyana to Argentina	Vulnerable to extinction
Cat, Andean	*Felis jacobita*	Chile, Peru, Bolivia, Argentina	Endangered
Chinchilla, short-tailed	*Chinchilla brevicaudata boliviana*	Bolivia	Endangered
Deer, pampas	*Ozotoceros bezoarticus*	Brazil, Argentina, Uruguay, Bolivia, Paraguay	Near threatened
Manatee, Amazonian	*Trichechus inunguis*	South America (Amazon River basin)	Vulnerable
Marmoset, buffy-headed	*Callithrix flaviceps*	Brazil	Endangered
Monkey, wooly spider	*Brachyteles arachnoides*	Brazil	Critically endangered
Otter, giant	*Pteronura brasiliensis*	South America	Endangered
Porcupine, thin-spined	*Chaetomys subspinosus*	Brazil	Vulnerable
Vicuña	*Vicugna vicugna*	South America (Andes)	Least concern

Source: U.S. Fish and Wildlife Service, U.S. Department of the Interior.

Birds that live in the Eastern Highlands include the James flamingo, which lives on Bolivia's frigid salt lakes, and the giant antshrike, which measures more than 12 inches (30 centimeters) in length.

Animals of Tierra del Fuego

On the islands that make up Tierra del Fuego, many unusual animals are found, including penguins and many other types of birds. Penguins cannot fly; they use their wings for swimming in the icy waters, and are insulated from the frigid sea water by three layers of short feathers and an underlying layer of fat. Other birds common to Tierra del Fuego are Magellanic cormorants, imperial cormorants, albatrosses, and various petrels. Sea lions also live on these islands.

Animals of the Galápagos Islands

In the Pacific Ocean, 600 miles (960 kilometers) off the coast of Ecuador, lie the Galápagos Islands. These islands, owned by Ecuador, are officially known as the Archipiélago de Colón. In the 1830s, the English naturalist Charles Darwin visited the Galápagos aboard the HMS *Beagle*. His interest in the diversity of animal life on the Galápagos—particularly the thirteen distinct species of finches

Sea lions at Isla de los Lobos in the Beagle Channel, near Ushuaia.

found there, each of which had adapted a distinct-shaped bill suitable for the type of seeds available on the island on which the particular species lived—eventually led to the development of his theory of evolution.

Many other birds live on the Galápagos. One is the flightless cormorant, which has lost its ability to fly because food is so abundant near shore that the bird, having no predators, has no need to fly. The flightless cormorant has powerful wings and webbed feet, but its wings are only about one-third of the wingspan that would be needed to lift its body.

On the Galápagos are found many flamingoes and Galápagos penguins, the species of penguins that lives closest to the equator. The frigatebird also lives there. It has long wings and a forked tail. Incapable of taking off from a flat surface, it must run downhill in order to fly. Galápagos hawks have a variety of feather colors, from white and brown to bright yellow and black. These hawks are much tamer than many of their cousins in other parts of the world.

The Galápagos are also home to the blue-footed booby, which lays its eggs directly on the ground, and the masked

*Male adult Galápagos penguin (*Spheniscus mendiculus*) swimming on Isabela Island off Moreno Point, Galápagos Islands. (Image by Charles J. Sharp)*

Two Galápagos tortoises engage in a dominance display in an enclosure on Santa Cruz Island.

booby, which lays two eggs, but raises only one chick to maturity. There is also the red-footed booby, which makes a nest of twigs and spends much of its time at sea hunting for food. The waved albatross, a rare seabird, nests on Head Island in the Galápagos. It cannot be said to live there because the only time it is not far out at sea is during the nesting period.

The Galápagos fur seal grows to about 5 feet (1.5 meters) in length and is the smallest of the fur seals. A special subspecies of the California sea lion lives in the Galápagos. It is 6 feet (1.8 meters) in length,

rarely goes far from shore, and is very intelligent. The Pacific waters around the Galápagos Islands also contain hammerhead sharks, marine iguanas, and Sally Light crabs. The marine iguanas there can swim as far as 50 feet (15 meters) below sea level. The Sally Light crab, yellowish-orange in color, was named after an eighteenth century English dancer.

One of the most famous inhabitants of these islands is the giant land tortoise, which can weigh up to 500 pounds (225 kilograms) and can live to be 150 years old.

Annita Marie Ward

FLORA AND FAUNA OF CENTRAL AMERICA

Because Central America is a land bridge that connects North America and South America, its plants and animals have similarities with the flora and fauna of both those continents. Northern species, such as white-tailed deer, raccoons, rattlesnakes, and mountain lions, may be seen in the northern part and the highlands of Central America. Tropical species, such as monkeys, coral snakes, jaguars, and sloths, abound in the coastal rain forests and southern regions of this geographic area.

Plant Life

Types of vegetation are influenced by climate, soil quality, land elevation, and human activity. Lowland tropical rain forests lie on the eastern half of Central America and typically have many tall,

HABITATS AND SELECTED VERTEBRATES OF CENTRAL AMERICA

broad-leaved evergreen trees that grow 130 feet (40 meters) or more in height, and 4 to 5 feet (1.2 to 1.5 meters) in diameter, forming a dense canopy. Shade-seeking plants, such as palms, figs, ferns, vines, philodendrons, and orchids, form the forest undergrowth beneath the trees.

Epiphytes such as orchids, ferns, bromeliads, and mosses cling to the branches of the trees in a dense mat of vegetation—these plants have no roots but grow by clinging to the trunks of trees and drawing moisture and nourishment from the air. Rain-forest trees that are harvested for their commercial value include mahogany, kapok, cedarwood, tagua, ebony, and rosewood for making furniture; breadfruit, palm, and cashew for food; sapodilla, used to make latex; and the rubber tree. Many brilliantly colored flowers also grow in Central America. The most

common of these are orchids (with nearly 1,500 species), heliconias, hibiscus, and bromeliads.

Elsewhere in the Caribbean lowlands, where the soil is porous and dry, extensive savanna grasslands with sparse forests of pines, palmettos, guanacastes, cedars, and oaks are found. Along the Caribbean coast (called the Mosquito Coast), mangroves and coconut palms flourish in swamps and lagoons.

The central mountains and highlands of Central America are cooler than the coastal lowlands, and the vegetation there is mainly deciduous hardwood trees such as walnuts, pines, oaks, and balsas. The eastern slopes of the mountains have abundant rainfall. "Cloud forests," which are 5,000 feet (1,525 meters) above sea level, are thickly choked with evergreen oaks, sweet gums, pines, and laurels. Laurels can grow to a height of about 65 feet (20

Plumeria (or frangipani) flowers are most fragrant at night in order to lure sphinx moths to pollinate them. The flowers yield no nectar, however, and simply trick their pollinators. The moths inadvertently pollinate them by transferring pollen from flower to flower in their fruitless search for nectar. (Image by Leon Brooks)

A red-eyed tree frog (Agalychnis callidryas), *photographed near Playa Jaco in Costa Rica. (Image by Carey James Balboa)*

meters) and be festooned with ferns, bromeliads, mosses, and orchids.

On the western side of the mountains, facing away from the moist Caribbean winds and receiving rain only seasonally, vegetation is sparse and semiarid, and soils are poor and unproductive. Tropical deciduous forests dominate there, and vegetation is characterized by evergreen herbs and shrubs, frangipanis, eupatorium pines, myrtles, and sphagnum mosses.

Animal Life

Wildlife in Central America generally has many similarities with that of South America, but species common to North America, or unique to Central America, also exist. Although the overall number of animal species is smaller than that of South America, Central America has a rich range of wildlife in relation to its small geographic area.

Competition among Central American animal species is intense; as a result, animals tend to be small and highly specialized as to where they live and what they eat. Mutualism between plant and animal species is also common, with each depending on the other for nourishment and survival. Ant acacias, for example, feed on the nectar of acacia trees, but also defend the trees vigorously from attacks by any other animals or invading plants. Similarly, hummingbird beaks are often shaped to allow them easy access to only one or two species of flowers—the flowers, in turn, have evolved special colors and shapes to encourage hummingbirds to drink their nectar and spread pollen from one blossom to another.

The forests of Central America support an especially rich array of bird species, with more than 1,000 species represented in total. The majority of these birds live in the tropical forests, where North

ENDANGERED SPECIES OF CENTRAL AMERICA

Common Name	Scientific Name	Range	Status
Cat, tiger	*Leopardus tigrinus*	Costa Rica to northern Argentina	Vulnerable
Jaguarundi, Panamanian	*Herpailurus yagouaroundi panamensis*	Nicaragua, Costa Rica, Panama	Least concern
Marmoset, cotton-top	*Saguinus oedipus*	Costa Rica to Colombia	Critically endangered
Monkey, red-backed squirrel	*Saimiri oerstedii*	Costa Rica, Panama	Vulnerable
Monkey, spider	*Ateles geoffroyi geoffroyi*	Costa Rica, Nicaragua	Critically endangered
Ocelot	*Leopardus pardalis*	U.S.A. (Arizona, Texas) to Central and S. America	Least concern
Puma, Costa Rican	*Puma concolor costaricensis*	Nicaragua, Panama, Costa Rica	Least concern
Tapir, Central American	*Tapirus bairdii*	Southern Mexico to Colombia and Ecuador	Endangered

Source: U.S. Fish and Wildlife Service, U.S. Department of the Interior.

American species of warblers and flycatchers (part of a large family of insectivorous birds called trogons) feed on the abundant insect life, along with more tropical species such as woodcreepers and flycatchers.

More than ninety species of hummingbirds feed upon the many colorful tropical flowers. Large-beaked birds such as toucans, parrots, and macaws feed upon the fruits and seeds of the numerous fruiting trees. Hawks, falcons, and eagles (including the world's largest eagle, the harpy eagle) hunt smaller birds, reptiles, and rodents in the forests and grasslands of Central America. Waterbirds and wading birds—ducks, egrets, storks, herons, ibis, kingfishers, and spoonbills—live along the riverbanks and lagoons. Ocean birds, including terns, gulls, and frigatebirds, live along the coastal areas. Central American birds display an extraordinary variety of shapes and brilliant colors.

Many species of animal life in Central America are also native to South America. Tapirs, capybaras, giant anteaters, peccaries, and small brocket deer live in the forest undergrowth, along with a variety of rodents, such as armadillos, agoutis, and pacas. These mammals are secretive and mainly nocturnal. Monkeys—howler, squirrel, woolly, spider, and capuchin—live in the forest canopy, along with tree sloths, tree squirrels, porcupines, tree rats, and tamarins. Many of Central America's predators are members of the cat family—jaguars, jaguarundis, margays, pumas, and ocelots. Other carnivores are coatimundis, kinkajous (raccoon-like animals), tayras (a large member of the weasel family), and vampire bats.

Central America is particularly rich in the many species of reptiles that inhabit it, including such venomous snakes as the fer-de-lance, bushmaster, pit viper, green palm viper, parrot snake, and coral snake, as well as many species of iguanas.

The warm rivers and coastal reefs of Central America also support many different species of fish, including catfish, barracudas, tarpons, groupers, and damselfish. Amphibians—tree frogs, true frogs, and toads—thrive in the wet environment of the rain forests, with the poison-arrow frog being perhaps the best known. Central America has five species of sea turtles—Ridley's, green, loggerhead, hawksbill, and leatherback—that nest on coastal beaches. The manatee (sea cow) is increasingly rare, but still lives in the rivers and lagoons of coastal areas.

ENDANGERED CENTRAL AMERICAN WILDLIFE

Deforestation is a major environmental concern throughout Central America, where population growth has led to the harvesting of trees for economic profit and the need to create new farmland. Typically, loggers remove the most valuable trees, and the land is then farmed or used for cattle ranching for three to seven years. As invasive weeds soon take over the newly cleared land, the land is quickly exhausted and human settlement then moves on to clear new areas of the forest. In recent years, all Central American countries have made efforts to create protected natural areas and to control illegal logging.

The wildlife of Central America is threatened not only by the loss of natural habitat through deforestation, but also by uncontrolled hunting, human population growth, and a world market for rare and exotic species. Many Central American species are endangered or nearly extinct—many parrot species, squirrel monkeys, and sea turtles are all becoming rare. As more and more land is deforested, only fragmented islands of forest remain, making it difficult for the remaining animals to forage and breed.

The rain forests of Central America support an enormous variety of insects and arachnids, with many species of ants, spiders, termites, millipedes, cockroaches, cicalas, beetles, chiggers, praying mantises, and scorpions, and such flying insects as bees, butterflies, blackflies, mosquitos, and sandflies. These animals feed on plant growth and animal life at all levels of the forest and the undergrowth, as well as in the marshy coastal areas. The insects are a food source for the many and varied reptiles, bats, birds, rodents, and amphibians that contribute to the rich diversity of life in Central America.

Helen Salmon

DISCUSSION QUESTIONS: FLORA AND FAUNA

Q1. What is the rain-forest canopy? How do scientists differentiate the various levels of the canopy? What plants grow on the floor of the rain forest?

Q2. Name the four members of the camel family that live in the Andes Mountains. How do they differ in size? Which has the most valuable fur? Compare and contrast other traits of the South American camelids.

Q3. What makes the animals of the Galápagos Islands so unique? How did the variations among the Galápagos finches lead Charles Darwin to his famous evolutionary theories?

HUMAN GEOGRAPHY

Overview

The Human Environment

No person lives in a vacuum. Every human being and community is surrounded by a world of external influences with which it interacts and by which it is affected. In turn, humans influence and change their environments: sometimes intentionally, sometimes not, and sometimes with effects that are harmful to these environments, and, in turn, to humans themselves. Humans have always shaped the world in which they live, but developments over the past few centuries have greatly enhanced this capacity.

Many people feel a sense of alarm about the consequences of widespread adoption of modern technology, including artificial intelligence (AI) and accelerating human population growth in the world. Travel and transportation among the world's regions have been made surer, safer, and faster, and global communication is virtually instantaneous. The human environment is no longer a matter of local physical, biological, or social conditions, or even of merely national or regional concerns—the postmodern world has become a true global community.

Students of human geography divide the human environment into three broad areas: the physical, biological, and social environments. The study of ecology describes and analyzes the interactions of biological forms (mainly plants and animals) and seeks to uncover the optimal means of species cooperation, or symbiosis. Everything that humans do affects life and the physical world around them, and this world provides potentials for and constraints on how humans can live.

As people acquired and shared ever-more knowledge about the world, their abilities to alter and shape it increased. Humans have always had a direct impact on Earth. Even 10,000 years ago, Neolithic people cut down trees, scratched the earth's surface with simple plows, and replaced diverse plant forms with single crops. From this basic agricultural technology grew more complex human communities, and people were freed from the need to hunt and gather. The alteration of the local ecosystems could have deleterious effects, however, as gardens turned eventually to deserts in places like North Africa and what later became Iraq. Those who kept herds of animals grazed them in areas rich in grasses, and animal fertilizer helped keep them rich. If the area was overgrazed, however, destroying important ground cover, the herders moved on, leaving a perfect setup for erosion and even desertification. Today, people have an even greater ability to alter their environments than did Neolithic people, and ecologists and other scientists as well as citizens and politicians are increasingly concerned about the negative effects of modern alterations.

The Physical Environment

The earth's biosphere is made up of the atmosphere—the mass of air surrounding the earth; the hydrosphere—bodies of water; and the lithosphere—the outer portion of the earth's crust. Each of these, alone and working together, affect human life and human communities.

Climate and weather at their most extreme can make human habitation impossible, or at least extremely uncomfortable. Desert and polar climates do not have the liquid water, vegetation, and animal life necessary to sustain human existence. Humans can adapt to a range of climates, however. Mild variations can be addressed simply, with clothing and shelter. Local droughts, tornadoes, hurricanes, heavy winds, lightning, and hail can have devastating effects even in the most comfortable of climates. Excess rain can be drained away to make habitable land, and arid areas can be irrigated. Most people live in temperate zones where weather extremes are rare or dealt with by technological adaptation. Heating and, more recently, air conditioning can create healthy microclimates, whatever the external conditions. Food can be grown and then transported across long distances to supply populations throughout the year.

The hydrosphere affects the atmosphere in countless ways, and provides the water necessary for human and other life. Bodies of water provide plants and animals for food, transportation routes, and aesthetic pleasure to people, and often serve to flush away waste products. People locate near water sources for all of these reasons, but sometimes suffer from sudden shifts in the water level, as in tidal waves (tsunamis) or flooding. Encroachment of salt water into freshwater bodies (salination) is a problem that can have natural or human causes.

The lithosphere provides the solid, generally dry surface on which people usually live. It has been shaped by the atmosphere (especially wind and rain that erode rocks into soil) and the hydrosphere (for example, alluvial deposits and beach erosion). It serves as the base for much plant life and for most agriculture. People have tapped its mineral deposits and reshaped it in many places; it also reshapes itself through, for example, earthquakes and volcanic eruption. Its great variations—including vegetation—draw or repel people, who exploit or enjoy them for reasons as varied as recreation, military defense, or farming.

The Biological Environment

Humans share the earth with over 8 million different species of plants, animals, and microorganisms—of which only about 2 million have been identified and named. As part of the natural food chain, people rely upon other life-forms for nourishment.

Through perhaps the first 99 percent of human history, people harvested the bounty of nature in its native setting, by hunting and gathering. Domestication of plants and animals, beginning about 10,000 years ago, provided humans a more stable and reliable food supply, revolutionizing human communities. Being omnivores, people can use a wide variety of plants and animals for food, and they have come to control or manage most important food sources through herding, agriculture, or mechanized harvesting. Which plants and animals are chosen as food, and thus which are cultivated, bred, or exploited, are matters of human culture, not, at least in the modern world, of necessity.

Huge increases in human population worldwide have, however, put tremendous strains on provision of adequate nourishment. Areas poorly endowed with foodstuffs or that suffer disastrous droughts or blights may benefit from the importation of food in the short run, but cannot sustain high populations fostered by medical advances and cultural considerations.

Human beings themselves are also hosts to myriad organisms, such as fungi, viruses, bacteria, eyelash mites, worms, and lice. While people usually can coexist with these organisms, at times they are destructive and even fatal to the human organism. Public health and medical efforts have eradicated some of humankind's biological enemies, but others remain, or are evolving, and continue to baffle modern science.

The presence of these enemies to health once played a major role in locating human habitations to avoid so-called "bad air" (*mal-aria*) and the breeding grounds of tsetse flies or other pests. The use of pesticides and draining of marshy grounds have alleviated a good deal of human suffering. Human efforts can also control or eliminate biological threats to the plants and animals used for food, clothing, and other purposes.

Social Environments

Human reproduction and the nurturing of young require cooperation among people. Over time, people gathered in groups that were diverse in age if not in other qualities, and the development of towns and cities eventually created an environment in which otherwise unrelated people interacted on intimate and constructive levels. Specialization, or division of labor, created a higher level of material wealth and culture and ensured interpersonal reliance.

The pooling of labor—both voluntary and forced—allowed for the creation of artificial living environments that defied the elements and met human needs for sustenance. Some seemingly basic human drives of exclusivity and territoriality may be responsible for interpersonal friction, violence and, at the extreme, war. Physical differences, such as size, skin, or hair color, and cultural differences, including language, religion, and customs, have often divided humans or communities. Even within close quarters such as cities, people often separate themselves along lines of perceived differences. Human social identity comes from shared characteristics, but which things are seen as shared, and which as differentiating, is arbitrary.

People can affect their social environment for good and ill through trade and war, cooperation and bigotry, altruism and greed. While people still are somewhat at the mercy of the biological and physical environments, technological developments have balanced the human relationship with these. Negative effects of human interaction, however, often offset the positive gains. People can seed clouds for rain, but also pollute the atmosphere around large cities, create acid rain, and perhaps contribute to global warming.

Human actions can direct water to where it is needed, but people also drain freshwater bodies and increase salination, pollute streams, lakes, and oceans, and encourage flooding by modifying riverbeds. People have terraced mountainsides and irrigated them to create gardens in mountains and deserts, but also lose about 24 billion metric tons of soil to erosion and 30 million acres (12 million hectares) of grazing land to desertification each year. These negative effects not only jeopardize other species of terrestrial life, but also humans' ability to live comfortably, or perhaps at all.

Globalization

Humankind's ability to affect its natural environments has increased enormously in the wake of the Industrial Revolution. The harnessing of steam, chemical, electrical, and atomic energy has enabled people to transform life on a global scale. Economically, the Western world still to dominates global markets despite effort of China to capture the crown, and computer and satellite technology have made even remote parts of the globe reliant on Western information and products. Efficient transportation of goods and people over huge distances has eliminated physical barriers to travel and commerce. The power and influence of multinational corporations and national corporations in international markets continues to grow.

Human environmental problems also have a global scope: Extreme weather, changes in ocean temperatures and sea level rise, global warming, and the spread of disease by travelers have become planetary concerns. International agencies seek to deal with such matters, and also social and political concerns once left to nations or colonial powers, such as population growth, the provision of justice, or environmental destruction within a country. Pessimists warn of horrendous trends in population and ecological damage, and further deterioration of human life and its environments. Optimists dismiss negative reports as exaggerated and alarmist, or expect further technological advances to mitigate the negative effects of human action.

Joseph P. Byrne

POPULATION GROWTH AND DISTRIBUTION

The population of the world has been growing steadily for thousands of years and has grown more in some places than in others. On November 2019, the total population of the earth had reached 7.7 billion people. The population of the United States in August 2019 was approximately 329.45 million. India's population in November 2019 was 1.37 billion, making it the world's second most populous country. China's population was about 1.45 billion—about 1 in 5 people on the planet.

How Populations Are Counted

The U.S. Constitution requires that a census, or enumeration, of the population of the United States be conducted every ten years. The U.S. Census Bureau mails out millions of census forms and pays thousands of people (enumerators) to count people that did not fill out their census forms. This task cost about US$5.6 billion in the year 2010, and estimates for the 2020 census have risen to over US$15 billion. Despite this great effort, millions of people are probably not counted in every U.S. census. Moreover, many countries have much less money to spend on censuses and more people to count. Therefore, information about the population of many poor or less-developed countries is even less accurate than that for the population of the United States.

Counting how many people were alive a hundred, a thousand, or hundreds of thousands of years ago is even more difficult. Estimates are made from archaeological findings, which include human skeletons, ruins of ancient buildings, and evidence of ancient agricultural practices. Historical records of births, deaths, taxes paid, and other information are also used. Although it is not possible to estimate the global population 1,000 years ago with great accuracy, it is a fascinating topic, and many people have participated in estimating the total population of the planet through the ages.

History of Human Population Growth

Ancient ancestors of humans, known as hominids, were alive in Africa and Europe around 1 million years ago. It is believed that modern humans (*Homo sapiens sapiens*) coexisted with the Neanderthals (*Homo sapiens neandertalensis*) about 100,000 years ago. By 8000 BCE (10,000 years ago) fully modern humans numbered around 8 million. If the presence of archaic *Homo sapiens* is accepted as the beginning of the human population 1 million years ago, then the first 990,000 years of human existence are characterized by a very low population growth rate (15 persons per million per year).

Around 10,000 years ago, humans began a practice that dramatically changed their growth rate: planting food crops. This shift in human history, called the Agricultural Revolution, paved the way for the development of cities, government, and civilizations. Before the Agricultural Revolution, there were no governments to count people. The earliest censuses were conducted less than 10,000 years ago in the ancient civilizations of Egypt, Babylon, China, Palestine, and Rome. For this reason, historical estimates of the earth's total population are difficult to make. However, there is no argument that human numbers have increased dramatically in the past 10,000 years. The dramatic changes in the growth rates of the human population are typically attributed to three significant epochs of human cultural evolution: the Agricultural, Industrial, and Green Revolutions.

Before the Agricultural Revolution, the size of the human population was probably fewer than 10 million people, who survived primarily by hunting and gathering. After plant and animal species were domesticated, the human population increased its growth rate. By about 5000 BCE, gains in food production caused by the Agricultural Revolution meant that the planet could support about 50 million people. For the next several thousand years, the human population continued to grow at a rate of about 0.03 percent per year. By the first year of the common era, the planet's population numbered about 300 million.

At the end of the Middle Ages, the human population numbered about 400 million. As people lived in densely populated cities, the effects of disease increased. Starting in 1348 and continuing to 1650, the human population was subjected to massive declines caused by the bubonic plague—the Black Death. At its peak in about 1400, the Black Death may have killed 25 percent of Europe's population in just over fifty years. By the end of the last great plague in 1650, the human population numbered 600 million.

The Industrial Revolution began between 1650 and 1750. Since then, the growth of the human population has increased greatly. In just under 300 years, the earth's population went from 0.5 billion to 7.7 billion people, and the annual rate of increase went from 0.1 percent to 1.1 percent. This population growth was not because people were having more babies, but because more babies lived to become adults and the average adult lived a longer life.

The Green Revolution occurred in the 1960s. The development of various vaccines and antibiotics in the twentieth century and the spread of their use to most of the world after World War II caused big drops in the death rate, increasing population growth rates. Feeding this growing population has presented a challenge. This third revolution is called the Green Revolution because of the technology used to increase the amount of food produced by farms. However, the Green Revolution was really a combination of improvements in health care, medicine, and sanitation, in addition to an increase in food production.

Geography of Human Population Growth

The present-day human race traces its lineage to Africa. Humans migrated from Africa to the Middle East, Europe, Asia, and eventually to Australia, North and South America, and the Pacific Islands. It is believed that during the last Ice Age, the world's sea levels were lower because much of the world's water was trapped in ice sheets. This lower sea level created land bridges that facilitated many of the major human migrations across the world.

Patterns of human settlement are not random. People generally avoid living in deserts because they lack water. Few humans are found above the Arctic Circle because of that region's severely cold climate. Environmental factors, such as the availability of water and food and the livability of climate, influence where humans choose to live. How much these factors influence the evolution and development of human societies is a subject of debate.

The domestication of plants and animals that resulted from the Agricultural Revolution did not take place everywhere on the earth. In many parts of the world, humans remained as hunter-gatherers while agriculture developed in other parts of the world. Eventually, the agriculturalists outbred the hunter-gatherers, and few hunter-gatherers remain in the twenty-first century. Early agricultural sites have been found in many places, including Central and South America, Southeast Asia and China, and along the Tigris and Euphrates Rivers in what is now Iraq. The practice of agriculture spread from these areas throughout most of the world.

By the time Christopher Columbus reached the Americas in the late fifteenth century, there were millions of Native Americans living in towns and villages and practicing agriculture. Most of them died from diseases that were brought by European colonists. Colonization, disease, and war are major mechanisms that have changed the composition and distribution of the world's population in the last 300 years.

The last few centuries also produced another change in the geography of the human population. During this period, the concentration of industry in urban areas and the efficiency gains of modern agricultural machinery caused large numbers of people to move from rural areas to cities to find jobs. From 1900 to 2020 the percentage of people living in cities went from 14 percent to just about 55 percent. Demographers estimate that by the year 2025, more than 68 percent of the earth's population will live in cities. Scientists estimate that the human population will continue to increase until the year

2050, at which time it will level out at between 8 and 15 billion.

Earth's Carrying Capacity

Many people are concerned that the earth cannot grow enough food or provide enough other resources to support 15 billion people. There is great debate about the concept of the earth's carrying capacity—the maximum human population that the earth can support indefinitely. Answers to questions about the earth's carrying capacity must account for variations in human behavior. For example, the earth could support more bicycle-riding vegetarians than car-driving carnivores. Questions about carrying capacity and the environmental impacts of the human race on the planet are fundamental to the United Nations' goals of sustainable development. Dealing with these questions will be one of the major challenges of the twenty-first century.

Paul C. Sutton

GLOBAL URBANIZATION

Urbanization is the process of building and living in cities. Although the human impulse to live in groups, sharing a "home base" probably dates back to cave-dweller times or before. The creation of towns and cities with a few hundred to many thousands to millions of inhabitants required several other developments.

Foremost of these was the invention of agriculture. Tilling crops requires a permanent living place near the cultivated land. The first agricultural villages were small. Jarmo, a village site from c. 7000 BCE, located in the Zagros Mountains of present-day Iran, appears to have had only twenty to twenty-five houses. Still, farmers' crops and livestock provided a food surplus that could be stored in the village or traded for other goods. Surplus food also meant surplus time, enabling some people to specialize in producing other useful items, or to engage in less tangible things like religious rituals or recordkeeping.

Given these conditions, it took people with foresight and political talents to lead the process of city formation. Once in cities, however, the inhabitants found many benefits. Walls and guards provided more security than the open country. Cities had regular markets where local craftspeople and traveling merchants displayed a variety of goods. City governments often provided amenities like primitive street lighting and sanitary facilities. The faster pace of life, and the exchange of ideas from diverse people interacting, made city life more interesting and speeded up the processes of social change and invention. Writing, law, and money all evolved in the earliest cities.

Ancient and Medieval Cities

Cities seem to have appeared almost simultaneously, around 3500 BCE, in three separate regions. In the Fertile Crescent, a wide curve of land stretching from the Persian gulf to the northwest Mediterranean Sea, the cities of Ur, Akkad, and Babylon rose, flourished, and succeeded one another. In Egypt, a connected chain of cities grew, soon unified by a ruler using Memphis, just south of the Nile River's delta, as his strategic and ceremonial base. On the Indian subcontinent, Mohenjo-Daro and Harappa oversaw about a hundred smaller towns in the Indus River valley. Similar developments took place about a thousand years later in northern China.

These first city sites were in the valleys of great river systems, where rich alluvial soil boosted large-scale food production. The rivers served as a "water highway" for ships carrying commodities and luxury items to and from the cities. They also furnished water for drinking, irrigation, and waste disposal. Even the rivers' rampages promoted civilization, as making flood control and irrigation systems required practical engineering, an organized workforce, and ongoing political authority to direct them.

Eurasia was still full of peoples who were not urbanized, however, and who lived by herding, pirating, or raiding. Early cities declined or disappeared, in some cases destroyed by invasions from such forces around 1200 BCE. Afterward, the cities of Greece became newly important. Their surrounding land was poor, but their access to the sea was an advantage. Greek cities prospered from fishing and trade. They also developed a new idea, the city-state, run by and for its citizens.

Rome, the Greek cities' successor to power, reached a new level of urbanization. Its rise owed more to historical accident and its citizens' political and military talents than to location, but some geographical features are salient. In some ways, the fertile coastal plain of Latium was an ideal site for a great city, central to both the Italian peninsula and the Mediterranean Sea. There, the Tiber River becomes navigible and crossable.

In other ways, Rome's site was far from ideal. Its lower areas were swampy and mosquito-ridden. The seven hills, with their sacred sites later filled with public buildings and luxury houses, imposed a crazy-quilt pattern on the city's growth. Romans built cities with a simple rectangular plan all over Europe and the Middle East, but their home city grew in a less rational way.

At its peak, Rome had a million residents, a population no other city reached before nineteenth century London. It provided facilities found in modern cities: a piped water supply, a sewage disposal system, a police force, public buildings, entertainment districts, shops, inns, restaurants, and taverns. The streets were crowded and noisy; to control traffic, wheeled wagons could make deliveries only at night. Fire and building collapse were constant risks in the cheaply built apartment structures that housed the city's poorer residents. Still, few wanted to live anywhere but in Rome, their world's preeminent city.

In the Early Middle Ages after the western Roman Empire collapsed, feudalism, based on land holdings, eclipsed urban life. Cities never disappeared, but their populations and services declined drastically. Urban life still flourished for another millenium in the eastern capital of Constantinople. When Islam spread across the Middle East, it caused the growth of new cities, centered around a mosque and a marketplace.

In the twelfth and thirteenth centuries, life revived in Western Europe. As in the Islamic cities, the driving forces were both religious—the building of cathedrals—and commercial—merchants and artisans expanding the reach of their activities. Medieval cities were usually walled, with narrow, twisting streets and a lack of basic sanitary measures, but they drew ambitious people and innovative forces together. Italy's cities revived the concept of the city-state with its outward reach. Venice sent its merchant fleet all over the known world. Farther north, Paris and Bologna hosted the first universities. The feudal system slowly gave way to nation-states ruled by one king.

Modern Cities

Modern cities differ from earlier ones because of changes wrought by technology, but most of today's cities arose before the Industrial Revolution. Until the early nineteenth century, travel within a city was by foot or on horse, which limited street widths and city sizes. The first effect of railroads was to shorten travel time between cities. This helped country residents moving to the cities, and speeded raw materials going into and manufactured goods coming out of the factories that increasingly dotted urban areas. Rail transit soon caused the growth of a suburban ring. Prosperous city workers could live in more spacious homes outside the city and ride rail lines to work every day. This pattern was common in London and New York City.

Factories, the lifeblood of the Industrial Revolution, were built in pockets of existing cities. Smaller cities like Glasgow, Scotland, and Pittsburgh, Pennsylvania, grew as ironworking industries, using nearby or easily transported coal and ore resources, built large foundries there. Neither industrialists nor city authorities worried about where the people working there would live. Workers took whatever housing they could find in tenements or subdivided old mansions.

Beginning in the 1880s, metal-framed construction made taller buildings possible. These skyscrapers towered over stately three- to eight-story structures of an earlier period. Because this technology enabled expensive central-city ground space to house many profitable office suites, up through the 1930s, city cores became quite compacted. Many people believed such skyward growth was the wave of the future and warned that city streets were becoming sunless, dangerous canyons.

Automobiles kept these predictions from fully coming true. As car ownership became widespread, more roads were built or widened to carry the traffic. Urban areas began to decentralize. The car, like rail transit before it, allowed people to flee the urban core for suburban living. Because roads could be built almost anywhere, built-up areas around cities came to resemble large patches filling a circle, rather than the spokes-of-a-wheel pattern introduced by rail lines. Cities born during the automotive age tend to have an indistinct city center, surrounded by large areas of diffuse urban development. The prime example is Los Angeles: It has a small downtown area, but a consolidated metropolitan area of about 34,000 square miles (88,000 sq. km.).

Almost everywhere, urban sprawl has created satellite cities with major manufacturing, office, and shopping nodes. These cause an increasing portion of daily travel within metropolitan areas to be between one edge city and another, rather than to and from downtown. Since these journeys have an almost limitless variety of start points and destinations within the urban region, mass transit is only a partial solution to highway crowding and air pollution problems.

The above trends typify the so-called developed world, especially the United States. Many cities in poor nations have grown even more rapidly but with a different mix of patterns and problems. However, the basic pattern can be detected around the globe, as urban dwellers seek to better their own cir-

URBANIZATION AND DEVELOPING NATIONS

The urban population, or number of people living in cities, in North America accounts for about 75 percent of its total population. In Europe, about 90 percent of the population lives in cities. In developing countries, the urban population is often less than 30 percent. The term "urbanization" refers to the rate of population growth of cities. Urbanization mainly results from people moving to cities from elsewhere. In developing countries, the urbanization rate is very high compared to those of North America or Europe. The high rate of urbanization of these countries makes it difficult for their governments to provide housing, water, sewers, jobs, schools, and other services for their fast-growing urban populations.

cumstances. Today, 55 percent of the world's population lives in urban areas, and that percentage is expected to rise to 68 percent by 2050. Projections show that urbanization combined with the overall growth of the world's population could add another 2.5 billion people to urban areas by 2050, with close to 90 percent of this increase taking place in Asia and Africa, according to a United Nations data set published in May 2018.

Megacities and the Future

In the year 2019 the world had thirty-three megacities, defined as urban areas with a population of 10 million or more. The largest was Tokyo, with an estimated 37.5 million people in 2018, predicted to grow to around 37 million by 2030. Second-largest was Delhi, with more than 28.5 million in 2018 and predicted to grow to around 38.94 million by 2030. Megacities in the United States include New York-Newark with a population of 18.8 million and Los Angeles at 12.5.

Megacities profoundly affect the air, weather, and terrain of their surrounding territory. Smog is a feature of urban life almost everywhere, but is worse where the exhaust from millions of cars mixes with industrial pollution. Some megacities have slowed the problem by regulating combustion technology; none have solved it. Huge expanses of soil pre-

URBAN HEAT ISLANDS

Large cities have distinctly different climates from the rural areas that surround them. The most important climatic characteristic of a city is the urban heat island, a concentration of relatively warmer temperatures, especially at nighttime. Large cities are frequently at least 11 degrees Fahrenheit (6 degrees Celsius) warmer than the surrounding countryside.

The urban heat island results from several factors. Primary among these are human activities, such as heating homes and operating factories and vehicles, that produce and release large quantities of energy to the atmosphere. Most of these activities involve the burning of fossil fuels such as oil, gas, and coal. A second factor is the abundance of heat-absorbing urban materials, such as brick, concrete, and asphalt. A third factor is the surface dryness of a city. Urban surface materials normally absorb little water and therefore quickly dry out after a storm. In contrast, the evaporation of moisture from wet soil and vegetation in rural areas uses a large quantity of solar energy—often more than is converted directly to heat—resulting in cooler air temperatures and higher relative humidities.

empted by buildings and pavements can turn heavy rains into floods almost instantly, and the ambient heat in large cities stays several degrees higher than in comparable rural areas. Recent engineering studies suggest that megacities create instability in the ground beneath, compressing and undermining it.

How will cities evolve? Barring an unforeseen technological or social breakthrough, the current growth and problems will probably continue. The process of megapolis—metropolitan areas blending together along the corridors between them—is well underway in many areas. Predictions that the computer will so change the nature of work as to cause massive population shifts away from cities have not been proven correct. Despite its drawbacks, increasing numbers of people are drawn to urban life, seeking the economic opportunities and wider social world that cities offer.

Emily Alward

PEOPLE

SOUTH AMERICA

The nature and pattern of the gradual habitation of South America by humans is similar to that of North America in some important aspects. The first South Americans were most likely the descendants of Asian hunters who dispersed throughout the Americas and physically and culturally evolved in accordance with the demands of their respective environments. There may have been other prehistoric migrants from Japan, Southeast Asia, and Europe.

Permanent European settlement began in 1492 with Christopher Columbus's arrival in the New World. In sharp contrast to most of North America, the southern half of the Western Hemisphere was dominated by the Portuguese and the Spanish. In more recent times, emigrants into South America have come from central and southern Europe, the Middle East, and Japan. The ethnic complexity created in South America rivals that of the United States.

Native Americans

There is no doubt that people from Asia populated the Americas, but which part of Asia is in dispute. Increasingly, the physical evidence suggests a migration of a non-Mongoloid human strain, possibly from Southeast Asia, as being the first significant human incursion into the Americas. Some genetic evidence suggests a relationship between the present-day Amerindians, or Native Americans, of Ecuador and southeast Asians, Japanese, and perhaps even Polynesians. There is similar genetic evidence for the Native Americans in Chile, Colombia, and Brazil. Some genetic and stone tool evidence points to a European presence, perhaps from France and Spain, around 20,000 and 16,000 BCE. However,

there is no fossil record to support that idea. Subsequent migrations into the Americas appear, instead, to have come from northeast Asia. This is the dominant theory as to the origins of Native Americans.

As to where the earliest humans entered into South America, two possible areas of initial settlement are likely. One area is located in Colombia and Ecuador, along the eastern slopes of the Andes and in the adjacent tropical lowlands. The other is central Amazonia. Some evidence suggests that initial movement into South America took place along the Pacific and Caribbean coastlines. The Native Americans eventually populated the entire continent.

Machu Picchu is a fifteenth-century Inca citadel, located in the East-ern Cordillera of southern Peru, on a 2,430-metre (7,970-ft) moun-tain ridge in the Cusco Region. Most archaeologists believe that Machu Picchu was constructed as an estate for the Inca emperor Pachacuti (1438–1472). The Incas abandoned it at the time of the Spanish conquest. Although known locally, it remained unknown to the outside world until American historian Hiram Bingham brought it to international attention in 1911. (Image by Diego Delso)

POPULATION DENSITIES OF SOUTH AMERICAN COUNTRIES

North
Atlantic
Ocean

GUYANA
VENEZUELA
SURINAME
FRENCH GUIANA

Bogotá
★

COLOMBIA

Galápagos
Islands

ECUADOR   ★Quito

Amazon Basin

Belem

PERU

BRAZIL

A
N
D
E
S

Lima ★

Brasilia
★

★ La Paz

Arica

BOLIVIA

South
Pacific
Ocean

M
O
U
N
T
A
I
N
S

PARAGUAY

Rio de Janeiro

Asunción

CHILE

ARGENTINA

URUGUAY

Santiago★ •Mendoza

Buenos Aires

★ Montevideo

South
Atlantic
Ocean

Falkland Islands

Fewer than 10
persons/sq. mi.

10–20
persons/sq. mi.

51–100
persons/sq. mi.

21–50
persons/sq. mi.

More than 100
persons/sq. mi.

The Cueva de las Manos, or Cave of the Hands, in Argentina. The art in the cave dates from 13,000 to 9,000 years ago and was calculated from the remains of bone-made pipes used for spraying the paint on the wall of the cave to create silhouettes of hands. Most of the hands are left hands, which suggests that painters held the spraying pipe with their right hand. (Image by Mariano)

The native ethnic groups that inhabited South America were diverse. The Arawak and Carib peoples inhabited the northern coastal areas such as present-day Venezuela, Colombia, Guyana, Suriname, and French Guiana. The Inca ruled much of Peru, Bolivia, and Ecuador before the Spanish *conquistador* Francisco Pizarro arrived. In Argentina, Native American groups included the Diaquita of the Andean northwest, the Querani, who inhabited the present-day region of Buenos Aires, the Tehuelche in Patagonia, and the Ona in Tierra del Fuego. Among the native peoples in Chile when the Europeans arrived were the Atacameños, Picunche, Cuncos, and the Araucanians, groups who still maintain a distinct identity.

The eventual arrival of the Portuguese and Spanish doomed the Amerindian civilizations and, to some extent, the people. Forced labor and European diseases decimated the native populations. In the coastal tropical areas, only about 5 percent of the native population was left ten years after the first significant contact with Europeans. The rate of population loss was less pronounced in higher, cooler areas, perhaps because disease did not spread so easily there. As a result, the native population in western and northern South America eventually recovered. In Brazil, the Amazonian tribes were almost eliminated by European diseases. In Argentina, Uruguay, and, to a lesser extent, Paraguay, the number of Amerindians was driven to low levels by disease and by being hunted down. The native populations became so small that the European population absorbed their remnants.

Iberians

Christopher Columbus led the first Europeans to the New World in 1492. He landed on an island in the Caribbean and claimed the lands he found for Spain. Conflicts at the end of the fifteenth century between the Portuguese and Spanish over their overseas claims led the Roman Catholic pope to negotiate the Treaty of Tordesillas in 1494. This treaty divided the New World between Portugal and Spain, with a line drawn such that what became Brazil eventually went to the Portuguese and the rest of the New World to the Spanish.

Until the early to mid-nineteenth century, when Iberian domination came to an end, only the Spanish and Portuguese were admitted to their South American colonies. The rigid exclusion of other foreigners had but few exceptions. Most of the Spaniards came from Castile and the southern regions. It is estimated that the total number of *licencias* (au-

thorizations to emigrate) granted by Spain was about 150,000 for the entire colonial period, which lasted from the sixteenth to the nineteenth century. It is possible that the number of illegal immigrants also approached this number. Of these, no more than two-fifths of the emigrants went to South America.

Little is known about the principal regions from which the Portuguese came. As many as 1 million Portuguese may have emigrated to Brazil, drawn primarily by a gold rush in Minas Gerais in the eighteenth century. Emigration from Portugal to Brazil continues.

Africans

Sugar production became important in northeastern Brazil in the early to mid-1540s. The Portuguese tried converting Amerindians to Christianity and paying them wages to work in sugar produc-

POPULATION, AREA, AND DENSITY IN SOUTH AND CENTRAL AMERICA

Country	Population	Area		Density	
		Kilometers	Miles	Per sq. km	Per sq. mi.
Argentina	44,780,677	2,736,690	1,056,642	16.36307985	42.380182
Bolivia	11,513,100	1,083,300	418,264	10.62780393	27.525886
Brazil	211,049,527	8,358,140	3,227,096	25.25077673	65.399211
Chile	18,952,038	743,532	287,079	25.48920289	66.016732
Colombia	50,339,443	1,109,500	428,380	45.37128707	117.51109
Ecuador	17,373,662	248,360	95,892	69.95354324	181.17885
French Guiana	290,832	82,200	31,738	3.53810219	9.1636426
Guyana	782,766	196,850	76,004	3.976459233	10.298982
Paraguay	7,044,636	397,300	153,398	17.73127611	45.923794
Peru	32,510,453	1,280,000	494,211	25.39879141	65.782568
Suriname	581,372	156,000	60,232	3.72674359	9.6522216
Uruguay	3,461,734	175,020	67,576	19.77907668	51.227573
Venezuela	28,515,829	882,050	340,561	32.32903917	83.731827
Belize	390,353	22,810	8,807	17.11323981	44.323088
Costa Rica	5,047,561	51,060	19,714	98.85548374	256.03453
El Salvador	6,453,553	20,720	8,000	311.4649131	806.69042
Guatemala	17,581,472	107,160	41,375	164.0674879	424.93284
Honduras	9,746,117	111,890	43,201	87.1044508	225.59949
Nicaragua	6,545,502	120,340	46,464	54.39174007	140.87396
Panama	4,246,439	74,340	28,703	57.12185903	147.94494

Source: Population Reference Bureau, 2019 World Population Data Sheet.

tion, but the primary means of extracting their labor was to turn them into chattel slaves.

In the 1560s, a major smallpox epidemic broke out among these previously unexposed populations of Amerindians. It has been estimated that 30,000 Amerindians under Portuguese control, either on plantations or in Jesuit mission villages, died of the disease. This decline in native slave labor led to the beginnings of mass importation of African slave labor after 1570. By the 1620s, sugar plantation labor was almost entirely African slaves. The sugar plantations in South America were concentrated mostly in the northeast part of the continent in present-day Guyana, Suriname, and northeastern Brazil.

The Dutch initially dominated the sugar trade in both the Caribbean and northeast South America, including that portion of Brazil, but the Portuguese had pushed the Dutch out by about 1660. The Dutch did much of the initial importation of African slaves into the region. Among the Dutch planters in the seventeenth and eighteenth centuries were an important minority of Jews. In Dutch Guiana (Suriname) by the 1760s, Jewish families owned 115 of the colony's 591 sugar estates and formed the largest number of native-born whites. A small mixed-race Jewish community developed and formed its own synagogue in 1759.

By 1780 there were 406,000 freed persons of slave descent and 1.5 million slaves in Brazil. African slaves in Brazil came from West Africa, the Congo, Angola, and Mozambique. The last slaves in Brazil were not freed until 1888. Brazil now contains the largest population of African origin outside of Africa. While African slaves were taken to other areas in South America, their numbers in other locales never approached those of Brazil. In Peru, for example, there were 90,000 African slaves in the eighteenth century.

By 1700 the native populations of Peru and other Spanish-controlled areas such as Colombia had adjusted to the European disease environment and were in the process of rapid population expansion. The free native and mestizo populations met labor needs in agricultural, mining, artisanal, and service areas; African slaves were not needed. In other ar-

eas of South America in the late eighteenth century, the African slave populations were 21,000 in Argentina; 12,000 in Chile; 54,000 in Colombia; 75,000 in Suriname; 8,000 in Ecuador; and 64,000 in Venezuela.

Indentured Laborers

In the 1830s slavery was abolished in much of the Caribbean and Central and South America. That left a vacuum in the labor market for the sugar plantations. In response to this need, indentured laborers were brought in from China and India. Between 1838 and 1918, 238,909 East Indians, primarily from Bengal, Bombay (now Mumbai), and Madras (now Chennai), went to Guyana. From 1853 to 1867, 11,282 Chinese went to Guyana. The numbers for French Guiana and the Dutch Caribbean were similar to those for Guyana. During the 1850s and 1860s, 100,000 Chinese went to Peru. Much smaller groups of Chinese went to Brazil, Chile, and Venezuela. At the same time, poor economic conditions in the Madeira Islands of Portugal caused approximately 10,000 Madeirans to immigrate to Guyana as indentured servants or subsidized labor, as it was often called.

Other Europeans

By 1800 Spanish immigration into most of the Americas had ceased. In 1810, during the period of French control of Spain under Napoleon, almost all of Spanish America declared its independence. Voluntary emigration of the Spanish into much of South America has been limited since the early 1800s.

Spanish governments typically have restricted emigration by controlling the numbers who leave and requiring the issuance of a permit to emigrate from Spain. Portugal, however, has never attempted to control emigration.

Only Brazil and Argentina have had significant immigration in post-colonial times. Argentina is quite European. Between 1857 and 1872, 221,000 immigrants entered Argentina and about 103,000 stayed in the country. Of that number, 63,825 were Italians (61 percent); 21,416 were Spaniards (20.5 percent); 7,055 were French (6.7 percent); and the rest of northwest European stock such as 2,604 English (2.5 percent). While Brazil has a strong African and even now an Amerindian presence, it also has had significant European immigration. Brazil is somewhat similar to the United States in its ethnic complexity and has a better history of race relations, despite Brazilian enslavement of Africans. Racial classifications in Brazil are not as sharply defined as in other nations. The Portuguese colonists who settled Brazil had a more relaxed attitude toward interracial relationships than other Europeans and often intermarried with Africans and Amerindians.

Argentina: Pressure-Relief Valve of Europe

Argentina, in particular, became a haven after World War II for anticommunists of all stripes, including Nazis and Nazi sympathizers. Argentina opened its doors to Italian immigrants on a large scale, with almost half a million arriving in the late 1940s. In opening its doors to immigration from Italy, Argentina also became a haven for Ukrainians, Croats, and Yugoslavians living in Italy who were unable or unwilling to be repatriated after World War II.

The United States Department of State also intervened on behalf of some 15,000 Yugoslavians who could not return to Josip Tito's Yugoslavia because they had fought under the promonarchist general Dragoljub Mihajlovic or were suspected of collaborating with the Nazis. Argentina was asked by the United States to take in 5,000 of them. The British government requested that Argentina absorb a group of Polish soldiers who had fought to liberate Italy under Wladyslaw Anders. Argentina complied in both cases, thereby bringing into the republic thousands of newcomers, some of them openly anti-Semitic. On October 18, 1948, Argentina signed an agreement with Spain, which, like Italy, was a preferred source of immigration. The agreement brought 140,000 Spaniards into Argentina during the next five years.

South America Today

Neither Colombia nor Venezuela had much net immigration in the past seventy years. Colombia does not encourage immigration. While Venezuela has had great influxes of people due to its petroleum economy, few have stayed. Bolivia has seen little immigration since the early 1800s, other than small groups of Japanese and Okinawan farmers who arrived in the 1950s and 1960s.

Paraguay has seen little immigration-its people are almost entirely mestizo. Around the beginning of the twentieth century, small numbers of displaced eastern European Jews and Christian Syrians and Palestinians fleeing the Ottoman Empire arrived in Chile. There was an unusual official encouragement of German and Swiss colonization in the Lake District of Chile during the second half of the nineteenth century. Ecuador has had immigration from a variety of foreign countries, including Lebanon, China, Korea, Japan, Italy, and Germany. Recent censuses in Ecuador have not inquired about ethnicity, language, religion, or origin. Thus, the number of individuals in each different group is not precisely known.

Bolivia has had an interesting small-scale influx of Europeans. Between the outer edge of the Amazon and the heart of the Chaco desert is a vast plain with some of the richest soil in the world. In eastern Bolivia, the jungle gives way to agricultural flatlands reminiscent of the American Midwest. Since the 1950s these plains have been colonized by the poor and religiously persecuted from all over the world. During the 1950s the Old Order Mennonites of German and Russian descent came to Bolivia from neighboring Paraguay in covered wagons. More recently, Mennonites from Canada, Mexico, and Belize have arrived to develop new colonies. The number of Mennonite colonies on the Bolivian frontier has been continually growing, and today totals more than sixty.

Today, South America receives few immigrants from abroad. Illegal immigration from less affluent countries, such as Peru and Bolivia, into wealthier Chile and, to a lesser extent Uruguay and Argentina, is far more common. The most prominent demographic trend is the movement of people from rural to urban areas.

Dana P. McDermott

CENTRAL AMERICA

Central America's first inhabitants were various groups of indigenous peoples now referred to as Native Americans or Amerindians. Originally thought to be migrants from the Asian continent, records of them have been found from Mesoamerica, now known as Mexico, to present-day Panama. The isthmus of Central America acted as a bridge for the migrations of peoples from South America and for the movement southward of peoples from North America.

Early History

Central America's most famous precolonial settlements were the Mayan city-states. The Maya had developed a complex society, skilled in astronomy and agriculture, and devoted to a complex set of reli-

POPULATION DENSITIES OF CENTRAL AMERICAN COUNTRIES

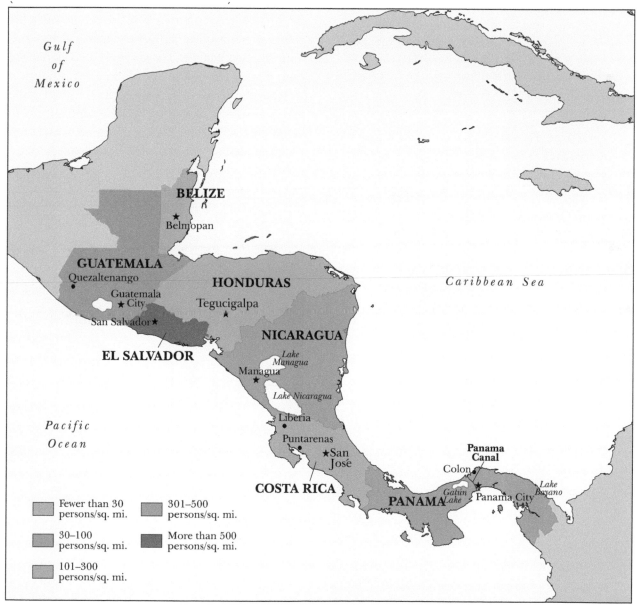

THE MAYA

The Mayan people inhabited much of Central America from the third to the ninth centuries. Archaeologists believe that during the height of their power, in what is called the Classic Period (around 1,800 to 1,200 years ago), the Maya flourished throughout most of Belize and Guatemala, the northwestern portion of El Salvador, northern Honduras, and much of southern Mexico, including the Yucatán.

The political geography of the Mayan world revolved around city-states, such as Tikal in Guatemala. Their architecture is some of the most impressive of any prehistoric peoples. Mayan culture began to deteriorate around 1,150 years ago, and when the Spanish entered their homeland in the 1500s, the Maya reflected but a shadow of their past glory. Today, over 8 million modern Maya inhabit the homeland of their ancestors, and continue to be a dynamic part of the region's human mosaic.

gious practices. They settled in modern Guatemala, Honduras, and El Salvador.

Another powerful Amerindian society, the Pipiles, came from Mexico to occupy western Central America, chiefly in modern El Salvador. They are believed to have been descendants of the Aztec peoples and spoke Nahuatl, a language common to both societies. Less-advanced indigenous cultures existed farther south, in modern Nicaragua, Costa Rica, and Panama.

Christopher Columbus, the Italian navigator in the service of the Spanish crown, opened up Central America to European exploration at the end of the fifteenth century. Afterward, Hernán Cortés, conqueror of Mexico, sent his lieutenant, Pedro de Alvarado, to invade Central America and claim the lands therein for the king of Spain. Alvarado moved south, defeating any Amerindians who attempted to oppose him, and established claim over much of the isthmus as part of New Spain, as Mexico was called during the sixteenth century.

Spanish Colonization

As the Spaniards began to colonize Central America, they introduced African slaves from the Caribbean to aid in the development of sugar plantations in Guatemala and Nicaragua. The slaves proved to be a source of effective and dependable labor. The Amerindian population was much less inclined to perform the demanding work in the fields and sugar mills that was involved in producing this cash crop. The Spaniards also brought in Africans to work in the mines in Honduras. Because of the high cost of African slaves, Indian labor was used instead wherever possible.

Slave Labor

In the seventeenth century, the British introduced African slave labor into British Honduras, which later became the independent country of Belize. The British had seized the area from Spain and began the export of hardwoods. Belize remained a British colony, despite claims to the territory by Guatemala, until it achieved independence in 1981. English is still the official language of the country, although the language spoken by roughly half the population is an English-based Creole. Peasants from Guatemala and El Salvador have moved into western Belize to escape the political unrest in those two countries. Belize, with a population of slightly fewer than 400,000 people, has more open land than any other Central American nation.

Public flogging of a slave in nineteenth-century Brazil. (Painting by Johann Moritz Rugenda)

Individual Nations

Guatemala, Central America's most populous country, is also the most ethnically Amerindian. The capital, Guatemala City, contains the majority of the country's ladinos—the people who adhere to

European culture. In the highlands, tribal languages and customs prevail over Spanish. Thousands of Guatemalans have attempted to emigrate to the United States in an effort to escape violence in their home country.

Honduras, Central America's second-largest country, had a population of roughly 9.2 million in 2018. Primarily an agricultural area, its major commercial operation is Chiquita Brands International, a Swiss enterprise. The country's major crop is bananas for export. Job opportunities on the banana plantations also have attracted a small minority of Caribbean blacks. Honduras received many immigrants from El Salvador during the latter's 1980s civil war.

In 1969 an attempt by the Honduran government to expel thousands of Salvadorans led to the so-called Soccer War, which lasted only a few days. The Organization of American States persuaded the smaller country to withdraw its forces from Honduran soil. The Salvadorans dispossessed by the war were precluded from returning, however. Relations between the two countries later were restored and the borderline more carefully drawn.

El Salvador, the smallest of the Central American countries, with only a Pacific coastline, has never had a major influx of Africans. During colonial times, the mountains dividing eastern and western Central America, together with the high costs of shipping, discouraged agriculture. Political oppression on the part of the government in the 1930s de-

The stone spheres of Costa Rica are an assortment of over 300 petrospheres, located on the Diquís Delta and on Isla del Caño. The spheres are commonly attributed to the extinct Diquis culture and are sometimes referred to as the Diquís Spheres. They are thought to have been placed in lines along the approach to the houses of chiefs, but their exact significance remains uncertain. (Image by Rodtico21)

terred the retention of native cultures; therefore, little evidence of the pre-colonial society exists. During one episode in 1936, government troops massacred 30,000 indigenous people, leading the latter to abandon tribal organization.

Violent civil wars plagued El Salvador during the 1980s. The prolonged unrest drove out nearly 20 percent of the country's people, with the United States receiving the largest number of refugees. El Salvador's population has recovered somewhat and now exceeds 6 million people.

Nicaragua's Mosquito Coast also became the focus of Caribbean African colonization. Although the country's population is heavily concentrated on the Pacific Coast and is predominantly Spanish-speaking, the more isolated Atlantic Coast has been influenced by the proximity of the island culture of the Caribbean. The central government in Managua has traditionally had a difficult time imposing its authority over the eastern shore. Nicaragua had a population of slightly more than 6 million in 2018.

Costa Rica, with a 2018 population of just under 5 million, is unusual in that the majority of its inhabitants are Caucasian. When Europeans arrived, they found an area sparsely settled by indigenous peoples. Costa Rica is a country of self-sustaining small farms as well as commercial agriculture.

"LOS TURCOS" IN CENTRAL AMERICA

During the late eighteenth and early nineteenth centuries, the countries of Central America experienced an influx of Near Eastern peoples, mainly from present-day Lebanon and Syria. Most practiced the Christian faith but came from the Turkish Empire, a Muslim state. Because they carried Turkish passports, they came to be identified as "Los Turcos," or "the Turks." Many emigrated to escape Turkish rule and to be able to practice their religion without interference. Descendants of these immigrants live throughout contemporary Central America. To some degree, they continue to maintain their ethnic heritage, socially and economically.

A view on globalization, titled Somos cultura que camina en un mundo globalizado *("We are a culture walking in a globalized world"). The mural is located in Humahuaca in the north of Argentina. (Image by Elemaki)*

There has been an influx of blacks from the Caribbean into the country's eastern port area, around Puerto Limón. For the most part, they work in Costa Rica's commercial agricultural industry. The country produces bananas, pineapples, coffee, melons, sugarcane, beef, and seafood for export.

Central America's southernmost country, Panama, came into existence in 1903, securing its independence from Colombia with help from the United States. At issue was the soon-to-be-built Panama Canal, a shortcut between the east and west coasts of the North and South American continents. The canal, built and operated by the United States, reverted to Panamanian sovereignty in 2000.

Panama's population, Caucasian, mestizo (mixed blood), and Amerindian at its inception, received a substantial influx of black workers from the Caribbean when construction on the canal began. Panama's modern population reflects that pattern of immigration.

Contemporary Central America represents a kaleidoscope of different races and ethnic backgrounds. Although primarily Latin in social structure, minorities from throughout the world have made these countries their home. In many cases, native Central Americans have returned to their countries of origin after spending years in political or economic exile. They have brought back ideas and practices from the countries that had furnished them with refuge, further broadening the social bases of their respective societies.

Carl Henry Marcoux

POPULATION DISTRIBUTION

The region of South and Central America is home to approximately 459 million people, about 6 percent of the world's total population. Its total land area is 7.1 million square miles (18.4 million sq. km.), giving the region a population density of about 65 people per square mile. In contrast to the world population density of approximately 38 people per square mile, the region appears to be relatively uncrowded. However, population density figures can be misleading.

There are few places in the world, and none whatsoever in South and Central America, where populations are spread evenly throughout an entire country. The actual distribution of people within a defined area is much more important than its numerical density. In South and Central America, population distributions are uneven. Some areas support extremely high densities, whereas vast areas have few people.

Many factors influence the distribution of people. The physical environment plays an important role in population density. In South and Central America, climate, land features, soils, and water strongly influence where people live, or in many cases, do not live. Cultural factors also are involved. For example, populations tend to cluster in places that offer at least the hope of achieving a better quality of life, and to avoid those places where living can be difficult.

Characteristics of Population Distribution

Population distribution in South and Central America has three primary characteristics: People are unevenly distributed, with most settlements bordering, or standing close to, coasts; the region is highly urbanized, with approximately three of every four people living in cities; and, finally, recent migrations have changed the population density of some areas.

Perhaps the most obvious feature of the region's settlement patterns is its uneven distribution. Some areas have high population densities, while others support few people. In South America, only Uruguay has nearly all of its territory settled and accessible by road or rail. In Central America, the same can be said only for Costa Rica and El Salvador. In all of the other countries in the region, large parts of their territory remain relatively isolated and thinly populated.

South America

The highest population densities in South America generally occur near the edge of the continent. Nearly 90 percent of that continent's people live within 100 miles (160 km.) of the coast. Densities ranging from 125 to more than 250 per square mile occur in several locations. The continent's greatest cluster of population is in southeastern Brazil. Major cities include São Paulo, a huge center of industry, agriculture, trade, commerce, and services; Rio de Janeiro, the former capital of Brazil, and still its center of culture and tourism; Santos, the port for São Paulo; and Belo Horizonte, a regional center for agriculture, mining, and industry. A thriving economy and pleasant climate have long attracted people to this area.

South America's second-largest cluster of people is in east central Argentina and southern Uruguay. There, the fertile soils and favorable climate of the Argentine Pampas and the rich grazing lands of Uruguay support one of the world's most productive agricultural regions. Buenos Aires (population, 15 million) is one of the continent's major centers of manufacturing and trade.

Secondary centers of high population density in South America occur in central Chile, central Colombia, northern Venezuela, and coastal Brazil. More than half of Chile's population lives in a large valley near its capital and primary economic center, Santiago. A mild Mediterranean climate and fertile soils help make Chile's Central Valley a productive agricultural region. Because it is in the Southern Hemisphere, Chile can grow summer crops for export to North American winter markets. The area's industry and agriculture are served by the coastal port city of Valparaiso.

In the northern Andean area, higher elevations provide relief from the hot, humid conditions of the tropical lowlands. In Colombia, an area of high population density is clustered in mountain and river valleys in an area roughly outlined by a triangle linking the capital, Bogotá, and the cities of Medellín and Cali. Historically, agriculture has been the primary source of income in this region.

A final area of high population density occurs in eastern coastal Brazil, from Rio de Janeiro northward to the cities of Salvador (Bahia), Recife (Pernambuco), and Fortaleza (Ceara). Fertile alluvial (stream-deposited) soils and a humid tropical climate have helped make this a productive agricultural region. Early Portuguese settlers, using African slaves as laborers, developed a plantation economy based on sugarcane. The narrow coastal strip has continued to be a major area for growing sugar and other tropical plantation crops.

Approximately half of South America has a population density of fewer than two persons per square mile. Vast areas are without roads, electricity, communications, or services. Such areas are not attractive to potential settlers. Geographers often refer to such areas as being outside the realm of effective national control. They contribute little to the national economies and, because of their isolation, are difficult to govern.

"Too Lands"

The concept of "too lands" can help to explain the areas of low population density both in South and in Central America. Such areas are places that are too high, too low (subject to flooding), too cold, too dry, too isolated, too infertile, or in some other way too difficult to inhabit and develop. The largest such area is the Amazon Basin, which spans nearly half the continent. Many factors explain why the region supports such a low population density. Brazil's Portuguese settlers avoided the interior at first. They were a coastal people and had come to the New World to grow sugarcane on the fertile soils of the narrow, tropical, Atlantic coastal plain.

When Brazilians finally turned their attention to the Amazon Basin, they learned that the rain-forest ecosystem presented many problems in terms of potential development. The humid tropical conditions were uncomfortable. Dense tropical rain forest was difficult to penetrate and clear. Most tropical soils, they soon learned, were infertile because their nutrients were leached by the region's high rainfall. Much of the area bordering the Amazon River and its many tributaries floods each year. Only recently has the vast Amazon frontier been opened by roads; as roads penetrate the interior, settlement and economic development follow.

To the south, in the continent's midlatitudes, Argentina's Patagonia region is a dry, cold, windswept plateau. Streams running from the Andes to the Atlantic Ocean have carved steep valleys that make north-south travel difficult. Livestock grazing is the primary economic activity of the region. Settlements and transportation routes are few and far between. To the west, in Chile's southern Andes Mountains and Pacific coastal region, elevation, cold, ruggedness, and lack of access make much of the area uninhabitable.

Central America

The population of Central America is also distributed unevenly. Only Costa Rica and El Salvador have most of their territory settled. Although there are some exceptions, most of the region's population and larger cities are in the cooler, more comfortable interior uplands. Hot, humid coastal plains are poorly developed, with low population densities and few communities. Exceptions are Panama's canal zone, with the coastal port cities of Colón and Panama, and Managua, Nicaragua's capital and largest city, which is on the shores of Lake Managua, near sea level. Elsewhere in Central America, most cities and their surrounding population clusters—such as San José, Costa Rica; Tegucigalpa, Honduras; San Salvador, El Salvador; and the city of Guatemala—are located in the interior uplands of their respective countries.

Urbanization

South and Central America stand apart among the world's regions in that they are highly urbanized, and are yet still not well developed industrially. Nearly four of every five people in the region live in a town or city. Of all South and Central American cities, only São Paulo, Brazil, ranks among the world's ten largest cities, at number nine. Other large cities include Buenos Aires, Argentina; Rio de Janeiro, Brazil; and Lima, Peru. No Central American city ranks in the top 100.

Geographers use the term "primate city" to refer to an individual city that is much larger and more important than all the other cities in a country. In South and Central America, most countries have one urban center that not only is much larger than others, but also serves as the country's economic, political, and cultural center. Exceptions include Belize, with Belize City the population, economic, and cultural center, and Belmopan the capital; Brazil, with population and economic functions, government, and culture served by São Paulo, Brasília, and Rio de Janeiro, respectively; and Ecua-dor, which has the Andean city of Quito as its capital and cultural center and the coastal port of Guayaquil as its largest city.

Because industry and commerce are not well developed in much of South and Central America, many urban people live in poverty. The cities in which they live lack a well-developed economy and enough jobs to support such large populations. As a result, huge slums—called *barrios* in Spanish-speaking countries and *favelas* in Brazil—surround nearly all South and Central American cities. It is estimated that more than one-third of the people in some cities live in poverty. Despite their size and the hardships that many residents endure, urban populations continue to swell throughout the region.

Recent Patterns of Migration

When people make a decision to move, two factors come into play: those that push, and those that pull. In South and Central America, as elsewhere in the world, the desire to improve one's economic well-being has been the major factor influencing a decision to move. Much of rural South and Central

Cityscape by Pinheiros river in São Paulo, Brazil. (Image by Tatian Sapateiro)

America is poorly developed. Jobs are scarce, living is difficult, and such important things as electricity, clean water, and social services are unreliable or nonexistent. Cities, despite the hardships often faced by first-generation migrants, offer many poor rural peasants at least the hope of a better living standard in the future. Over the past century, millions of people have moved to cities, making rural-to-urban migration the single most important cause of the region's changing distribution of population.

Other patterns of migration have influenced population distribution in both South and Central America. In Brazil, the government has taken deliberate steps to encourage population growth in the isolated and sparsely populated interior. In the 1960s, Brazil's capital was moved from Rio de Janeiro to a new city, Brasília, designed and built 900 miles (1,450 km.) inland. In 1966 the interior Amazon city of Manaus was made a free port in which trade can be conducted without tariffs. Networks of roads were built, making it much easier to reach large areas of the once-isolated Amazon Basin. Many settlers, particularly from the poor and densely populated northeast Brazil, have followed these routes into the interior. The population boom in the Amazon has caused much concern about its environmental consequences. The Amazon rain forest is being cut and burned at an alarming rate. Scientists fear that many plants and animals native to the region will disappear altogether.

In the central and northern Andean region, from Bolivia northward to Venezuela, many people are moving from the highlands into the eastern wet tropical lowlands. Here, in the once almost impenetrable upper basins of the Orinoco and Amazon Rivers and their tributaries, the lure of land and opportunity has attracted people to the hot, humid, and isolated rain-forest region. Venezuela however, has experienced major population growth and urban development in its tropical lowlands. Rich deposits of iron ore and bauxite (the ore from which aluminum is made) near the Orinoco River have helped this once-remote area to grow in importance.

Not all internal migrations are prompted by economic motives. In the 1980s, for example, the brutal tactics of a revolutionary movement called Shining Path (*Sendero Luminos*), centered mainly in the Peruvian Andes, drove perhaps a million people from their mountain villages to the country's capital, Lima, and to other coastal cities. By 1991, Shining Path controlled much of the country, but the next year its leader was captured. Today, although activity by the group has declined substantially, Shining Path is still being implicated in sporadic terrorist activity around the country.

Most Central American countries have experienced internal migrations resulting primarily from economic influences or, in some instances, civil unrest. For the region as a whole, the major migration of recent decades has been northward to the United States and Canada, in search of jobs and a better life.

Future Population Distribution

A population distribution map of South and Central America shows huge areas of low population density. Some areas of low population can be explained by their harsh natural environment, but environment alone cannot fully explain the region's distribution of people. History and culture both played an important role. Brazil's early European settlers, for example, were a coastal, seafaring people who came to raise sugarcane. Plantations and, later, cities flourished on the fertile alluvial coastal plains. There was little desire or need to explore, develop, and settle the country's remote interior, where the tropical rain-forest environment presents many challenges.

Not all economic and population growth is influenced by climate, soil, and rain forest. Manaus, located nearly 1,000 miles (1,600 km.) up the Amazon River, grew from a small town to a booming city with more than 1 million people after it was made a free port in 1966; today, its population exceeds 2 million. By allowing international trade to be conducted without payment of tariffs, the Brazilian government was able to overcome environmental obstacles and encourage population growth in this remote location.

Elsewhere in South and Central America, Spaniards came in search of mineral wealth, particularly gold and silver. Partly because of the region's distribution of precious metals, and partly because of the Spaniards' desire to have easy access to the sea, their settlements rarely occurred more than 100 miles (160 km.) from the coast. The Spanish settlers strongly preferred city living, so once the urban centers were established, they grew at the expense of rural development. Outside the cities, vast outlying areas remained remote and poorly developed.

Many areas of South America are capable of supporting much greater populations than they do today. For much of the region, the problem is not a challenging environment. Much of the continent's empty areas have environmental conditions that elsewhere in the world support large populations. To attract settlers to areas of potential development, many things will need to be done.

The areas must be made accessible: transportation and communication linkages must be built. Utilities and other public services must be provided. New communities must be built, and they must provide the kinds of things that young and old alike have become accustomed to having in the twenty-first century. Of greatest importance, economic growth and development must take place, providing jobs and an acceptable standard of living. If these goals for rural development can be reached, the existing flow of migration from country to city can be reversed. Should this happen, the region's existing cities will become more livable. There will be a more even distribution of population and development throughout the region. Areas that today are of little importance will begin to make valuable contributions to the economies of their nation and the region.

Charles F. Gritzner

DISCUSSION QUESTIONS: PEOPLE

Q1. Who were the ancestors of the indigenous people of South and Central America? Where did they come from? When do scientists think the first humans arrived in South America?

Q2. Discuss the history of slavery in South and Central America. Which European countries imported slaves into the region? When was slavery abolished? What roles do the descendants of slaves play in their societies today?

Q3. How did the settlement of South and Central America by the Spanish and Portuguese differ from the ways the English settled parts of North America?

CULTURE REGIONS

SOUTH AMERICA

The cultural patterns of South America might appear to be simple reflections of Iberia. Portuguese and Spanish are the official languages of almost every nation, Roman Catholicism is the dominant faith, the buildings in most urban areas have a distinctive Mediterranean style, and the landscape is filled with reminders of colonial links to Europe. First impressions can be misleading, however.

The cultural patterns of South America are the result of a complex mixture of many elements derived from several ethnic groups over the centuries. In some areas, the mixing has resulted in a blended culture with contributions from various sources. A person playing the *carioca* on the streets of Rio de Janeiro, for example, is predominantly of an African racial background, may speak Iberian Portuguese; practice a form of Christianity laced with animist symbolism; play local samba music that features foreign melodies and rhythms; and exhibit the independent spirit and easygoing style considered by many to be inherited from indigenous ancestors. However, the *carioca* player will claim to be nothing other than Brazilian.

In other regions of South America, the landscape is occupied by ethnic groups that have not mixed. They coexist side by side, accommodated to each other and the physical environment. A town in the highlands of Ecuador or Peru, for example, may include an upper class of people who trace their pure-blood ancestry directly back to Europe. The type of work they do, their family relationships, prevalent patterns of courtship, traditions of land ownership, the style of their clothing, and many other features may be similar to those of distant Spanish cousins. In contrast, not far away on the outskirts of town, Amerindians till the soil and live out their lives in ways little different from those of their Incan ancestors. Although the present European and indigenous cultures occupy the same geographical region, their traditions remain separate.

Cultural Realms

Four principal ethnic groups have contributed to the cultures of South America: Amerindians; colonial-era Europeans; Africans; and postcolonial immigrants. These groups have intermingled and merged to different degrees throughout the continent, creating a multispectral ethnic tapestry with great diversity. Although conditions can vary significantly over a small area, South America exhibits four broadly generalized cultural realms. The first is the Amerindian/mestizo highlands of Colombia, Ecuador, Peru, and Bolivia. There, the descendants of the Amerindians and Spanish, usually mixed in a racial type referred to as mestizo, are prevalent. Predominantly Amerindian communities occupy rural and highland areas while the cities and coastal regions have a greater proportion of colonial European descendants.

The second cultural realm is a discontinuous set of African/multiracial coastal settlement regions, including Esmeraldas in northern Ecuador; Cartagena, Barranquilla, and Santa Marta in coastal Colombia; north-central Venezuela; Northeast (Nordeste) Brazil; and Bahia and Rio de Janeiro in coastal east central Brazil. In some of these areas, large numbers of slaves were brought from Africa, and many of their cultural traits have persisted. In other areas, especially along the Caribbean coast, the prevalence of black racial groups and cultures spread from the islands to mainland northern South America. The racial composition is dominated by Africans from sub-Saharan tribes, often mixed with European blood in a racial type identified as mulatto.

The third cultural realm is focused in the southern nations of the continent: Uruguay, Chile, Argentina, and southeast Brazil. In these areas, the population has been most heavily influenced by the large number of immigrants who arrived well after the colonial period, especially during the late nine-

teenth century. Many came from southern Europe to work in the alfalfa fields of Argentina or the coffee plantations of Brazil, providing labor that was no longer available from slavery. Throughout the region, the traits of the new immigrants, mostly Europeans, have interacted with traditions from the older colonial European culture. This interaction has led to new and unique lifestyles. At times, such as when the Peronistas wrestled political power from the conservative groups in Argentina, the interactions have caused conflict and violence.

Afro-Colombian fruit sellers in Cartagena. (Image by Luz A. Villa)

Crowds watch the Carnaval parade in the Sambadrome Marquês de Sapucaí, in Rio de Janeiro. (Image by Agência Brasil/Marco Antonio Cavalcanti)

Finally, a broad interior region, occupied by persons of mixed Amerindian and African blood includes dispersed farmworkers scattered across the land. They herd cattle on the savannas or move about the forest practicing subsistence-level shifting cultivation much as earlier Amerindians did—clearing the land, burning the debris, planting and harvesting crops, and moving on when the nutrients of the field were exhausted. These people often present a variety of cultural traits that can be traced to each of the principal ethnic ancestral roots of South American society.

Amerindians

When Christopher Columbus arrived in the New World, he found that the region was already occupied by humans. These people were identified as "Indians," because leaders of the expedition were certain that the goal of reaching the Far East (Indies) had been accomplished. Furthermore, the na-

tive people exhibited racial features similar to those of Asiatic groups.

Although already present when Columbus reached the New World, the Amerindian peoples were themselves relatively recent arrivals in the Western Hemisphere. The estimated span of humans, or humanlike creatures, on Earth stretches back at least 4 million years, but the Amerindians of North and South America probably arrived no earlier than 12,000 to 14,000 years ago. Traditional theories regarding the populating of the Americas postulate a migration from northeast Asia into North America across the Bering Strait during a glacial period around 11,500 years ago. This was followed by a slow diffusion into the rest of the northern and southern portions of the continent. Recent evidence from archaeology sites at Monte Verde, Chile, and Pedra Furada, Brazil, indicates that these South American sites are probably older than any so far discovered in North America. Theories regarding the method of penetration and settlement have been revised, and many modern scientists support the idea that people moved rapidly from Asia along the seashore or in boats all the way into South America.

By 1492 numerous cultural groups had emerged in South America, ranging from simple hunting and gathering societies in the extreme south to the sophisticated civilizations of the Incas and Chibchas of the Andes. The bulk of the territory in South America was occupied by simple farming communities, such as the Tupi-Guaraní of Southeast Brazil or the Arawak linguistic communities of central and northern South America. They survived by cultivating their crops using a slash-and-burn technique and supplementing the food supply with protein from hunting or fishing. Many of the Amerindians, mestizos, and mulattos continue this practice throughout the rain forest today.

At the time of European contact, the most important communities lived in the mountains of the west. In Colombia, the Chibchas dominated the Cundinamarca Basin where Bogotá is presently located. Farther south, the Incas controlled the land from southern Colombia to northern Chile. Although it is difficult to estimate, the Incas may have numbered from 9 million to 13 million around 1500. They had made remarkable advances in farming technology, crop domestication, irrigation, metallurgy, and social organization, but their military capability was far inferior to that of the Spaniards.

Following their conquest by Spain, the Amerindians of the Andes experienced centuries of subjugation to Spanish rule, great loss of life from the introduction of Old-World diseases such as smallpox and influenza, backbreaking work in the fields and mines under the *encomienda* and *mit'a* systems, and continued oppression even after the South American nations gained political independence. Nevertheless, the cultural fabric in the rural highlands of South America still reflects Incan traditions. Large numbers of people speak Quechua, the language of the Incas; many in southeast Peru and northwest Bolivia use the even older Aymara tongue. Religion, the style of dress, music, farming methods, and many other cultural traits are dominated by the Amerindian and mestizo traditions of the ancient groups in the Andes.

Amerindian groups have left their mark in other areas. Jesuit missionaries brought European ways to the Guaraní tribes of the upper courses of the Paraná and Paraguay Rivers in a region known as Misiones. Although frequently raided by Portuguese expeditions and later slaughtered in the Paraguay War of the 1860s, the Guaraní continue to influence Paraguayan society. Their language is still widely spoken in that country. In isolated pockets throughout the interior of the continent, remnants of the many South American indigenous groups can still be found. Some governments have offered modest protection through agencies devoted to preserving indigenous culture, but most of the Amerindians continue to suffer as a result of contact with the outside world. Many die when exposed to diseases against which they have little immunity, and the people are heavily exploited by local ranchers, miners, and farmers.

Colonial Europeans

The second group to leave its mark on South America is perhaps the most obvious: Iberian Europeans. From the initial contact in 1492 until the conclusion of the wars of independence in the 1820s, Spain and Portugal maintained political control over most of the lands of South America. During this period, the Amerindians were forced to adopt European customs, religion, and languages. Africans were imported to work in the fields and also were compelled to take on European ways.

Migration from Iberia to the New World during the early colonial period was composed primarily of men seeking adventure and fortune. In Spain, the armies that had recently defeated the North African Moors were available to launch new conquests for God and king. Estates in Spain were inherited by the eldest son, so siblings born later were left without land. This provided manpower for the conquest of the Americas. Tales of El Dorado, the legendary city of gold, enticed others, as did the hope for recognition and land grants in payment for heroic deeds. Portuguese men also arrived in America, and, more accustomed to contact with foreign races, the Europeans in Brazil intermarried with Amerindian and African women. Racial mixture in Brazil began very early. Today, in fact, the Brazilian population is one of the most genetically intermingled groups in the world.

The modern influence of the colonial European culture is most evident in cities and in areas of early and continuous Spanish or Portuguese settlement. For example, the Spanish came to the Cauca Valley of Colombia in significant numbers during the early colonial period. Here their population grew and prospered. Their descendants maintain a conservative political and religious outlook. Their farms and towns are among the most productive in the nation. Throughout the continent, the Spanish and Portuguese mixed with the Amerindians and Africans to form the pervasive mestizo and mulatto groups. In the south, the descendants of the colonial-era Europeans have mixed with the more-recent immigrants from Europe.

While some of the legacy influences of the Europeans from the colonial period are obvious, others are not readily evident. Among the more visible are the official languages—Spanish and Portuguese—of almost every country; the prevalent Roman Catholic religion; the street layout of cities (especially in Spanish-speaking countries); the coastal orientation of the transportation systems that were used to ship minerals and farm products to the coast; and a peripheral population pattern by which most large cities are located near the coast, where there was easy access to ships from Europe.

Other effects are real but less obvious. The mercantile economic system created export economies in the New World, and those patterns have persisted. Elements of the Iberian patriarchal society, where men dominated, were transferred and have persisted, although they are disappearing in many areas. Paternalism also helped to produce political traditions of authoritarian leadership, whereby the actions of the president, or *caudillo*, were not subject to question, just as those of the *hacienda patron* were beyond dispute. The traits of colonial Europeans are probably best preserved among the mestizo and mulatto inhabitants of highland and backcountry

communities. They are part of a traditional way of life that is often resistant to change.

Africans

Africans were brought to South America as slaves at the beginning of the sixteenth century. The earliest arrivals were household servants of Portuguese aristocrats, but soon shipments were made for purely economic reasons. Slavery was authorized by Spain for the first time in 1518, and the heaviest concentrations served the plantations of the Caribbean region. The largest numbers of Africans in South America, an estimated 4 million, came to Brazil. This was more than the estimated 3 million who came to Spanish America (including Central America and Mexico) and constitute approximately two-fifths of all of the slaves brought to the New World.

Several conditions in Brazil contributed to the heavy influx. The Portuguese became involved in hot-weather plantation agriculture in the northeast, and many African societies were well adjusted to hard farmwork. The Amerindians discovered by the Portuguese in eastern Brazil were not adaptable; most of those who did not escape died soon after capture. This was in contrast to the native populations in Spanish South America, where the people were capable of providing the required labor. Finally, by the time of their arrival in Brazil in 1500, the Portuguese had a long history of contact with Africans, thanks to the efforts by Prince Henry the Navigator's sailors to discover a passage to the Indies. For decades, ship captains sailed along the coast of Africa, so the cultures of African societies to the south were familiar to the Portuguese.

Slavery persisted in the New World until 1888, when the final emancipation declaration was issued by Dom Pedro II, Emperor of Brazil. By the second half of the nineteenth century, slavery was a dying institution around the world, although it persisted in Brazil longer than elsewhere because of the considerable political power of the landed elite. The use of African slaves in Brazil was concentrated in the coastal northeast and center-east regions, stretching from the states of Piauí to Espírito Santo and Rio de Janeiro. Throughout these areas, the cultural influence of Africans has been great. The Portuguese language is altered into local dialects that include a large number of words from the Angolan and Guinea regions of Africa. Food preferences in places such as Recife and Salvador reflect African cuisine, and the local visual arts express African themes. The samba, frevo, and other forms of traditional Brazilian music have roots in Africa.

Religion in Brazil also has been affected by African culture. Sects such as Macumba and Candomblé incorporate the worship of traditional spirits into traditional Christian beliefs. Much of the activity of these faiths focuses on establishing contact with spirits that can be made to temporarily take possession of the physical bodies of the ceremony participants.

Outside of Brazil, African influences have not been as significant and are restricted to regional pockets of northern South America and a few southern urban enclaves. Descendants of Africans in Montevideo and Buenos Aires are credited with inventing the tango, Argentina's best-known contribution to music and dance. Around the city of Esmeraldas in northwest Ecuador, residents of African descent have farmed the forest lands for many years. They comprise the most important racial segment in the area. Similarly, the northern coastal regions of Colombia and Venezuela reflect the heavy influence of African culture. Although slaves were brought to these areas, much of the impact has been due to the proximity to black-dominated areas of the Caribbean region.

Recent Immigrants

The latest contributions to the culture of South America were introduced by several immigrant groups during the late nineteenth and twentieth centuries. The majority were Europeans attracted to agricultural regions of southern Brazil, Uruguay, and Argentina; but Asians and North Americans were also among the arrivals. The largest cluster, including approximately 6 million Italians, Spanish, Germans, and other Europeans, settled in Argentina, where most became tenant farmers on the rich soils of the Pampas. They raised alfalfa to feed the cattle of the landowners and produced wheat, corn,

and flax for their own profit. Following the collapse of meat prices in the 1920s, most moved to Buenos Aires and other large cities, where they gained employment in factories and shops.

The immigrant population had a tremendous impact on Argentina, and at one time three-fourths of the total population were either born overseas or were a first-generation descendant of an immigrant. From the Spanish language, which is spoken with a definitive Italian-sounding accent, to clothing styles, Argentine culture reflects the recent European heritage. The architecture of Buenos Aires is reminiscent of the stately cities of southern Europe, and the picturesque Boca area near the port preserves the Italian urban landscape.

Though less dominant than in Argentina, European immigrants also had a strong influence in Uruguay, southern Brazil, and Chile. In southeast Brazil, Italians were attracted by work opportunities on the coffee *fazendas*, . As in Argentina, many accumulated wealth that was later invested in factories or other businesses. In extreme south Brazil and in the regions of Chile south of the Bío-Bío River, German influence has been strong. Cities are named for locations in Bavaria, and the modest-sized, family-operated farms of these areas contrast with the larger plantations and ranches of São Paulo and Middle Chile.

Other immigrant groups have left their mark in South America. The Guiana region, later occupied by French Guiana and the independent nations of Guyana and Suriname, received additional settlers from the colonial and former colonial powers of France, Great Britain, and the Netherlands. In order to support farming activity, the British imported large numbers of laborers from South Asia (referred to as East Indians) into Trinidad (an island off the coast of Venezuela) and British Guiana (later Guyana). In some areas, South Asians comprise a substantial part of the total population.

Chinese immigrants have been a significant presence in several of the large cities, and Levantine (Palestine, Lebanon, and Syria) groups have established themselves in commercial enterprises. A large number of Japanese colonized several rural areas of interior south Brazil during the early twentieth century, and they have had an important presence in the Amazon region as well. In fact, there are more Japanese living in Brazil than in any other country outside of Japan. Elsewhere in South America, Alberto Fujimori, a Peruvian of Japanese descent, served as president of Peru from 1990 to 2000. Fujimori was the first person of Asian descent to lead a Central or South American country.

Mennonites from the United States and Canada were drawn to the forbidding Chaco of Paraguay near the remote city of Filadelfia, and Americans from the U.S. South moved to interior southeast Brazil following the U.S. Civil War. There, near the city that would eventually take on the name of Americana, they established an enclave that had a remarkable influence on local Brazilian society. They introduced new farming technology and con-

tributed toward the establishment of mainline Protestant religions in that country.

The cultural panorama of South America is diverse. In many areas, the peoples and cultures have mixed and blended to create a way of life that is uniquely South American. In other areas, much of the original culture of the dominant groups has been preserved. As communication increases between South Americans and people elsewhere in the world, the cultural mixing is accelerating. Especially evident is the influence of the United States and other Western societies. Computer technology, Internet usage, and social media have become pervasive. North American music, film, and other forms of entertainment compete with traditional styles. Even so, South America maintains its unique cultural identity.

Cyrus B. Dawsey

CENTRAL AMERICA

Central America has a culture that is largely stratified according to perceived racial background. The primary races for the inhabitants of these lands are European, Amerindian, mestizo (mixed European and Amerindian), mulatto (mixed European and African), zambo (mixed African and Amerindian), and African. In general, the more European one appears to be, the higher one's status is. The culture that has developed in these regions is a blending of the diverse traditions of the cultures of origin of these peoples.

The blending of the cultures is evidenced in a number of areas. One obvious area is that of religion. The dominant religion in the region is Roman Catholicism. However, the Catholicism practiced varies greatly. The role of saints in Catholic theology was adapted by the indigenous populations to include the worship of some of their primary deities; thus, feast days to these gods have crept into

the calendars of most Amerindian societies. The church initially allowed the practice, given that the process of conversion was underway. These syncretistic events have now become the core of festival days for these societies. The Mayan practice of offerings to the gods is preserved in the practice of erecting cave shrines to saints, who are given fruit and flowers as offerings. African elements have also crept into the Catholicism practiced by the Caribbean descendants of the region. More recently, liberation theology, a Marxist-influenced movement with an emphasis on social and political activism, came into conflict with mainstream Catholicism. The movement and the church have resolved most of their differences in recent years.

European Culture
Central America's cities and larger estates are the areas that traditionally have been populated pri-

Altun Ha was a Mayan temple and burial site in Belize. Evidence shows that it may have been used as a trading center with other Mesoamerican societies. (Image by Michael Lazarev)

marily by Europeans and mestizos. Europeans came to those areas to exploit the resources of the region, primarily the volcanic soils of the mountains. Estates were situated on the foothills, river valleys, and along the coasts of these lands. In Guatemala, Nicaragua, Panama, and El Salvador, more than 10 percent of the population is unambiguously classified as European. However, Europeans generally form the core of the elite in all of Central American societies, including in Honduras, where Europeans comprise less than 1 percent of the population. The perceived homogeneity of Costa Rica, where nearly 90 percent are European or mestizo (most claim to only be the former), has led to a society in which the majority has enjoyed a more egalitarian existence, because a middle class emerged here long before anywhere else in the region.

Mestizo and Ladino Culture

This category represents the majority of the population in every country in the region with the exception of Belize; there, it is the largest racial group, at

49 percent. A sizable number of mestizos in Belize are refugees from Guatemala. Mestizos make up 90 percent of the population of Honduras, 86 percent of El Salvador, 70 percent of Panama, 69 percent of Nicaragua, and 41 percent of Guatemala.

The classification "ladino" is found in the cities and larger villages of Guatemala, Honduras, and the southern Mexican state of Chiapas. Ladinos are mestizos who consciously deny their Amerindian heritage. They practice more mainstream Roman Catholicism and tend to shun traditional medicine and dress. Nevertheless, the Mayan imprint upon these areas and this class is unmistakable.

Amerindian Culture

Every Central American country has a significant population of Amerindians. Most live in either the marginal agricultural areas (the better soils having been reserved for European *encomienda* or for mestizos) of the hills or in tropical rain forests. The largest Amerindian ethnic group is the Maya, numbering over 6 million, distributed in southern

Mexico, Guatemala, Belize, Honduras, and El Salvador. Indigenous groups of sizeable numbers inhabit the coasts of Honduras, Nicaragua, and Panama.

The key to the preservation of these cultures has been their ability to maintain a continuous homeland. Many of the governments in the region have attempted to set aside these traditional lands as preserves. The Maya have survived due, in part, to isolation. In many isolated villages throughout the region, the "day keepers," or priests, still exist, responsible for keeping track of Mayan rituals and other events of the Mayan calendar. Cultural awareness among the Maya has led its leaders to try to resist assimilation to European culture.

The preserves of the tropical rain forest that have been established by governments for the Maya, Garifuna, Kuna, Emberá, and other ethnic groups are a means of protecting the forests themselves.

The peoples who inhabit these regions live in symbiosis with the land. To abuse nature is to lose resources for the people. Their lifestyles promote uses for plants that have yet to be employed by Western cultures.

Many of the indigenous people still engage in horticulture or hunting and gathering (although most now hunt with shotguns and rifles) when they are not doing menial labor for outsiders who are exploiting the resources of the forests. Overall, they are not integrated into the general economies of Central American nations.

Mulatto Culture

Mulatto regions are generally found on the Caribbean coasts. This classification results from Caribbean migrations onto the mainland. The mulatto population of Belize comprises 31 percent of the population, while 14 percent of Panama's popula-

An Amerindian woman and child in the Sacred Valley, Andes, Peru. (Image by Thomas Quine)

151

A Kuna woman displays a selection of molas *for sale in the San Blas Islands of Panama. (Image by Johantheghost)*

tion is mulatto. Fewer numbers are found in the other Central American countries. The similarities of these communities with mulatto regions in South American countries reflects the impact this group has had on their host cultures.

Zambo Culture

The primary region of zambo culture is in the Atlantic coastal areas of Honduras and Nicaragua. The most prominent group is the Miskits—the offspring of escaped slaves, mostly from Jamaica—and a number of local indigenous peoples, who assimilated other nearby groups. The Amerindian groups that escaped assimilation by moving inland into the forests are known collectively as the Sumu, a group that formed (as did the Miskits) 250 years after the arrival of Europeans.

African Culture

African culture comes by way of the Caribbean, having undergone a cultural transformation before entering Central America. Most Africans were subsumed into mulatto status. Sizeable populations still remain in Nicaragua and Belize.

Africans often endure more prejudice than other cultures, given that lightness of skin color often is a prerequisite for prestige. For example, the Costa Rican port of Limón is one-third African (who make up about 3 percent of the population as a whole), the descendants of late nineteenth century railroad workers. For years, Africans were restricted to living in Limón. Today, although no longer subject to official discrimination, many of the country's blacks have chosen to move from Limón to other Costa Rican cities or to the United States.

Mark Anthony Phelps

EXPLORATION

Europeans were first drawn to South America during the so-called Age of Exploration, a part of the broader cultural movement known as the Renaissance. Initial contact and exploration stemmed from several often contradictory motives: desire to find a sea route to Asia, greed for native wealth, the Christian mission to convert non-Christians, conquest and colonization for national expansion, curiosity, and a desire for adventure. As Europeans staked their claims, competition for territorial control sparked further exploration and colonization. The early exploration of South America cannot be viewed apart from its attendant activities of conversion, colonization, and exploitation. More recent exploration has been accompanied by economic development and ecological destruction.

Probing the Eastern Coast

On his third voyage for Spain, Christopher Columbus landed on the mainland near the mouth of the Orinoco River, in 1498. He coasted along westward to Margarita Island and returned to Hispaniola. At about the same time, the experienced Portuguese captain and scientist Duarte Pacheco Pereira may have been the first European to sail along Brazil's coast, for he reported a vast continent stretching from 70 degrees north to 28 degrees south latitude. His later accounts of Portuguese exploits are valuable records of his adventures in Africa and India, as well as the western Atlantic.

Between May 1499 and June 1500, the Spanish navigator Alonso de Ojeda and Florentine businessman Amerigo Vespucci landed in what would become French Guiana and sailed south to the mouth of the Amazon River, which they described for the first time. They then moved farther south to Cape St. Augustine (about 6 degrees south latitude), then north and west to the Magdalena River. As Spain's chief navigator, Vespucci was responsible for preparing the official maps of newly discovered and surveyed areas.

In January 1500, Vicente Yáñez Pinzón , the Spaniard who had captained Columbus's ship *Niña*, explored around Cape St. Roque and coasted north and westward. He probed the Amazon estuary and sailed up to the Gulf of Paria. Diego de Lepe, also from Spain, at about the same time traveled south from Cape St. Roque to near 10 degrees south latitude. Portuguese captain Pedro Cabral, with a fleet of thirteen caravels, landed and formally took possession of Brazil for the king of Portugal, following the provisions of the Treaty of Tordesillas of 1494. He was on his way to India on a trading voyage and stayed only ten days, naming the region Tierra de Santa Cruz.

Woodcut probably depicting Amerigo Vespucci's first voyage (1497-98) to the New World, from the first known published edition of Vespucci's 1504 Letter to Soderini, entitled "Lettera di Amerigo Vespucci delle isole nuovament trovate in quattro suoi viaggi," published by Pietro Pacini in Florence c. 1505. (Photographic facsimile reproduced from 1893 First Four Voyages of Amerigo Vespucci, *London: B. Quaritch)*

EXPLORATION OF SOUTH AMERICA, 1498–1547

Amerigo Vespucci sailed again in May 1501, this time for Portugal, and reached Guanabara Bay (Rio de Janeiro), the Rio de la Plata, and somewhat farther south. His published account of this adventure led German mapmaker Martin Waldseemüller to name the region "America," a name that eventually encompassed both continents. From the Panamanian isthmus, Rodrigo de Bastidas explored the coast southward to Port Manzanillo (1500–1502). Vicente Pinzón may have sailed down the coast from the Bay of Honduras to the easternmost tip of Brazil in 1508.

La Plata Region

The La Plata River region was first explored by Juan Díaz de Solís of Spain, who was inspired by the reports of Vespucci to seek a passage to the Pacific Ocean. He left Spain in October 1515, with three ships and about seventy men. He charted the coast from near Rio de Janeiro to La Plata, naming the river's mouth the "Sweet Sea" (Mar Dulce). They sailed up the Uruguay River and disembarked, only to be killed and eaten by the native Guarani people in 1516. The lone survivor told his tale to Sebastian Cabot, who returned in 1526. Convinced that the area held a wealth of silver, Díaz named the river La Plata, as in "silver-plate."

On his famous circumnavigational voyage (1519–1522), Ferdinand Magellan first reached Brazil near Pernambuco, then sailed into La Plata estuary. This was formally charted by Sebastian del Cano. Convinced that this was no passage to the Pacific Ocean, Magellan wintered at Port St. Julian, proceeded through the straits that now bear his name, then up the Pacific coast to about 40 degrees south latitude. From there he headed westward to Guam and the Philippines, where he was killed by the local people.

In 1525 a group of Sevillian merchants hired the adventurous Sebastian Cabot to lead three ships to establish trading ties with Molucca in the Indian Ocean. Rather than proceed directly, he explored South America's Atlantic coast (1526–1529) in search of the "City of the Caesars." He was drawn by the promise of silver that Díaz de Solís posthumously held out to him. His men probed La Plata,

as well as the Paraná and Paraguay Rivers, establishing a small colony on the lower Paraná River that he named Sancti Spíritu. This was abandoned in September 1529. Upon his return to Spain, Cabot, he was banished to Africa for his failure to carry out his original mission. For a variety of reasons, Cabot's sentence was never carried out.

Within a few years, Portuguese developments in Brazil and Spanish successes in Peru persuaded Spanish authorities there was a need for an official presence in the area that would become Argentina. In 1535 the Spanish crown sent Pedro de Mendoza with thirteen ships and 2,000 men to establish formal colonies in the region. In La Plata estuary, he founded a settlement that he named Santa Maria del Buen Aire. After less than a year, the colonists were moved to a fort at Asunción; Buenos Aires would be refounded by Juan de Garay in 1580. Mendoza oversaw a number of expeditions inland that finally succeeded in linking with the Spanish presence in Peru.

Moving Inland

Early exploration resulted in few European settlements, since few resources of interest to the explorers were found. Even the attempt by Bartolomé de las Casas—famed for his religious work among the people of Mexico—to establish a nonexploitative religious colony at Cumaná failed. Beginning in 1529, greed resulted in a better record of colonial success. King Charles I of Spain (also known as Emperor Charles V) was in debt to the Welser family of bankers in Augsburg, and provided them with a broad concession to explore, colonize, and exploit the region of the Orinoco River and beyond. Their colonists pushed through the *llanos*, up the river, across the Apure and Meta Rivers, and into the Andes, establishing a presence so violent that King Charles revoked their rights between 1546 and 1556.

In New Granada, the earliest permanent settlement was Santa Marta, founded by Rodrigo de Bastidas in 1525. From here and Cartagena, established by Pedro de Heredia in 1533, the Spanish could rule and explorers could move inland. Gonzalo Jiménez de Quesada proceeded up the Magdalena River onto the Bogotá Plateau, where he founded Santa Fe de Bogotá in 1538.

Conquest of Peru and the West Coast

Spanish exploration and expansion down the west coast from Panama began in the early 1520s under Pascual de Andagoya, who penetrated the northern regions of Biru (Peru) in 1522. Pascual fell ill, but intrigued Francisco Pizarro with tales of the great wealth of the region, something the Europeans had failed to find elsewhere on the continent. Pizarro initially pushed to the San Juan River, and in another trek, as far as the Gulf of Guayaquil and Túmbez. Convinced now of Pascual's claims, he obtained from the Spanish crown titles of captain general and governor and permission to conquer up to 600 miles (965 km.) south of Panama. In 1531 he set off with 180 men, twenty-seven horses, two cannons, and his two brothers. They seized the Incan ruler Atahualpa in his capital city of Cuzco on November 16, 1532.

After the Incan ruler was executed in 1533, Spanish authority and exploration stretched out from Cuzco through the Andes. Lima was founded in 1535, and the rich silver mines of Potosí were worked by native slaves for Spain from 1545. Sebastián Belalcázar, one of Pizarro's lieutenants, moved through the Quito region and farther north, founding Cali and Papayán (1535–1536). He pushed onto the Bogotá Plateau in 1539, and

linked with the followers of Gonzalo Jiménez de Quesada.

In 1539 Pizarro, now governor of Quito, led an expedition across the Andes to the headwaters of the Amazon. From here Francisco de Orellana and his followers worked their way down the Amazon River to the sea. He encountered a great wealth of resources, and fierce female warriors after whom he named the river. He died on his return trip to Spain, and no claims for Spain were made by his men.

Diego de Almagro spearheaded Spanish movement south from Peru into Chile (1535–1537), reaching the Maule River valley. Pedro de Valdivia, with 150 followers, explored the Maule Valley further and established Santiago in 1541. Fierce opposition from the local native peoples halted further expansion until the late 1550s. Between 1557 and 1561, Hurtado de Mendoza reached the Strait of Magellan and moved into the Cuyo region. He founded the town of Mendoza.

Exploration and the Native Peoples

Early explorers learned from the indigenous people to survive on native foodstuffs that were often strange to them and to travel in native canoes where their larger vessels could not penetrate. Contact with Europeans often resulted in demographic catastrophes, as smallpox, measles, and influenza ran unchecked through the local populations. Peru's population dropped by 90 percent and Brazil's by perhaps 60 percent in the first century of contact. Since exploration usually meant settlement and control by Europeans, road-building and deforestation for agriculture often altered the local landscape.

The establishment of plantation economies in the secured territories of the north required slave labor: native at first and where possible, later augmented by huge numbers of enslaved Africans. Christian missionaries, especially the Jesuits, sought to protect the native peoples from the explorers and created huge settlements for their charges. In Brazil, slave-hunting expeditions called *bandeiras* began early in 1628 under António Raposo Tavares from São Paulo. These expeditions, controlled by wealthy entrepreneurs, often swept

deep into uncharted jungle, preying on the unde-
fended mission villages. These incursions forced
the Jesuits and their followers farther into unex-
plored territory. As many of these peaceful European Jesuits were men of learning and curiosity,
they made excellent observations and valuable
discoveries.

Other Patterns

The coastal region of Guyana, known then as "the
wild coast," was generally avoided by explorers, but
Dutch settlers began moving up the rivers beginning about 1580. By the seventeenth century, they
had large sugar plantations worked by African
slaves. The Pacific coast of South America is far
more forbidding than the Atlantic coast, and coastal

exploration developed only after the wealth of Peru
had been discovered.

Exploration for Scientific Discovery

The earliest attempt to disclose the nature of the
South American interior apart from the river valleys resulted in the *Relaciones geográficas* of
1579–1585. In the seventeenth century, Jesuit missionaries kept meticulous records as they moved
through the mountains and jungles and established
villages for converts. They and their followers surveyed and charted the entire Paraná River Basin. In
the 1630s, the Portuguese Pedro Teixeira explored
the Amazon Basin from Belem to Quito.

The Spanish and Portuguese authorities kept
tight control on access to South America in the eigh-

Entrance into Guadalajara, Jalisco. The Tlaxcalans are carrying their traditional obsidian-tipped war clubs. Tlaxcalan forces accompanied the Span-
iards on post-conquest explorations of northern Mexico. Shown here is a scene from the 1522 exploration led by Cristóbal de Olid, one of Cortés' most
trusted lieutenants.

teenth century. Nevertheless, several important expeditions of discovery and research occurred in the 1700s. In 1743 the French scientist Charles-Marie de La Condamine traveled down the Amazon River, from Quito to its mouth on rafts for four months. He took copious notes on both the natural plant and animal life, and the native cultures he encountered in the Andes and the Amazon Basin. He made the first serious observations on the uses of rubber by indigenous people. He published his findings both in an official scientific form and in a more popular work titled *Journal of a Voyage to the Equator Made by Order of the King*.

At the end of the eighteenth century, the Prussian naturalist and explorer Alexander von Humboldt sailed from Marseilles with the French botanist Aimé Bonpland on a self-financed expedition of research and discovery in northern South America. Well prepared to conduct sophisticated research—even by the standards of the Enlightenment—Humboldt had obtained the support of the Spanish prime minister Mariano de Urquijo. Between 1799 and 1804, the pair traveled over some 6,000 miles (9,650 km.) of the roughest mountain and jungle terrain on Earth. Beginning in Caracas, they explored and mapped the drainage network of the northern Andes, discovering that the upper reaches of the Orinoco and Amazon systems were connected by the Casiquiare River. After a rest in Cuba, they continued into the Andes range, traveling from Bogotá to Trujillo, Peru. Poking into volcanoes and climbing extraordinary peaks, they eventually ended up in Quito at the end of 1802.

The remainder of their time was spent in Central America. Collation and publication of the data they gathered on plant and animal life, geology, climate, and the effects of high-altitude oxygen deprivation on the travelers occupied Humboldt from his return to Paris in 1804 to 1827. Eventually, thirty well-illustrated volumes presented his immense range of findings to the world.

With independence from Spain and Portugal in the early nineteenth century, new national boundaries between the former colonies needed to be surveyed, and restrictions on access by scientists were loosened. Among foreign visitors of a scientific bent

Portrait of Alexander von Humboldt by Friedrich Georg Weitsch, 1806.

one should include Charles Darwin, the English naturalist, who explored the natural world of the various anchorages along the Pacific coast that his ship visited, and Henry W. Bates, the English entomologist. Beginning in 1848, Bates spent eleven years crisscrossing the Amazon Basin collecting various species of insects, eventually accumulating 14,712 specimens, of which nearly 8,000 were newly discovered. He published his findings in London, supporting Darwin's ideas of natural selection.

Recent Times

The Amazon Basin is the largest drainage system in the world and covers about 2.3 million square miles (6 million sq. km.). It continues to draw explorers who seek to understand its mysteries and exploit its vast resources. Expeditions are funded by many sources, including regional governmental agencies; international organizations, such as the United Nations Educational, Scientific and Cultural Organization (UNESCO); private foundations, such as the National Geographic Society; and multinational private industries, such as pharmaceutical companies researching new materials for medicines. The great diversity of life-forms—both plant and animal—draws scientists from around the world, who seek to better understand the area's ecology and the threats to it.

Since World War II, all the region's nations have looked to their portions of the Amazon Basin for economic development. Private companies in Brazil and surrounding countries extend human activity ever deeper into the rain forest of the basin. At the same time, indigenous peoples move farther west and south to maintain a buffer between them and Western influences. The alternative is destruction of their culture.

Mining companies stretch pathways deep into the forest in search of lucrative ores, and governments create networks of roads to integrate towns and villages. Cattle grazing, oil exploration, timber exploitation, the demand for natural rubber from the native trees, and even cultivation of the coca leaf for cocaine production have all served to push back the boundaries of the unknown territories. The building of the Brazilian capital city of Brasília, beginning in 1956, and the gold rush of the 1980s both were accompanied by an increase in human understanding of the area and by tragic destruction of its natural elements. As in centuries past, exploration today is a complex phenomenon with many motivations and even more consequences.

Joseph P. Byrne

DISCUSSION QUESTIONS: PEOPLE

Q1. How did the racial type known as mestizo arise? Where in South and Central America do mestizos live? What roles do they play in their societies?

Q2. What impacts did the Spanish have on the indigenous peoples of South and Central America? How have European religious practices meshed with those of the local people? Why in all of South America does only Brazil recognize Portuguese as its official language?

Q3. Which South American countries have received the most recent waves of European and Asian immigrants? How have these newcomers assimilated into the fabric of their new homelands?

POLITICAL GEOGRAPHY

SOUTH AMERICA

South America's approximately 414 million people live in twelve independent nations and one overseas department of France. Brazil dominates the continent in land area, size of its economy, and population. The remaining nations, in descending order of population size, are Colombia, Argentina, Venezuela, Peru, Chile, Ecuador, Bolivia, Paraguay, Uruguay, Guyana, and Suriname. French Guiana, the last European colony on the continent, has the smallest population.

Cultural Diversity

South America's great cultural diversity is complex and the source of much political injustice and sometimes violence. Throughout the continent's history, distinctions in race and culture have determined a person's status and access to political power, land, military rank, and business resources.

After founding their first colonies in the sixteenth century, the Spanish and Portuguese soon gained control of most of the wealth and nearly all of the political power. Their descendants still consider themselves the ruling elite. Later immigrants, especially European Christians, have been next in status, followed by mestizos, people of mixed white-Amerindian parentage; people of African descent; people of mixed racial heritage; and Amerindians.

Revolutions

By the mid-eighteenth century there existed a large number of South American-born whites, known as Creoles. European officials and military officers often denied Creoles the same status and political power as European-born whites. This discrimination caused great discord and led to revolts in which the Creoles allied themselves with mestizos and mulattos in their common contempt of native Europeans.

Spain and Portugal successfully repressed the revolts until the early nineteenth century. Then, led by brilliant military leaders, such as Simón Bolívar, José de San Martín, and Bernardo O'Higgins, the Spanish colonies gained independence. By 1825 the political map of the continent was divided into essentially the same countries that exist today, although the borders were somewhat different. Brazil became technically independent of Portugal in 1822 but remained close to its parent country and had its own Portuguese-based monarchy until 1889.

During the new nations' early years, their leaders tried to imitate the democracy of the United States or modified parliamentary systems, such as in Great Britain. In fact Bolívar dreamed of uniting South America under one large federal government, a United States of South America. However, democracy never took a firm hold.

Regional rivalries over land and resources among wealthy Creoles prevented continental union. Led by economic and military strongmen, called caudillos, the elites ruled their territories like private clubs; nonwhites had few rights. Revolutions occurred, but until the late twentieth century they usually only replaced one group of elite white rulers with another. Small communities of Amerindians in remote areas were free to follow their traditional ways of life. Most, however, lived as farmworkers on large estates or in segregated villages.

Intra-American Wars

South American countries have fought several wars over disputed territory. For example, Paraguay lost the War of the Triple Alliance (1865–1870) to the combined forces of Argentina, Brazil, and Uruguay. The War of the Pacific (1879–1884) pitted Chile, the eventual victor, against Peru and Bolivia. Bolivia and Paraguay fought each other to a bloody draw in the Chaco War (1932–1935), in which each sought to gain control of a valuable oil field that had not been correctly mapped.

A long quarrel between Peru and Ecuador over their eastern border erupted into a brief war in 1941. Ecuador repudiated its border treaty with Peru in 1960. Another war between them broke out in 1981, which Ecuador also lost. In 1995 the two countries skirmished over a small territory, but after a cease-fire was arranged by other countries, the dispute remained unsettled. Tensions between Peru and Ecuador subsided in 1998 when El Niño-spawned storms devastated both countries, and their leaders pledged to help each other recover. That same year the two nations signed a comprehensive treaty that finally brought peace to the border. However, the legacy of historical grievances and mutual distrust led to renewed friction between Ecuador and Peru in the 2010s over Ecuador's plan to build a mile-long border wall, purportedly as part of an urban planning project and a measure to control smuggling. Peru vigorously protested, insisting that the wall would disrupt local trade and prevent Peru's access to a canal that runs along the border. In order to avoid further escalation, Ecuador, in 2017, suspended the construction work, but strains on the border have remained.

Brazil and Argentina, long major commercial rivals, for years engaged in disputes over each other's trade policies. However, since the creation of the Organization of American States (OAS) and regional trading blocs, a high volume of trade and migration has fostered increasingly close ties. Brazil is now Argentina's largest export and import market; Argentina is Brazil's third-largest. Both countries claim each other as strategic economic and political partners.

Rebels and Insurgent States
Until the mid-twentieth century, it was nearly impossible, or difficult, for outsiders to win their way into the ruling elites who controlled the economies

Simón Bolívar, Liberator of Venezuela, Colombia, Ecuador, Bolivia, Peru, and Panama. (By Arturo Michelena, 1863-98, Galería de Arte Nacional)

and politics of South American countries. Amerindians, Africans, and poor people of all ethnic groups had especially hard times. They were the disenfranchised peoples with no meaningful roles in political decision-making. One way for the disenfranchised to gain some control over their own destinies was to rebel against established governments.

South America has a long history of rebellions, most of them unsuccessful. It has been typical for left-wing intellectuals to gain support of the disenfranchised by promising them more social services and greater participation in government. Leftists formed guerrilla armies, manned largely by the poor, and battled the right-wing elements in the government. The latter have usually controlled the national armies and wished to maintain the traditional power structure. When leftist guerrillas have won and taken over governments, right-wing forces have often formed guerrilla groups of their own, starting new cycles of rebellion.

Between the 1950s and the 1990s twenty-eight major revolutionary organizations conspired to overthrow one or more of the South American governments. Nearly all these organizations were left-wing and wanted to create socialist or communist states. Most relied upon support from the underprivileged and disenfranchised. Guerrilla warfare was especially widespread in Venezuela, Colombia, Bolivia, and Peru. By the beginning of the twenty-first century most guerrilla groups were quashed, petered out, or had reinvented themselves as political parties. Notable exceptions included the FARC (Revolutionary Armed Forces of Colombia), which signed a peace agreement with the Colombian government only in 2016, and the *Sendero Luminoso* (Shining Path) in Peru, whose remaining factions are still active in rural areas.

Social Unrest and Ethnic Conflict

According to sociologists, social conflict is nearly inevitable in a class-divided society. Moreover, when large segments of population are impoverished, as is the case throughout South America, the chances for conflict are still greater. Even in prosperous times, such as the second decade of the twenty-first century, the income of a large number of South Americans remained below the poverty line (between 20 and 50 percent in most countries, with the exception of the more prosperous Argentina, Chile, and Uruguay). Many rural and indigenous South Americans are desperately poor, especially in the Andes.

Large numbers of rural poor have migrated to South American cities in hopes of finding jobs. Rings of shantytowns have grown up around industrial cities, notably the *pueblos jóvenes* of Lima in Peru and the *favelas* of São Paulo and Rio de Janeiro in Brazil. Because of high unemployment rates and low in-

comes among the working poor, poor South Americans often turn to crime, such as theft and drug dealing, to support themselves. Police forces in major South America cities are hard pressed to control it.

Military Coups

South American nations have had hundreds of coups d'etat in their history, many by right-wing leaders and some by left-wing leaders. Paraguay, an especially unstable country, alone has had more than 250 coups. Between 1950 and 1990 coups put military officers in power at least once in almost every South American country.

At issue in most coups was the question of how to develop national economies and who should control them: whether to have free-market systems, as in the United States and Western Europe, or to have centralized economic control under a socialist or communist system. During the period of military dictatorships, juntas suppressed political opponents, often arresting and jailing them without trial.

In the 1980s, however, the process of democratization began spreading throughout the continent. One by one, all South American nations did away with their caudillos and military juntas, adopted new constitutions, and held democratic elections. This wave of democratization soon turned into what became known as the "pink tide"—the election of leftist and center-left populist political leaders. Their promises of better jobs and social welfare attracted the large segments of the population suffering from pervasive economic and social inequalities. Starting in Venezuela, the pink tide enveloped Peru, Brazil, Argentina, Chile, and other countries. By the beginning of the 2020s, the tide was showing signs of receding. Disillusioned by failed socioeconomic policies and government corruption, in 2015 Argentine voters opted for a conservative president, a center-right candidate came to power in Chile in 2017, and in 2019 a right-wing populist won the presidential election in Brazil.

Environmental Conflict

Governments with territory in the Amazon Basin are under great international pressure to end destruction of the rain forest. Brazil, Peru, and Ecuador have had difficulty in complying: Each had based much of its economic growth on exploitation of Amazon resources, such as gold, oil, and timber; and each had encouraged its citizens to establish communities in the region. These settlers, in turn, cut down the forests to create farmland. As international criticism mounted in recent years, local attempts were implemented to restrict field burning, logging, and hunting. These were either ignored or met with harsh opposition by Amazon residents. Today, the situation in the Amazon Basin remains a key focus of international environmental groups.

Roger Smith

CENTRAL AMERICA

Most violent conflicts in Central America have resulted from economic issues. These include a long history of unequal land ownership and use, unfair exploitation of workers, and abuses of political power. Political power has derived from combinations of land ownership and control over its use and its workers. The powerful presence of aggressive, profit-seeking foreign companies has also contributed to aggravating local economic problems. Fears of the spread of communism caused some Central American nations to support unpopular governments. Opposition groups within Central America have fought civil wars over land reform and resource control, civil rights violations, and political control.

Since the European conquest began five centuries ago, Central America has never been totally isolated from foreign influence. However, its tropical jungles and mountains make travel difficult or impossible in many areas.

Modern Cultures

Central America's peoples are descended from numerous Amerindian groups, as well as people of European, African-Caribbean, and Asian roots. A majority of the people in most of the region's countries are mestizos—persons of mixed ancestry. The populations of Guatemala and Honduras are heavily Amerindian. Costa Rica's population, by contrast, is mostly European by descent, with many African-Caribbeans and Asians living on the country's east coast. Belize, a former British colony, has large numbers of people of British descent.

Many Central Americans deny that discrimination is ethnic or racial. They say that it is economic: the few rich discriminate against the many poor. However, due to historical oppression, disproportionate numbers of poor are of either Amerindian or African-Caribbean descent.

Economic Conflicts

Economic problems based on land ownership and use have been a root cause of violent conflicts in Central America since the sixteenth century. Immediately before the Spanish arrived, the peoples of many Central American societies shared their land and lived simple rural lives. To encourage aggressive exploration and colonization, the Spanish crown granted explorers both huge tracts of land and the right to use forced Indian labor.

Most Amerindians eventually converted to Roman Catholicism, while often retaining elements of their ancestral religious beliefs. Though European and Amerindian religious traditions blended over the centuries, the socioeconomic practices and the legal system did not foster intermarriage between groups. Central Americans of European descent formed the elite. Though the Spanish explorers never found as much gold in Central America as they had hoped to discover, the Spanish prospered. They became landowners of huge estates worked by forced Amerindian labor (and later imported African labor). In general, Central Americans of European descent prospered, while those of Amerindian descent remained poor. Racial distinctions were officially recognized by the colonial government.

Independence

The process of winning independence in Central America began in the early nineteenth century and ended when British Honduras gained its complete independence, as Belize, from Great Britain in 1981. The other Central American countries won their independence from Spain more than a century earlier, when virtually all of Latin America rebelled against colonial rule.

Immediately after winning their independence from Spain in 1823–1824, the nations now known as Guatemala, El Salvador, Honduras, Nicaragua, and Costa Rica rejected Mexican rule and attempted to form a union of their own, with a constitution similar to that of the United States. Due to violent uprisings, it failed. Each of the countries then gained independence on its own. Since the early nineteenth century, each of them has by and large been ruled by dictators.

Economic Issues

Resentment about foreign exploitation of Central American resources and people has led to much violence, both in protest of the exploitation and in attempts to maintain control. Central America's rich land—which is especially suited for plantation agriculture—began attracting foreign companies soon after the countries became independent.

Foreign-owned businesses have played an important role in influencing both Central America's governments and U.S. policy toward these countries. In return for monetary gain and political support from the foreign companies and the U.S. government, many local dictators have granted favors to the companies and the United States. The dependence of Central American countries on foreign companies contributed to their being disparagingly dubbed "banana republics."

Ecological Issues

Ecological destruction threatens much of Central America. A major problem is deforestation, which has caused the land to lose its nutrients and produce fewer crops. Deforestation has been a byproduct of excessive logging—mostly by foreign companies—and the indigenous system of slash-and-burn agriculture, which Central American farmers have long used to clear and fertilize crop land. Exacerbating the situation are the growing populations in many areas, which have created an urgent need for more farmland.

Debra D. Andrist

URBANIZATION

SOUTH AMERICA

In South America, 360 million people live in cities—a number greater than the entire population of the United States. The populations of South American cities have grown rapidly as people leave rural areas to move to cities. By 2015, thirty-three cities, most on or near seacoasts, had more than 1 million residents each. São Paulo, the largest, has about 12 million people—roughly the population of Pennsylvania. Lima and Bogotá are about the same size as New York City (8 million) while Rio de Janeiro and Santiago are about the same size as Los Angeles and Houston combined (6 million).

The gap between the few rich and the many poor is visible throughout South American cities. There are glistening skyscrapers and grim shantytowns. Outside fashionable restaurants and shops, children sell gum and cigarettes. While the gap between the rich and the poor has somewhat narrowed in the twenty-first century, urban poverty remains widespread and cities are struggling to cope with the pressures caused by rapid population growth and poverty-driven crime.

Pre-Columbian Urbanization

Before Christopher Columbus opened the Americas to the Old World, most of South America was sparsely populated. Only two areas supported larger populations: the dry valleys of coastal Peru, where there were rivers for irrigation; and the high valleys of the Andes, where there was good farmland. The first urban settlements emerged in these two areas more than 2,000 years ago.

By the fifteenth century, the high valleys of the Andes were part of the Incan Empire. The largest empire in the Americas, it stretched 2,500 miles (4,000 km.) from southern Colombia down to central Chile. The Incas built a network of cities that served as centers of trade, religion, and government. Spain conquered the Incas in the sixteenth century. In the process, the Spanish destroyed Incan cities, reducing the buildings and monuments to rubble. Afterward, Spain built new cities on top of the old, often using stones from previous buildings.

Colonial Heritage

The number of cities in South America grew rapidly during the colonial period. Spain controlled most of the native population from new cities in the Andes. However, Spain needed many more cities to establish trade. The Spanish, and later the Portuguese, added three main types of cities: mining

---

THE CITY IN THE CLOUDS

Around 1460, the Incas built a city high on a mountaintop 320 miles (500 km.) southeast of what is now Lima, Peru, as an estate for the Inca emperor Pachacuti (1438-1472). They called the city Machu Picchu. Most of the city's 200 buildings were constructed of huge granite stones. The Incas did not use wheels or draft animals, yet somehow they moved these heavy stones up steep slopes from quarries several miles away.

The mystery of Machu Picchu may remain unsolved forever. Around 1530, smallpox killed half of the city's residents. The rest fled. The city remained abandoned and mostly forgotten until Hiram Bingham, a Yale professor and explorer, rediscovered it in 1911. In 1983, UNESCO designated Machu Picchu a World Heritage site, describing it as "an absolute masterpiece of architecture and a unique testimony to the Inca civilization."

MAJOR URBAN CENTERS IN SOUTH AMERICA

towns to extract gold and other minerals; ports to send and receive goods; and forts to protect their settlements from pirates and from hostile Amerindian groups.

The Spanish Laws of the Indies dictated the layout of colonial cities. These laws required a central plaza surrounded by a regular grid of east-west and north-south streets. Regulations controlled the location of most activities. Near the central plaza were churches, government buildings, office buildings, and stores. The homes of the elite were also near the city center, where they could easily enjoy the pleasures of urban life. The poorest people lived on the outskirts of town.

Portuguese colonial cities were different. Some had town squares, but they lacked a rectangular street pattern. Early Brazilian cities looked like medieval European cities, with outer walls surrounding cramped buildings overlooking narrow, winding roads.

By the end of the colonial period, population distribution changed in two important ways. The first change was dramatic. The center of population shifted from the high valleys of the Andes to cities on or near the coasts. The largest cities ringed the northern two-thirds of the continent from Santiago, Chile, to Buenos Aires, Argentina. The second change was more gradual. As Europeans seized control, native people lost their land. Without land, they needed work. They could work as farmworkers for European landlords or they could find work in a city. So began a trickle of rural-to-urban migration that would turn into a flood three centuries later.

Independence and Urbanization

The number of cities increased sharply after countries in South America achieved independence from the colonial powers. Immigration, the railroads, and changes in exports spurred city building. Between 1880 and 1930, thousands of people migrated to southern South America. They came from southern and eastern Europe, Germany, and Switzerland. Most of these immigrants moved to cities, where they started small businesses and worked as craftspeople and builders.

Immigrants also helped build the railroads. Railroad lines radiated out from port cities toward the interior. Waves of people followed the railroads, establishing new towns and cities along the way. The railroads opened up the interior, making more land available for farming. At the same time, advances in agricultural technology improved yields. Exports of agricultural goods rapidly expanded. Cities like Argentina's Buenos Aires and Uruguay's Montevideo became important agricultural ports. In the interior, new cities were established as service centers for farm communities.

By the late nineteenth century, coal, petroleum, and rubber became major South American exports, prompting new city-building. Maracaibo, Venezuela, was an oil town and Manaus, Brazil, handled rubber. Industrial towns also grew. São Paulo, for instance, became the most important industrial center of Brazil and the largest city in South America.

Medellín, Colombia, is a good example of the impact of these economic changes on South American urbanization. Founded in 1675, it was originally a small, isolated center for gold and silver mining. In the late nineteenth century, the city expanded rapidly for several reasons. First, Medellín was located near important deposits of coal. Second, the railroads connected the city with the coast. With the railroads and coal from nearby mines, Medellín became the chief manufacturing center of Colombia.

Rural-to-Urban Migration

Approximately 80 percent of South Americans live in cities. Every year this percentage increases as people leave the countryside and move to urban areas. People leave rural areas in part because much of the work on farms has become mechanized. As a result, there are too many people competing for too few jobs. Looking for work, young people move to cities where they hope to find employment. Cities, overwhelmed by this internal migration, are unable to provide regular jobs or affordable housing for the flood of new arrivals.

Primate Cities

In most South American countries, one city is much bigger and more important than the rest. These are called primate cities. Usually, they are several times bigger than the second-largest city. For example,

Lima, Buenos Aires, and Santiago are many times bigger than the next-largest cities in Peru, Argentina, and Chile, respectively. In Uruguay, nearly half the population lives in Montevideo. Brazil and Ecuador have dual primacy, because their two largest cities are much bigger and more important than the third. Only Bolivia and Venezuela do not have primate cities.

The Spanish and Portuguese set the stage for primate cities. The colonies were easier to control from a single large capital, which was often a port. As capitals, they were politically powerful; as ports, they were economically powerful. Their status did not change after independence. The largest cities were favored by government policies and attracted the lion's share of public and private investment.

Primate cities are a sign of uneven economic development. With investment concentrated in one or two cities, other regions of the country fall behind economically. This is a major cause of rural-to-urban migration. A stubborn cycle emerges. Immigrants move to the cities with the best job op-portunities. New immigrants require services, which require more businesses, which require more jobs, which require more immigrants. Coping with rapidly increasing urban populations becomes almost impossible.

To relieve the pressure on coastal cities and to encourage the settlement of the sparsely populated interior, Brazil and Venezuela took a creative approach: in the 1960s, each country designed and built new cities in poor regions of the Amazon. Brazil took the further step of moving its capital to the new city, which it named Brasília. Venezuela built steel and aluminum factories in remote Ciudad Guayana. Although far from other major cities, both cities prospered. But in the end, this model proved costly and controversial, and similar plans in other countries were abandoned.

Urban Structure

The heart of every South American city is the downtown area. There, sedate central plazas have evolved into lively central business districts (CBDs)

Buenos Aires, Argentina. (Image byt Falk2)

View of the City of Rio de Janeiro, from the Chacara do Ceu Museum in Santa Teresa. At the center of the image, the Metropolitan Cathedral and the Arcos da Lapa (Carioca Aqueduct). (Image by Victor Tarcitano)

with modern skyscrapers. People work and play in the CBD. There are commercial offices, department stores, government buildings, fine restaurants, movie theaters, and dance clubs. CBDs do not shut down at six o'clock when businesses close for the day. Activity stretches long into the night as young and old converge on the city center.

A commercial "spine" often extends from the CBD to the outer edge of the city. The spine has gracious, tree-lined boulevards, expensive office buildings, fine shops, and large parks with museums, theaters, golf courses, and zoos. Wealthy neighborhoods lie on both sides of the spine, forming a wedge-shaped residential area that becomes wider farther away from the CBD. The most expensive houses lie farthest from the city center, where lots are large and houses have all the modern conveniences. The city makes sure these neighborhoods have modern plumbing and sewer systems, paved roads, and electric, cable, and telephone lines.

Outside the elite wedge, residential areas form concentric rings around the CBD. More than 95 percent of the urban population live in one of these concentric rings. Better neighborhoods lie closest to the CBD. Neighborhood quality gets worse with increasing distance from the CBD. Most cities also have a *periferico*, a ring road similar to the beltways found in the United States. Squatter settlements

and large industrial parks generally lie outside the *perifericos*.

Shantytowns

Squatter settlements ring most cities of South America These shantytowns are called *callampas* in Chile, *barriadas* in Peru, *favelas* in Brazil, and *villas miserias* ("miserable towns") in Argentina. New shantytowns are like refugee camps. Residents clear vegetation for building material and fuel, leaving bare soil. In the dry season, dust coats everything; in the rainy season, unpaved roads are mired in mud. The houses, no more than rickety shacks, do not have access to fresh water, electricity, trash collection, or sewers. Shantytowns have reputations as dangerous places where even the police have little authority.

Shantytowns do not remain like refugee camps. Residents slowly build their own homes on land they rent, buy, or simply take over. At first, houses may be just one room made of cardboard, plastic, or tree branches. Later, as owners save more money, they add rooms or rebuild with brick and concrete. Residents help their friends and neighbors. They start community associations. They demand fresh water, sewers, schools, and health clinics. Eventually, the refugee camps become established neighborhoods.

171

Rio's Santa Teresa neighborhood features favelas *(right) contrasted with more affluent houses (left). The Christ the Redeemer, shrouded in clouds, is in the left background. (Image by chensiyuan)*

Rocinha, a *favela* of Rio de Janeiro, is the largest shantytown in South America. More than 100 years ago, squatters claimed an unwanted mountainside with a beautiful view of the Pacific. Rocinha has between 150,000 and 300,000 people, two schools, and two health clinics. Houses in the older sections are made of brick and most have water, sewers, and electricity.

Rocinha lacks standard middle-class amenities, however. There is no street garbage collection because city garbage trucks cannot squeeze through the narrow streets. On the upper edge of the *favela*, newer houses have no city services. Trash and garbage accumulate. Sewage runs through the narrow alleys and then down the mountainside. Flies, mosquitoes, and rats flourish.

Urban Problems

Unemployment, homelessness, crime, and pollution are problems in all cities. In South American cities, where poverty and overcrowding worsen every year, these problems are widespread.

Unemployment is one of the biggest problems in South American cities. Some families often survive by putting everyone to work on the streets. The poorest eke out a living shining shoes or picking through garbage for items to sell or recycle. Others work as street vendors, beggars, and prostitutes. The situation is particularly acute for homeless young children, who lack proper food, health care, and education, and may be the victims of physical, sexual, and emotional abuse. Rising poverty contributes to rising crime. Murder is the number one cause of death among poor men between fifteen and thirty-nine. Many crimes are linked to illegal drugs: People steal to get money to buy drugs; drug dealers fight over territory and business deals. Like the worst inner-city neighborhoods in the United States, in some shantytowns gunshots ring out in the night and children are afraid to go to school.

Rapid population growth means more pollution. Cars and factories pollute the air. Cities dump untreated sewage directly into rivers, canals, and bays. Throughout South American cities, trees and buildings are damaged by acid rain, and respiratory diseases are increasing. Laws have been passed to reduce pollution levels, but they are often ignored or unevenly enforced.

Prospects for the Future

Population growth and poverty are the most pressing problems for South American cities. Unfortunately, cities cannot effectively eliminate either one. Cities can neither build fences to keep out poor people, nor can they narrow the gap between rich and poor. Improving the conditions in South American cities requires healthy economies and well-implemented policies aimed at reducing poverty.

National governments must be able to invest in basic infrastructure-schools, hospitals, roads, sanitation measures, and telecommunications systems. This is particularly true in rural areas, where the challenge is focused on slowing the rural-to-urban

---

BRASILIA: A PLANNED CAPITAL CITY

Since gaining independence from Portugal in 1822, Brazilians had wanted to build a new capital in the Amazon. The government finally chose a modern design by Lucio Costa, and construction was begun in 1956. The city was supposed to look and work like a machine. Unlike other Brazilian cities, which have winding, narrow roads, Brasilia has wide boulevards and large open spaces. Older cities have mixed land use, with offices, stores, and homes all in the same block; Brasilia has a separate zone for each. Critics complain that Brasilia is cold and charmless and that the city has not aged well. In spite of these criticisms, Brasilia has been a success. Originally planned for a half million people, its population was six times that size by 2017.

---

migration. Everywhere, the poor need education, health care, and jobs that pay a decent wage. Throughout South America, governments are instituting policies to help alleviate poverty, albeit with mixed results.

Virginia Thompson

CENTRAL AMERICA

In many rural parts of Central America, living conditions are poor. Each year, thousands of Central Americans migrate to cities in search of better jobs, housing, and education. Rural-to-urban migration is influenced by both push and pull factors. Push factors that persuade Central Americans to leave rural areas include loss of agricultural jobs, lack of educational facilities, and rising poverty. Migrants are attracted to urban areas through pull factors, including employment opportunities and access to education. These factors are also at play in many South American countries.

Degree of Urbanization

In contrast to South America, the percentage of people living in Central America's cities is relatively low. The most urbanized country in Central America is Costa Rica, whose urban population grew from 43 percent in 1980 to 78 percent in 2016. The urban populations of El Salvador and Panama are slightly over 65 percent, and Guatemala and Honduras barely exceed 50 percent. In Belize, less than half the population lives in cities.

The majority of Central America's people are Native Americans, or Amerindians, who make their

MAJOR URBAN CENTERS IN CENTRAL AMERICA

living off the land. They may be involved in seasonal work as laborers on large plantations or farms, or they may cultivate their own food and crops as sharecroppers—people who work the land for a share of the crop—or on land that they own or rent. Despite the large number of people living in rural villages, many do not have access to enough cultivable land. This is because the elite minority holds the majority of the land on huge ranches, plantations, or haciendas. The shortage of farm-land for the poor in rural areas has been a major factor contributing to urban migration.

Dissatisfied with political and economic circumstances, many Central Americans have attempted to migrate, legally or illegally, to cities in other countries. New migrants are often poor and unable to read or write and must compete for limited jobs. Women may find domestic work, and men may find low-skill jobs in construction, maintenance, and small-scale manufacturing. They often depend on day work that is frequently short-term, or earn money through jobs in construction or landscaping. Significant numbers of Hondurans, Salvadorans, and Guatemalans in particular continue to migrate to the United States.

Unless poverty diminishes, violence decreases, and living conditions improve, the trend of rural-to-urban migration will likely continue. And beyond internal migration, many Central Americans are willing to make the dangerous trek northward to the United States in search of a better life.

Anne Galantowicz

DISCUSSION QUESTIONS: URBANIZATION

Q1. How did the layouts of Spanish and Portuguese colonial cities differ? How did social status determine where in and around a city people lived? How have these patterns carried through to today?

Q2. Why did Brazil create a new capital, Brasília, in the Amazon, remote from the urbanized coastal cities? What challenges are involved in creating a new city on the scale of Brasília?

Q3. What factors have led to the internal migration of South and Central Americans from rural areas to cities? How are the cities dealing with their rapidly growing populations?

ECONOMIC GEOGRAPHY

OVERVIEW

TRADITIONAL AGRICULTURE

Two agricultural practices that are widespread among the world's traditional cultures, slash-and-burn and nomadism, share several common features. Both are ancient forms of agriculture, both involve farmers not remaining in a fixed location, and both can pose serious environmental threats if practiced in a nonsustainable fashion. The most significant difference between the two forms is that slash-and-burn generally is associated with raising field crops, while nomadism as a rule involves herding livestock.

Slash-and-Burn Agriculture

Farmers have practiced slash-and-burn agriculture, which is also referrred to as shifting cultivation or swidden agriculture, in almost every region of the world where the climate makes farming possible. Humans have practiced this method for about 12,000 years, ever since the Neolithic Revolution. Swidden agriculture once dominated agriculture in more temperate regions, such as northern Europe. It was, in fact, common in Finland and northern Russia well into the early decades of the twentieth century. Today, between 200 and 500 million people use slash-and-burn agriculture, roughly 7 percent of the world's population. It is most commonly practiced in areas where open land for farming is not readily available because of dense vegetation. These regions include central Africa, northern South America, and Southeast Asia

Slash-and-burn acquired its name from the practice of farmers who cleared land for planting crops by cutting down the trees or brush on the land and then burning the fallen timber on the site. The farmers literally slash and burn. The ashes of the burnt wood add minerals to the soil, which temporarily improves its fertility. Crops the first year following clearing and burning are generally the best crops the site will provide. Each year after that, the yield diminishes slightly as the fertility of the soil is depleted.

Farmers who practice swidden cultivation do not attempt to improve fertility by adding fertilizers such as animal manures but instead rely on the soil to replenish itself over time. When the yield from one site drops below acceptable levels, the farmers then clear another piece of land, burn the brush and other vegetation, and cultivate that site while leaving their previous field to lie fallow and its natural vegetation to return. This cycle will be repeated over and over, with some sites being allowed to lie fallow indefinitely while others may be revisited and farmed again in five, ten, or twenty years.

Farmers who practice shifting cultivation do not necessarily move their dwelling places as they change the fields they cultivate. In some geographic regions, farmers live in a central village and farm cooperatively, with the fields being alternately allowed to remain fallow, and the fields being farmed making a gradual circuit around the central village. In other cases, the village itself may move as new fields are cultivated. Anthropologists studying indigenous peoples in Amazonia, discovered that village garden sites were on a hundred-year cycle. Villagers farmed cooperatively, with the entire village working together to clear a garden site. That garden would be used for about five years, then a new site was cleared. When the garden moved an inconvenient distance from the village, about once every twenty years, the entire village would move to be closer to the new garden. Over a period of approximately 100 years, a village would make a circle through the forest, eventually ending up close to where it had been located long before any of the present villagers had been born.

In more temperate climates, individual farmers often owned and lived on the land on which they practiced swidden agriculture. Farmers in Finland, for example, would clear a portion of their land, burn the brush and other covering vegetation, grow grains for several years, and then allow that land to remain fallow for from five to twenty years. The individual farmer rotated cultivation around the land in a fashion similar to that practiced by whole villages in other areas, but did so as an individual rather than as part of a communal society.

Although slash-and-burn is frequently denounced as a cause of environmental degradation in tropical areas, the problem with shifting cultivation is not the practice itself but the length of the cycle. If the cycle of shifting cultivation is long enough, forests will grow back, the soil will regain its fertility, and minimal adverse effects will occur. In some regions, a piece of land may require as little as five years to regain its maximum fertility; in others, it may take 100 years. Problems arise when growing populations put pressure on traditional farmers to return to fallow land too soon. Crops are smaller than needed, leading to a vicious cycle in which the next strip of land is also farmed too soon, and each site yields less and less. As a result, more and more land must be cleared.

Nomadism

Nomadic peoples have no permanent homes. They earn their livings by raising herd animals, such as sheep, cattle, or horses, and they spend their lives following their herds from pasture to pasture with the seasons. Most nomadic animals tend to be hardy breeds of goats, sheep, or cattle that can withstand hardship and live on marginal lands. Traditional nomads rely on natural pasturage to support their herds and grow no grains or hay for themselves. If a drought occurs or a traditional pasturing site is unavailable, they can lose most of their herds to starvation.

THE HERITAGE SEED MOVEMENT

Modern hybrid seeds have increased yields and enabled the tremendous productivity of the modern mechanized farm. However, the widespread use of a few hybrid varieties has meant that almost all plants of a given species in a wide area are almost identical genetically. This loss of biodiversity, or, the range of genetic difference in a given species, means that a blight could wipe out an entire season's crop. Historical examples of blight include the nineteenth century Great Potato Famine of Ireland and the 1971 corn blight in the United States.

In response to the concern for biodiversity, there has been a movement in North America to preserve older forms of crops with different genes that would otherwise be lost to the gene pool. Nostalgia also motivates many people to keep alive the varieties of fruits and vegetables that their grandparents raised. Many older recipes do not taste the same with modern varieties of vegetables that have been optimized for commercial considerations such as transportability. Thus, raising heritage varieties also can be a way of continuing to enjoy the foods one's ancestors ate.

In many nomadic societies, the herd animal is almost the entire basis for sustaining the people. The animals are slaughtered for food, clothing is woven from the fibers of their hair, and cheese and yogurt may be made from milk. The animals may also be used for sustenance without being slaughtered. Nomads in Mongolia, for example, occasionally drink horses' blood, removing only a cup or two at a time from the animal. Nomads go where there is sufficient vegetation to feed their animals.

In mountainous regions, nomads often spend the summers high up on mountain meadows, returning to lower altitudes in the autumn when snow begins to fall. In desert regions, they move from oasis to oasis, going to the places where sufficient natural water exists to allow brush and grass to grow, allowing their animals to graze for a few days, weeks, or months, then moving on. In some cases, the pressure to move on comes not from the depletion of food for the animals but from the depletion of a water source, such as a spring or well. At many natural desert oases, a natural water seep or spring provides only enough water to support a nomadic group for a few days at a time.

In addition to true nomads—people who never live in one place permanently—a number of cultures have practiced seminomadic farming: The temperate months of the year, spring through fall, are spent following the herds on a long loop, sometimes hundreds of miles long, through traditional grazing areas; then the winter is spent in a permanent village.

Nomadism has been practiced for millennia, but there is strong pressure from several sources to eliminate it. Pressures generated by industrialized society are increasingly threatening the traditional cultures of nomadic societies, such as the Bedouin of the Arabian Peninsula. Traditional grazing areas are being fenced off or developed for other purposes. Environmentalists are also concerned about the ecological damage caused by nomadism.

Nomads generally measure their wealth by the number of animals they own and so will try to develop their herds to be as large as possible, well beyond the numbers required for simple sustainability. The herd animals eat increasingly large amounts of vegetation, which then has no opportunity to regenerate, and desertification may occur. Nomadism based on herding goats and sheep,

DESERTIFICATION

Desertification is the extension of desert conditions into new areas. Typically, this term refers to the expansion of deserts into adjacent nondesert areas, but it can also refer to the creation of a new desert. Land that is susceptible to prolonged drought is always in danger of losing its vegetative ground cover, thereby exposing its soil to wind. The wind carries away the smaller silt particles and leaves behind the larger sand particles, stripping the land of its fertility. This naturally occurring process is assisted in many areas by overgrazing.

In the African Sahel, south of the Sahara, the impact of desertification is acute. Recurring drought has reduced the vegetation available for cattle, but the need for cattle remains high to feed populations that continue to grow. The cattle eat the grass, the soil is exposed, and the area becomes less fertile and less able to support the population. The desert slowly encroaches, and the people must either move or die.

for example, has been blamed for the expansion of the Sahara Desert in Africa. For this reason, many environmental policymakers have been attempting to persuade nomads to give up their roaming lifestyle and become sedentary farmers.

Nancy Farm Männikkö

COMMERCIAL AGRICULTURE

Commercial farmers are those who sell substantial portions of their output of crops, livestock, and dairy products for cash. In some regions, commercial agriculture is as old as recorded history, but only in the twentieth century did the majority of farmers come to participate in it. For individual farmers, this has offered the prospect of larger income and the opportunity to buy a wider range of products. For society, commercial agriculture has been associated with specialization and increased productivity.

Commercial agriculture has enabled world food production to increase more rapidly than world population, improving nutrition levels for millions of people.

Steps in Commercial Agriculture

In order for commercial agriculture to exist, products must move from farmer to ultimate consumer, usually through six stages:

181

1. Processing, packaging, and preserving to protect the products and reduce their bulk to facilitate shipping.

2. Transport to specialized processing facilities and to final consumers.

3. Networks of merchant middlemen who buy products in bulk from farmers and processors and sell them to final consumers.

4. Specialized suppliers of inputs to farmers, such as seed, livestock feed, chemical inputs (fertilizers, insecticides, pesticides, soil conditioners), and equipment.

5. A market for land, so that farmers can buy or lease the land they need.

6. Specialized financial services, especially loans to enable farmers to buy land and other inputs before they receive sales revenues.

Improvements in agricultural science and technology have resulted from extensive research programs by government, business firms, and universities.

International Trade

Products such as grain, olive oil, and wine moved by ship across the Mediterranean Sea in ancient times. Trade in spices, tea, coffee, and cocoa provided powerful stimulus for exploration and colonization around 1500 CE. The coming of steam locomotives and steamships in the nineteenth century greatly aided in the shipment of farm products and spurred the spread of population into potentially productive farmland all over the world. Beginning with Great Britain in the 1840s, countries were willing to relinquish agricultural self-sufficiency to obtain cheap imported food, paid for by exporting manufactured goods.

Most of the leaders in agricultural trade were highly developed countries, which typically had large amounts of both imports and exports. These countries are highly productive both in agriculture and in other commercial activities. Much of their trade is in high-value packaged and processed goods. Although the vast majority of China's labor force works in agriculture, their average productivity is low and the country showed an import surplus in agricultural products. The same was true for Rus-sia. India, similar to China in size, development, and population, had relatively little agricultural trade. Australia and Argentina are examples of countries with large export surpluses, while Japan and South Korea had large import surpluses. Judged by volume, trade is dominated by grains, sugar, and soybeans. In contrast, meat, tobacco, cotton, and coffee reflect much higher values per unit of weight.

The United States

Blessed with advantageous soil, topography, and climate, the United States has become one of the most productive agricultural countries in the world. Technological advances have enabled the United States to feed its own residents and export substantial quantities with only 3 percent of its labor force engaged directly in farming. In the 2020s there are about 2 million farms cultivating about 1 billion acres. They produced about US$133 billion worth of products. After expenses, this yielded about US$92.5 billion of net farm income—an average of only about US$25,000 per farm. However, most farm families derive substantial income from nonfarm employment.

There is a great deal of agricultural specialization by region. Corn, soybeans, and wheat are grown in many parts of the United States (outside New England). Some other crops have much more limited growing areas. Cotton, rice, and sugarcane require warmer temperatures. Significant production of cotton occurred in seventeen states, rice in six, and sugarcane in four. In 2018, the top 10 agricultural producing states in terms of cash receipts were (in descending order): California, Iowa, Texas, Nebraska, Minnesota, Illinois, Kansas, North Carolina, Wisconsin, and Indiana Typically the top two states in a category account for about 30 percent of sales. Fruits and vegetables are the main exception; the great size, diversity, and mild climate of California gives it a dominant 45 percent.

Socialist Experiments

Under the dictatorship of Joseph Stalin, the communist government of the Soviet Union established a program of compulsory collectivized agriculture in 1929. Private ownership of land, buildings, and other assets was abolished. There were some state farms, "factories in the fields," operated on a large scale with many hired workers. Most, however, were collective farms, theoretically run as cooperative ventures of all residents of a village, but in practice directed by government functionaries. The arrangements had disastrous effects on productivity and kept the rural residents in poverty. Nevertheless, similar arrangements were established in China in 1950 under the rule of Mao Zedong. A restoration of commercial agriculture after Mao's death in 1976 enabled China to achieve greater farm output and farm incomes.

Most Western countries, including the United States, subsidize agriculture and restrict imports of competing farm products. Objectives are to support farm incomes, reduce rural discontent, and slow the downward trend in the number of farmers. Farmers in the European Union will see aid shrink in the 2021–2027 period to 365 billion euros (US$438 billion), down 5 percent from the current Common Agricultural Policy (CAP). Japan's Ministry of Agriculture, Forestry and Fisheries (MAFF) has requested 2.65 trillion yen (roughly US$24 billion) for the Japan Fiscal Year (JFY) 2018 budget, a 15 percent increase over last year. The budget request eliminates the direct payment subsidy for table rice production, but requests significant funding for a new income insurance program, agricultural ex-port promotion, and underwriting goals to expand domestic potato production. In 2019, trade wars with China and punishing tariffs have led to increased subsidies by the U.S. government, totaling US$10 billion dollars in 2018 and US$14.5 billion dollars in 2019.

Problems for Farmers

Farmers in a system of commercial agriculture are vulnerable to changes in market prices as well as the universal problems of fluctuating weather. Congress tried to reduce farm subsidies through the Freedom to Farm Act of 1996, but serious price declines in 1997-1999 led to backtracking. Efforts to increase productivity by genetic alterations, radiation, and feeding synthetic hormones to livestock have drawn critical responses from some consumer groups. Environmentalists have been concerned about soil depletion and water pollution resulting from chemical inputs.

Productivity and World Hunger

Despite advances in agricultural production, the problem of world hunger persists. Even in countries that store surpluses of farm commodities, there are still people who go hungry. In less-developed countries, the prices of imported food from the West are too low for local producers to compete and too high for the poor to buy them.

Paul B. Trescott

MODERN AGRICULTURAL PROBLEMS

Ever since human societies started to grow their own food, there have been problems to solve. Much of the work of nature was disrupted by the work of agriculture as many as 10,000 years ago. Nature took care of the land and made it productive in its own intricate way, through its own web of interdependent systems. Agriculture disrupts those systems with the hope of making the land even more productive, growing even more food to feed even more people. Since the first spade of soil was turned over and the first plants domesticated, farmers have been trying to figure out how to care for the land as well as nature did before.

Many modern problems in agriculture are not really modern at all. Erosion and pollution, for example, have been around as long as agriculture.

However, agriculture has changed drastically within those 10,000 years, especially since the dawn of the Industrial Revolution in the seventeenth century. Erosion and pollution are now bigger problems than before and have been joined by a host of others that are equally critical—not all related to physical deterioration. Modern farmers use many more machines than did farmers of old, and modern machines require advanced sources of energy to unleash their power. The machines do more work than could be accomplished before, so fewer farmers are needed, which causes economic problems.

Cities continue to grow bigger as land—usually the best farmland around—is converted to homes and parking lots for shopping centers. The farmers that remain on the land, needing to grow ever more food, turn to the research and engineering industries to improve their seeds. These industries have responded with recombinant technologies that move genes from one species to another; for example, genes cut from peanuts may be spliced into chickens. This creates another set of cultural problems, which are even more difficult to solve because most are still "potential"—their impact is not yet known.

Erosion

Soil loss from erosion continues to be a huge problem all over the world. As agriculture struggles to feed more millions of people, more land is plowed. The newly plowed lands usually are considered more marginal, meaning they are either too steep, too thin, or too sandy; are subject to too much rain; or suffer some other deficiency. Natural vegetative cover blankets these soils and protects them from whatever erosive agents are active in their regions: water, wind, ice, or gravity. Plant cover also increases the amount of rain that seeps downward into the soil rather than running off into rivers. The more marginal land that is turned over for crops, the faster the erosive agents will act and the more erosion will occur.

Expansion of land under cultivation is not the only factor contributing to erosion. Fragile grasslands in dry areas also are being used more intensively. Grazing more livestock than these pastures can handle decreases the amount of grass in the pasture and exposes more of the soil to wind—the primary erosive agent in dry regions.

Overgrazing can affect pastureland in tropical regions too. Thousands of acres of tropical forest have been cleared to establish cattle-grazing ranges in Latin America. Tropical soils, although thick, are not very fertile. Fertility comes from organic waste in the surface layers of the soil. Tropical soils form under constantly high temperatures and receive much more rain than soils in moderate, midlatitude climates; thus, tropical organic waste materials rot so fast they are not worked into the soil at all. After one or two growing seasons, crops grown in these soils will yield substantially less than before.

Tropical fields require fallow periods of about ten years to restore themselves after they are depleted. That is why tropical cultures using slash-and-burn methods of agriculture move to new fields every other year in a cycle that returns them to the same place about every ten years, or however long it takes those particular lands to regenerate. The heavy forest cover protects these soils from exposure to the massive amounts of rainfall and provides enough organic material for crops—as long as the forest remains in place. When the forest is cleared, however, the resulting grassland cannot provide the adequate protection, and erosion accelerates. Grasslands that are heavily grazed provide even less protection from heavy rains, and erosion accelerates even more.

The use of machines also promotes erosion, and modern agriculture relies on machinery: tractors, harvesters, trucks, balers, ditchers, and so on. In the United States, Canada, Europe, Russia, Brazil, South Africa, and other industrialized areas, machinery use is intense. Machinery use is also on the rise in countries such as India, China, Mexico, and Indonesia, where traditional nonmechanized methods are practiced widely. Farming machines, in gaining traction, loosen the topsoil and inhibit vegetative cover growth, especially when they pull behind them any of the various farm implements designed to rid the soil of weeds, that is, all vegetation except the desired crop. This leaves the soil more exposed to erosive weather, so more soil is carried away in the runoff of water to streams.

Eco-fallow farming has become more popular in the United States and Europe as a solution to reducing erosion. This method of agriculture, which leaves the crop residue in place over the fallow (nongrowing) season, does not root the soil in place, however. Dead plants do not "grab" the soil like live plants that need to extract from it the nutrients they need to live, so erosion continues, even though it is at a slower rate. Eco-fallow methods also require heavier use of chemicals, such as herbicides, to "burn down" weed growth at the start of the growing season, which contributes to accelerated erosion and increases pollution.

Pollution

Pollution, besides being a problem in general, continues to grow as an agricultural problem. With the onset of the Green Revolution, the use of herbicides, insecticides, and pesticides has increased dramatically all over the world. These chemicals are not used up completely in the growth of the crop, so the leftovers (residue) wash into, and contaminate, surface and groundwater supplies. These supplies then must be treated to become useful for other purposes, a job nature used to do on its own. Agricultural chemicals reduce nature's ability to act as a filter by inhibiting the growth of the kinds of plant life that perform that function in aquatic environments. The chemical residues that are not washed into surface supplies contaminate wells.

As chemical use increases, contamination accumulates in the soil and fertility decreases. The microorganisms and animal life in the soil, which had facilitated the breakdown of soil minerals into usable plant products, are no longer nourished because the crop residue on which they feed is depleted, or they are killed by the active ingredients in the chemical. As a result, soil fertility must be restored to maintain yield. Chemical replacement is usually the method of choice, and increased applications of chemical fertilizers intensify the toxicity of this cyclical chemical dependency.

Chemicals, although problematic, are not as difficult to contend with as the increasingly heavy silt load choking the life out of streams and rivers. Accelerated erosion from water runoff carries silt particles into streams, where they remain suspended and inhibit the growth of many beneficial forms of plant and animal life. The silt load in U.S. streams has become so heavy that the Mississippi River Delta is growing faster than it used to. The heavy silt load, combined with the increased load of chemical residues, is seriously taxing the capabilities of the ecosystems around the delta that filter out sediments, absorb nutrients, and stabilize salinity levels for ocean life, creating an expanding dead zone.

This general phenomenon is not limited to the Mississippi Delta—it is widespread. Its impact on people is high, because most of the world's population lives in coastal zones and comes in direct contact with the sea. Additionally, eighty percent of the world's fish catch comes from the coastal waters over continental shelves that are most susceptible to this form of pollution.

Monoculture

Modern agriculture emphasizes crop specialization. Farmers, especially in industrialized regions, often grow a single crop on most of their land, perhaps rotating it with a second crop in successive years: corn one year, for example, then soybeans, then back to corn. Such a strategy allows the farmer to reduce costs, but it also makes the crop, and, thus, the farmer and community, susceptible to widespread crop failure. When the crop is infested by any of an ever-changing number and variety of pests—worms, molds, bacteria, fungi, insects, or other diseases—the whole crop is likely to die quickly, unless an appropriate antidote is immediately applied. Chemical antidotes can do the job but increase pollution. Maintaining species diversity—growing several different crops instead of one or two—allows for crop failures without jeopardizing the entire income for a farm or region that specializes in a particular monoculture, such as tobacco, coffee, or bananas.

Chemicals are not the only modern methods of preventing crop loss. Genetically engineered seeds are one attempt at replacing post-infestation chemical treatments. For example, splicing genes into varieties of rice or potatoes from wholly unrelated species—say, hypothetically, a grasshopper—to prevent common forms of blight is occurring more often. Even if the new genes make the crop more resistant, however, they could trigger unknown side effects that have more serious long-term environmental and economic consequences than the problem they were used to solve. Genetically altered crops are essentially new life-forms being introduced into nature with no observable precedents to watch beforehand for clues as to what might happen.

Urban Sprawl

As more farms become mechanized, the need for farmers is being drastically reduced. There were more farmers in the United States in 1860 than there were in the year 2000. From a peak in 1935 of about 6.8 million farmers farming 1.1 billion acres, the United States at the end of the twentieth century counted fewer than 2.1 million farmers farming 950 million acres. As fewer people care for land, the potential for erosion and pollution to accelerate is likely to increase, causing land quality to decline.

As farmers are displaced and move into towns, the cities take up more space. The resulting urban sprawl converts a tremendous amount of cropland into parking lots, malls, industrial parks, or suburban neighborhoods. If cities were located in marginal areas, then the concern over the loss of farmland to commercial development would be nominal. However, the cities attracting the greatest numbers of people have too often replaced the best cropland. Taking the best cropland out of primary production imposes a severe economic penalty.

James Knotwell and Denise Knotwell

WORLD FOOD SUPPLIES

All living things need food to begin the life process and to live, grow, work, and survive. Almost all foods that humans consume come from plants and animals. Not all of Earth's people eat the same foods, however, nor do they require the same caloric intakes. The types, combinations, and amounts of food consumed by different peoples depend upon historic, socioeconomic, and environmental factors.

The History of Food Consumption

Early in human history, people ate what they could gather or scavenge. Later, people ate what they could plant and harvest and what animals they could domesticate and raise. Modern people eat what they can grow, raise, or purchase. Their diets or food composition are determined by income, local customs, religion or food biases, and advertis- ing. There is a global food market, and many people can select what they want to eat and when they eat it according to the prices they can pay and what is available.

Historically, in places where food was plentiful, accessible, and inexpensive, humans devoted less time to basic survival needs and more time to activities that led to human progress and enjoyment of leisure. Despite a modern global food system, instant telecommunications, the United Nations, and food surpluses at places, however, the problem of providing food for everyone on Earth has not been solved.

According to the United Nations Sustainable Development Goals that were adopted by all Member States in 2015, an estimated 821 million people were undernourished in 2017. In developing countries, 12.9 per cent of the population is undernourished. Sub-Saharan Africa has the highest prevalence of hunger; the number of undernourished people increased from 195 million in 2014 to 237 million in 2017. Poor nutrition causes nearly half (45 per cent) of deaths in children under five—3.1 million children each year. As of 2018, 22 per cent of the global under-5 population were still chronically undernourished in 2018. To meet challenge of Goal 2: Zero Hunger, significant changes both in terms of agriculture and conservation as well as in financing and social equality will be required to nourish the 821 million people who are hungry today and the additional 2 billion people expected to be undernourished by 2050.

World Food Source Regions

Agriculture and related primary food production activities, such as fishing, hunting, and gathering, continue to employ more than one-third of the world's labor force. Agriculture's relative importance in the world economic system has declined with urbanization and industrialization, but it still plays a vital role in human survival and general economic growth. Agriculture in the third millennium must supply food to an increasing world population of nonfood producers. It must also produce food and nonfood crude materials for industry, accumulate capital needed for further economic growth, and allow workers from rural areas to industrial, construction, and expanding intraurban service functions.

Soil types, topography, weather, climate, socioeconomic history, location, population pressures, dietary preferences, stages in modern agricultural development, and governmental policies combine to give a distinctive personality to regional agricultural characteristics. Two of the most productive food-producing regions of the world are North America and Asia. Countries in these regions export large amounts of food to other parts of the world.

Foods from Plants

Most basic staple foods come from a small number of plants and animals. Ranked by tonnage produced, the most important food plants throughout the world are corn (maize), wheats, rice, potatoes, cassava (manioc), barley, soybeans, sorghums and millets, beans, peas and chickpeas, and peanuts (groundnuts).

More than one-third of the world's cultivated land is planted with wheat and rice. Wheat is the dominant food staple in North America, Western and Eastern Europe, northern China, and the Middle East and North Africa. Rice is the dominant food staple in southern and eastern Asia. Corn, used primarily as animal food in developed nations, is a staple food in Latin America and Southeast Africa. Potatoes are a basic food in the highlands of South America and in Central and Eastern Europe. Cassava (manioc) is a tropical starch-producing root crop of special dietary importance in portions of lowland South America, the west coast countries of Africa, and sections of South Asia. Barley is an important component of diets in North African, Middle Eastern, and Eastern European countries. Soybeans are an integral part of the diets of those who live in eastern, southeastern, and southern Asia. Sorghums and millets are staple subsistence foods in the savanna regions of Africa and south Asia, while peanuts are a facet of dietary mixes in tropical Africa, Southeast Asia, and South America.

Food from Animals

Animals have been used as food by humans from the time the earliest people learned to hunt, trap, and fish. However, humans have domesticated only a few varieties of animals. Ranked by tonnage of meat produced, the most commonly eaten animals are cattle, pigs, chickens and turkeys, sheep, goats, water buffalo, camels, rabbits and guinea pigs, yaks, and llamas and alpacas.

Cattle, which produce milk and meat, are important food sources in North America, Western Europe, Eastern Europe, Australia and New Zealand, Argentina, and Uruguay. Pigs are bred and reared for food on a massive scale in southern and eastern Asia, North America, Western Europe, and Eastern Europe. Chickens are the most important domesticated fowl used as a human food source and are a part of the diets of most of the world's people.

Sheep and goats, as a source of meat and milk, are especially important to the diets of those who live in the Middle East and North Africa, Eastern Europe, Western Europe, and Australia and New Zealand.

Water buffalo, camels, rabbits, guinea pigs, yaks, llamas, and alpacas are food sources in regions of the world where there is low consumption of meat for religious, cultural, or socioeconomic reasons. Fish is an inexpensive and wholesome source of food. Seafood is an important component to the diets of those who live in southern and eastern Asia, Western Europe, and North America.

The World's Growing Population

The problem of feeding the world is compounded by the fact that population was increasing at a rate of nearly 82 million persons per year at the end of second decade of the twenty-first century. That rate of increase is roughly equivalent to adding a country the size of Germany to the world every single year.

Also compounding the problem of feeding the world are population redistribution patterns and changing food consumption standards. In the year 2050, the world population is projected to reach approximately 10 billion—4 billion people more than were on the earth in 2000. Most of the increase in world population is expected to occur within the developing nations.

Urbanization

Along with an increase in population in developing nations is massive urbanization. City dwellers are food consumers, not food producers. The exodus of young men and women from rural areas has given rise to a new series of megacities, most of which are in developing countries. By the year 2030, there could be as many as forty-one megacities (cities with populations of 10 million people or more).

When rural dwellers move to cities, they tend to change their dietary composition and food-consumption patterns. Qualitative changes in dietary consumption standards are positive, for the most part, and are a result of copying the diets of what is considered a more prestigious group or positive educational activities of modern nutritional scientists working in developing countries. During the last four decades of the twentieth century, a tremendous shift took place in overall dietary habits as Western foods became increasingly available and popular throughout the world. While improved nutrition has contributed to a decrease in child mortality, an increase in longevity, and a greater resistance to disease, it is also true that conditions including morbid obesity, Type II diabetes, and hypertension are on the rise.

Strategies for Increasing Food Production

To meet the food demands and the food distribution needs of the world's people in the future, several strategies have been proposed. One such strategy calls for the intensification of agriculture—improving biological, mechanical, and chemical technology and applying proven agricultural innovations to regions of the world where the physical and cultural environments are most suitable for rapid food production increases.

The second step is to expand the areas where food is produced so that areas that are empty or underused will be made productive. Reclaiming areas damaged by human mismanagement, expanding irrigation in carefully selected areas, and introducing extensive agrotechniques to areas not under cultivation could increase the production of inexpensive grains and meats.

Finally, interregional, international, and global commerce should be expanded, in most instances, increasing regional specializations and production of high-quality, high-demand agricultural products for export and importing low-cost basic foods. A disequilibrium of supply and demand for certain commodities will persist, but food producers, regional and national agricultural planners, and those who strive for regional economic integration must take advantage of local conditions and location or create the new products needed by the food-consuming public in a one-world economy.

Perspectives

Humanity is entering a time of volatility in food production and distribution. The world will produce enough food to meet the demands of those who can afford to buy food. In many developing countries, however, food production is unlikely to keep pace with increases in the demand for food by growing populations.

Factors that could lead to larger fluctuations in food availability include weather variations such as those induced by El Niño and climate change, the growing scarcity of water, civil strife and political in-stability, and declining food aid. In developing countries, decision makers need to ensure that policies promote broad-based economic growth—and in particular agricultural growth—so that their countries can produce enough food to feed themselves or enough income to buy the necessary food on the world market.

William A. Dando

AGRICULTURE

SOUTH AMERICA

Most people of South America, like people of the United States, live in large cities. About 15 percent of South Americans are considered rural, compared to about 19 percent of the U.S. population. The real contrast is in the number of people who make their living from agriculture. About 12.5 percent of the population of South America receives most or all of its income from agriculture, compared to less than 2 percent in the United States. Only about 40 million acres (96 million hectares) of the continent is farmable (arable)—considerably less than the U.S.'s 922 million acres (373 million hectares).

The agricultural economy of South America has undergone dramatic changes in recent years. South America has become much more urbanized and industrialized. For example, Brazil now contains the ninth largest economy in the world. This presents a tremendous challenge to people in rural communities who have depended on agriculture as a way of life for generations. Because it is so difficult to make a living in the farming areas, more South Americans are leaving their rural homelands to live in large cities. As the economy of South America expands through its industrial base, poverty nevertheless continues to rise. Throughout the continent, a wide gap exists between the wealthy and the poor. Even during good economic times, the poor benefit very little.

Today, 83 million South Americans live in rural areas and about 331 million live in urban areas. Cities such as Brazil's São Paulo (21.7 million) and Rio de Janeiro (population 5.8 million) and São Paulo (13.4 million) continue to grow quickly, and at a pace higher than the overall industrialization rate of the continent. Vast shantytowns surround the cit-

ies, but people keep coming. The poor migrants leave behind impoverished rural agricultural areas, where many still barely get by on what they can grow. This migration to the city has had a tremendous impact on the farmland that is left behind, idle and nonproductive. At the same time, cities are faced with the dual challenges of uncontrolled growth and of lack of jobs for newcomers.

Agriculture and the Tropical Forest

Some of the most remote parts in South America are also its fastest-developing areas. The northern interior regions of Brazil in the Amazon Basin face

---

HIDDEN AGRICULTURE

Nearly 75 percent of the total world supply of the illegal drug cocaine comes from Peru, Colombia, and Bolivia. This drug is manufactured from the coca plant, which is grown mostly along the east side of the Andes. It grows well in the tropical lowland areas of Peru and Bolivia. The coca plant is historically embedded in the culture of South America. The Incas cultivated this plant and still chew its leaves and drink it as a tea to help them cope with the high altitudes of the area.

Coca cultivation has impacted agriculture in the region. Many farmers, seeing how much money can be made growing coca instead of a conventional crop like maize, leave their traditional farming to raise coca. In this case, farmers do not leave their farm for the city but changes their crop to make more money. The government has tried to help farmers by giving them funds to help bring up the prices of traditional food crops, but these programs have not worked. So long as markets for high-priced cocaine exist, it will be grown in these countries.

SELECTED AGRICULTURAL PRODUCTS OF SOUTH AMERICA

tremendous pressure from development and population growth. This area is covered with the world's largest tropical rain forest, as vast as the forty-eight contiguous U.S. states. Within some areas of this region, gold and huge iron ore deposits have been discovered. Development programs to promote mining, hydroelectric projects, ranching, and farming require that the forests be cleared and encourage a large migration of people into this otherwise remote, isolated region.

Competition for the land and its resources has been fierce, with the detrimental consequence of thousands of square miles of tropical forest being removed every year. The World Wildlife Fund (WWF) estimates that 27 percent of the Amazon rain forest will be destroyed by 2030 if the current rate of deforestation continues.

Although the rain forest might appear to be a highly productive growing region for crops, it is not.

Soils in these forests are thin and lacking in nutrients. To grow crops and grasses for cattle, slash-and-burn agricultural practices are used. Crops such as maize (corn) and rice are planted on the newly cleared ground. But after about three years, the nutrients have been washed out of the soil by the heavy tropical rains. Farmers and ranchers move on and find a new forest area to cut, and the cycle goes on. Why conserve these forests? Rain forests cover only about 6 percent of the total land area of Earth but contain more than half of the plant and animal species on Earth. These forests provide lumber products, medicines, and food. No one knows what amazing products they might yet hold for human good.

South American Agriculture as a System

Agriculture is a complex system comprising both human and natural variables. Components include

Burning forest in Brazil. The removal of forest to make way for cattle ranching was the leading cause of deforestation in the Brazilian Amazon rain-forest from the mid-1960s. (Image by NASA)

the type of landforms, soil, water, and climate, and the technological possibilities that can be applied by humans to an agricultural economy. These variables interact to produce a variety of agricultural productions specific to the region, in South America as elsewhere.

South America can be divided into three general landform regions: the Andes Mountains, the plateaus of the interior of the continent, and the river lowlands.

The Andes Mountains

The Andes Cordillera reaches from Venezuela in the north to Tierra del Fuego at the southern tip of the continent. The Andes are composed of both folded mountains and volcanic peaks, some exceeding 20,000 feet (6,100 meters). The soils there are rocky and mostly found on steep grades or hillsides. This makes farming difficult but not impossible. Farmers use terrace-type systems, building up steplike fields carved into the side of a mountain. Because of the limitation of these soils and the difficulty in farming them, the region has only noncommercial, subsistence-type agricultural settlements. Native peoples in the highlands grow small plots of corn, barley, and especially potatoes on the high-altitude soils. Llamas and alpacas—adapted to the rugged terrain—are raised there as well.

The Highland Growing Regions

In the central eastern region of Brazil are the Brazilian Highlands. To the north of the Amazon Basin lies another plateau region called the Guiana Highlands. Both plateaus, which are not much higher than 9,000 feet (2,743 meters), are old geologic structures with relatively rough surfaces to farm. The Brazilian Highlands is the world's primary coffee-growing region. More than one-third of the world's coffee is grown here, along with soybeans and oranges. The Guiana Highlands, which stretches through southern Venezuela and Guyana, is another plateau of geologically old soils. This area is covered in savanna (grasslands with trees and shrubs). Native peoples here use the slash-and-burn system of agriculture, cutting down trees and brush on the land, burning them, then working them into the soil. This enriches the soils for a time and allows for limited production of corn and other subsistence crops.

The River Lowlands

The river valleys of South America are some of the largest in the world. The Orinoco River drains the Guiana Highlands and the *llanos* region of Venezuela, flowing out into the Atlantic Ocean. The *llanos* are a large, expansive plain of grassland between the Andes in Venezuela and the Guiana Highlands. During the rainy seasons, soils are flooded in this area, providing for rich grass development and the support of large cattle ranches.

The Amazon River drains the north and western regions of Brazil. Although more water moves through the Amazon River than any other river in the world, its location in dense tropical forests under a humid tropical climate makes agriculture there a challenge. Amerindians of the region practice subsistence agriculture, growing yams and bananas and raising some small animals. Manioc (a root crop from which tapioca is made) and sugarcane are also grown here in slash-and-burn fashion.

The Paraná, Paraguay, and Uruguay Rivers, south of the Amazon lowlands, cross through a great grassland region known as the Pampas and a forested region known as the Gran Chaco. The Pampas of Argentina is not unlike the Midwest of North America. It is a huge grassy plain nearly 400 miles (640 km.) long in the central part of Argentina. This is a land of gauchos (cowboys), cattle,

The Orinoco River delta. (Image by Milito)

sheep, and barbecues. Historically, ranching began in this region because grazing animals did not require a great number of people, and early settlements in the area by the Spanish were few. Also, the products from cattle and sheep, such as hides and wool, could be shipped long distances without damage.

Agriculture is a primary industry in Argentina and encompasses 54 percent of the country's land. Cattle is king here, and large *estancias* (ranches) are common in the region. The soils are well suited to wheat and other grains as well as alfalfa, a grass grown for cattle and horse feed. The level character of the land makes it easy to work, but it is difficult to drain and is prone to flooding.

In northern Argentina and the Pampas and through the central region of Paraguay lies the Gran Chaco. Paraguay's agricultural zones are divided by the Paraguay River. Tobacco, rice, and sugarcane grow to the east of the river in the more humid climates. To the west, in the drier climates and the Chaco region, the unique growing area of the Quebracho Forest is found. Quebracho is a hardwood that grows only in the Chaco. The wood contains tannin, which is used to produce tannic acid, a chemical used for tanning leather. This area is also suited for cattle, but because of its rocky and steep topography, the population of goats and sheep rises as one moves farther west toward the Andes. This region is irrigated with mountain streams from the eastern slopes of the Andes. Grapes also grow well here and are made into raisins and wine.

The Hot, Wet Lowlands

The coastal lowlands up to an altitude of about 2,500 feet (750 meters) encompass an area known as the hot land (*tierra calienta*). Temperatures there average between 75 and 80 degrees Fahrenheit (22

LEADING AGRICULTURAL PRODUCTS OF SOUTH AMERICAN COUNTRIES

Country	Products
Argentina	sunflower seeds; lemons; soybeans; grapes; corn; tobacco; peanuts; tea; wheat; livestock
Bolivia	soybeans; quinoa; Brazil nuts; sugarcane; coffee; corn; rice; potatoes; chia; coca
Brazil	coffee; soybeans; wheat; rice; corn; sugarcane; cocoa; citrus; beef
Chile	grapes; apples; pears; onions; wheat; corn; oats; peaches; garlic; asparagus; beans; beef; poultry; wool; fish; timber
Colombia	coffee; cut flowers; bananas; rice; tobacco; corn; sugarcane; cocoa beans; oilseed; vegetables; shrimp; forest products
Ecuador	bananas; coffee; cocoa; rice; potatoes; cassava (manioc; tapioca); plantains; sugarcane; cattle; sheep; pigs; beef; pork; dairy products; fish; shrimp; balsa wood
Guyana	sugarcane; rice; edible oils; beef; pork; poultry; shrimp; fish
Paraguay	cotton; sugarcane; soybeans; corn; wheat; tobacco; cassava (manioc; tapioca); fruits; vegetables; beef; pork; eggs; milk; timber
Peru	artichokes; asparagus; avocados; blueberries; coffee; cocoa; cotton; sugarcane; rice; potatoes; corn; plantains; grapes; oranges; pineapples; guavas; bananas; apples; lemons; pears; coca; tomatoes; mangoes; barley; medicinal plants; quinoa; palm oil; marigolds; onions; wheat; dry beans; poultry; beef; pork; dairy products; guinea pigs; fish
Suriname	rice; bananas; seabob shrimp; yellow-fin tuna; vegetables
Uruguay	Cellulose; beef; soybeans; rice; wheat; dairy products; fish; lumber; tobacco; wine
Venezuela	corn; sorghum; sugarcane; rice; bananas; vegetables; coffee; beef; pork; milk; eggs; fish

Source: Central Intelligence Agency; The World Factbook; 2019

and 24 degrees Celsius), and plantation agriculture abounds. Plantations are huge commercial farming operations that grow large quantities of crops that are usually sold for export. Because of their easy access to port facilities, coastal lowlands historically have been linked with the markets of Europe and North America.

The banana is one of the best-known examples of a plantation crop. It grows well in the wet, hot climate of this altitude zone and has been cultivated here for U.S. and European markets since 1866. In 2017, more bananas were traded on the international fruit market than any other commodity except for watermelons. North America, the third-largest importing market for bananas, brings in more than 8 billion pounds every year. The closeness of the banana-growing regions to the markets of North America is good for U.S. and Canadian consumers, since the bananas can reach the market quickly, eliminating spoilage owing to travel. In South America, Ecuador and Brazil are the leading banana producers, with Colombia third.

A vineyard in Luján de Cuyo, province of Mendoza, Argentina. Argentina is the fifth-largest producer of wine in the world. The most important wine regions of the country are located in the provinces of Mendoza, San Juan and La Rioja. The Mendoza province produces more than 60 percent of the Argentine wine and is the source of an even higher percentage of the total exports. (Image by Juan Pelizzatti)

Cacao, the bean pods from which cocoa and chocolate are made, are also grown on plantations in this zone. Brazil and Ecuador are sixth and seventh in world yearly production. (The largest producing area for cacao is Côte d'Ivoire in West Africa.)

Although sugarcane is grown in almost every South American country, it does especially well as a plantation crop in the lowlands of eastern Brazil, the world's largest exporter. There the crop is not used just to produce sugar but also to produce gasohol, an alcohol-based gasoline that is the mandated fuel for automobiles in Argentina, Brazil, Colombia, Paraguay, and Peru. This type of commercial agriculture has made agricultural businesses the third-largest component of the Brazilian economy.

Yams, cassava (manioc), and other root crops used as staple foods grow well in this humid, hot climate. Cassava is a highly productive native crop of the region, containing more starch than potatoes. Cassava root can be made into bread and tapioca. Some rice is also grown in this zone.

The Tierra Templada

Just above the *calienta* zone lies a zone of cooler temperatures that extends to about 6,000 feet (1,850 meters). Temperatures there range from 65 to 75 degrees Fahrenheit (17 to 22 degrees Celsius), and the commercial crop that dominates the landscape is coffee. It is grown on large plantations called *fazendas*. Brazil, the largest South American coffee producer, exports about one-third of the world's coffee, producing nearly 2.7 million metric tons annually. Colombia, South America's second-largest coffee producer, produces about 810,000 metric tons per year. Coffee was once the leading export from Colombia but, as a result of a coffee-worm infestation and lower world prices, other products have taken the lead.

Other commercial crops from this zone include fresh fruit. The central valley of Chile contains grape vineyards and apple orchards. Fresh produce from this region enters stores in the United States and elsewhere as the growing seasons of the domestic products are finishing up. Chilean grapes, apples, peaches, and plums are sold worldwide. Fine vineyards and low labor costs have helped make Argentina and Chile major players in the world's wine industry. Argentina is the leading producer of wine in all of South America and ranks sixth in the world in wine production. Chile is the world's fifth-largest exporter of wine and its seventh leading producer.

Corn and wheat are also grown in this zone. These staple foods are produced for local consumption and sold only at local markets. However, corn production is common throughout South America. In 2017, Brazil and Argentina ranked third and fourth, respectively, in world corn production, but most of this was for domestic consumption.

In Colombia and Ecuador, certain higher-elevation areas have become important centers of flower production, thanks to both climate and low labor costs. Ecuador is now the world's leading producer and exporter of roses. Colombia, with a floral industry dating back nearly a half-century, ranks second among the world's flower exporters.

In 2012, the United States and Colombia signed a free-trade agreement that eliminated tariffs on flowers, therefore lowering the cost of Colombian flowers to U.S. consumers. As a result, Colombia is now the source of approximately 80 percent of all flowers imported to the United States (the U.S. imports 82 percent of its flowers).

The Tierra Fria

This is the cold land that extends from 6,000 feet (1,850 meters) to about 15,000 feet (4,570 meters). Average temperatures there range from 55 to 64 degrees Fahrenheit (12 to 17 degrees Celsius). This zone exists throughout the Andes Mountains and can maintain only such plant life that can withstand the limited soils and the cold climate conditions. In this region of subsistence-type farming, crops and animals are grown and raised mostly for family use.

In the lower reaches of this zone, barley grows. Because it requires a short growing season, it grows well at high elevations. In the cooler portions of this zone, the potato evolved. The potato of the Andes is much smaller in size than the potato common in the United States, but they are relatives. A tuber, the potato can flourish in cold conditions with moderate moisture. The loose soils of this zone are perfect for its production, but it requires a considerable amount of cultivation. Although the potato is used throughout South America, potato production for the continent constitutes only a tiny percentage of the world total production.

Throughout Peru, Bolivia, and Chile, flocks of alpaca are commonly found at elevations around 12,000 feet (3,650 meters). Each year the alpaca is shorn for its fine wool, which is naturally white but can also be black or brown. The fibers are excellent for sweaters and other clothing items. The llama is mainly found in Peru, but its popularity has spread around the world. Llamas make excellent pack animals and also produce a long, smooth coat that is sheared and used for clothing.

Near-Term Outlook

South America has the potential to develop its agricultural economy even further if it refocuses on dealing with its internal infrastructure challenges. This is a continent of economic contrasts. Can rain forests and cattle ranches find common ground? Can the cities survive the influx of people from rural areas? What about the land the migrants leave behind? The challenge in South America is to find a way to unite the agricultural potential of the land with wise usage and the development that is imminent.

M. Mustoe

CENTRAL AMERICA

Agriculture is generally understood to be concerned with the production of food; however, in Central America, ornamental plants and flowers, forest products, and fibers are also important agricultural commodities. In 2019, the agricultural sector employed about 22 percent of the available labor force in Central America, much of which is engaged in subsistence agriculture. This percent-

age is also higher than those of the neighboring developing countries of Mexico (13 percent) and Colombia (16 percent). The Central American percentage is far higher than those in more developed countries, such as the United States and Canada, each of which is below 3 percent. The percentage of suitable land (9 percent) is less than that in Mexico (12 percent) but significantly more than in Colom-

SELECTED AGRICULTURAL PRODUCTS OF CENTRAL AMERICA

bia (1.5 percent). Arable land in the United States is about 17 percent.

Early Agriculture

Considerable archaeological evidence supports the existence of sedentary agriculture in the region for more than 2,000 years. The early Maya farmed raised fields in lowland swamp areas and constructed irrigation systems in areas with a dry season. In highland areas, steep slopes were terraced. The most prominent terrace agriculture in the Americas was developed in the Andean cultures, but Central Americans also used this practice. Agriculture was based mainly on maize (corn), but other crops were widely grown, including squash, beans, and chili peppers. Nonfood crops such as cotton and tobacco were grown for both domestic use and trade. These two crops continue to be important.

Exactly what group of Central Americans established the various agricultural practices, or when, is debatable. However, it is known that agriculture supported large communities of people early in the first millennium. The cities of Tikal, Copán, Caracol, and others had populations of 35,000 or more.

Raised field agriculture had several benefits. The muck dredged from channel bottoms was added to the fields, raising the surface above the surrounding swamp, creating dry land. This material was rich in nutrients from decaying plant matter and wastes from fish and other aquatic creatures. Channels of water dividing the dry land provided habitats for fish and turtles, which supplied protein to the diet.

Slash-and-burn agriculture was the process of stripping forests and burning the debris in place. Trees too large to be cut with primitive stone implements were girded, that is, a circle of bark was removed from around the tree and the tree left to die. The burned debris added nutrients to the top two to three inches (5 to 8 centimeters) of soil. Because the soil was generally poor, the fields, usually known as *milpas* or cornfields but sometimes referred to as swidden, were abandoned after two or three years of production and left fallow for up to twenty years. This process is still practiced today.

Intercropping or polyculture was a practice that helped ensure a harvest. The planting of several crops and different varieties provided a harvest even if one crop failed because of climate, insect, or disease problems. This practice is also in use today.

Traditional and Nontraditional Crops

Certain crops are raised in Central America as export crops and others principally for domestic consumption. Many of the so-called traditional crops grown are not native to the region. They are termed exotics, that is, plants not native to the region but

LEADING AGRICULTURAL PRODUCTS OF CENTRAL AMERICAN COUNTRIES

Country	Products
Belize	bananas; cacao; citrus; sugar; fish; cultured shrimp; lumber
Costa Rica	bananas; pineapples; coffee; melons; ornamental plants; sugar; corn; rice; beans; potatoes; beef; poultry; dairy; timber
El Salvador	coffee; sugar; corn; rice; beans; oilseed; cotton; sorghum; beef; dairy products
Guatemala	sugarcane; corn; bananas; coffee; beans; cardamom; cattle; sheep; pigs; chickens
Honduras	bananas; coffee; citrus; corn; African palm; beef; timber; shrimp; tilapia; lobster; sugar; oriental vegetables
Nicaragua	coffee; bananas; sugarcane; rice; corn; tobacco; cotton; sesame; soya; beans; beef; veal; pork; poultry; dairy products; shrimp; lobsters; peanuts
Panama	sugarcane; rice; edible oils; beef; pork; poultry; shrimp; fish
Paraguay	bananas; rice; corn; coffee; sugarcane; vegetables; livestock; shrimp

Source: Central Intelligence Agency; The World Factbook; 2019

introduced by European settlers. Bananas, coffee, and sugarcane are the three principal exotic crops, with corn being a fourth. Most of the production of introduced plants is intended for export, although native corn is for local use.

Bananas are grown extensively in the Caribbean and Pacific lowlands, but most prominently in the Sula Valley of Honduras, a leading world exporter of this fruit. The banana industry flourished under the control of North American growers, especially the United Fruit Company. Beginning in the latter half of the nineteenth century, the banana export business grew and enjoyed large markets in the United States and Europe. The United Fruit Company also exerted strong influence over governmental policies in the region. (United Fruit became United Brands in 1970, and in 1984, United Brands was transformed into Chiquita Brands International, based in Switzerland.)

Hurricanes are a constant threat to the banana crop. Hurricane Mitch, in late 1998, devastated the banana plantations in Honduras. Chiquita Brands and Dole Fruit replanted, and the industry recovered by 2001.

Coffee beans being sorted and pulped by workers and volunteers, on an organic, fair-trade, shade-grown coffee plantation in Guatemala. (Image by rohsstreetcafe)

Coffee is grown extensively in the highland areas of all seven Central American countries. A slow-ripening crop, coffee requires as long as two months to harvest. Small-scale growers who sell their product through cooperatives produce much of the area's coffee. The best-quality coffee is shade-grown, and so banana trees often are interspersed throughout the small fields.

Sugarcane, first introduced by Christopher Columbus to the island of Hispanola, is another plant grown over a wide area. Sugarcane is labor-intensive during harvest but requires little attention at other times. The harvest of sugarcane begins with the burning of the fields. This practice reduces the volume of foliage and leaves standing only the stalks, or canes, which are the source of sugar. After the burning—which has the side benefit of chasing out the snakes that inhabit the cane fields—teams of workers with machetes march through the fields cutting the cane.

Corn is not grown for export. Along with regionally grown rice, it is for domestic consumption. Corn meal is used in the preparation of tortillas, which are eaten at nearly every meal. Rice is commonly served with red or black beans.

Export crops have had peaks and valleys in their economic value to the region. A disease of banana plants nearly ruined the industry in the 1930s. The Great Depression in that same decade sharply reduced exports to North America. Import quotas imposed by the United States on sugar and the U.S.

Banana "tree" showing fruit and inflorescence. An inflorescence is a group or cluster of flowers arranged on a stem that is composed of a main branch or a complicated arrangement of branches. (Image by Daniela Kloth)

ban on importing Cuban sugar provided both a low and a high for Central American sugar producers. Overproduction of coffee by South American producers has led to depressed prices several times. Free-trade agreements with the United States and among the nations of Central America have helped boost the local economies.

Nontraditional Crops

Vegetables, high-value crops, and ornamental plants and flowers are being grown at an increased rate. The leading crops are bananas, coffee, cacao, broccoli, cauliflower, snowpeas, melons, strawberries, and pineapples. Palm oil from a nonnative tree is another high-value farm product. Nontraditional crops are labor-intensive and also affect the environment because of the heavy requirements for chemical insecticides and disease control. There are health risks to workers because of these chemical applications, but employment is high. In Costa Rica, the government encourages investment in reforestation using teak from Southeast Asia.

Livestock

Every country in Central America raises cattle, with swine the second-ranking livestock. Cattle are mostly range-fed; as a result, the beef is of inferior quality compared to North American beef. Most cattle are improved breeds of zebu, a type of domes-

---

CACAO: A FAVORITE CENTRAL AMERICAN CROP

Cacao beans, from which chocolate is made, have been cultivated in Central America for centuries. Once considered the drink of the gods, chocolate was reserved for royalty. Today, millions around the world enjoy chocolate, especially mixed with sugar and milk.

The beans are actually the berries of the small cacao tree, which grows in shade and is rarely more than 20 feet (6 meters) tall. The tree's football-shaped pods are 6–8 inches (15-20 centimeters) in length. When ripe, the pods can be red, yellow, or orange, depending on the variety. The tree's flowers are tiny, inconspicuous white blossoms that emerge singularly from the lower branches or trunk, not from stem ends. The cacao seeds, or beans, are surrounded by a whitish, gelatinous mass. Cacao beans must be fermented, dried, and cleaned before the chocolate aroma and taste develops.

---

tic ox, with some dairy herds of other breeds. The dominant area of cattle production is the Guanacaste region of Costa Rica and generally the western slope of the Pacific highlands. Once forest is removed for timber or subsistence agriculture, the area frequently transitions to pasture after a few years.

Donald Andrew Wiley

INDUSTRIES

SOUTH AMERICA

South America is well known for the number of minerals and other natural resources existing there. While it lacks the variety found in North America and Russia, it nevertheless has strategic and historically important reserves of silver, petroleum, and other minerals.

Silver

During Spanish colonial rule, silver was a major export from the region. Today, Peru ranks first in South America and third in the world (after Mexico and China) in the production of silver. Peru's Cerro de Pasco and surrounding areas have produced a large portion of the world's silver from colonial times to the present.

Perhaps the most famous South American source of silver is the Potosí mine in Bolivia. During the sixteenth century, Potosí produced nearly half the world's total supply of silver. Bolivia remains an important source of silver, and now ranks seventh in the world. The more recent discovery and exploitation of silver deposits in Chile have moved Chile into sixth place among silver-producing nations.

Labor for silver mines was hard to come by in colonial times. Working conditions were poor, the work itself dangerous, and diseases ran rampant in the mines. Native Peruvians were not willing to work as wage laborers in great numbers. As a result, colonial mine managers and owners relied on coercion to draw enough laborers. The system used was known as the *mita*, or "turn," whereby local villages were required to supply workers to take their turn in the mines.

Gold and Diamonds

The other major precious metal, gold, is not as common in South America as is silver, although southeastern Brazil and western Colombia were originally settled because of the presence of gold. Brazil also mines a considerable amount of diamonds.

The largest area in which gold was found was in southeastern Brazil. Gold was discovered in the highlands of São Paulo and Minas Gerais states in the late sixteenth century by Paulista *bandeiras*, bands of men from São Paulo who were trying to capture native peoples to sell as slaves. Larger deposits were later discovered in Rio de Janeiro state, leading to a gold rush in the early eighteenth century. Further discoveries to the west in Goiás and Mato Grosso do Sul states, and the discovery of diamonds in the same areas, helped establish southeastern Brazil and particularly the city of Rio de Janeiro as the center of Brazilian wealth and settlement.

A minor gold deposit was found in the 1530s in the Cordillera Occidental in Colombia, near the city of Medellín, leading to settlement of that area. Gold and diamonds are also found in the Amazon rain forest in northern Brazil and the Guianas. The exploitation of these reserves has led to significant deforestation and conflicts between independent miners, or *garimpeiros*, and native peoples of the region.

Industrial Minerals

Precious metals and stones are valuable in their own right. Modern mining of industrial minerals, however, depends on factories to process the minerals, since iron ore, petroleum, and bauxite are not useful in their natural states. Some of the most important industrial minerals in the world today are found in large quantities in South America. Many mines are owned and operated by North American, European, and Asian investors with the main purpose of exporting these minerals to overseas markets.

Petroleum is essential for modern industry, since it is the raw material for the production of fuels, plastics, and other petrochemicals. The world's largest petroleum reserves are found in Venezuela in and around Lake Maracaibo, offshore in the Orinoco Oil Belt, and in the sparsely settled eastern part of that country. Venezuelan petroleum production peaked in the 1960s and again in 1977 and in 2006, before falling dramatically due to Venezuelan government policies. By 2018, oil production in Venezuela was less than half the 2006 level. Petroleum nevertheless remains the largest Venezuelan export and has been the basis for development policy in that country. Other large reserves are found off the southeast coast of Brazil. Smaller reserves of petroleum are found in Brazil, Colombia, Ecuador, and Argentina.

Tin traditionally has been associated with South America. Only Asia as a continent produces more tin than South America. South American mines produced just under 20 percent of the world's tin in 2017. Peru is the largest producer in the region and third in the world, after China in the total number of tons mined. Bolivia and Brazil hold the fourth and fifth spots, respectively, Bolivia relies on tin exports more than does Brazil. The Brazilian mines are concentrated in the Amazon River valley and the state of Rondônia in the far west. Pitinga, thought to be the largest underdeveloped tin deposit in the world, is located in Amazonas state in Brazil. Elsewhere in South America, deposits are found in Bolivia, primarily in the region between La Paz and Sucre; in southern Peru; and in northern Argentina.

Bauxite, from which aluminum is made, is another major mineral found in South America. The leading South American producer and the fourth-largest in the world is Brazil. Bauxite mining is concentrated in the north, especially in Pará state. The largest mine is found at Trombetas, about 150 miles (240 km.) northeast of Manáos. Bauxite also is mined in Suriname, Guyana, and Venezuela.

Tuyeres of Blast Furnace at Gerdau, Brazil. A tuyere *or* tuyère *is a tube, nozzle or pipe through which air is blown into a furnace or hearth. (Image by Mohanrajnp)*

Many other important minerals are found in South America. Copper-mining is important in northern Chile and Peru, the world's leading copper-producing countries. Iron ore can be found in Brazil, which is the third-largest producer in the world (after China and Australia), especially in Minas Gerais state and in Serra dos Carajás in Pará state; iron also is mined in Venezuela and Colombia. Coal-mining is an important industry in Colombia (the world's ninth-biggest producer), and to a lesser extent in Brazil, Chile, Peru, and Argentina. Uranium reserves are found in Brazil, Argentina, and Paraguay, although they have not been exploited to any great extent. Finally, phosphates and nitrates for use as fertilizers can be found in coastal Peru, the Atacama Desert in Chile, and in the Pampas in Argentina.

Manufacturing

Modern economies are dependent on manufacturing. In the manufacturing process, minerals and agricultural products, most of which are useless in their original state, are turned into useful goods that people can consume. Because of this, manufacturing is said to add value to raw materials by making them into finished products, which are worth more than the raw materials alone.

South American industry traditionally has been dominated by the exportation of minerals and raw materials and the importation of finished products. Since most of the value of products is added in the manufacturing process, and since the value of manufactured products has risen as compared to raw materials, South American economies usually have been at a disadvantage in terms of trade—if they sell inexpensive products and buy expensive products, they get less back than they sell. This disadvantage became more pronounced during the Great Depression, when the demand for South American agricultural products and minerals declined in Europe and North America.

One strategy for overcoming this disadvantage after World War II was the encouragement of import-substituting industrialization (ISI). Import substitution means replacing previously imported finished products with domestically produced alter-

natives. This was achieved by placing higher tariffs (taxes on imports) on finished products than on raw materials or intermediate goods. Producers were encouraged to import parts and machinery and do final assembly in South American countries, the goal being to provide highly paid manufacturing jobs.

ISI was largely unsuccessful in bringing development to South America and eventually was abandoned for several reasons. First, the domestic industries were too small and inefficient, and assembling the finished products was more expensive than if done by foreign counterparts. This left South American consumers worse off than they had been before ISI. Second, foreign manufacturers moved only the final assembly work to South America, bringing only a few low-paying jobs. Third, since the manufacturing plants that were built required heavy investment in machinery that had to be imported, South Americans found themselves unable to purchase other imports.

Export-Oriented Manufacturing

South American governments have also encouraged export-oriented manufacturing (EOM). Traditionally, it has been difficult for South American economies to export manufactured goods to the industrialized countries of North America, Europe, and Japan. South American firms could not compete with the efficiency and productivity of North American, European, and Japanese firms. Also, the types and quality of goods that the people of South America demanded were different from the types and quality of goods that North Americans, Europeans, and Japanese people demanded. Thus, it was difficult for South American economies to develop based on manufacturing, since they could not gain access to major markets.

Two EOM promotion strategies were followed by South American governments beginning in the late 1950s. The first strategy was intended to address a significant challenge: North Americans and Europeans did not purchase the same sorts of goods as Brazilians or Peruvians. Since most South Americans demanded similar types and qualities of goods, if a factory produced for domestic consumption, it could also export its output to places that demanded similar goods. South American governments, therefore, encouraged trade with other South American economies by forming trade agreements and common markets. For example, a factory in Argentina could increase its market beyond the relatively small Argentine population by selling also to Chileans and Venezuelans, becoming more efficient in the process. Although there was some progress, this strategy was not successful in the long run, as disagreements about how to coordinate these development efforts emerged between national governments.

The second and more successful strategy addressed the problem of competing with more industrialized economies. Since South American firms were not as efficient as North American and European firms, they had to concentrate on manufacturing methods that took advantage of their strengths. Wage rates in South America have been significantly lower than those in North America, Europe, or Asia. Therefore, South American firms attempted to use more workers and fewer machines in their factories than a U.S. or German factory would.

South American firms also concentrated on manufacturing goods that needed a great deal of labor in the manufacturing process, especially less-skilled labor. Thus, South American manufacturers have concentrated on producing things like textiles and apparel (especially footgear and underwear), radios and televisions, automobile parts, and toys. Usually, EOM factories use imported parts, and only assembly takes place in the South American factory. The goods produced in many of these factories are too expensive for the local people and workers to purchase. Most EOM goods are produced for export to the United States and Europe.

Manufacturing Regions

Despite the plentiful natural resources, South America as a whole has not emerged as a major manufacturing region. Certain regions of South America have developed significant manufacturing, and industry remains concentrated in those regions. The rest of South America relies on agriculture and mining. One reason for this is that in the colonial period, manufacturing was discouraged in South America by the Spanish and Portuguese governments. Since independence from Spain and Portugal in the first half of the nineteenth century, South American industrial development has been relatively slow and highly concentrated.

Southeastern Brazil

Brazil is the dominant manufacturing country in South America, producing more than one-third of the total manufacturing output of the continent. The largest area of manufacturing is in the southeast, centered in the cities of São Paulo, Rio de Janeiro, and Belo Horizonte, and the areas in between them. São Paulo by itself is home to more than half of all Brazilian manufacturing.

Before the 1950s, there was little manufacturing in Brazil. Since then, the southeast region has emerged as the center of Brazilian industry for historical and geographical reasons. Historically, the coffee industry, a tremendous wealth-generator for Brazil, has been concentrated in São Paulo state, providing a significant amount of financial capital for investment in factories and also providing Brazil's earliest and most highly developed transportation system. That, combined with the eighteenth century gold rush in the area and the selection of Rio de Janeiro as Brazil's capital (before the capital was relocated to Brasília), tended to concentrate wealth and population in this area. This area is home to a large population that provides both labor and consumers. Southeastern Brazil also contains important natural resources such as iron ore, and electricity is cheaper and more easily available in this area than in other parts of Brazil.

As a result, southeastern Brazil is home to many industries, both heavy and light. Primary among heavy industry is steel and iron production. Brazil is a significant world producer of steel and pig iron, producing about 1.1 percent of the world's steel. Given its extensive iron ore reserves, Brazil has the potential to greatly expand that share.

Brazil is home to the world's sixth-largest automobile market based on new car registrations. In 2017 alone, the Brazilian automotive industry gen-

Nuclear power plant of Angra 1 and Angra 2. (Image by Rodrigo Soldon)

erated a revenue in new car sales of approximately US$60 billion. Brazil also ranks among the top ten countries for automotive aftermarket and parts sales. The industry produced more than 2.5 million cars in 2018, and employed more than 130,000 workers.

South America's aerospace industry is almost entirely centered in Brazil, thanks in large part to Embraer, an aerospace conglomerate headquartered in Sao Jose dos Campos. Embraer, which produces commercial, military, executive, and agricultural aircraft, is the third-largest aircraft manufacturer in the world, after Boeing and Airbus. In 2019, Embraer's stockholders approved a joint venture with Boeing to be called Boeing Brasil-Commercial. Details about branding and other issues are still being worked out.

Brazil has a small space industry. The Agência Espacial Brasileira (AEB) oversees its space projects, including satellites and their applications as well as related research and development. In 2006,

a Brazilian astronaut, Marcos Pontes, the only South American citizen to travel into space, spent 10 days on the *International Space Station*.

Since 1968, France has operated the Guiana Space Centre in Kourou, French Guiana. Since 1975, France has shared the facility with the European Space Agency (ESA). The spaceport has been the site of dozens of successful launches through the years.

Finally, chemical and pharmaceutical industries are becoming more important in the region. Light industries in São Paulo, Rio de Janeiro, and Minas Gerais states include textiles, apparel (especially footwear), food processing, and light consumer goods.

Río de la Plata

Manufacturing in the La Plata River area (including Buenos Aires in Argentina and Montevideo in Uruguay) began as an outgrowth of the agricultural exports from the fertile, productive Pampas region.

Argentina and Uruguay are well known for sheep and cattle ranching. These animals are in great demand in Europe and Japan as food. It is difficult to ship live animals, so the meat must be prepared for export. Industry developed around the mouth of the Paraná River, where the agricultural goods would have been loaded for export. Animals first were driven to Buenos Aires or Montevideo or later brought by rail, butchered in slaughterhouses, and prepared for export. In the beginning, the meat was salted in *saladeros*, but with the invention of refrigeration in the 1870s, it now can be shipped frozen. Other related industries that emerged in this time included leather goods and textiles.

In the 1950s and after, import-substituting industries were established in response to the urging of long-time Argentine leader Juan Perón, who was president from 1946 to 1955, and again in 1973 and 1974. Consumer appliances, such as refrigerators, stoves, and radios, began to be produced in Argentina, primarily in Buenos Aires. More recently, industry has begun to decentralize from Buenos Aires proper into its suburbs, although it remains highly concentrated in the Pampas region. Most of the manufacturing in the country is conducted in industrial parks, chiefly based in the Buenos Aires area.

CLEANING UP CUBATÃO

Cubatão, Brazil, is an industrial town 49 miles (80 km.) from São Paulo. In the mid-1980s Cubatão's air stank and residents complained of headaches and nausea. Each year, two dozen factories released a quarter-million tons of toxins into the air. In nearby shantytowns one-third of the babies died before their first birthdays and eight out of ten children had respiratory problems. In 1984 an oil pipeline exploded, killing more than 500 people. The following year, there were mass evacuations when an ammonia pipeline broke. In the face of this environmental nightmare, Brazil passed stricter pollution laws. By 1990, factories in Cubatão had cut emissions by more than three-fourths. Although conditions in the city have improved, many large industries continue to work in such a small area. It is likely that there will always be some pollution in the soil and groundwater.

Today, manufacturing represents about 15 percent of Argentina's economy. More than half of the country's industrial exports are agriculture in nature. Argentina has a thriving automotive industry, producing around 800,000 cars per year. Other products include chemicals and pharmaceuticals, electronics and home appliances, plastics and tires, and textiles, among many others. The far-southern city of Ushuaia dominates the country's production of computers, laptops, servers, and cell phones.

Uruguay continues to make its mark in a number of other areas, aside from its traditional agricultural industries. The country has emerged as South America's most important manufacturer of plastics products. Thanks to its well-educated workforce, Uruguay has become a player in the information technology (IT) field, most notably for software and consulting. Tourism and related services account for more than 10 percent of Uruguay's economy. And its stable banks and financial sector have helped Uruguay earn the nickname "Switzerland of the Americas."

Central Chile

Chile was more dependent on just a few commodities—copper and nitrates—than were most South American economies and thus was harder hit by the Great Depression than most places. As a result, the Chilean government encouraged industrialization in the 1930s and 1940s, earlier than did most governments in South America. Chile now has several well-established industries, including textiles and apparel, construction materials, and food processing. Some attempts at establishing industry have failed, for example, building an automobile assembly industry in the middle of the desert in the north of the country.

Today, Chile is considered South America's most successful economy. Mining represents about 60 percent of the country's exports. Chile is noted for its strong services and tourism industries.

Other Regions

Manufacturing is found in several other regions of South America. Northern Venezuela is dominated by petroleum refining and petrochemicals as a result of its plentiful petroleum in the Lake Maracaibo region. It also has seen the rise of manufacturing for the domestic market in the region from Caracas westward to Valencia. But by the mid-2010s, the country's manufacturing sector has experienced extreme difficulties amid the country's dramatic economic collapse.

Colombia has been manufacturing home appliances for nearly 90 years, but only in the past few decades has been exporting them to its neighboring countries. Some foreign companies, including Whirlpool and General Electric, manufacture various lines of refrigerators and other appliances in Colombian factories. Colombia holds the distinction of being the only country in South and Central America to domestically produce a 4K television set, and it's second only to Brazil in overall production of appliances.

Colombia is now the second-leading producer of electronics manufactured by domestic companies in all of South and Central America. Today, these companies have vigorously worked to create strong export markets for their products. The Colombian government enthusiastically supports these efforts. Its investment in technology education and innovation centers around the country serves to recognize the great potential of the high-tech industries.

Colombia's longstanding food-processing industry is concentrated around Cali. Textile production is centered near Medellín. In Peru, similar industries are found near Lima.

Timothy C. Pitts

CENTRAL AMERICA

Central America has exported raw materials and agricultural products to Europe and other countries in the Western Hemisphere from its earliest colonial days. Its warm weather and rich volcanic soil make the isthmus an ideal environment for raising crops and livestock. Its traditional agricultural exports—coffee, bananas, cotton, and shrimp—are still major sources of income for each of the seven Central American independent countries.

Historical Perspective

In the 1950s, the United Nations Commission for Latin America, led by Raúl Prebisch of Argentina, suggested that groups of countries such as those in Central America should reorganize their economies. The commission recommended a plan to develop local industry and establish trade regulations designed to help one another. Private and public investment in industry in each Central American country would be encouraged, and tariff walls would be erected around the region to protect incipient local industrial development. The result would be an economic trading unit with a population base in the tens of millions rather than one restricted to that of a single country.

Five Central American countries—Guatemala, El Salvador, Nicaragua, Honduras, and Costa Rica—formed the Central American Common Market (CACM), designed to integrate the economies of its members. In 1960 planners began establishing local industries to take advantage of the larger Central American market by reducing or eliminating tariffs among its members. The member countries, at the same time, established tariff barriers on goods from outside Central America that would compete with the newly formed domestic economic units. This process is known as import-substituting industrialization (ISI). The initial agreement called for free trade within the group of 239 Central American products, as well as a ten-year phase-in on all the remaining goods produced in Central America. The largest group of commodities that began to flow among the members was consumer goods, mostly processed foods.

During the 1960s and 1970s, trade within the CACM increased dramatically. Unfortunately, most of Central America's industrial development remained in the consumer goods sector and did not expand into the area of extraregional exports designed to compete in global markets, initially a major goal for the CACM.

While the tariff walls set up to protect local industry from foreign competition increased trade among the CACM members, they failed to provide any incentive for Central American industry to break into established markets outside of the isthmus itself. Both the price and quality of their goods faced real competition in open world markets. Capital goods required to establish local industries had to be imported from abroad as well. Again, the countries in the region faced a stagnation of their respective economies as the market within the CACM became saturated, and local industry failed to enter the more competitive foreign markets. Only the agricultural exports, broadened to include

cotton and seafood, proved attractive to buyers from outside of the CACM.

By the 1980s, various tensions and conflicts among CACM's members led to the suspension of the group's activities. The CACM was resurrected in the 1990s as the Central American Integration System (SICA), which has helped expand regional trade. In 2004, SICA members signed a free-trade agreement with the United States, further promoting the isthmus's products.

Globalization has resulted in the expansion of foreign industry into Central America. *Maquiladoras,* in effect assembly plants for unfinished goods, brought into the area by foreign entrepreneurs, have been established throughout the region to take advantage of an increasingly large labor pool of skilled workers willing to perform at competitively low wages. Foreign manufacturers have also received favorable tax treatment for the goods assembled in Central America. The arrangement reduces the rate of unemployment locally, decreasing, to some degree, the pressure on Central American governments to solve local job shortages.

Industries Today

Despite these efforts, by the early 2020s only two Central American countries—Panama and Costa Rica—can be said to have diversified industries. And of these two countries, only Costa Rica, with the strongest economy in Central America, has been able to translate its economic performance into a general prosperity for its people. In Panama and throughout the remainder of Central America, income distribution remains very uneven, with large percentages of each country's population living below the poverty line.

Panama's operation of the Panama Canal continues to be a reliable source of revenue, with the United States and China being the principal users. Both Panama and Costa Rica have established enormously successful free-trade zones (FTZs). Prominent foreign companies have established facilities in these FTZs, producing pharmaceuticals, medical equipment, computer components, building materials, and dozens of other products for export.

THE INTRODUCTION OF COFFEE TO EL SALVADOR

After a synthetic indigo dye was developed by German scientists in the mid-1800s, El Salvador looked for a new cash crop to replace indigo, on which the country's agriculture had been based. The coffee bean had already been adopted in Costa Rica, demonstrating that the fertile soil of the Central American isthmus provided an ideal environment for coffee cultivation. The introduction of the coffee bean in El Salvador in the late 1800s changed the whole economic, political, and social structure of the country. It gave rise to a new economic elite and destroyed the system of subsistence farming carried on by the country's majority indigenous population. The tiny country, together with its larger Central American neighbors, became principal participants in the global coffee economy, reaching a peak in 1980 with a revenue of more than US$615 million. Coffee exports, once the backbone of El Salvador's economy, have fallen by more than half since 2010, due to political and economic turmoil. Today, some growers are hoping that a focus on specialty coffee trees will help the industry rebound.

Agriculture plays a significant role in the economies of all the Central American countries. This has led to small-scale food-processing industries throughout the isthmus. Forest products remain important, especially in Honduras and Belize. Nicaragua has a small but growing gold-mining industry. Apparel and textile production is important throughout the region.

Tourism has almost infinite potential—not surprising given the isthmus's vast stretches of fine beaches and nearby rain forests. On both coasts, substantial investment has already been made in resorts and other tourist facilities to draw visitors from North America, Europe, and Asia. Costa Rica, in particular, has been a pioneer in the development of ecotourism as an industry. Throughout Central America, it is widely recognized that the tourism industry will benefit when more Central Americans gain fluency in English and other non-Spanish languages. Tourism and, indeed, virtually every industry, would benefit from improved transportation and communication infrastructure.

Carl Henry Marcoux

DISCUSSION QUESTIONS: INDUSTRIES

Q1. What role did mineral resources play in drawing early European explorers to South and Central America? Which metals remain important to the local economies? How has the discovery of petroleum reserves transformed various countries?

Q2. What is meant by export-based manufacturing? How does it differ from import-substituting industrialization? Compare and contrast the two strategies. What roles do they now play in the South and Central American economies?

Q3. Which countries in South and Central America are considered to be the most industrialized? How did these countries achieve this status? In which countries does the economy depend primarily on agricultural products?

TRANSPORTATION

SOUTH AMERICA

Waterways, highways, railroads, and airways are crucial facilitators of trade. Throughout the world, economic growth is almost always accompanied by an expansion of transportation systems. Unlike North America and Western Europe, South America does not possess fully integrated transportation networks. Goods and people do not move easily to and from the farthest corners of the continent, and many regions can be reached only by circuitous and lengthy routes. Sometimes, several changes in the mode of travel are required. A traveler setting off from the Brazilian metropolitan center of São Paulo to a small village of the Amazon region may first need to travel on an airplane, then ride a bus, switch to a horse cart, and complete the trek in a dugout canoe.

Historical Background

The earliest engineered transport systems in South America were footpaths maintained in the Andes mountains by the Incas. These routes included hanging bridges across deep gorges, stone stairways up steep slopes, and well-maintained roadbeds through some of the most difficult terrain in the world. The roads were used by couriers of the Inca emperor and were a vital mechanism by which he exercised his authority over a region that stretched from present-day Chile to central Ecuador. After the Spanish conquest, the Inca roads were abandoned. The Spaniards who conquered the Incas had little use for these routes connecting the Amerindian communities in the highlands. Their priority was to establish roads from the highlands to the coast.

The colonial period brought with it an economic system known as mercantilism. Under this system, the main function of the South American colonies was to supply Spain and Portugal with raw materials. The European nations, in turn, were responsible for providing the manufactured items that were needed in the colonies. All trade relationships, therefore, involved connections from each of the South American colonial areas directly to Spain (or Portugal for the Brazilian colony). During the sixteenth and seventeenth centuries, the South American colonies produced minerals (gold, silver, and gems) and agricultural items (sugar, spices, and tobacco) that were shipped to Spain and Portugal. The colonial areas produced commodities that were desired back in Europe, not items that could be exchanged with their South American neighbors. As a result, there was very little intracontinental trade.

The transportation routes constructed during the colonial period were designed to accommodate the flow of trade. Most of the mining areas in Spanish South America were located near the Pacific coast, so roads were built directly from the highlands and across the coastal plains to port cities. Minerals were brought down the mountains on the backs of mules from mining centers such as Cerro de Pasco and Potosí. When the coast was reached, the cargo was loaded aboard ships destined for Panama. There, the freight was unloaded and carried overland across the isthmus to the Caribbean shore, where it was shipped on vessels in large convoys across the Atlantic Ocean to Spain.

The Portuguese likewise built roads from the cane fields of the northeast and later from the gold mines of Ouro Preto in the southeast, directly to the coast. Port cities such as Recife and Rio de Janeiro flourished because of the activities linked to loading the sailing vessels destined for Lisbon. As in Spanish South America, trade and the associated economic activity prompted the building of roads and paths to link the interior to the coast.

Impact of Independence

For most South American countries, the independence movements of the early nineteenth century altered neither the pattern of trade nor the layout of transportation routes. South American republics continued to provide raw materials to the rest of the world, although not necessarily for Spain and Portugal. During the nineteenth and early twentieth centuries, for example, Brazil and Colombia became the world's major suppliers of coffee, Argentina and Uruguay produced meat and grain for Europe, Chile sent nitrate and copper to North America, Bolivia shipped tin, and Peru exported guano. In most areas of South America, this pattern continues to the present.

The majority of the countries derive revenue from exports and use the money to purchase needed manufactured goods from abroad. Only the larger economies—Argentina, Chile, and especially Brazil—have been able to develop a substantial internal industrial base and market for their products. In those countries, transportation networks that interlink important centers within the national territory have evolved. However, much attention is still focused on moving commodities to the coast.

Railroads

The earliest railroads in South America date from the first half of the nineteenth century. As with the colonial routes, the South American railroads were designed to facilitate the exploitation of export

Santiago Metro is South America's most extensive metro system. (Image by Osmar Valdebenito)

The La Paz cable car system in Bolivia is home to both the longest and highest urban cable car network in the world. (Image by EEJCC)

commodities. In mining areas such as northern Chile and mountainous Peru, a single line led to the sea; in predominantly agricultural regions such as southeast Brazil and the Pampas of Argentina, a more substantial network was needed because there the production areas were dispersed. Construction was financed from abroad, especially by British investors.

Several rail lines showcased imaginative strategies for overcoming the engineering difficulties posed by the mountainous terrain. Peru's line from Lima to Cerro de Pasco featured a series of sharp switchbacks, and the climb up the southeast Brazilian escarpment from Santos to São Paulo used a cable pulley system.

Later lines linked interior Colombia to the coast, and Bolivia to northern Argentina and Brazil. One of the most interesting railroad projects was the construction of a line that provided a detour around the falls of the Madeira-Mamoré route was built to facilitate the export of natural rubber from regions of northern Bolivia during the rubber boom of the

early twentieth century. The railway, which became known as the "Devil's Railroad," was never profitable, and its use was discontinued in the 1960s.

By the early twentieth century, dense rail networks had been laid out in southeast Brazil and in the Pampas of Argentina. Established to support the export of coffee from Brazil and meat and grains from Argentina, these lines also provided the urban centers along the coast with access to the interior. As the economies of these two countries evolved toward greater self-sufficiency, the rail lines helped support the needs of the manufacturing plants of São Paulo, Buenos Aires, and neighboring centers. For these countries, rail lines played an important role in both stimulating industrialization and facilitating commodity exports. Although in modern times much of the track of the networks has fallen into disuse (as in the United States), Brazil and Argentina still have the densest rail network in South America.

Even in these countries, the governments' active programs of highway construction and subsidies offered to truck manufacturing have diverted traffic away from the rail lines. Almost all the railroads in South America were run by government agencies; graft, corruption, inefficiency, and a lack of safety were constant problems. By the twenty-first century, most rail lines had been privatized with the hope that management operations would improve. Today, they are operated by a variety of public and private companies.

The South American railway sector may regain its importance if a long-discussed multinational plan for building a transcontinental railroad is put into place. If built, the Central Bioceanic Railway Corridor (Corredor Ferroviario Bioceánico Central) will run from Peru through Bolivia to Brazil, connecting the continent's Atlantic and Pacific coasts. Due to its projected trade benefits and impressive length—between 2,360 and 3,290 miles (3,800 and 5,300 km.)—this megaproject has been dubbed "the Panama Canal of the twenty-first century."

Highways

Many of the highways of South America occupy routes that date back to colonial and even pre-colonial times. During the nineteenth and early twentieth centuries, railroads took over the bulk of the freight traffic in some areas, but roads continued to support vital local and regional travel. Since the second half of the twentieth century, highway construction and traffic have been rapidly growing while the railroads came to play a secondary role. In some countries, such as Venezuela, railroads never achieved a major role, and highways have always been dominant.

In the densely populated areas around Caracas in Venezuela, the Cauca Valley of Colombia, the Guayas lowland of Ecuador, the Peruvian capital of Lima, middle Chile, the Pampas of Argentina, and southeast Brazil, dense road networks have evolved. Most roads in South America are dirt or gravel, but a significant percentage in the larger and more developed countries has been paved.

The construction of highways and the proliferation of vehicles has led to serious problems in most countries. Though paved, many routes are in dire

need of repair. Two-lane highways, clogged with slow-moving trucks and speeding cars operated by inexperienced drivers, combine to create serious safety hazards. South American countries have some of the highest accident rates in the world. Urban areas are greatly congested and polluted by the fumes from trucks, buses, and automobiles. Little urban planning has occurred, and vehicles often travel on roads that were built for pedestrian or animal traffic. Residents of the larger centers of Rio de Janeiro, São Paulo, Santiago, Caracas, and Buenos Aires have access to subway systems, but commuters in these and other cities may nevertheless spend hours traveling to their jobs. Highways provide the principal access to interurban public transportation for most South Americans. Buses serve most communities of the continent, and travel is relatively inexpensive. Many of the bus companies offer comfortable accommodations on overnight sleepers linking major cities. Brazil and Argentina have large transportation manufacturing industries that turn out a great variety of automobiles, buses, trucks, and motorcycles.

While most roads are built in response to economic or social need, over the past decades, several South American nations have used highway construction as a tool for economic development. The first of these efforts was the Pan-American Highway

Stretch of the Pan-American Highway in Argentina. (Image by Ale4110)

project, initiated after a conference of the American states in 1923. This highway, which links all the nations of the Western Hemisphere over a distance of 16,031 miles (25,800 km.), has been completed with the exception of a 54-mile (87-km.)-long stretch, called the Darién Gap, on the border between Panama and Colombia.

The Pan-American Highway has contributed greatly toward interlinking the nations of South America, thus modifying the historical down-to-the-coast orientation of the road and rail systems. Engineering difficulties and environmental concerns have delayed construction the Darién Gap, which is covered by mountainous rain forest on the Panamanian side and vast wetlands on the Colombian side. Governments also have used road-building projects to accomplish policy goals. Most of these routes lead directly into the rain forest, built to encourage economic development and establish sovereign control over remote territory. During the late 1960s, the government of Peru undertook the construction of the Carretera Marginal de la Selva (Jungle Border Highway), which runs along the eastern flank of the Andes. By 1968 it had been connected with other routes across the mountains, and the project opened large areas of the Amazon rain forest to settlement.

The monumental Interoceanic Highway, built between 2005 and 2011, links Peru's Pacific ports to Brazil's Atlantic coast. Spanning 1,615 miles (2,600 km.), it runs from the Pacific coastline across the Andes mountains and the Peruvian part of the Amazon rain forest to the Brazilian border, where it connects with a network of existing highways leading to the Atlantic.

From the late 1960s through the 1980s, the Brazilian government diverted a substantial amount of government funding to the construction of three highways: the Belém-Brasília, the Trans-Amazonian, and BR-364. Although initially little more than dirt tracks that were impassible during the rainy season, the now-paved routes have become heavily traveled by trucks and buses. The first ran directly north from Brazil's new (at the time) capital

SOUTH AMERICAN EXPORTS AND MAJOR MARKETS

Many South American economies rely on exports for the bulk of their activity. Compare how much they make and where their exports go.

Country	Total Exports (U.S. dollars)	Major Markets
Argentina	$61.5B	EU, 30%; Latin America, 23%; North America, 14%; Eastern Europe & former USSR, 12%
Bolivia	$8.9B	Latin America, 58%; EU, 22%; North America, 17%
Brazil	$239B	North America, 28%; EU, 28%; Latin America, 7%; Japan, 7%
Chile	$75.5B	EU, 37%; North America, 21%; Latin America, 13%; Japan, 13%
Colombia	$41.8B	North America, 45%; EU, 26%; Latin America, 16%; Japan, 4%
Ecuador	$21.6B	North America, 61%; Latin America, 18%; EU, 9%; Japan, 3%
Guyana	$3.6B	Latin America, 53%; EU, 20%; North America, 14%; Japan, 4%
Paraguay	$9B	Latin America, 49%; EU, 35%; other Western Europe, 8%; North America, 4%
Peru	$47.2B	North America, 31%; EU, 23%; Latin America, 18%; Japan, 10%
Suriname	$3B	EU, 39%; North America, 26%; other Western Europe, 18%; Latin America, 14%
Uruguay	$7.6B	Latin America, 28%; EU, 26%; North America, 12%; Eastern Europe & former Soviet Union, 8%
Venezuela	$34.9B	North America, 56%; Latin America, 17%; EU, 17%; Japan, 5%

Source: The Observatory of Economic Complexity, 2020. https://oec.world/

city to the important port at the mouth of the Amazon River. The second extended more than 2,500 miles (4,000 km.) from the overpopulated northeast into the interior forest region. The third linked the urbanized southeast with the remote Amazon states of Rondônia and Acre. The BR-364 has been especially successful in facilitating the migration of subsistence farmworkers into the heart of the Amazon, where they have contributed significantly to rain-forest destruction by using the slash-and-burn method of agriculture. Rondônia has become one of the most deforested parts of the Amazon.

These highways were expected to facilitate the migration of people from overcrowded coastal regions to relatively empty interior areas and expedite the movement of government forces for political control. They also were designed to provide a network over which newly produced forest products might be shipped to urban areas along the coast. Reducing the cost of transportation was expected to create economic viability for many activities and thus promote development.

Although successful in encouraging migration and the extraction of forest resources, these frontier road-building schemes have not brought growth and development to the region. The most important consequences have been the rampant deforestation and endangerment of indigenous peoples, animals, and plant communities, which are threatened by encroaching civilization. The combination of poor soil conditions, limited economic resources, and remoteness from population centers has made the interior Amazon region one of the poorest areas in the Western Hemisphere.

Inland Waterways

Traffic on inland waterways is handicapped by several problems. Many regions of South America experience heavy rainfall during the summer months (December, January, and February, south of the equator), followed by dry conditions during the winter, so water levels in the rivers fluctuate dramatically. Many of the streams include rapids, waterfalls, and artificial dams that block navigation. Nevertheless, some of the rivers provide important routes of access.

The Amazon River is wide and deep, so ocean-going ships can easily navigate its course from the mouth to the city of Iquitos in Peru. The Paraná system in the south is not as favorable a channel for shipping, but because the rivers flow through important urban and industrial areas, they are economically more important than the Amazon.

The countries of the Common Market of the South (Mercosur)—Brazil, Argentina, Paraguay, and Uruguay—have undertaken two important projects aimed at extending inland shipping in southeastern South America. The basin of the Paraná and its tributaries has been dredged and otherwise improved to accommodate large vessels from the mouth near Buenos Aires into the interior of southeast Brazil near São Paulo. A substantial amount of investment has been made to build lock systems that can elevate boats around the massive dams that block several of the rivers. The Tietê-Paraná waterway has become Brazil's major shipping channel between its southern and southwestern regions. Its planned extension to Buenos Aires in Argentina will create a streamlined bulk-cargo shipping route through the heart of the Mercosur region.

The second major project—the Paraguay-Paraná Hidrovía—involves an attempt to alter (through dredging, rock removal, and channeling) the Paraguay and Paraná riverbeds from Corumbá on the Bolivia-Brazil border to Cuiabá in the Brazilian state of Mato Grosso. Although beneficial to shipping, the creation of this industrial waterway will jeopardize wildlife in the Pantanal, the world's largest tropical wetlands. For this reason, environmental groups have been lobbying the Brazilian government to change its plans, and the project has been put on hold.

Waterborne navigation is also important in northern South America. The streams of Guyana and Suriname are used to transport bauxite from inland mines to the coast, where it is smelted and exported. In Venezuela, the Orinoco River has sustained navigation as far back as the colonial period and is used to ship iron ore to the sea, where it will be loaded on vessels destined for the United States.

Roll-on/roll-off ships, such as this one pictured here at Miraflores locks, are among the largest ships to pass through the Panama Canal. The canal cuts across the Isthmus of Panama and is a key conduit for international maritime trade. (Image by Dozenist)

The Magdalena and Cauca Rivers of Colombia have provided access into the interior, and the Magdalena was the principal route into the central area near the city of Bogotá until the construction of a railroad. Across South America, the rivers of the interior offer a network of transportation routes for rural inhabitants, riparian tribes, prospectors, and tourists who travel the streams in canoes, flatboats, rafts, and other small craft.

Maritime Transportation

Because of the coastal distribution of the population of South America and the poor conditions of overland routes, several nations make heavy use of ocean shipping. Ports were developed to service the export economies, but the docking facilities also support local traffic from one coastal city to another. Some countries have significant merchant fleets. Especially notable is the large fleet of Brazil and some twenty large cargo ships possessed by Paraguay, a landlocked nation. As those of many other nations, the ships of the South American countries crisscross the globe with little reference to the home nations. Brazil also has a major shipbuilding industry. Brazilian-manufactured ships make up the bulk of the nation's fleet, and they are sold to other nationsas well. Large cargo and passenger ports operate in Brazil, Peru, Colombia, Argentina, and Chile.

Air Travel

For the more remote areas of South America, air travel is the principal means of access. Communities in the eastern regions of the Andean countries, the central and upper Amazon basin of Brazil, and the extreme south of Argentina and Chile depend on air connections as links to the outside world. Many villages, *estancias*, *fazendas*, and haciendas include an airstrip. Local air travel has been especially important in Colombia, where the mountainous terrain hampers land travel.

South America is serviced by many commercial airlines. Airports in Buenos Aires, Argentina; Lima, Peru; São Paulo, Rio de Janeiro, and Brasília, Brazil; Caracas, Venezuela; and Santa Cruz, Bolivia, are major hubs for international travelers. The facilities are modern, and high standards of aircraft maintenance and air traffic control are adhered to. Brazil is one of the world's major producers of aircraft, supplying a substantial portion of the commuter aircraft market in the United States. At the same time, commercial air travel in South America, except for in the Andean region, remains quite expensive. Only Brazil and Colombia have their own low-cost airlines, which serve major South American cities.

Although several areas of South America have developed dense, interlinked networks, many regions remain largely unserved by modern transportation. They still exhibit a pattern that is reminiscent of that of the colonial period, with the major routes connecting interior mining or farming areas directly to the coast.

Cyrus B. Dawsey

CENTRAL AMERICA

The seven nations of Central America cover an area of about 202,000 square miles (523,200 sq. km.). The highly diverse landscape ranges from coastal lowlands outlined with islands, to uplands, to mountain ranges of active volcanic peaks. Earthquakes and landslides are common. Occasional volcanic eruptions of ash and mud cover towns. Hurricanes, high winds, and flooding also add to the challenging environment of Central America. All of these factors and conditions contribute to making the development of state-of-the-art transportation infrastructure extremely difficult.

Central America comprises four physical regions. The Nicaraguan lowlands stretch through the central part of the region for about 250 miles (400 km.) and divide up the rugged highland region running through most of the center of the isthmus. These highland areas can exceed 1,000 feet (300 meters). The Caribbean lowland stretches from Belize in the north through Panama in the south and includes some navigable rivers. Cays (islands) fringing the coastline and shallow water ports keep larger ships from entering harbors along the Caribbean east coast. The Pacific lowlands is a narrow strip of land bordering the western side of the isthmus. The Nicaraguan lowlands serves as a passageway between the Gulf of Fonseca on the Pacific side to the Caribbean Sea.

The challenge of transportation has been a key to cultural and political control of this region. The Maya, Carib, and Arawak were adept at making their way through the rough terrain. They used forest paths and rivers to develop trade routes north to Mexico and perhaps farther. The Island Caribs (indigenous people of the Lesser Antilles in the Caribbean) were skilled at open-water navigation and sailed among islands in boats capable of carrying as many as fifty people. In the seventeenth century, as more Europeans filtered into the region, British pirates knew that the eastern coastal waters contained many treacherous barriers for Spanish sailing ships. Pirates hid in the keys and attacked the Spanish as they sailed beyond the reefs, unable to enter the shallower safe harbors. In modern times, the region has been a battleground for controlling access to the world's great oceans via the Panama Canal.

Rough Roads

Travel continues to be challenging in Central America. It is difficult to construct roads there, not only because of the extreme physical conditions but also due to social issues. Military conflicts, economic troubles, soaring populations, and environmental issues have all contributed to the burden of developing the region.

By the third decade of the twenty-first century, there were about 120,000 miles (193,000 km.) of roadway throughout Central America. Of this total, only about 30 percent was paved—the remainder were gravel and dirt roads. Nearly all highways outside of the major cities are two-lane undivided roadways without shoulders. It is not unusual to find a single automobile traveling on a road next to an ox cart, a loaded donkey, or a pickup truck crammed with people in the back. Many local buses in rural areas are old, modified American school buses whose passengers commonly use them to transport produce, chickens, and other small livestock. In lieu of bus stations, these so-called "chicken buses" make their stops near markets on the outskirts of towns and villages.

In the wake of Hurricane Mitch (the deadliest hurricane in Central American history), which hit Honduras and Nicaragua in November 1998, the magnitude of the rural road problem became clear. Transportation on roads that were already in bad condition, and in some cases practically nonexistent, before the hurricane became an even greater logistics obstacle. Now they were totally wiped out, making it nearly impossible to get aid to victims in remote areas. Even in the twenty-first century, the inadequacy of Central America's roadways continues to negatively affect the region's rural development and market competitiveness of its agricultural products.

The region's longest roadway is the Inter-American Highway, which is part of the Pan-American Highway that runs across the American continents from Alaska to the lower reaches of South America. The Inter-American Highway stretches 1,500 miles (2,400 km.) through Central America. While this highway links all Central American countries except Belize, lengthy border crossing procedures impede car and truck traffic among them.

Railroads

The rail systems in Central America are short, isolated lines that were built by sugar, coffee, and banana plantations to haul their commodities to the coast for shipment. By the beginning of the twenty-first century, most of them went into decline and were decommissioned. The remaining ones operate only in Honduras, Costa Rica, and Panama. The Panama Canal RailwayRailroad, which runs along the canal, is 47 miles (76 km.), but nevertheless holds the distinction of being the world's first transcontinental railroad. Opened in 1855, it was instrumental in the building of the Panama Canal half a century later; the canal itself was constructed parallel to the railway. The railroad was completely rebuilt in 2001 to accommodate double-stack container railcars. It continues to run several historic passenger cars as a tourist attraction.

Maritime Water Transport

The least expensive way to move large quantities of commodities is by water. The closeness to and accessibility of Central American seaports to markets in the United States and Canada helped to develop the banana, coffee, and sugar trade in the region. Because Central America is a land block between the Atlantic and Pacific oceans, it is convenient for ships to call on ports there. This blockage was also a hindrance to trade between the two oceans, however. In 1914 that blockage was removed and a new nation emerged in the region as a result.

The Panama Canal was constructed between 1904 and 1914 by the United States. Panama, which after gaining independence from Spain became a part of Colombia, was not a sovereign state at the time. After the French attempted but failed to build a sea-level canal in the 1880s, the United States wanted the chance. When petitioned with a treaty by the United States to build the canal, Colombia's government refused. A general uprising occurred in the area of the isthmus. The Panamanians declared their independence from Colombia, and the United States was given full authority to build the canal. This time, an alternative canal-building scheme was adopted that involved creating a large reservoir above sea level and using a series of locks to raise and lower ships. The area for the canal was leased from the Republic of Panama, and the completed canal and adjacent land were controlled and operated exclusively by the United States as the Panama Canal Zone until 1977. In that year, joint American-Panamanian control over the Canal Zone was established. On January 1, 2000, the United States completely withdrew from its operations.

The Panama Canal represents one of the largest and most difficult engineering projects ever undertaken. The canal cuts through the isthmus of Panama, allowing ships to avoid going around the southern tip of South America—a distance of some 8,000 nautical miles (14,800 km.). The canal carries ships from the Atlantic Ocean, at Limôn Bay at Colón, to the Pacific Ocean, at the Bay of Panama at Balboa, and back. The canal is slightly longer than 50 miles (80 km.). Ships entering it from the Atlantic side pass across the large artificial Gatun Lake and are raised and lowered 85 feet (26 meters) through three sets of locks before reentering sea level on the Pacific side.

It takes ships seven to eight hours to pass through, or transit, the canal. When the canal was planned, it accommodated the largest ships then made. By the end of the twentieth century, cruise ships, supertankers, and container ships with deeper and wider hulls would not fit through the canal, which was only 41 feet (12.5 meters) deep at its deepest point and 110 feet (33.5 meters) wide at its widest. Those ships therefore had to make the long voyage around the tip of South America. Between 2007 and 2016, a major expansion of the canal was undertaken, doubling its capacity. Today, nearly 14,000 vessels pass through the canal yearly, including both cargo and tourist ships.

Panama also has over 8,000 merchant marine vessels in its ship registry (the world's largest). Most ships belong to foreign owners, who fly the Panamanian "flag of convenience," and are thus not subject to the stricter maritime regulations and higher operating costs of their own countries.

Inland Waterways

River transportation in the region is limited to shallow-hulled boats and by seasonal high-water conditions. The Belize River (also called Old River) in Belize runs for 180 miles (290 km.) through the heart of the country and flows into the Caribbean. It can be navigated almost all the way to Guatemala with small craft. El Salvador's Lempa River is also navigable. In Guatemala, about 155 miles (250 km.) of rivers can be navigated and an additional 290 miles (470 km.) can be used during the high-water season. Nicaragua has two large lakes and over 1,240 miles (2,000 km.) of waterways. Honduras contains about 290 miles (465 km.) of navigable waterways.

Costa Rica has an abundance of small, seasonally navigable rivers. Panama's Chagres River forms part of the Panama Canal system.

Air Transportation

Over 900 airports and developed airstrips are found in Central America, but only about 160 of them have paved runways and can handle all sizes of airplanes. The remainder service small planes that can penetrate into the remote regions. For some communities, airplanes are the only link to the outside world. Air traffic into the larger cities has been increasing, partly to meet tourist needs. Seventeen national air carriers serve Central America, and planes from Europe, the United States, Mexico, and Canada make daily flights to Panama City, Guatemala City, Belize City, and San José.

The Future

Economic development in Central America is dependent on developing transportation links. Cooperation among nations to build these links is essential. Its geographic position between the two American continents gives Central America the potential to improve its accessibility to world markets. Although much of the region's environment is challenging, given its extreme weather and topography, the Panama Canal is a good example of the progress that can be made by connecting this region to the rest of the world.

M. Mustoe

TRADE

SOUTH AMERICA

South America historically has played a peripheral role in the global economy. By contrast, the global economy has always played a major role in the economies of South American countries. Whereas the United States exports about 12 percent of what it produces, some South American countries export more than twice that proportion. For example, Chile exports over 28 percent of its total production.

Major Trading Goods

Ever since the sixteenth century, and the beginnings of European colonization of the region, South America has exported primary products to the more industrialized countries of Europe and North America and imported finished goods from them. South America still fills this role, though there is a growing emphasis on developing exports of manufactured goods from the region.

Raw materials that are typically not used in their original state, such as agricultural goods and minerals, are considered to be primary products. One example is iron ore, which must be smelted and then manufactured into pig iron or steel before it is useful. Another example might be cattle, which must be slaughtered, butchered, and packed before becoming meat that can be eaten. Primary products have always constituted the major exports from South America.

Agriculture

Argentina and Uruguay are best known for their exports of agricultural products. Argentina is a major producer of wheat and the world's seventh-largest wheat exporter. Most of Argentina's wheat goes to Asia, although some is shipped to Brazil. Brazil is the world's largest beef exporter. Both Argentina and Uruguay export significant amounts of cattle and sheep products, usually after only minimal processing and packaging. Beef exports once went mostly to Europe, the United Kingdom in particular, but in the twenty-first century, China and Hong Kong became the major destinations. Cleaned wool is the major product from sheep exported from these countries.

Brazil, although a net importer of many agricultural products, is the origin of many tropical crops exported to countries around the world. The most notable crops are soybeans, cotton, tobacco, corn, and sugarcane. Other crops grown primarily for export from Brazil include cocoa, beans, and natural rubber. Brazil exports large quantities of orange juice concentrate, primarily to the United States. It is also known for its coffee, shipped to the United States and Europe. Finally, Brazil exports large amounts of wood products, either sawn lumber or wood pulp for paper manufacturing.

Other South American countries export tropical fruits, seafood, and specialized crops. Colombia is known for its coffee, as is Ecuador to a lesser extent. Ecuador also exports cocoa beans and shrimp. Chile is one of the world's leading exporters of farm-raised salmon and trout. Both Colombia and Ecuador export bananas. Chile and Argentina are known for apples, grapes, and wine. Fruits are shipped to Northern Hemisphere destinations during South America's autumn months (March through June).

Minerals

Many major industrial minerals are exported from South America. One of the most important is bauxite, the ore from which aluminum is made. Brazil, Suriname, Guyana, and Venezuela all export large amounts of raw bauxite. Since aluminum smelting requires heavy inputs of electricity, the process tends to take place near cheap hydroelectric power sources. Canada, the United States, Europe, and Russia import most of South America's bauxite.

Another important industrial mineral is iron ore, used to produce steel. Brazil is the world's largest exporter of iron ore, most of which is shipped to Europe and Japan. Chile is the only other country in South America that exports a significant amount of iron ore. Petroleum is found in Venezuela, Brazil, and Ecuador, and in smaller amounts in Colombia. Most petroleum exports go to the United States and China.

South America exports several other minerals. Precious metals include gold from Brazil and silver from Peru (the world's second-largest silver producer) and Bolivia. Lead, zinc, and tin come from Peru and Bolivia. Raw and refined copper are exported from Peru and Chile; both countries are among the world's top three producers of the metal.

Manufactured Goods

Few South American countries export significant amounts of manufactured goods; most import them. Major imports include high-tech products such as computers and other electronics; capital goods such as machine tools; and plastics. Few manufactured consumer goods and appliances are imported, however. Of all the South American countries, only Argentina and Brazil export significant amounts of manufactured goods, including automobiles, other transport equipment, and their parts.

Brazil is the major exporter of manufactured goods in South America. One of its most important products is steel. Steel exports from Brazil find their way primarily to Southeast Asia, although Nigeria and the United States also import Brazilian steel. Brazil also exports other iron products, plus some aluminum and various processed ores.

Textiles, apparel, and automobile parts are other important exports from Brazil, which is well known for making shoes and sneakers. Most products of this type are produced to be exported to North America or Europe.

Chile and Argentina compete for the place of the second-largest South American exporter in terms of total monetary value, but most of their exports are in the primary sector. In the secondary sector, Argentina exports automobiles, often to other South American countries. Some textiles, mostly leather derived from its large livestock industry, are also shipped out. Steel products represent most of the remaining manufactured exports.

Brazil has tried to expand its export sector to help bring development to the poorer and depressed regions of the country. The most successful attempt has been around Manaus in the Central Amazon River valley. Manaus, the former center of the rubber industry, declined after synthetic rubber from petroleum replaced natural rubber. It was de-

Detail of hairsash being brocaded on a Jacaltec Maya backstrap loom.

Trading panel of the São Paulo Stock Exchange is the second-biggest in the Americas and 13th in the world. (Image by Rafael Matsunaga)

clared a free-trade zone, meaning that goods can be imported and exported from the region without tariffs being paid. This has brought some well-paying factory jobs to the people of the region. Similar attempts have been made in the northeast of Brazil, but without the same promising results.

Venezuela, which used to be one of the top exporters in South America, has suffered a socioeconomic and political crisis that began during the presidency of Hugo Chávez and has continued into the presidency of Nicolás Maduro. Venezuela has the largest proven reserves of petroleum, and its major exports are petroleum and petroleum-based products. Venezuelan petroleum production has decreased significantly in recent years due to economic mismanagement, failed social policies, and widespread corruption. The economic impact on the country has been compounded by the worldwide drop in oil prices. Venezuela has traditionally relied on earnings from the petroleum sector to pay for its many imports, including food and medicine. Oil-production decline led to the dramatic drop in both exports and imports, which in turn caused catastrophic shortages in essential goods for the Venezuelan population.

South America's Major Trading Partners

Historically, the countries in the north of South America have had close ties with the United States, while those in the southern portions have had closer ties to Europe. Colombia, Ecuador, and Venezuela all count the United States as their leading trading partner. The European Union is the second-largest partner for trade in goods.

In recent years, there has also been a growing role for China in trade with the region. China has surpassed the United States as the top export destination for countries such as Chile, Peru, and Brazil. Other developing countries, especially in the Middle East and East Asia, have begun buying South American goods. Finally, trade between and among South American countries is rising in importance. The smaller countries such as Bolivia, Paraguay, and Guyana, have traded with their neighbors for many years. But historically, most South American countries have looked to the outside world, at least in part because they have difficulty interacting with their neighbors, given the tall mountains, thick jungles, and poor transportation infrastructure. This trend toward intracontinental trade has been further boosted by regional trading blocs. Brazil and Argentina now claim each other as major trading partners, and both of these countries are top export destinations for Bolivia.

U.S. Trade with South America

South America is an important source of imports into the United States and an even more important destination for exports from it. Brazil and Colombia are the two biggest trading partners, in that order, but Argentina and Venezuela also have a significant share of the United States' trade.

Major U.S. exports to South America include telecommunications equipment, electronic devices, and various types of industrial machinery. South American countries are largely self-sufficient in the production of consumer goods. However, the production of consumer goods requires machinery for the factories that produce those goods. Since South America does not produce enough of this machinery, it must import it from abroad. Machines that are used to make any goods are known as capital goods, and they often come from the United States, Canada, Europe, or China.

The United States imports a significant amount of products from South America. Brazil supplies a great deal of iron and steel and also shoes and sneakers. Venezuela and Colombia sell petroleum to the United States. Other products that the United States gets from South America include coffee, tea, spices, and various types of apparel.

Trade Associations

South American countries, because of their reliance on exports, have a long history of participating in trade associations. That participation expanded as new organizations were being developed.

LAFTA/ALADI

In 1960 seven South American countries, along with Mexico, formed the Latin American Free Trade Association (LAFTA), whose stated purpose was the lowering of tariff barriers among member states. Unfortunately for LAFTA, its members were all following import-substituting industrialization policies that tried to keep out foreign goods as a means to encourage the manufacture of domestic

KC-390 is the largest military transport aircraft produced in South America by the Brazilian company Embraer. (Image by Ministério da Defesa - Apresentação KC-390)

alternatives. Since a free-trade agreement requires members to open their markets, those policies proved incompatible. In addition, the far-flung members shared few things in common and found it difficult to agree on tariff levels. Thus, trade barriers were lowered on only a few goods, and LAFTA generally failed. In 1980, the defunct LAFTA was replaced with the Latin American Integration Association (Asociación Latinoamericana de Integración, or ALADI), which currently has 13 member countries. ALADI members adopted a regional tariff preference scheme, which they later expanded. These and other ALADI actions helped increase intraregional trade flows.

Andean Community

In 1969, after it was evident that LAFTA was not working, five countries in the Andes—Colombia, Ecuador, Peru, Bolivia, and Chile—formed a customs union with the goal of boosting the economic competitiveness of its member nations with respect to the more developed South American economies. Venezuela later joined what became known as the Andean Group. The member states neither could agree on such basic things as where the boundaries between their countries were nor could they agree on more complicated issues, such as the tariff levels for a variety of goods. Chile withdrew its membership in 1976, as did Venezuela in 2006—the same year Chile came back as an associate member. By that time, the organization had transformed itself

and, in 1996, was renamed the Andean Community (Comunidad Andina, or CAN). Currently, it has four full members (Colombia, Ecuador, Peru, and Bolivia) and five associate members—Chile and the four countries (Argentina, Brazil, Paraguay, and Uruguay) that make up Mercosur, the other of the two main trading blocs in South America. Together, CAN and Mercosur are spearheading economic integration of South American nations.

Mercosur

The Common Market of the South (Mercado Común del Sur, usually shortened to Mercosur), was founded in 1991 and includes Argentina, Brazil, Paraguay, and Uruguay (Venezuela is also a member but has been suspended since 2016 for failure to maintain democratic norms). A common market is a trade organization that incorporates lower trade barriers among members (like a free-trade association) and common trade barriers with nonmembers (like a customs union), and places no restrictions on the movement of workers and money among its members. Since the formation of Mercosur trade among its member countries increased significantly. In 2019, Mercosur concluded a landmark trade deal with the European Union. If ratified, it will be the largest free trade agreement in the world, creating a market of some 800 million people and involving a quarter of the global economy.

Timothy C. Pitts

CENTRAL AMERICA

During the early nineteenth century, Guatemala, El Salvador, Honduras, Nicaragua, and Costa Rica briefly joined in a single nation—the United Provinces of Central America,(also called the Federal Republic of Central America), following the region's successful war of independence from Mexico.

Bickering broke out among the participants soon after, and ultimately each country went its separate way. Since that time, there have been several attempts to reunite the original members of the United Provinces, but without any meaningful success.

Economic Unity

The attempt on the part of the Central American nations to achieve some form of economic union has proven to be much more popular than efforts aimed at creating a political union. Recognizing the need for creating a stronger vehicle for the exchange of goods and services among themselves, the five original nations belonging to the old federation, and—to a lesser extent—Panama, joined together in the early 1960s to form the Central American Common Market (CACM). Under its provisions, goods could move among the signatory nations either without tariffs or with greatly reduced ones. In this manner, a larger market for domestic products could be developed, strengthening the economies of all the members.

The CACM proved to be unsatisfactory to Honduras and Nicaragua, whose leader believed that their countries did not share equally in the benefits that the agreement was supposed to produce. These complaints, together with the 1969 "Soccer War" between El Salvador and Honduras, led to the suspension of CACM activities in the 1980s. The organization revived in the 1990s with the creation of the Central American Integration System (Sistema de la Integración Centroamericana, or SICA). The SICA includes the seven Central American nations and the Dominican Republic. Within the SICA system, the CACM has removed customs duties on most products throughout the member countries and unified external tariffs, which helped expand regional trade.

Central America's principal exports consist of natural resources such as lumber and metals; crops such as coffee, bananas, sugar, and cotton; and livestock, mostly beef. Much of the isthmus is covered with fertile volcanic soil that lends itself to the raising of plantation crops. While there is wide variety in the emphasis that each country places on items exported, the countries compete with one another, as well as neighboring countries, in finding markets for their major agricultural produce.

Maquiladora Industries

The existence of a substantial labor pool of skilled and semiskilled workers, available at comparatively

THE RISE OF BANANA REPUBLICS

The term "banana republic" was first used by O. Henry in 1901. In political science, it describes a politically unstable country with an economy dependent upon the exportation of a limited-resource product, such as bananas. It is widely used as a pejorative descriptor for a dictatorship, for kickbacks, and the exploitation of large-scale plantation agriculture, especially banana cultivation.

Bananas were introduced to the U.S. public by late nineteenth century shipowners, who included small amounts of the fruit as supplements to their regular shipments. Bananas proved to be so popular that some U.S. entrepreneurs decided to expand the trade. Banana cultivation at that time was limited to the small output of a few Caribbean islands, hardly adequate for the presumed demand that existed in the United States. The investors decided to establish banana plantations elsewhere on a much grander scale.

The east coast of the Central American isthmus offered an ideal location for the establishment of extensive banana cultivation, providing a broad geographical production base that could weather the frequent storms that plagued the Caribbean. Central American governments involved eagerly sought potential investors in the largely undeveloped lowlands. Railroad builders in the area saw business opportunities in the development of the new agricultural industry. Soon, swamps were drained, modern harbor facilities built, and large tracts of land put into cultivation. The banana industry provided jobs for native labor and furnished health and educational facilities for their employees.

The banana business became so large and profitable that the major U.S. companies, such as United Fruit Company (now Chiquita Brands International) and Standard Fruit Company (now Dole Food Company), could virtually dictate policy to the local Central American governments. Throughout the early twentieth century, the fruit companies dominated the economies of Honduras, Costa Rica, Panama, and, to a lesser extent, Guatemala. After World War II, however, as both the governments and the economies of these countries grew more sophisticated, the economic and political power of the fruit companies diminished. Moreover, the U.S. government ceased to support the claims of private companies to the same degree that it did in the nineteenth and early twentieth centuries.

low wage rates, brought business enterprises to Central America. The governments of the isthmus also have been willing to provide tax incentives to encourage companies to set up factories (*maquiladoras*) for the processing of raw materials into a variety of finished products. Members of the SICA produce around 10 percent of all apparel goods purchased in the United States; of these goods, about 70 percent are produced in Guatemala, El Salvador, and Honduras. The apparel export industry in Central America is concentrated in the hands of a few multinational companies, including Fruit of the Loom and Hanes.

In the meantime, the European Union maintained trade barriers to the importation of Central American agricultural produce such as bananas and sugar. These tariffs were designed to protect certain Caribbean markets—dominated by Europeans—from Central American competition. This kind of economic discrimination prompted Central Americans to seek economic and political aid from the United States, the home base of many companies with operations in the isthmus. In the twenty-first century, trade relations of the region with the European Union improved. An association agreement between the EU and six Central American countries, designed to open up markets on both sides, was signed in 2012, creating a free-trade area between the EU and Central America.

Patterns

Central America's trade continues to be dominated by the export of agricultural products. The combination of rich soil, tropical climate, and low labor costs have made bananas, coffee, cotton, and sugar the region's primary commodities for shipment abroad. In colonial times, most parts of Central America developed as tropical crop-producing monoeconomies. The trend continued later, when young independent countries were dominated by U.S. corporations, such as the United Fruit Company (now Chiquita Brands International, based in Switzerland). In the case of bananas, particularly, foreign investment has been the major impetus in the production and exportation of the fruit. Foreign corporations supported servile local dictatorships

that allowed them to make huge profits by exploiting local labor. The pejorative term "banana republic" was coined by American writer O. Henry in 1904 with respect to Honduras and its neighboring countries. Current social and political problems of these countries are deeply rooted in this economic history.

Today, Central America's seven republics vary widely in terms of the levels of economic progress. During the 1980s, El Salvador, Nicaragua, Guatemala, and to a lesser extent Honduras, had their productive capabilities disrupted by civil strife. (A great deal of El Salvador's infrastructure—its roads, bridges, communications system, and transportation equipment—suffered extensive damage during its brutal civil war.) In the 1990s, these disruptions were eliminated, making production for foreign trade possible, but residual poverty, high crime rates, and political instability would hinder its development for decades to come.

Nicaragua and Honduras suffered the most grievous losses to their respective economies as a result of Hurricane Mitch in 1998. Vast areas of both countries suffered extensive flood damage to their economic infrastructures. Nicaragua and Honduras required substantial outside help to resume normal economic expansion.

Costa Rica, Belize, and Panama have not faced the same economic dislocations as their neighbors. Costa Rica and Panama have become the most advanced Central American countries in terms of economic development and foreign trade. Politically stable Costa Rica has a robust economy; besides traditional bananas and coffee, it now exports medical instruments, pharmaceuticals, and software as well as offering financial and ecotourism services. Panama is the only Central American nation whose economy is based primarily on the services sector (it accounts for nearly 80 percent of its GDP). It trades in high-income services centered around the Panama Canal, banking, the Colón Free Trade Zone, container ports, and flagship registry. At the opposite end of the spectrum, Belize is still seeking to establish its industries and diversify its trade after centuries of colonial administration by Great Britain, which maintained it essentially as a mahogany plantation. It is the least advanced country in the area. Future success for Central American trade will depend on each country's ability to improve its economic development. Critical issues include the broadening of trade relations within the area and with the rest of the world. In 2004, members of the Central American Integration System signed a wide-ranging agreement (Dominican Republic-Central America Free Trade Agreement, or CAFTA-DR) with the United States, the first free-trade agreement between the United States and a group of smaller developing economies. It boosted the economies of South American countries. Still, the full trade potential of Central America remains to be realized.

Carl Henry Marcoux

COMMUNICATIONS

SOUTH AMERICA

While most South American countries have well-developed communication systems, although they are less sophisticated than those in Europe, North America, Australia, and Asian countries such as Japan, South Korea, and Singapore. Communication systems and devices differ in sophistication and prevalence from one South American country to another, and from the continent's coastal cities to its interior regions.

Early History
During the colonial period, South American cities such as Buenos Aires and Montevideo were likely to have closer communications with Europe than with communities in the interiors of their own territories. This was because people of the coastal regions were generally of European birth or descent and had a language and cultural background similar to that of Europeans. They spoke the European languages, primarily Spanish, but also French, Dutch, and Portuguese. The people of the continent's interior were predominantly Amerindians, and they seldom spoke these languages. Nor did they share the European culture, and it was difficult for these disparate groups of people to cross cultural barriers and communicate effectively.

Complicating communication matters was the South American topography. The Andes Mountains, rising to above 20,000 feet (6,100 meters) on the west, the Guiana Highlands, rising to nearly 10,000 feet (3,000 meters) on the north, and the Brazilian Highlands, also rising to nearly 10,000 feet on the east, made crossing the continent, or reaching people in its interior, extremely difficult.

Climate differences from one area to another also made travel and communication between the coastal regions and the interior difficult. The hot, humid climate of the Amazon Basin differs greatly from the cold climate of the mountains and of the Tierra del Fuego archipelago near Antarctica. The isolation of people within their own geographic areas and climate zones exacerbated the communication barriers.

Modern Developments
Certainly, communication from one region of South America to another, from one South American country to another, and from the coastal areas to the interior has improved enormously since colonial times. Technology has made communications easier and faster, and education and migration have broken down many of the cultural barriers that existed among the various groups of South Americans. Still, telecommunication in South America is far less effective and less extensive than it is in North America.

Monetary considerations play an important role. The combined gross national product of the countries of South America is significantly lower than the gross national product of the United States or Canada. Therefore, the governments and industries in South America have less money to devote to developing communications technology and infrastructure. The personal income and purchasing power of the citizens of South America also are much lower than those of North Americans. This is why fewer South Americans, especially in rural regions, can afford costly computers, sophisticated cellular phones, and paid Internet and cable TV subscriptions compared to residents of the United States or Canada.

Nevertheless, the continent has made great strides in the growth of communications. South Americans have hugely benefited from the technological revolution brought on by the development of the Internet. Submarine cable systems now provide direct links between the countries as well as to Central and North America, Europe, and Africa. Almost all countries have domestic systems of communications satellites, and all participate in international satellite networks.

Languages and Communication

Linguistic diversity has hampered communications within and among many South American nations. Spanish is the official language of most countries of South America, but some, such as Peru and Bolivia, have more than one official language, and many countries have mixes of unofficial languages. This makes daily communications difficult, and communication over the Internet was almost impossible until the recent advent of machine translation software and voice translation apps. While most computers in South America have Spanish or Portuguese user interfaces, there are practically none that accommodate the many languages spoken and written by South America's indigenous peoples.

Literacy levels also present problems in communication. In 2020, over 90 percent of the citizens of South America could read and write. Despite this, the functional illiteracy rate was high—more than 20 percent in Brazil, for example. This has made it difficult for people to communicate effectively and to use instruments of communication such as the printed media and Internet publications.

Venezuela, Colombia, Ecuador

Located in northern South America, these three countries are similar in their communications systems and their overall lifestyles. Spanish is the official language of all three countries, but all have many people who speak indigenous languages such as Quechua. The mix of languages and cultures affects the overall communication levels. All three

countries have high poverty levels, despite their rich resources. Venezuela had a 2019 population of about 32 million, and there were some 20 landline and 80 cellular telephones per 100 inhabitants. The phone system is relatively modern and expanding, especially in rural areas. The country has a national inter-urban fiber-optic network capable of digital multimedia services. There are two domestic satellite systems with three Earth-based stations and two international satellite stations. The use of satellite TV is widespread, and the prices are low. However, as a result of the deepening economic crisis in recent years, the state of fixed-line equipment and the quality of service in many areas have deteriorated. Venezuela, which imports all of its telecommunications equipment, has struggled with paying foreign vendors for it.

Venezuelan broadcast and print media have been under increased government control and supervision in recent years. Today, the country has more than 100 national television channels, including public service networks, privately owned networks, and state-sponsored community TV broadcasters. The government controls over 55 percent of TV broadcasting, compared to 12 percent in 1998. The state-run radio network includes over 60 news stations and another 30 stations targeting specific audiences, including speakers of indigenous languages. More than 230 radio stations are state-sponsored community broadcasters. The number of private broadcast radio stations has been declining, but many still remain in operation.

In 2013, 90 newspapers were in circulation in Venezuela. Following the election of President Nicolás Maduro, most newspapers stopped circulation due to government censorship as well as paper and toner shortages. The last oppositional newspaper ceased its print edition in 2018. By 2019, the number of newspapers had dropped to 28. Because freedom of the press is lacking in mainstream media, the Internet has become a lifeline for Venezuela's independent media. Yet in 2019, less than 60 percent of the population had reliable Internet access. As a result of underdeveloped infrastructure and insufficient funding, the country has the slowest Internet speed in Latin America, and blackouts are common. Nevertheless, the growing popularity of social networks, boosted by the political upheaval, has led to the rapid expansion of mobile data traffic.

The current communications situation in Colombia is better than the situation in Venezuela. Colombia had less diversity of language and culture and a lower poverty rate. Telephone usage and ownership are greater in Colombia, which has a more modern system. Mobile cellular telephone subscribership is about 130 per 100 persons. Colombia has a nationwide microwave radio relay system for domestic calls and ten Earth stations as part of its satellite system. Submarine cable systems provide links to other parts of South America, Central and North America, and the Caribbean.

Despite economic instability and the challenges posed by leftist insurgencies and narcotrafficking, the country maintains democratic institutions and freedom of the press, which allow for political diversity in its print and broadcast media. Today, there are more than 500 radio stations, including three nationwide networks, as well as five national television channels and many regional and local TV stations. Broadband Internet access has been available in Colombia since 1997. In 2019, the Internet penetration of the population exceeded 60 percent.

Ecuador is one of the least developed and most ethnically and culturally diverse nations in South America. Face-to-face communication among its citizens can be difficult due to the variety of languages spoken there, although Spanish is the official language. While the majority of the population is mestizo, a combination of Spanish and Amerindian heritage, there are several other distinct ethnic groups that include Afro-Ecuadorians, white Ecuadorians (a mixture of European immigrants from different countries), and fourteen Amerindian nations. Many Amerindians languages are spoken, but Quechua is the most common.

The telecommunication structure in Ecuador is influenced by topographical challenges associated with the Andes Mountains. It is dominated by the mobile cellular telephone sector. Its subscribership has surged in recent years and by 2018 exceeded 90 per 100 persons; still, it is among the lowest in South America. The country has landing points for several submarine cables and one satellite Earth station. There are multiple television networks and many local channels, and more than 300 radio stations. Radio remains the most prevalent type of telecommunications media, especially in mountainous areas. Many TV and radio stations are privately owned; however, broadcast media are legally required to provide the government free airtime to broadcast its programs. While there is generally no censorship, the government, being the largest advertiser, tends to grant its advertising contracts to broadcast and print media outlets that provide favorable coverage. Some TV and radio broadcasting, as well as print publications, are in Quechua and other indigenous languages. The Internet penetration in 2018 was around 60 percent of the population. There are multiple Internet television channels.

Guyana, Suriname, and French Guiana

These three entities—two small countries and an overseas region of France—are located on the northeast coast of South America. Their historical backgrounds and their cultural makeup are unlike the rest of South America. Because the languages here are different from those spoken in other parts of South America, communication between them and other parts of South America can be difficult. In Suriname, Dutch is the official language, with English being widely spoken. French is the official language of French Guiana, and English is the official language in Guyana, which was formerly British Guiana.

In the 1990s, Guyana was one of the world's most heavily indebted countries. While in later years the debt was gradually reduced, it impeded the development of Guyana's infrastructure, which relied on foreign investment. In 2020, the communications infrastructure remains underdeveloped. The country has a submarine cable connecting it to the neighboring Suriname and one satellite Earth station. While Guyana has an entirely digital telephone network, many regions outside the two densely populated areas lack fixed-line telephone services. There are about ninety cellular telephones per 100 inhabitants. Broadcast media are government-dominated. Almost all radio and television broadcasting occurs within the state-run National Communications Network. Out of the three nationwide newspapers, one is government-owned. The Internet penetration is slightly above 40 percent, the second-lowest level in South America (after French Guiana). Despite the country's still sizeable external debt, budget allocations were made in 2019 for the development of information and communications technology.

Suriname's cultural diversity is among the highest in the world. Its population in 2019 was composed of 37 percent Afro-Surinamese (including the Creoles and the Maroons), 27 percent Hindustani (East Indians), 14 percent Javanese, and smaller groups of Amerindians, Chinese, Brazilians, and other ethnicities. Such diversity creates problems with face-to-face communications among the country's 580,000 citizens. Besides the official Dutch, there are eight regionally recognized languages and over a dozen other widely spoken languages. The domestic phone system is based on microwave relay capability, and there are two Intelsat stations. Suriname has good international telephone facilities, and by 2019, there were over 130 cellular phone subscriptions per 100 inhabitants. The Internet penetration is about 60 percent. Broadcast media include two state-owned television stations, one state-owned radio station, and multiple private radio and TV stations. The two leading daily newspapers (both in Dutch) are privately owned; numerous print and online publications are published in different languages.

French Guiana is tied economically to France and, therefore, its communication equipment is supplied predominantly by that country. The Guiana Space Centre at Kourou, used by the European Space Agency as its primary launch site, is of great importance to the economy and local infrastructure development. In 2019, French Guiana had a microwave radio relay system and one satellite Earth station. Several commercial broadcasters operate alongside one public broadcaster, which is part of RFO, a French network of radio and television stations servicing France's overseas departments and territories. There are about 130 cellular phones per 100 residents. The Internet penetration is only slightly over 40 percent of the total population, although the Internet and social media usage are predictably much higher among the young French Guianese (in 2019, more than 40 percent of the residents were younger than 20 years old). Currently, social media was used by almost 90 percent of that large and growing population segment.

Brazil, Bolivia, and Paraguay

Located in the central part of South America, these countries have significant communication problems, created by linguistic diversity, cultural differences, and physical barriers. Although French, Spanish, and English are widely spoken in Brazil, Portuguese is its official language. Thus, there are problems with formal communication between Brazil and its neighbors Bolivia and Paraguay, where the official languages are different. In Bolivia, the official languages are Spanish, Quechua, and Aymara; in Paraguay, the official languages are Spanish and Guarani. The Brazilian Highlands and the Andes Mountains in Bolivia contribute to transportation and communication problems.

Brazil, with a population of over 211 million (as of 2019), is the largest country in both size and population in South America. In that year, about 20 percent of its population lived below the poverty level. Its functional literacy rate of about 80 percent was also relatively low. The largest economy in South America, Brazil has an extensive microwave radio relay system. With its sixty-four domestic and five international satellite Earth stations, Brazil uses satellites as its main communications platform, as it is difficult and costly to lay fiber-optic cables in the dense rain forest that covers much of its territory. Numerous submarine cables connect Brazil to other parts of the Americas, the Caribbean, Africa, and Europe. As of 2019, Brazil had one of the largest broadband markets in Latin America and a teledensity of 114 cell phones per 100 persons. There are over 1,000 radio stations, more than 100 TV channels, numerous print and online periodicals, and several news agencies. Television remains the most-used type of media. The ownership of broadcast and major print media is highly concentrated in the hands of media conglomerates. The Internet penetration of the population exceeds 70 percent. Brazilians rank among the world's top users of social media. However, there remains a large "digital divide" between urban and rural areas.

The deficient communications structure in Bolivia is a result of the extreme poverty in that country, fiscal and trade deficits, and state-oriented policies that deter foreign investment. As of 2019, there were only eight fixed-line telephones per 100 people, with landlines centered in the capital and other cities. Mobile phone subscriptions were over ninety per 100 inhabitants, and the number was growing. The Internet penetration is around 50 percent of the population. Landlocked Bolivia has no direct access to submarine cable networks and has to connect to the rest of the world either via its one Intelsat Earth station or through terrestrial links across neighboring countries. Bolivia has several hundred radio and television stations. Most are privately operated, but the number of state-owned community broadcasters is rapidly growing. Radio and TV broadcasts are predominantly in Spanish, with some coverage in the Aymara and Quechua languages. There are several national newspapers and many local ones, some in Aymara and Quechua, but newspaper readership is limited by low functional literacy. Since television and Internet reception is poor in many areas of the country, radio remains the leading information source. Several telecommunications companies, which had been privatized in the 1990s, were re-nationalized in 2007–2008.

The citizens of Paraguay also had limited fixed-line phone service in 2020. There are only about four landline phones every 100 people. Paraguay has a fairly well-developed microwave radio relay network and one Intelsat station, but the services are concentrated around the capital district of Asunción and a few more populated areas within the country's seventeen departments. The people of Paraguay generally rely on mobile cellular phones for personal communication. There are some 110 cell phone subscriptions per 100 people. Because of Paraguay's landlocked position, it de-

pends on neighboring countries for interconnection with submarine cable networks, making the cost of broadband services higher. The Internet is available in most of the country, but reports on actual Internet usage are somewhat unreliable, ranging from 50 to 70 percent. Radio remains the leading news and information medium. The country has more than seventy commercial and community radio stations and several national newspapers.

Peru, Chile, Argentina, and Uruguay

These four countries share communications problems that are common throughout South America. Face-to-face communications are impeded by language barriers. The development of land-based communications systems is difficult because of physical barriers, most notably the Andes Mountains. Yet all these countries, which have experienced stable economic development in recent decades, have made significant progress in communications technology, and two of them—Chile and Argentina—have created high-quality telecommunications infrastructures.

Peru has two official languages, Spanish and Quechua, and several other languages, such as Aymara, that are widely spoken. Today, Peru has more than 2,000 radio stations, including a substantial number of indigenous language stations. Of Peru's ten major television networks, nine are private and one is state owned; multi-channel cable TV services are available. The newspaper market is dominated by one media company. The country's telecommunications infrastructure is based on two submarine cable systems, a nationwide radio relay system, a domestic satellite system with twelve Earth stations, and two international (Intelsat) satellite Earth stations. Work has begun on providing fiber-optic connections to remote areas. There are over 120 cellular telephone subscriptions per 100 inhabitants. The Internet penetration exceeds 50 percent.

Chile has created the most advanced telecommunications infrastructure in South America, based on an extensive microwave radio relay network, a satellite system with three domestic Earth stations and two Intelsat stations, and a large submarine cable system providing links to other parts of the Americas and the Caribbean. The Chilean government is working on plans to install a submarine line with Asia and deploy a nationwide fiber-optic network. The country has one of the highest broadband speeds and mobile penetration rates in the region, with more than 130 cellular phones per 100 people, and an Internet penetration of over 80 percent. Ownership of Chile's numerous privately owned television channels, radio stations, and newspapers is concentrated in the hands of several large media groups.

Argentina has an extensive and advanced telecommunications system and one of South America's biggest media markets. By 2019, its major telephone networks were entirely digital. They are served by microwave radio relay, a large domestic satellite system with 40 Earth stations, numerous international satellite stations, and several submarine cable systems. A fiber-optic ring connects cities and is being extended into rural areas. The country has the second-largest (after Uruguay) percentage of cellular phone subscriptions in South America (more than 140 per 100 people); the second-highest (after Brazil) Internet and social media usage; and one of the highest rates of cable TV subscriptions in the world. There are dozens of television channels and hundreds of radio stations. With more than 150 daily newspapers, Argentina's print media industry is highly developed and independent of the government.

While most people in Uruguay speak Spanish, the official language, on the Brazilian border many people speak Portuñol or Uruguayan Portuguese, a combination of Spanish and Portuguese that allows speakers of either language to communicate with one another. By 2019, Uruguay had modern landline telephone facilities with high broadband penetration, but they were concentrated in the capital, Montevideo. People outside the metropolitan area have much poorer access to the fixed-line phone network. Because of this, most Uruguayans turned to cellular phones. Uruguay has the largest number of cell phone subscriptions in South America (more than 150 per 100 inhabitants). The Internet penetration exceeds 70 percent of the population. There are about twenty state-run and private television channels, more than 100 commercial radio stations, and numerous community radio and TV stations. Many newspapers are owned by, or linked to, the main political parties.

Conclusion

While South America has made significant achievements in the development of its communications systems, there are still barriers to such development. These barriers include linguistic and cultural differences that create communication problems, physical obstacles that make the delivery of services difficult, and political and economic constraints. There is, predictably, a notable disparity in the availability and quality of modern communications media between richer and poorer countries. The subscription cost differed greatly between urban and rural areas: while it was relatively cheap in large cities, in rural areas it was significantly higher. Community radio remains a leading source of news and information in remote regions. In most countries, broadcast and print media are dominated by large media conglomerates. In 2020, the Internet in South America reached, on average, some 70 percent of the population, compared to the world average of about 50 percent. Still, in most countries of the continent, it was much lower than in North America, where the Internet penetration exceeds 90 percent.

Annita Marie Ward

CENTRAL AMERICA

Central America has the same types of communications systems that the rest of the world enjoys. It has strong radio presence; television is very popular but is not available everywhere; cellular telephones are numerous; newspapers are varied and easily available; and the Internet and social media have taken firm hold, especially among the young populations of the cities and urban areas.

Radio

Central America had more than 620 AM and hundreds of FM radio stations in 2020. Radio continues to play an important role in Central American society, especially in remote areas, where television and the Internet are not readily available and where radio remains the main news source. With the exception of Nicaragua, each Central American country has only one government-run station or network, the remainder being privately operated.

Television

Television plays an important role in Central America's society and has done so for decades. In urban areas, television has been a major source of entertainment. Readily available to urban residents, cable and satellite television networks carry international channels. The rural areas of all seven countries have less access to television. Terrestrial television stations in Central America are supported by numerous repeater stations that catch television signals from the original stations and retransmit them. The repeater stations broaden the main signal base out to the people who live in remote areas, especially in mountainous regions. Almost all television stations are privately owned.

Telephones

The telephone systems in the seven Central American countries are diversified. They include microwave radio relay networks connected to the Central American Microwave System and Intelsat Earth stations as well as fiber-optic submarine and terrestrial networks. Submarine cables link Central American countries to each other, South America, the Caribbean, and the United States. In 2019, the fixed-line teledensity (the number of landline phones per 100 inhabitants) differed significantly among Central American countries—from five to six in Nicaragua, Belize, and Honduras to sixteen in Costa Rica, Panama, and Guatemala. Broadband penetration lagged behind many South American countries; Costa Rica's broadband market was the most advanced in the region. The number of mobile cellular phone subscriptions was high and growing. Less expensive and farther-reaching than landline phone systems, cellular phones became the main communication means for large lower-income and remote-area populations. In El Salvador, for example, 95 percent of all telephones were mobile in 2019. Belize, with some seventy cell phones per 100 people, had the lowest number.

Newspapers

Central America has followed the trend of the rest of the world, where print media has been losing ground to the Internet and social media. Still, the newspaper scene in Central America differs among individual countries. In 2010, Belize had no daily newspapers, while Costa Rica had over a dozen dailies and weeklies. Freedom of the press was also unequal: in Costa Rica, it is largely unrestricted; in Panama, media workers risked legal action when they criticized the government; in El Salvador, the press is controlled by business interests who exercised editorial influence; and in Guatemala, Honduras, and Nicaragua, investigative journalists face harassment and death threats.

Internet

Most Central American governments play an active role in promoting Internet use in their countries, largely so that Central America would not miss out on the benefits of business globalization. Thanks to the Panamanian government's "Internet for All" project, in 2010 Panama became one of the first countries in the world to offer free wireless broadband access nationwide. Predictably, the Internet penetration per 100 inhabitants differs widely among the region's less developed and more developed countries—from around forty in Honduras and Nicaragua to over eighty in Costa Rica.

Mexico has assisted the Central American region in the development of the Web and the delivery of e-commerce data to other Spanish-speaking nations in the Americas, as well as to international markets using other languages. Initially, it was thought that the Spanish language would cause a problem in doing online business with Europe, the United States, and other non-Spanish-speaking parts of the world, but that has proven not to be the case.

The Internet gives Central Americans unprecedented access to the world of information, education, and entertainment. As in many places, the use of the Internet and social media is largely concentrated among younger, more educated people, and centered in metropolitan areas.

Earl P. Andresen

DISCUSSION QUESTIONS:
COMMUNICATIONS

Q1. What languages are spoken in South and Central America? Which languages are dominant and where? What challenges do language barriers continue to present to the development of effective communication systems?

Q2. What geographical obstacles in South and Central America have impeded the development of efficient systems of communication? How have these challenges been met? What challenges remain?

Q3. How has the digital revolution impacted South and Central America? How has Panama been particularly innovative in this regard? What demographic group has been most apt to embrace the Internet and social media?

ENGINEERING PROJECTS

Engineering projects in South and Central America are closely linked to the economic growth of the regions as well as to those of the individual countries. Economic and political fluctuations have often prevented these countries from undertaking expensive infrastructure projects. Central America has been plagued by devastating hurricanes such as Hurricane Mitch in 1998 and Hurricane Nate in 2017. These and other natural disasters with their human toll and property destruction, can set a nation's economy back decades. In this region, the effects of these tragic events are particularly exacerbated by deficient infrastructure, poor living conditions, and inadequate land management.

Constraints on Investment

Because of their economic instability and weak rule of law, the least developed nations of South and Central America lack the confidence of foreign investors, who are reluctant to sink their money into expensive infrastructure projects. South and Central Americans also generally have low personal-savings rates, which mean they contribute little toward investment within their own economies. As a result, the region has been heavily dependent on cash flows from foreign nations and international institutions.

The growing trend toward privatization of businesses and industries has gradually improved economic conditions in Chile, Peru, and Argentina. Reforms in the banking and finance sectors have made these countries more attractive to international financial institutions and foreign corporations, which are now more willing to invest in development projects. Such initiatives include dams, industrial plants, transportation facilities, and telecommunications. By contrast, the economic development of Bolivia, Ecuador, Venezuela, and most Central American countries has been ham-pered by social unrest, widespread corruption, and state-oriented policies that deter foreign investment.

Hydroelectric Projects

The South American dam-building boom, which started in the 1970s, has continued well into the twenty-first century. Brazil is the second-largest producer of hydroelectricity in the world. Most other South American countries also rely on hydroelectric power plants as their main source of electricity. Dam-building has allowed South American countries to industrialize and urbanize rapidly and relatively cheaply. As a result, remote areas of the continent have been opened up to farming and mining.

However, there has been much criticism of the environmental impact of these projects. Another problem with these large dam projects has been the huge numbers of people who have been directly affected. In Brazil alone, it has been estimated that more than 1 million people were forced to relocate when their lands were seized to make way for dams, reservoirs, and the industrial developments that accompany dams. Many of those who have been relocated feel they were inadequately compensated for their losses. Resettlement provisions have been paid, but they are rarely enough to cover the losses of those displaced by dams. Spokespeople for the indigenous people and others involved in these resettlements contend that constitutional guarantees of land rights have been violated.

Despite these problems, Brazilian officials have stated that energy alternatives are still years away from being technically and economically practical. Brazil produces around 80 percent of its electricity from hydroelectric power plants and likely will remain dependent on them for the foreseeable future. Critics assert that Brazil should develop its

The Yacyretá Dam or Jasyretâ-Apipé Hydroelectric Power Station (from Guaraní jasy retã, "land of the moon") is a dam and hydroelectric power plant built over the waterfalls of Jasyretâ-Apipé in the Paraná River, between the Argentine Province of Corrientes and the Paraguayan City of Ayolas.

vast potential in other alternative energy sources, such as wind, biomass, and solar energy, and should also be more aggressive at managing electricity consumption by energy-intensive industries.

Other Controversies

Controversies surrounding dam construction have occurred in Chile, Paraguay, Argentina, Colombia, Panama, and Guatemala. In Guatemala, where indigenous anti-dam movements have been particularly active, members of the dam-resistance movement were even attacked and killed by the military back in the 1980s. Nevertheless, they eventually succeeded in halting construction of several dams.

Concern over the effects of dam projects in Central and South America has received scientific support from major research institutions. For example, in 1990, Brazil's National Institute for Amazonian Research published a report showing that the country's Tucuruí Dam—the first large-scale dam built

in a tropical rain forest—emits more greenhouse gases than all the cars and industries in São Paulo, the country's largest city. Emission calculations reveal that tropical dams generate more greenhouse gases than those in the temperate and boreal zones and can be as harmful to the environment as fossil fuel emissions. Similar alarming research findings have been published for decades, but they have not significantly influenced dam-building decisions.

Another environmental impact of hydroelectric dams is siltation in the reservoirs behind them. If allowed to go unchecked, silt buildups can greatly reduce water storage capacity and the power-generating ability of the facilities themselves. Some South American dam reservoirs have silted up within twenty to twenty-five years. One of the worst cases of siltation ever documented occurred in the Amaluza reservoir behind Ecuador's Paute River hydroelectric complex. Deforestation and soil erosion were so extreme in the surrounding highlands that the reservoir was virtually filled with sediment by 1992 and had to be dredged in order to remain operational only ten years after its opening. In 2009, low water levels at the reservoir were the primary cause of an electricity crisis in Ecuador.

Other environmental threats posed by hydroelectric projects include destruction of rain forests and the loss of fish, plant, and animal species owing to downstream water changes brought about by the projects. A greater occurrence of serious diseases such as malaria, yellow fever, and South American sleeping sickness, or Chagas disease, are also threats caused by the increased populations of mosquitoes, black flies, snails, and rodents in the areas affected by the dams.

The largest hydroelectric projects in South America include Brazil's new Belo Monte Dam as well as its Tucuruí, Jirau, and Santo Antônio dams; Venezuela's Guri and Macagua dams; and the Argentine-Paraguayan Yacyretá Dam. The Brazilian-Paraguayan Itaipú Dam on the Paraná River is the world's largest hydroelectricity project in terms of power output.

Peru's Water Projects

During the first decades of the twenty-first century, Peru has undertaken several large engineering projects aimed at creating a cleaner supply of drinking water and building additional irrigation facilities for its agricultural sector. An ambitious investment plan for the water and sanitation sector called "Water for everyone" (*Agua para todos*) was launched in 2006. While in 1980 only 30 percent of Peru's people had access to a central water supply, by 2010 that number had increased to 85 percent. Access to public sewer systems, which was less than 50 percent, increased to almost 80 percent. Work continues to improve access to clean water and sewerage in rural areas. Those helping in the upgrades of these systems include the World Bank, the Inter-American Development Bank, and the Japanese and German governments.

Reforms in the agricultural sector have led to laws improving the establishment of private water rights. Peru has invested a great deal of money in its ongoing irrigation projects in the country's fertile but arid coastal areas. Major new irrigation systems,

Most of the continent's energy is generated through hydroelectric power plants, but there is also an important share of thermoelectric and wind energy. The Jepírachi wind farm in the Guajira Peninsula is shown here. (Image by Jorge Mahecha)

delivering water from the Andean highlands and Amazon regions, included Chira-Piura, Olmos, Chavimochic, Chinecas, Majes, and Pasto Grande. These projects alone have already provided water to more than 120,000 acres (48,500 hectares) of coastal farmlands, which produce high-value export crops such as premium-quality cotton, coffee beans, avocados, asparagus, and quinoa.

Pipeline Projects in South America

A major pipeline project to move natural gas from eastern Bolivia to São Paulo in Brazil, 1,950 miles (3,150 km.) away, was completed in 2000. The Bolivia-Brazil pipeline (GASBOL) is now the longest natural gas pipeline in South America. Brazil has also completed the first section of its National Unification Gas Pipeline (GASUN), which will be even longer. After its second section is built, the 3,100-mile (5,000 km.)-long pipeline will bring Bolivian gas to Brazil's northeast states and the northern Amazon.

Space launch at the Kourou Space Center in French Guiana. (Image by DLR German Aerospace Center)

Several natural gas pipelines, designed to transport Argentina's natural gas to Chile, were constructed between 1997 and 2004, and Chile built large gas-fired power plants to be fueled by cheap Argentine gas. However, as a result of a serious economic crisis and a deterioration in the country's energy sector, Argentina stopped natural gas exports by 2007, and Chile switched to liquefied natural gas shipped from other countries.

The governments of Peru and Bolivia have developed plans to build a pipeline that would transport natural gas from Bolivia to Peru's southern port of Ilo. This pipeline is considered key to the economic development of southern Peru. It will also provide crucial access to the Pacific Ocean for landlocked Bolivia.

Ports, Roads, and Urban Infrastructure

In recent years, many ports in South America have undergone renovation and modernization, including the installation of container terminals. São Paulo's port of Santos, which in the past was widely condemned as the least efficient and most expensive big port in the Western Hemisphere, is now Brazil's most modern. It is the busiest container port in all of Latin America. In Central America, the Panama Canal was expanded in 2007–2016. New locks and deeper approach channels were built, and parts of the canal were widened, doubling the canal's capacity.

Highway systems are being improved to provide much-needed transportation links within and between the countries. The Interoceanic Highway between Peru and Brazil was inaugurated in 2011. The project entailed renovation and construction of 1,615 miles (2,600 km.) of roads and 22 bridges on Peruvian territory. It is one of the largest infrastructure projects in the history of Peru and part of the country's ambitious national road-building plan that involves the construction of three longitudinal highways and twenty transversal highways.

Central America's urban population is projected to double by 2050 due to migration from rural areas, where infrastructure and jobs are lacking. In

General Rafael Urdaneta Bridge, by Italian engineer Riccardo Morandi, showing the view from the lake to Cabimas side. (Image by Orlando Pozo)

2017, the Inter-American Development Bank joined the countries of the Northern Triangle of Central America (El Salvador, Guatemala, and Honduras) in an unprecedented initiative to channel massive funding into infrastructure projects in their cities, which are experiencing accelerated urbanization. It is part of a broader international effort (which also includes Mexico and the United States) to promote economic development and create more employment opportunities in this part of Central America.

Carol Ann Gillespie

GAZETTEER

Places whose names are printed in SMALL CAPS *are subjects of their own entries in this gazetteer.*

Aconcagua, Mount. Highest peak in the WESTERN HEMISPHERE, at 22,842 feet (6,962 meters). Located at 32°39′ south latitude, longitude 70°1′ west, in a group of Andean summits that exceed 21,000 feet (6,400 meters). Perpetual snow and ice cover these extinct volcanoes and provide meltwater for irrigation on both the Chilean and Argentine sides.

Acre. State in NORTH region of BRAZIL. Covers an area of 58,912 square miles (152,581 sq. km.), with a 2010 population of more than 733,000. Capital is RIO BRANCO. Land descends in west from foothills of sub-Andean range to eastern lowlands. Has an equatorial climate. Business includes rubber extraction, some agriculture, and ranching.

Alagoas. State in NORTHEAST region of BRAZIL. Total area of 10,721 square miles (27,767 sq. km.), with a 2010 population that exceeds 3.1 million. Capital is MACEIÓ. Land descends from high, arid interior in west to tropical Atlantic coast. Lowlands in the south, along SÃO FRANCISCO RIVER. Cattle are raised in upper region scrubland; tropical agriculture; tourism is substantial.

Altiplano. With Lake TITICACA at its center, South America's *altiplano* is a high plain in the ANDES MOUNTAINS of southern PERU and western BOLIVIA. Situated at more than 12,000 feet (3,650 meters) above sea level, this cold, dry, windswept plain is inhabited by farmers who eke out a living cultivating small plots of potatoes and raising sheep and llamas. Spanish for "high plain."

Amapá. State in NORTH region of BRAZIL. Total area of 55,141 square miles (142,814 sq. km.), with a 2010 population of nearly 670,000. Capital is MACAPÁ. From Brazil's western border with FRENCH GUIANA, land descends gradually eastward to Atlantic Ocean and AMAZON RIVER delta. Mostly equatorial rain forest. Extensive mining for manganese.

Amazon Basin. Drainage area of the AMAZON RIVER system, covering about 2.25 million square miles (5.83 million sq. km.) of South America. *See also* AMAZONIA.

Amazon River. South American river, almost 4,000 miles (6,500 km.) long. Forms the largest water basin in world, fifth of world's fresh water. Begins in Peruvian ANDES as APURÍMAC RIVER near Lake TITICACA, becomes the Ucyali River in eastern PERU, then meets the Marañon River and becomes the Peruvian Amazon River. After entering BRAZIL, the river's name changes to Solimões River until it meets NEGRO RIVER, below MANAUS, and becomes the Brazilian Amazon River. Its major tributaries are the Negro River to the north; and to the south and progressing downriver, the JURUÁ, PURUS, MADEIRA, TAPAJÓS, XINGÚ, and TOCANTINS Rivers. These tributaries are each more than 1,000 miles (1,600 km.) long.

Amazonas. Largest state in BRAZIL. Located in its NORTH region, with a total area of 606,468 square miles (1,570,745 sq. km.) and a 2010 population of 3.5 million. Capital is MANAUS. It occupies the western half of the Amazon plain, bisected from west to east by the AMAZON RIVER, into which drain waters of the GUIANA and BRAZILIAN HIGHLANDS and the ANDES MOUNTAINS. Equatorial climate and dense rain forest. Rubber extraction, lumber, mining, cattle raising, ecotourism.

Amazonia. Largest rain forest in world, covering more than 2 million square miles (5.5 million sq. km.), two-thirds situated in BRAZIL. Dominates entire northern part of Brazil and extends into eight other countries and territories in South America. Equatorial climate with short dry season in the eastern half. Nutrients are primarily in vegetation, not soil.

Ambergris Caye. See CAYES (CENTRAL AMERICA).

Americas. Collective term for the lands of the WESTERN HEMISPHERE, including North America, CENTRAL AMERICA, South America, and the islands of the Caribbean.

Andes Mountains. Forming the western spine of South America, the Andes rise from the waters of the Caribbean in eastern VENEZUELA and extend southward for 4,500 miles (7,240 km.) to the southern tip of the continent and TIERRA DEL FUEGO. The Andes have twelve peaks over 20,000 feet (6,100 meters), including the highest peak in the WESTERN HEMISPHERE, Mount ACONCAGUA. Climate, vegetation, settlement, and agriculture are highly influenced by elevation. The dramatic vertical rises of the mountain slopes contain almost every climate and environment found on Earth, from rain forest at sea level to tundra above the tree line. The Andes are part of the Pacific Ring of Fire, a zone around the Pacific Rim that is actively mountain-building. Earthquakes and volcanic activity are common in this zone. The mountains are still inhabited by many Amerindian people, who live by subsistence agriculture. The Andes are the place of origin of potatoes, llamas, and guinea pigs.

Angel Falls. World's highest waterfall, with a vertical drop of 3,212 feet (979 meters), almost the height of three Empire State Buildings and about sixteen times higher than Niagara Falls. Located in southeastern VENEZUELA, at 5°44' north latitude, longitude 62°27' west. Named after U.S. bush pilot and gold seeker Jimmie Angel, who first sighted the falls in 1933.

Angostura. See CIUDAD BOLÍVAR.

Antigua. Colonial city in GUATEMALA,; served as capital from 1543 until an earthquake struck in 1773, when the capital was moved to GUATEMALA CITY. Population was 45,669 in 2013. Surrounded by three volcanoes—Fuego, Acatenango, and Agua —evidence of the restless land on which the city lies. Colonial churches and other architectural displays are remnants of Spain's golden age in the AMERICAS. Casa K'ojom is a museum dedicated to the culture of the Maya, who have played such a vital role in the development of Guatemala.

Antioquia. Hilly department in northwest COLOMBIA, known for the high-quality coffee it produces. Settled by Spanish farmers, it is a picturesque area known for conservative, Old World, small-farm traditions, and the largest percentage of residents of European ancestry in Colombia. MEDELLÍN is the capital city and an important commercial and industrial center. The 2013 population was nearly 6.3 million.

Antofagasta. The principal port in CHILE for mineral exports, especially copper and nitrates. Located in ATACAMA DESERT 700 miles (1,100 km.) north of SANTIAGO, with a 2012 population of about 323,000. The world's largest copper mine, Chuquicamata, is located nearby. Antofagasta was part of BOLIVIA until Chile won it in the War of the Pacific (1879–1884), but it is still the terminus of the LA PAZ/ORURO railway from Bolivia.

Apurímac River. Headwaters of the AMAZON RIVER, the second-longest river in the world and the largest by volume. Beginning in the high ANDES MOUNTAINS of southern PERU, the Apurímac River eventually merges with the Amazon River and flows 4,000 miles (6,440 km.) to the Atlantic Ocean.

Aracaju. Capital of SERGIPE state, BRAZIL, with a 2010 population of 571,000. Port city, located on the right bank of Sergipe River and Atlantic Ocean in terrain of dunes and lagoon. City was planned on a grid pattern in 1858. It has textile manufacturing, as well as growing tourism owing to its beaches.

Arenal Volcano National Park. Park in the northern lowlands of COSTA RICA containing a 4,000-year-old volcano. The perfectly conical volcano rises 5,356 feet (1,624 meters). Arenal remained dormant for 400 years, then erupted in 1968, destroying several square miles of farmland around it; it erupted several more times, most recently in 1998. The park preserves eight of the twelve natural life habitat zones found in Costa Rica. Also includes Arenal's sister volcano, Chato.

Arequipa. Second-largest city in PERU, with a 2017 population of more than 1 million inhabitants. Founded on an Inca site in 1540, at an elevation of 7,636 feet (2,325 meters) in an earthquake zone in southern Peru. Known as the White City because many of the buildings are made of a light-colored volcanic rock called sillar.

Argentina. Also known as the Argentine Republic, a South American nation that occupies most of the southern portion of the continent, along the South Atlantic coast. The eighth-largest country in the world, it covers an area of 1.07 million square miles (2.78 million sq. km.) and has an undulating coastline some 2,900 miles (4,700 km.) in length. In the east, the country is separated from neighboring Uruguay and Brazil by the Uruguay River. In the west, the Andes Mountains separate it from Chile. Argentina also borders Bolivia and Paraguay to the north. To an extent greater than was the case in most other South American countries, Argentina attracted immigrants from Europe in the late nineteenth century and much of the twentieth century. In 2018 its population was 44.7 million, about 79 percent of whom were European in descent. The country still attracts European immigrants. Its capital is Buenos Aires.

Arica. Northernmost seaport and international duty-free trade zone in Chile. Located in the bone-dry Atacama Desert, Arica was part of Peru until the War of the Pacific (1879–1884) between Chile, Peru, and Bolivia. At that time, Bolivia lost access to the Pacific Ocean and became a landlocked country. The population exceeded 220,000 in 2017.

Artigas. Capital of the department of the same name in Uruguay. This small town is Uruguay's northernmost important town. It has excellent road and rail connections with Brazil and with the cities of Argentina and Uruguay that lie along the Uruguay River. Remote from the national core, Artigas services a large pastoral hinterland.

Asunción. Capital of Paraguay, and its first Spanish settlement. Population of the metro area was more than 2 million in 2017. Strategically located where the Pilcomayo and Paraguay Rivers meet, both flowing into the Paraná River.

Atacama Desert. Located in South America along the coast of northern Chile and southern Peru, this is one of the driest places on Earth and almost totally devoid of vegetation. Some locations go for decades without measurable rainfall. Human settlements are generally situated along rivers that originate in the Andes Mountains to the east and transverse the desert to the Pacific Ocean.

Atitlán, Lake. Beautiful, famous lake in the highlands of Guatemala and one of the most impressive sites in Central America. Located 90 miles (150 km.) west of Guatemala City. Covers 50 square miles (31 sq. km.); more than 11 miles (17 km.) long, almost 8 miles (13 km.) wide, and more than 1,000 feet (305 meters) deep in places, making it the deepest lake in Central America. Surrounded by volcanoes and granite cliffs, with deep blue water. Name is Mayan for "abundance of waters."

Avellaneda. Suburb of Buenos Aires, Argentina, in South America, and port city, with a 2010 population of nearly 343,000. Originally a community for the working poor, but modern industries of meat packing, wool, and hides have brought a measure of prosperity. Its factory workers were an important power base for Juan Perón in the 1940s and 1950s.

Ayacucho. Located in the Andes Mountains between Lima and Cuzco, Peru. Famous for its Holy Week religious processions and colorful local market. In the 1980s and early 1990s, it was a center of Amerindian resistance and the antigovernment movement Shining Path (*Sendero Luminoso*).

Bahia. State in Northeast region of Brazil. Total area of 218,431 square miles (565,733 sq. km.) with a 2010 population of 14 million. Capital is Salvador. Dry uplands in west descend to fertile, tropical Atlantic coastal area in east. Cattle are raised in upper region scrubland; tropical agriculture is practiced along the coast. Automotive manufacturing has grown substantially.

Bahia Blanca. Port city located at the southwesternmost shore of Buenos Aires province, Argentina. Originally settled as a fort to protect the settled cities in the north, this city with a population of 300,000 in 2010 is an important shipping point for products from the Pampas and Patagonia. Grain, wool, and fruit are sent here by rail, then shipped to final markets.

Bananal Island. Largest river island in BRAZIL, comprising 7,600 square miles (20,000 sq. km.). Located in the Araguaia River where MATO GROSSO borders TOCANTINS. Extending on a north-south axis, the island is almost 200 miles (320 km.) long. The island is inhabited solely by four tribes of Amerindians.

Baños. Popular resort on the eastern slope of the ANDES MOUNTAINS in ECUADOR. Known for its hot springs and as a gateway for climbers of the volcanic peak Tungurahua (16,460 feet/5,016 meters) and visits to the AMAZON BASIN.

Bariloche. See SAN CARLOS DE BARILOCHE.

Bay Islands. Three islands off the north coast of HONDURAS: Guanaja, Roatan, and Utila. Located about 31 miles (50 km.) off the coast, they are a continuation of the barrier reefs found off the coast of BELIZE. Christopher Columbus was the first European to land on the islands (Guanaja) in 1502; they were not populated by Europeans until the 1520s, when pirates occupied them. The islands remain Caribbean-like, and their people speak a Caribbean style of English. They are popular tourist destinations because of the excellent diving and snorkeling.

Beagle Channel. Waterway along the southern border of TIERRA DEL FUEGO Island, located at 54°53′ south latitude, longitude 68°10′ west. English naturalist Charles Darwin passed through this channel on the HMS *Beagle* on his famous scientific voyage in 1832–1834. Observing the perpetual fires built by the scantily clad Amerindians in their attempt to stay warm, he named the place *Tierra del Fuego*, or "land of fire." An ARGENTINE national park is on the channel just west of USHUAIA.

Belém. Capital of PARÁ state in BRAZIL. Located on the eastern edge of the mouth of the AMAZON RIVER, with a 2015 metro population nearing 2.5 million. Founded early in the colonial period. Major port and a regional commercial and services area. Noted for annual procession to Our Lady of Nazareth.

Belize. Small Central American nation located at the base of Mexico's Yucatán Peninsula, where its much larger neighbor, GUATEMALA, touches the Caribbean. Belize's land area is 8,867 square miles (22,965 sq. km.). It has a pronounced tropical climate, and more than 60 percent of its land is covered by forest. Its population was estimated to be 385,854 in 2018. In contrast to the rest of CENTRAL AMERICA, Belize was colonized by Great Britain—which named it BRITISH HONDURAS—and uses English as its official language. It became independent in 1981, joined the Commonwealth of Nations, and continues to recognize the British monarch as head of state.

Belize City. Largest city and former capital of BELIZE, and the country's economic and transportation hub. Population was about 61,000 in 2018. A bustling Caribbean coastal city, with brightly colored homes, tin roofs, and hurricane shutters; not a strong tourist destination. The new capital, BELMOPAN, is in the country's interior to avoid damage from hurricanes.

Belmopan. Capital of BELIZE since 1970. Located 41 miles (66 km.) southwest of BELIZE CITY, the former capital, in an interior location less subject to flooding from hurricanes. Had an estimated population in 2016 of 20,600.

Belo Horizonte. Capital of MINAS GERAIS state in BRAZIL. Located in central southeast of state. Brazil's sixth-largest city, with a metro area population of over 5 million in 2015. Founded in 1893, this planned city replaced the colonial capital of Ouro Preto. Industrial, commercial, and regional services center.

Bermudian Landing Community Baboon Sanctuary. Animal and bird sanctuary located about an hour's drive west of BELIZE CITY in BELIZE. The community of Bermudian Landing has set aside forest land to preserve native species and the endangered black howler monkey. Has a visitor information center to explain preservation efforts.

Bluefields. Town and export center of NICARAGUA. Located on the Caribbean side of Nicaragua, which was not colonized and remained a British protectorate until the late nineteenth century. A distinctly Caribbean town with a mix of Caribbean blacks, mestizos, and Amerindians; population in 2005 was just over 45,000. Bluefields can only be reached by taking a boat down the

Escondido River, since no roads connect it to MANAGUA

Boa Vista. Capital of RORAIMA state in BRAZIL. In 2018, its estimated population was 375,000. Formerly located on the right bank of the RIO BRANCO, it lies in the central northeast part of state. Center of cattle-raising district.

Bocas del Toro Archipelago. Islands in the CARIBBEAN SEA northwest of PANAMA that make up the Bastimentos National Park. The islands and park offer excellent snorkeling and diving. Sea turtles use the islands as nesting grounds.

Bogotá. Capital of COLOMBIA; officially known as Santa Fe de Bogotá. Located in the northern ANDES MOUNTAINS at an elevation of 8,563 feet (2,610 meters). Founded in 1538, it is a sprawling city with a metro-area population in 2018 of more than 10 million. Built in a long valley beneath the spectacular peaks of Monserrate and Guadalupe, Bogotá's splendid colonial architecture and wealthy modern districts contrast with its vast shantytowns. Crime has fallen dramatically in recent years.

Bolívar. See CIUDAD BOLÍVAR.

Bolivia. Landlocked country in Andean South America known for Lake TITICACA, people maintaining their traditional culture, llamas, and traditional Andean music. LA PAZ is the seat of government, but SUCRE is the constitutional capital. Bolivia has a land area of 424,165 square miles (1,083,301 sq. km.), approximately the size of the U.S. states of Texas and California combined. The population in 2018 was 11.3 million. The official languages are Spanish, Quechua, and Aymara. The second-poorest country in South America, Bolivia has a dual economy: Local farmers grow potatoes and other traditional subsistence crops and graze sheep and llamas; the export sector produces minerals (zinc, lead, and tin), natural gas, soybeans, and sugar. It is the second-largest producer of coca leaf, for cocaine production, after PERU.

Branco River. See RIO BRANCO.

Brasília. Capital of BRAZIL, located in the Distrito Federal (FEDERAL DISTRICT), surrounded by state of GOIÁS, except for the southeast tip, which touches MINAS GERAIS. The metropolitan population was almost 4.3 million in 2015. Inaugurated in 1960 as the new national capital, it represented an attempt to draw population to the center of the country and project modernization and development of Brazil through images of cultural modernism. The United Nations declared it a World Cultural Heritage Site in 1987. Numerous satellite cities also occupy the Federal District.

Brazil. Fifth-largest country in the world in terms of area (3.28 million square miles/8.5 million sq. km.) and population (208.8 million in 2018). Capital is BRASÍLIA. Spanning four time zones and occupying the central eastern half of South America, Brazil extends from under 5 degrees north of the equator to more than 30 degrees south of it and 10 degrees below the Tropic of Capricorn. Its border is more than 14,000 miles (23,000 km.) long, less than one-third of which is along the Atlantic coast; the rest touches all the countries of South America except CHILE and ECUADOR. Highly humid equatorial climate predominates in upper western half, tropical to semiarid climates in upper eastern half. Coast has tropical Atlantic climate, central south a highlands tropical climate, and far south a subtropical climate. Naturally rich soils for productive agriculture exist primarily along the eastern coast and in the southern interior. The AMAZON RIVER drains waters west to east from GUIANA and BRAZILIAN HIGHLANDS and the ANDES MOUNTAINS; in the south, waters drain north to south through the PANTANAL and the PARAGUAY, PARANÁ, and Plata Rivers. The SÃO FRANCISCO RIVER drains eastern central waters from south to north.

Brazilian Highlands. High plain of the interior of northern BRAZIL, which descends west-northwest from coastal mountains to the AMAZON RIVER floodplain. Together with the central and northern ANDES MOUNTAINS and GUIANA HIGHLANDS, they form the continental semicircular system of elevations from where waters drain into the Amazon River. Soil varies from *caatinga* to *cerrado* and even *terra roxa*. Cattle raising and agriculture dominate.

British Guiana. Colonial-era name of the South American nation of GUYANA.

British Honduras. Colonial era name of Central American nation of BELIZE, which became independent in 1981.

Buenos Aires. Capital of ARGENTINA, and the nation's cultural, economic, and political core. Population of the metropolitan area in 2015 was nearly 17 million. So much power in Argentina is concentrated in Buenos Aires that the outlying regions suffer. *Porteños*, as the people who live in Buenos Aires are known, tend to regard themselves as cosmopolitan and sophisticated and the rest of their countrymen as rustic.

Cajamarca. Mountain town in northern PERU, with a 2017 population of 201,000. The site where Spanish conquistador Francisco Pizarro captured and executed Inca emperor Atahualpa in 1533. Has interesting colonial architecture and a colorful local culture.

Cali. Third-largest city in COLOMBIA, with a 2012 population of 2.4 million. Located in the CAUCA RIVER valley, a rich agricultural region that produces sugar, cotton, coffee, and cattle. The principal urban center in the southwest of the country, Cali is a center for salsa and Caribbean music, and a world center for cocaine trafficking and the Cali Cartel.

Campo Grande. Capital of MATO GROSSO DO SUL state in BRAZIL. Located in the center of the state on a level peak of low-lying Serra de Maracaju, lying east of PANTANAL, with a 2010 population of 774,000. Railroad and transportation hub, and a center of agricultural and cattle production.

Canal Zone. Territory in PANAMA that includes the PANAMA CANAL. Was acquired by the United States from COLOMBIA, without the agreement of the latter, before the Panama Canal was built. Ceased to exist as a political entity in 1979.

Cape Horn. Southernmost point of South America. Located at 55°59′ south latitude in CHILE. Ships prefer to pass through the Strait of MAGELLAN rather than navigate around the treacherous waters of the cape.

Caracas. Capital of VENEZUELA and the country's largest urban agglomeration. Founded in 1567, Caracas grew to nearly 2 million inhabitants in 2017, fueled by the oil boom that began early in the twentieth century. Set in the hills at 2,950 feet (900 meters), just 8 miles (13 km.) from the Caribbean coast. Has an agreeable climate with a mean temperature of 70° Fahrenheit (21° Celsius). One of LATIN AMERICA's most modern cities, with high-rise buildings, affluent commercial districts, world-class museums, and a web of highways. Migration has led to sprawling slums and a marked contrast between wealth and poverty. In 2016, Caracas had the world's highest homicide rate for a city not in a war zone: nearly 120 per 100,000 people.

Caribbean Sea. Portion of the western Atlantic Ocean bounded by CENTRAL and SOUTH AMERICA to the west and south, and the islands of the Antilles chain on the north and east. Separated from the Gulf of Mexico on the west by the Yucatán Channel, which runs from the north tip of the Yucatán Peninsula to the southern tip of Florida. Covers about 1.05 million square miles (2.7 million sq. km.) and has a maximum depth of about 25,000 feet (7,620 meters) in the Cayman Trench.

Cartagena. Seaport on the northern coast of COLOMBIA, and one of the most beautiful colonial cities in LATIN AMERICA. Founded in 1533, it was the Spanish Empire's most important Caribbean port during the seventeenth century and the gateway to northern South America. The old walled city center has splendidly preserved colonial architecture. It is Colombia's most important petroleum port and processing center. Its population in 2016 was 971,600.

Cartago. Colonial city and first capital (until 1823) of COSTA RICA. Located 16 miles (25 km.) southeast of SAN JOSÉ, with a 2017 population of 160,000. Founded in 1563. Plagued by earthquakes over the centuries, the largest occurring in 1841 and 1910, and at times showered with ash from the nearby Irazu volcano. Home of the patron saint of Costa Rica, La Negrita. A small statue of the saint is housed in the Basilica de Nuestra Señora de los Ángeles; every August, Costa Ricans make pilgrimages to pay homage to the sacred statue.

Catamarca. An oasis in the northwest dry belt of ARGENTINA. Founded by colonizers from CHILE in 1558. Irrigation from Andean streams enables farmers to grow abundant crops of grapes, cotton, and fruits; planted pastures are the basis of the cattle industry. Many religious pilgrims visit an old church in the city that contains a revered statue of the Virgin Mary.

Catedral, Mount. Highest point in URUGUAY, at only 1,683 feet (513 meters). Located 47 miles (75 km.) north of PUNTA DEL ESTE, it emphasizes the nation's flat terrain. Sweeping grassy plains stretch from the Atlantic seaboard and URUGUAY RIVER to the Brazilian border. Cattle raising has been a dominant economic activity since the arrival of Europeans.

Cauca River. Major river in northwest COLOMBIA and a major tributary of the MAGDALENA RIVER; flows north from its headwaters in the ANDES MOUNTAINS for 838 miles (1,348 km.). The cities of CALI and MEDELLÍN are situated in the Cauca River valley.

Caulker Caye. See CAYES (CENTRAL AMERICA).

Cayenne. Capital, principal port, and largest city of FRENCH GUIANA. Founded in 1637 by French merchants. Had a 2016 population of about 60,500. Its interesting architecture, colorful outdoor cafés and food stalls, and ethnic diversity give it the feel of "France in the tropics."

Cayes (Central America). Barrier reef on BELIZE's Caribbean coast in CENTRAL AMERICA; the longest such reef in the WESTERN HEMISPHERE, at 180 miles (290 km.). Cayes (pronounced "keys") on the west side of the reef have clear, warm waters ideal for scuba diving and snorkeling, and they have become a popular tourist destinations. Two of the most popular cayes are Ambergris and Caulker. AMBERGRIS CAYE, located 36 miles (60 km.) north of BELIZE CITY, had a 2015 population of 16,000 people. It is Belize's longest caye, connected to the Mexican mainland to the north, and the site of the only coral atolls in the Western Hemisphere: Half Moon Caye, Turneffe Islands, and Blue Hole. Caulker Caye, 20 miles (33 km.) north of Belize City, is noted for its crystal-clear water, with visibility to a depth of 200 feet (60 meters).

Ceará. State in NORTHEAST region of BRAZIL. Total area of 57,147 square miles (148,000 sq. km.), with a 2014 population of about 8.8 million. Capital is FORTALEZA. Low mountains in the south gradually descend to northern Atlantic coast. Ranching is practiced in arid southern interior, tropical agriculture along the coast. There is some manufacturing and growing tourism.

Center-West. Region of BRAZIL, comprising states of GOIÁS, MATO GROSSO, and MATO GROSSO DO SUL, and the FEDERAL DISTRICT. Covers more than 622,000 square miles (1.6 million sq. km.), with a 2018 population of more than 16 million. Dominating feature in western portion is PANTANAL. Tropical environment.

Central America. Definitions of this region vary, but it is most generally understood to constitute the irregularly shaped neck of land that links North America and South America, containing the seven independent nations between Mexico and COLOMBIA: BELIZE, GUATEMALA, HONDURAS, EL SALVADOR, NICARAGUA, COSTA RICA, and PANAMA. These countries collectively cover 228,578 square miles (592,014 sq. km.) and were home to 44.5 million people in 2015. Also sometimes call Meso-America.

Cerro de Pasco. Area of extensive mineral deposits in the central Andean highlands of PERU. Once a major site of Spanish colonial silver mining, today it exports copper, lead, zinc, and gold. At an elevation of 14,200 feet (4,330 meters), Cerro de Pasco has some of the highest mines in the world.

Cerro Verde National Park. Home of Izalco and Santa Ana volcanoes in EL SALVADOR. Izalco began to grow more than 200 years ago, until it formed a cone 6,400 feet (1,950 meters) high. It erupted periodically until 1966, then stopped. Adventuresome travelers can hike to the top.

Coatepeque Caldera, El Salvador (Image by JMRAFFi)

Santa Ana volcano is known for its deep blue crater lake, Coatepeque Lake, located on the east side of the volcano. Name is Spanish for "green hill."

Chacaltaya. Highest ski run in the world, at 17,130 feet (5,221 meters) above sea level. Located in the ANDES MOUNTAINS of BOLIVIA just north of LA PAZ. From the peak, one can see Lake TITICACA, the ALTIPLANO, and the volcanic peaks of the eastern range of the Andes. Its famous glacier melted completely between 1940 and 2009.

Chaco. Immense alluvial plain, stretching from just south of the PANTANAL to the middle latitudes of northern ARGENTINA, extending 700 miles (1,125 km.) in a north-south direction. A dry tropical/semitropical region.

Chan Chán. Ruin of the largest adobe city in the world. Located on the north coast of PERU, near TRUJILLO, Chan Chán was the imperial city of the Chimú people who were conquered by the Incas in the late fifteenth century. The city wall encompasses 11 square miles (28 sq. km.) of palaces, temples, royal burial mounds, and ruins of 10,000 other structures.

Chichicastenango. Mayan city in GUATEMALA. Located in a high mountain valley 90 miles (150 km.) west of GUATEMALA CITY. Known for its colorful markets, located in the main plaza on Thursdays and Sundays, selling wares from food to cultural items. In its churches, one can witness the blending of Roman Catholicism and ancient Mayan beliefs. Nowhere is this more evident than at the shrine of Pascual Abaj, which pays homage to the Mayan god of the earth.

Chile. An elongated country in southwest South America bounded by the ANDES MOUNTAINS and the Pacific Ocean. It is 2,650 miles (4,265 km.) long, but never more than 220 miles (355 km.) wide. Its tremendously varied natural environments range from the bone-dry ATACAMA DESERT in the north, to the Mediterranean climate in the center, to the glaciated fjordlike landscape of the far south. The estimated 2018 population is 18 million, of whom most live in the temperate central region of the country where the capital, SANTIAGO, is located. At the beginning of the twenty-first century, Chile had one of the most dynamic economies in LATIN AMERICA and was a major world supplier of fresh fruits (grapes, pears, plums, apples, peaches, and citrus) and the world's largest exporter of copper. In 1970 Salvador Allende became the first democratically elected Marxist leader in Latin America. He was overthrown by the military and Augusto Pinochet in 1973. Chile returned to democracy in 1989.

Chiloé. Island off southern CHILE (42°30′ south latitude, longitude 73°55′ west) known for picturesque forest and farmland and traditional Spanish culture. The descendants of early settlers and Jesuit converts inhabit small agricultural and fishing villages. More than 150 wooden churches dot the island. The island is 155 by 32 miles (250 by 50 km.) Chiloé National Park is noted for its marine fauna, most notably the highly endangered blue whale.

Chimborazo, Mount. Highest mountain in ECUADOR, at 20,700 feet (6,310 meters); once thought to be the highest mountain in the world. In his celebrated work *Kosmos*, German explorer and geographer Alexander von Humboldt described the relationship between elevation and changing vegetation and life zones based on observations on the slopes of Chimborazo and other Andean peaks.

Chocó. Lowland province in the extreme northwest of COLOMBIA, with dense tropical rain forests bordering the DARIÉN GAP; sparsely populated, mainly by persons of African descent. With more than 160 inches (400 centimeters) of annual rainfall, it is wetter than the AMAZON BASIN and one of the world's richest areas for plants and animals. Gold and platinum mines are also found there.

Ciudad Bolívar. Capital of Bolívar, the largest state in VENEZUELA. Located on bank of the ORINOCO RIVER 250 miles (400 km.) upstream from the Atlantic Ocean, with a 2015 population of 400,000. The jumping-off point for visits to ANGEL FALLS. Originally named Angostura, Ciudad Bolívar is where liberator Simón Bolívar, in 1817, based the military operations against Spanish colonial control in his effort to form Gran Colombia. The state of Bolívar is the site of a major industrial complex based on hydroelectric power and steel and aluminum production.

Cochabamba. Fourth-largest city in BOLIVIA, with a 2012 population of 630,000. Set in a valley in the ANDES MOUNTAINS at an elevation of 8,430 feet (2,570 meters), it has a pleasant climate and is an important agricultural center for grains and vegetables. The surrounding rural areas are dominated by Quechua-speaking Amerindians.

Colca Canyon. Canyon in southern PERU, near AREQUIPA, that is said to be twice as deep as the Grand Canyon. Both sides are beautifully terraced in agriculture and dotted with traditional villages. The region also contains many majestic volcanic peaks.

Colombia. Country located in northwest South America, bounded by the CARIBBEAN SEA, the Pacific Ocean, VENEZUELA, PANAMA, ECUADOR, PERU, and BRAZIL. BOGOTÁ is the capital. Total area of 439,735 square miles (1,138,914 sq. km.); the population was 48.2 million in 2018. Spanish is the principal language. The physical environment is varied: the dissected terrain of the northern ANDES MOUNTAINS, deep river valleys of the MAGDALENA and CAUCA, Pacific coastal lowlands, and the isolated eastern lowlands in the AMAZON BASIN. In the twentieth century, Colombia was plagued by fractious regional divisions, political violence, guerrilla warfare, and drug wars. In the years since 2005, security and stability improved significantly. Today, the country has a relatively diversified and prosperous legal economy. Its principal exports are petroleum products, coal, emeralds, nickel, cut flowers, and coffee, of which it is the world's third-largest producer.

Colonia del Sacramento. City in URUGUAY. Located on LA PLATA RIVER ESTUARY across from BUENOS AIRES. Founded in 1680 by Portuguese from BRAZIL to solidify their southern boundary against expansion of Spanish settlements from Buenos Aires. Ferry services link it to Buenos Aires.

Colorado River. River in ARGENTINA, marking the start of PATAGONIA. Its valley has a broad flat bottom and steep banks, because the Andean meltwater has cut deeply into the steplike plateaus that make up most of Patagonia. The region began to be developed near the end of the nineteenth century at the end of the war against the Amerindians. Irrigation is necessary to allow the cultivation of grain, animal feed, and fruit.

Comayagua. Capital of HONDURAS, 1537–1880. Population was 152,000 in 2015. Known for its colonial architecture, including numerous churches such as La Merced, built between 1550 and 1558. The first university in CENTRAL AMERICA was founded there in 1632, in the Casa Cural, which houses the Colonial Museum.

Comodoro Rivadavia. Seaport in ARGENTINA and center of the oil industry in Argentina's PATAGONIA. Located in Chubut Province, originally as a port to ship the region's agricultural production, it took on new importance when oil was discovered nearby. Tankers and pipelines send the crude oil to refineries near BUENOS AIRES. Population was 180,000 in 2010.

Concepción. City in PARAGUAY. Located on the PARAGUAY RIVER, with a 2008 population of 76,000. The most important commercial center for the northern part of the country, and a free port for trade with southwest BRAZIL. Sawmills, flour mills, tanneries, and sugar refineries are major industries.

Concepción. Third-largest city in CHILE and a center for wood products. Located on the Bío-Bío River 310 miles (500 km.) south of SANTIAGO, with a 2017 population of 221,000. The surrounding region has a marine west coast climate much like coastal Oregon and Washington in the U.S. Large tracts of temperate rain forest contribute to expanding exports of timber products. The area has many Spanish and Italian immigrant families.

Copán. Ruling center of the Maya from about 1,400 to 1,200 years ago. Prehistoric peoples began moving into the Copán Valley, part of present-day HONDURAS, about 2,000 years ago. The Maya and their rulers left behind impressive artifacts depicting their culture. Its acropolis has carved reliefs of the sixteen Mayan rulers of Copán, and the Stelae of the Great Plaza portray these rulers and their reigns.

Córdoba. Second-largest city in ARGENTINA. Located at the eastern margin of the picturesque Córdoba Mountains, with a 2019 population of nearly 1.4 million. Boasting the nation's oldest university, it is an important agricultural and industrial center. A transportation hub between the coast and interior, and a center for leather, textile, glass, automobiles, and food processing.

Corrientes. City in ARGENTINA. Located on the PARANÁ RIVER near the confluence of the PARAGUAY RIVER, with a 2010 population of 346,000 people. Its warm, wet climate allows a diversified agriculture of citrus, sugar, cotton, tobacco, and fruit, as well as cattle and hides. Excellent river, road, and rail transportation link the region with domestic and foreign markets. The 430-year-old city is also a cultural center with a university, museums, and a monastery.

Costa Rica. Central American country bordered by NICARAGUA, PANAMA, the Caribbean, and the Pacific; its area is 19,730 square miles (51,100 sq. km.), with a population of 4,987,000 in 2018. Though within the tropics, its mountains reach as high as 12,530 feet (3,820 meters) at Mount Chirripó, the country's highest peak. Costa Rica differs from other Central and South American countries in having a much smaller gap between its richest and poorest classes. Historically it has shown a greater propensity for social and governmental stability. Its capital is SAN JOSÉ.

Cuenca. Third-largest city in ECUADOR (2010 population 330,000). A pleasant colonial town located at an elevation of 8,300 feet (2,530 meters) in the southern highlands of Ecuador, about 56 miles (90 km.) from Ingapirca, Ecuador's most important Inca ruin. Because of its high-elevation equatorial location, the climate is springlike. Surrounding region is home to several groups of indigenous peoples, and the place where the original Panama straw hats are woven by hand.

Cuiabá. Capital of MATO GROSSO state in BRAZIL. Located on left bank of Cuiabá River in the central southern portion of state. Population in 2019 was 613,000. A colonial gold-mining area, a cattle-raising region, and gateway to the PANTANAL. Considered the geographical center of South America. One of the hosts of the 2014 FIFA World Cup.

Curitiba. Capital of PARANÁ state in BRAZIL. Located on east central edge of state at top of escarpment connected by train to coastal port of Paranaguá. Had a 2017 population of 1.9 million. Noted for its exceptional urban administration. One of the hosts of the 2014 FIFA World Cup.

Cuzco. Magnificent Spanish colonial city in southern PERU and the ancient capital of the Inca Empire. Located at an elevation of 10,860 feet (3,310 meters) in the ANDES MOUNTAINS. Population in 2010 was 428,000. Most residents are Amerindians. It has many examples of Spanish colonial architecture and churches, as well as beautifully preserved Inca stonework and ruin sites. Cuzco means "navel" or "center" in the language of Quechua.

Darién Gap. The area of dense jungle, wetlands, and mountains that straddles the Panama-Colombia border. The only section of the Pan-American Highway yet to be built is the 100-mile-long (160-km.-long) stretch through the Darién Gap. Given the difficult terrain, environmental considerations, and the lack of political consensus, the near-term outlook for completing the highway remains unclear.

Devil's Island. Notorious prison in FRENCH GUIANA that was a penal colony from 1852 to 1953. Located 8 miles (13 km.) offshore on one of the Safety Islands (Îles du Salut), it was nearly impossible to escape from because of shark-infested waters and strong currents. A tourist attraction today.

Distrito Federal. See FEDERAL DISTRICT.

Dutch Guiana. Colonial-era name of SURINAME.

Easter Island. Located in the South Pacific 2,350 miles (3,780 km.) off the coast of CHILE. Known for hundreds of monolithic stone figures, it originally was inhabited by Polynesians. The first European to land there was the Dutch explorer Jacob Roggeveen, on Easter Sunday in 1722. Annexed by Chile in 1888. Entire island is a UNESCO World Heritage Site.

Ecuador. Small Andean country located on the equator in western South America. It has an area of 109,483 square miles (283,561 sq. km.), about the size of the U.S. state of Colorado. Its population was 16.5 million in 2018. QUITO is the capital; the predominant languages are Spanish and Quechua. Ecuador has three geographic regions. The Sierra, located in the ANDES MOUNTAINS, is characterized by large estates and by small farms worked by farmers who raise potatoes and corn and graze animals. The Costa—the lowlands between the Andes Mountains and the Pacific Ocean—is dominated by commercial farms producing bananas, rice, cocoa, and shrimp for export. The Oriente is in the AMAZON BASIN in the east of the country. Oil was discovered there in the 1960s. Ecuador is an oil-exporting country and the world's leading exporter of bananas.

El Chapare. A frontier zone in the eastern foothills of the ANDES MOUNTAINS between COCHABAMBA and SANTA CRUZ, and the largest coca-growing region in Bolivia. Most of the coca is processed into cocaine, and the region has been the focus of U.S. crop substitution and eradication programs.

El Salvador. Smallest and most densely populated of CENTRAL AMERICA's seven countries, with a land area of 8,124 square miles (21,041 sq. km.) and a population of 6,187,000 (2018). The only Central American country not to have a Caribbean coast, El Salvador faces the Pacific Ocean and is bordered by GUATEMALA on the northwest and HONDURAS on the east. Its economy is primarily agricultural; its most important cash crop is coffee. Its capital is SAN SALVADOR.

Encarnación. City in southeastern PARAGUAY. Located on the PARANÁ RIVER, with a 2002 population of 67,000 people. It has road and rail links with ASUNCIÓN; connections with Posadas, ARGENTINA, are by ferry across the river. A major port on the river, trading yerba maté tea, cattle, timber, hides, and grains.

Esmeraldas. City on the northern coast of ECUADOR, where Spanish conquistadors first landed in Ecuador. Its population in 2010 was 162,000. Site of the terminal point of the trans-Ecuadorian Pipeline, an oil refinery, and a port. Surrounding region is known for shrimp fishing and a Creole population descended from African slaves.

Espírito Santo. State in SOUTHEAST region of BRAZIL. Total area of 17,658 square miles (45,734 sq. km.) with a 2014 population of 3.9 million. Capital is VITÓRIA. High mountainous western interior descends to tropical Atlantic eastern coast. Industries include cattle raising, coffee production, tropical agriculture, mining, some industry.

Falkland Islands. Also known as the Islas Malvinas, a windswept archipelago in the South Atlantic Ocean, about 320 miles (500 km.) off the southeastern coast of ARGENTINA, on almost the same latitude as the Strait of MAGELLAN. The archipelago consists of two large islands, East and West Falkland, and more than 340 small islands and islets. Their land area is about 4,699 square miles (12,175 sq. km.)—slightly smaller than the U.S. state of Connecticut. Most of the approximately 3,400 (2016) residents of the islands live on East Falkland, where the capital, Stanley, is located. Great Britain has administered the islands since the 1830s. Argentina's long-standing claims to the islands led to a brief war with

Great Britain in 1982 that did not change their status.

Federal District (Brazil). Special administrative district of BRAZIL containing the headquarters of the national government—a territory similar in function to Mexico's Federal District and the District of Columbia of the United States. Standing on a savanna at an elevation of about 3,300 feet (1,000 meters) above sea level, it has a mild, dry climate. It is located inside the state of GOIÁS in south central Brazil and contains the national capital city, BRASÍLIA. The district and capital were dedicated in 1960.

Fernando de Noronha. Brazilian archipelago more than 200 miles (320 km.) northeast of RIO GRANDE DO NORTE. Largest of islands, comprising tips of ancient volcanoes, is Fernando de Noronha. The territory is an ecological reserve with a small military contingent, administered from PERNAMBUCO.

Florianópolis. Capital of SANTA CATARINA state in BRAZIL. Population in 2016 was 478,000. Located on western side of the Atlantic coastal island of Santa Catarina, connected by bridge to mainland. Eastern side of island is dunes and ocean beaches, popular with tourists from BRAZIL, ARGENTINA, and URUGUAY. Formerly known as Desterro.

Fortaleza. Capital of CEARÁ state in BRAZIL. Located on Atlantic Ocean, with a 2018 population of 2.6 million. Growing regional commercial, port, banking, manufacturing, and tourist center. Some fishing; beaches popular with tourists despite high crime rate.

Fray Bentos. Major river port in URUGUAY for navigation on the URUGUAY RIVER of oceangoing vessels. An international bridge crosses the river to ARGENTINA. Uruguay's first large-scale meatpacking plant was built here in 1861. With a deep-water port and good rail, road, and air connections, this city of 24,000 (2011) people is economically important despite its small size.

French Guiana. Overseas Department (state) of France located on the north coast of South America, bordered by SURINAME to the west and BRAZIL to the east and south. Total area of 35,126 square miles (90,976 sq. km.) with a 2019 population of 297,000. Capital is CAYENNE. Was a French penal colony, including the notorious DEVIL'S ISLAND, from 1852 to 1953. Population is mostly Creole (people of African ancestry), with notable Vietnamese, Indonesian, and European minorities. Most of the settlement is on the coast, with a third of the people living in Cayenne. The interior is largely uninhabited. French is the official language, but a local Creole dialect is spoken. Principal exports are gold, shrimp, rice, and timber. The territory is also used as a launch site for commercial satellites at the French Space Center near Kourou.

Galápagos Islands Archipelago. Archipelego of nineteen volcanic islands located on the equator 597 miles (960 km.) west of South America, a territory of ECUADOR. Visited by English naturalist Charles Darwin, sailing on the HMS *Beagle*, in 1835. Darwin's observations on the islands contributed to his theories about evolution by natural selection. With a land area of about 3,000 square miles (8,000 sq. km.), the Galápagos are slightly less than half the size of the Hawaiian Islands. In 1979 the United Nations designated the islands as a World Heritage Site, recognizing them as one of the world's most magnificent natural areas and home to many rare and unique plant and animal species, including tortoises, marine iguanas, and blue-footed boobies. Tourism is closely monitored, but the Galápagos are threatened by introduced plants and animals, poorly regulated fishing, and excessive migration from the mainland.

Georgetown. Capital, principal port, and only large city of GUYANA. Located at the mouth of the Demerara River on the Atlantic Ocean, with a 2016 population of 200,000. Founded by the English in 1781. Parts of the city are notable for remnants of old sugar plantations, streets lined with flowering trees, and white wooden nineteenth century houses.

Goiânia. Capital of GOIÁS state in BRAZIL. Population in 2010 exceeded 1.3 million. Founded as planned city in 1933 to replace colonial capital of Goiás Velho. Growing commercial and services center for dynamic rural economy.

Goiás. State in CENTER-WEST region of BRAZIL, extending over the central plateau of country. Total area of 131,339 square miles (340,168 sq. km.) with a 2017 population of 6.7 million. Capital is GOIÂNIA. Combines tropical and scrub vegetation; southern half of state productive in agriculture and ranching.

Gran Chaco. Alluvial plain covering western PARAGUAY, northwestern ARGENTINA, southwest BRAZIL, and southeastern BOLIVIA, formed by sediments washing down from the ANDES MOUNTAINS. Covers 280,000 square miles (725,000 sq. km.). Large and small rivers cross the plain flood in the hot, wet summers, creating vast swamps, and dry out in the cool winters, making irrigation waters scarce.

Guanabara Bay. Major bay off the Atlantic coast of South America, in Brazilian state of RIO DE JANEIRO. Along south edge of bay stretch the port areas of city of RIO DE JANEIRO; along northeast edge is port city of Niterói. A major bridge spanning the bay connects the two cities. There are many islands in the bay. Its water was once so clean that dolphins swam in it, but today it is polluted, especially from oil spills and sewage.

Guatemala. Third-largest country in CENTRAL AMERICA; bordered by Mexico on the north and west, EL SALVADOR and HONDURAS on the east, and BELIZE—whose territory it claims—on the northeast. It has a 150-mile (240-km.)-long coastline on its Pacific Ocean side, but its Caribbean coastline is only a fraction of that length. Guatemala is largely tropical, with low-lying coastlands, but does have some drier areas. Cool temperatures exist at high elevations. Its capital is GUATEMALA CITY. Its land area is 42,042 square miles (108,889 sq. km.). Its population was 16,581,000 in 2018.

Guatemala City. Capital and cosmopolitan hub of GUATEMALA. Population in 2018 was 2.4 million. Impressive for its size, but little remains of its colonial heritage. The city's retail district is located in the Plaza Mayor, and the central market caters mostly to the needs of tourists.

Guayaquil. Largest city in ECUADOR, with 2.7 million inhabitants in 2019; the country's principal seaport and commercial and industrial center. Located on the ocean near the equator, at the heart of the coastal region that produces bananas, rice, sugarcane, cacao, and shrimp. Known for its oppressive heat and humidity.

Guiana Highlands. Elevated uplands in north of BRAZIL bordering COLOMBIA, VENEZUELA, GUIANA, and SURINAME. Waters from this region drain to the lower AMAZON RIVER. Brazil's highest peaks are located here: PICO DE NEBLINA (about 10,000 feet/3,000 meters), Pico de Trinta e Um de Março (about 9,800 feet), and Mount Roraima (just over 9,400 feet).

Guianas. Region of northern South America comprising GUYANA (formerly BRITISH GUIANA), SURINAME (formerly Dutch Guiana), and FRENCH GUIANA (a territory of France). These territories have interesting colonial histories, highlighted by sugar plantations and imported laborers. This has left them with ethnically diverse populations, including Creoles, Asians (Indians, Indonesians, Chinese, and Vietnamese), and Europeans. Most of the settlement is along a narrow coastal fringe, leaving the undeveloped interior with some of South America's most pristine rain forests. Economies rely heavily on agriculture and mineral extraction.

Guyana. Country on the north coast of South America, bordered by VENEZUELA to the west, SURINAME to the east, and BRAZIL to the south. Total area of 83,000 square miles (214,970 sq. km.), with a 2018 population of 741,000, concentrated on the coastal fringe. Capital is GEORGETOWN. The poorest country in South America. First settled by the Dutch in 1613, became a British colony in 1831 (then known as British Guiana), achieved independence in 1966. The land area is roughly the size of the U.S. state of Idaho; approximately 77 percent is covered by dense tropical rain forests and savanna, largely in the undeveloped interior. Its ethnically diverse population is divided between Creoles (people of African ancestry) and East Indians (people of southeast Asian ancestry). English is the official language, but Hindi, Urdu, and Amerindian dialects are spoken also. Principal exports are sugar, gold, rice, bauxite, and alumina.

Honduras. Republic in CENTRAL AMERICA. Covering an area of 43,433 square miles (112,492 sq. km.), it is encompassed by GUATEMALA on the west and NICARAGUA on the east. It has a long coastline on the CARIBBEAN SEA, but EL SALVADOR stands between it and the Pacific Ocean on the southwest, leaving Honduras its only outlet to the Pacific through the tiny Gulf of Fonseca. Its capital and largest city is TEGUCIGALPA. The tropical climate of Honduras is tempered by the higher elevations in the interior. Its population was 9,182,766 in 2018.

Huascarán Nevado. The highest mountain in PERU. Located at 9°6′ south latitude, longitude 77°50′ west, at an elevation of 22,133 feet (6,746 meters). Part of the picturesque Cordillera Blanca (white range), it is part of a national park that protects an area of high mountain cloud forest and tundra in the ANDES MOUNTAINS.

Huayaga Valley. Zone on the eastern flank of the ANDES MOUNTAINS in PERU, known for cultivating coca leaf, the raw material for cocaine. Poor farmworkers there have been caught between U.S. antidrug efforts and Peruvian antigovernment terrorist activities.

Humboldt Current. See PERU CURRENT.

Iguaçu Falls. Located in southwest corner of state of PARANÁ, BRAZIL, these stunning falls are formed by a deep, semicircular depression in the Iguaçu River as it approaches the PARANÁ RIVER. Rushing down numerous tiers of rock molded in a massive arc, the spewing waters glisten and roar through an exotic subtropical scenario of butterflies, hummingbirds, sunlight, gleaming flowers, and rainbows. Also known as Iguazú Falls.

Iguazú National Park. A favorite tourist destination in ARGENTINA, located along the Iguazú River. One can view tropical vegetation, animals, and spectacular waterfalls where the Iguazú River plunges over the PARANÁ PLATEAU in a series of high cascades.

Ipanema. Famed beach and neighborhood in the Brazilian city of RIO DE JANEIRO, nestled between the Atlantic Ocean and a lagoon. Its sophistication and physical allure are celebrated in the song "The Girl from Ipanema." It is the second of the first three ocean-beach neighborhoods of Rio, preceded by the highly commercialized Copacabana and followed by the residential Leblon.

Iquitos. PERU's most important city in the AMAZON BASIN of South America. Located on the river Marañón, a tributary of the AMAZON RIVER, with a 2017 population of 378,000. Accessible to oceangoing vessels that come 2,240 miles (3,600 km.) up the Amazon River from the Atlantic Ocean. In the late nineteenth century, the region experienced a rubber boom; later, cacao became the dominant crop. Since the 1960s, oil has driven the development of the region.

Isla de Margarita. Venezuelan island in the CARIBBEAN SEA 25 miles (40 km.) north of the mainland. Total area is about 350 square miles (900 sq. km.). VENEZUELA's number one tourist destination for Caribbean atmosphere, white beaches, snorkeling, and visits to nearby national parks.

Itaipú Dam. Hydroelectric dam spanning the PARANÁ RIVER between BRAZIL and PARAGUAY in South America. Opened in 1984 as the largest dam and one of the largest reinforced concrete structures in the world. Twenty generators produce 700 megawatts each. The largest generator of renewable, clean energy in the world.

João Pessoa. Capital of PARAÍBA state in BRAZIL. Located between the right bank of Sanhaú River and the Atlantic Ocean. Popular tourist city with 2018 population of 800,000. Formerly Filipéia (under Spain), Frederikstad (under Netherlands), and Paraíba.

Juruá River. A major tributary of the AMAZON RIVER in South America, draining from the Peruvian ANDES. About 1,500 miles (2,400 km.) long, it crosses the Brazilian states of ACRE and AMAZONAS, emptying into the upper Solimões River. It has many tributaries of its own.

La Amistad International Park. International Peace Park shared between COSTA RICA and PANama. Along with several reserves and some Amerindian reservations, the combined landmass is about 124,000 acres (500,000 hectares) and forms the Amistad Biosphere Reserve, recognized as a World Heritage Site. The region protects thousands of species of plants and animals, including more than 600 species of birds. One of the best places in Costa Rica to witness Amerindians in more traditional settings.

La Paz. The seat of government of BOLIVIA. Located 11,900 feet (3,630 meters) above sea level, La Paz is the highest capital city in the world. The La Paz metropolitan area had a 2012 population of 2.3 million people, which also makes it the world's most populous city at such a high elevation. The average annual temperature is a comfortable 64° Fahrenheit (18° Celsius), resulting from its high-altitude location near the equator.

La Plata. Secondary port for BUENOS AIRES on the LA PLATA RIVER ESTUARY in ARGENTINA. Buenos Aires attracts so many ships that the traffic congestion causes many of them to be rerouted to La Plata's deep-water harbor. A major industrial center for meat packing, oil refining, flour milling, and textiles. With almost 765,000 people (2010), the well-planned city has a pleasant urban environment with many plazas and parks.

La Plata River Estuary. The place where the great URUGUAY-PARAGUAY-PARANÁ RIVER system meets the south Atlantic Ocean. BUENOS AIRES, MONTEVIDEO, and many other cities are located on the estuary's banks to take advantage of the region's outstanding navigation potential. Large amounts of silt are washed downriver from the vast continental interior basin of southern South America. Shipping channels near Buenos Aires must be continually dredged to keep them open to large ships.

Lagoa dos Patos. Largest coastal lagoon in BRAZIL, stretching along half the coast of the state of RIO GRANDE DO SUL. It is about 175 miles (280 km.) long and 40 miles (60 km.) wide. At the northwest tip of the lagoon is the Guaíba River, where the port city of PORTO ALEGRE is located. At its south end, the lagoon connects to the Atlantic Ocean and Lagoa Mirim.

Lamanai. Mayan site located near the remote village of Indian Church in BELIZE. Although undergoing restoration and excavation, it is an impressive work in progress. The nearly sixty structures and ruins include a ball court, a small temple, and a late pre-classical Mayan structure that rises more than 112 feet (34 meters) above the jungle. Was occupied until the arrival of the Spanish, as evidenced by the two ruins of Mayan churches on the site. Name is Mayan for "submerged crocodile."

Las Piedras. Suburb northwest of MONTEVIDEO, Uruguay. Located on the floodplain of the URUGUAY RIVER, with a 2011 population of 71,000. Food and wine are major products; it is a major rail hub for the large cities along the river. The Uruguayan patriot and nationalist José Gervasio Artigas won an important battle against the Spaniards here in 1811.

Latin America. Loosely used term for countries and territories in the AMERICAS that are culturally predominantly Hispanic or Portuguese in their backgrounds. By this definition, Latin America includes BRAZIL, Mexico, and all the Spanish-speaking nations of Central and South America and the Caribbean, while excluding such English- and Dutch-speaking lands as GUYANA, SURINAME, BELIZE, Jamaica, and other English-speaking Caribbean islands. Because French—like Spanish and Portuguese—is a descendant of the Latin language, Latin America has also been construed as including such French-speaking lands as FRENCH GUIANA, Haiti, and other Caribbean islands.

León. The radical, liberal, and intellectual center of NICARAGUA, and its second-largest city, with a 2016 population of 168,000. Located near the Pacific coast, 50 miles (80 km.) northwest of MANAGUA. Has many examples of fine Spanish colonial architecture but is best known for its monuments to the Sandinista political and military movement, including large wall murals.

Lima. Capital and largest city of PERU, with a 2017 population of 9 million. Founded in 1535 by the

Spanish and the principal city of Spain's colonial holdings in Andean South America. Designated by the United Nations as a World Heritage Site, Lima's center has impressive architecture and numerous museums holding colonial and indigenous treasures. Since 2014, Lima has become a major destination for Venezuelan refugees fleeing the turmoil in their country.

Llanos. Vast plains in northern South America. Drained by the ORINOCO RIVER and its tributaries in VENEZUELA and COLOMBIA, forming the world's fourth-largest river by volume of water. Humid, often marshy, and sparsely populated; principal land use is cattle ranching. Petroleum is produced in the eastern *llanos* of Venezuela.

Macapá. Capital of AMAPÁ state in BRAZIL. Located on western delta of AMAZON RIVER, with the equator running just to its south. Population in 2018 was nearly 500,000. Originated as a Portuguese fortress to protect the Amazon from foreign penetration; a port for shipping manganese.

Maceió. Capital of ALAGOAS state in BRAZIL. Located on the Atlantic coast on an elevated sandy area of lagoon region, with a 2013 population of nearly 1 million. Commercial center with light industry, offshore oil drilling, and growing tourism.

Machu Picchu. Famous lost city of the Incas, located near CUZCO in PERU. Set on a high mountain terrace on the eastern flank of the ANDES MOUNTAINS, it was abandoned in the fifteenth century before being rediscovered by U.S. explorer Hiram Bingham in 1911. Machu Picchu is a United Nations-designated World Heritage Site and a major tourist attraction in South America. Plans to build a cable car system to the site have brought concerns that too much tourism might be detrimental to its preservation.

Madeira River. The largest tributary of the AMAZON RIVER in South America. Flows southwest/northeast through Brazilian states of RONDÔNIA and AMAZONAS. About 2,000 miles (3,200 km.) long, it drains from the Bolivian ANDES and enters the Amazon below MANAUS.

Magdalena River. Largest river in COLOMBIA. Flows from its headwaters in the ANDES MOUNTAINS north 950 miles (1,540 km.) into the CARIBBEAN SEA near Barranquilla. Navigable most of the way, it is an important transportation corridor in a country that is fragmented by north-to-south mountain ranges and river valleys.

Magellan, Strait of. Strait passing between the southernmost point of the mainland of South America and TIERRA DEL FUEGO. The Portuguese explorer Ferdinand Magellan discovered the passage in 1520 as an alternative to the longer route around CAPE HORN. The straits connect the Atlantic and Pacific Oceans.

Malvinas, Islas. See FALKLAND ISLANDS.

Managua. Capital of NICARAGUA. Had a 2016 population of 1 million people. Natural disasters such as earthquakes have destroyed much of its architecture and heritage. A large earthquake in 1972 virtually destroyed the downtown area, little of which has been rebuilt. The volcanoes and craters in Masaya Volcano National Park, less than a half hour's drive from the city center, are one of Nicaragua's top tourist draws.

Managua, Lake. See NICARAGUA, LAKE.

Manaus. Capital of AMAZONAS state in BRAZIL. Located on the southern bank of the NEGRO RIVER just above where that river joins the Solimões River to form the AMAZON RIVER. Population in 2019 was more than 2 million. A historic river port, especially for export of rubber; a tariff-free zone, with some industrialization and ecotourism.

Manu National Park. Largest nature reserve in PERU. Located on the Madre de Dios river near Puerto Maldonado in the southeast corner of Peru, it encompasses 4.6 million acres (1.9 million hectares). The zone of the AMAZON BASIN along the eastern flank of the ANDES MOUNTAINS is one of the world's centers of biodiversity, with an almost unmatched diversity of plant and animal species.

Manuel Antonio National Park. Second-most-visited park in COSTA RICA. Located on the Pacific side of Costa Rica near the town of Quepos, it has about 150,000 visitors annually. Magnificent beaches and rain forests offer hiking to view the numerous species of plant and animal life.

Mar Chiquita Lake. Largest lake in ARGENTINA. The Dulce, Primero, and Segundo Rivers flow into the lake, but there is no outlet to the sea. Like the Great Salt Lake in the United States, this salty lake is surrounded by salt flats and nearly barren land.

Mar del Plata. Fashionable seaside resort in ARGENTINA. Population of 614,000 in 2010. Clear blue ocean, mild temperatures, beautiful beaches and parks, and a large casino draw many summer vacationers from BUENOS AIRES. City also has a naval base, a university, and many cultural institutions.

Maracaibo. Second-largest city in VENEZUELA, with a 2015 population of 1.6 million. The center of Venezuela's oil-rich Maracaibo lowland region. With an average annual temperature of 82.5° Fahrenheit (28° Celsius), it is one of the warmest, most humid places in LATIN AMERICA.

Maracaibo, Lake. Brackish lake in northwestern VENEZUELA that opens into the CARIBBEAN SEA. Its area is 75 by 130 miles (121 by 209 km.). The lake and surrounding lowlands are the site of one of Earth's largest petroleum deposits (discovered in 1917) and a forest of oil derricks.

Marajó Island. The largest river delta island in the world; larger than Ireland. Situated in the mouth of the AMAZON RIVER in South America. The name also describes a type of lower Amazon indigenous ceramic.

Maranhão. State in the farthest western part of NORTHEAST region of BRAZIL. Total area of 127,242 square miles (329,557 sq. km.) with a 2010 population of 6.5 million. Capital is SÃO LUÍS. From south to north, the land gradually descends to the Atlantic coast; from west to east, it varies from equatorial rain forest to tropical savanna. Agriculture predominates, especially cultivation of *babaçú* and cattle raising; also some mining.

Mato Grosso. State in CENTER-WEST region of BRAZIL. Total area of 352,400 square miles (912,716 sq. km.) with a 2010 population of 3

million. Capital is CUIABÁ. Low-lying central highlands form headwaters for rivers flowing north to the AMAZON BASIN and south to PARAGUAY, with extensive wetlands in the southwest along border with BOLIVIA. Tropical agriculture and ranching; some gold and diamond mining.

Mato Grosso do Sul. State in CENTER-WEST region of BRAZIL. Total area of 140,219 square miles (363,167 sq. km.) with a 2010 population of 2.4 million. Capital is CAMPO GRANDE. Land gradually descends from east to west, the western half forming vast wetlands (the PANTANAL) along border with BOLIVIA and PARAGUAY. Subequatorial climate with tropical agriculture and extensive cattle raising.

Pantanal jaguar. (Image by Wolves201)

Medellín. Second-largest city in COLOMBIA (2019 population, 2.6 million); capital of ANTIOQUIA province. Founded in 1616 by Spanish settlers and farmers, its growth was fueled by an early twentieth century coffee boom. The city is a dynamic commercial and industrial city, producing most of Colombia's textiles, but received international notoriety for its former drug cartel.

Mendoza. Capital of Mendoza province in ARGENTINA. Metropolitan area had 1 million people in 2010. Transportation hub on the major international route from BUENOS AIRES to SANTIAGO, Chile. Nestled at the base of the ANDES MOUNTAINS, the city is known for its natural beauty, five universities, and as an important economic center. Alluvial soils and irrigation water from the Andes enable the fertile region to produce fruits, vegetables, and grains. As ARGENTINA's major grape-growing area, it has developed wine-making and food-processing industries. Metalworking and oil refining also are economically significant. Earthquakes and droughts are hazards.

Mesopotamia. Region of the northeast of ARGENTINA, between the PARANÁ and URUGUAY Rivers. Good soils, warm climate, and ample rainfall make this an important agricultural area for citrus, yerba maté tea, sugarcane, and vegetables. Jesuit missions dotted the area in the seventeenth century and formed the basis for productive agriculture.

Minas Gerais. State in SOUTHEAST region of BRAZIL. Total area of 226,707 square miles (587,168 sq. km.) with a 2012 population of 19.9 million. Capital is BELO HORIZONTE. Most mountainous state in Brazil; headwaters of SÃO FRANCISCO RIVER form in the northwest. Industries include agriculture and cattle raising; mining and extensive mineral reserves, especially iron; and manufacturing. Historic site for gold and diamonds.

Missiones Province. Province in the northeast arm of ARGENTINA. Located between the PARANÁ and URUGUAY Rivers. Lush, semitropical vegetation covers the area; the climate is suitable for the production of yerba maté tea, citrus, maize (corn), and other fruits and vegetables that need warm temperatures and much rain. The name of the province comes from the colonial missions founded by the Jesuits. Agricultural productivity was improved greatly in the mission stations.

Mitad del Mundo. Exact line of the equator, as determined by Charles-Marie de la Condamine during the French expedition of 1735. Located 14 miles (23 km.) north of QUITO, ECUADOR. One can visit the monument and straddle the equator, having one foot in the Northern Hemisphere and the other in the Southern Hemisphere. Spanish for "middle of the world."

Montecristo National Park. Cloud forest in CENTRAL AMERICA. Located at the convergence of EL SALVADOR, GUATEMALA, and HONDURAS, several hours north of SAN SALVADOR, at an elevation of 7,900 feet (2,400 meters). Because of the altitude, it receives about 80 inches (200 centimeters) of rain a year, resulting in a wet, cloud forest environment. Flora include ferns, mushrooms, and moss-covered forest floor canopied by laurel and oak trees. Fauna include the anteater, puma, toucan, and spider monkey.

Monteverde. Small community northwest of SAN JOSÉ in the cloud forest of COSTA RICA. Founded in 1951 by North American Quakers seeking a country that did not require registration for the military. Home to the Monteverde Cloud Forest Reserve, a research center for the preservation of the cloud-forest plants and animals. Name is Spanish for "green mountain."

Montevideo. Capital and primate city of URUGUAY. Located on the LA PLATA RIVER ESTUARY on trade routes between the south Atlantic Ocean and PARANÁ-LA PLATA BASIN, with a population in 2011 of about 1.7 million people—one-third of the nation's total. The nation's political, economic, and social core.

Nahuel Huapí Lake. Lake in the ANDES MOUNTAINS of South America. Located at 40°58' south latitude, longitude 71°30' west. Formed by glaciers in a deep valley, it has the typical long, narrow shape of mountain valley lakes. SAN CARLOS DE BARILOCHE resort on the lake is a popular vacation spot for skiers, fishers, hikers, and those who enjoy glaciated mountain landscapes.

Napo River. Largest river in the AMAZON BASIN region of eastern ECUADOR and part of the AMAZON RIVER drainage. In 1542, Spanish explorer Francisco de Orellana, beginning on the Napo, sailed from the eastern foothills of the ANDES MOUNTAINS to the mouth of the Amazon River at the Atlantic Ocean, traversing the continent of South America.

Natal. Capital of RIO GRANDE DO NORTE state in BRAZIL. Located between the right bank of the Potenji River and the Atlantic Ocean, with a 2018 population of 878,000. Important trans-Atlantic air hub to North Africa for Allies during World War II; a growing tourist city.

Nazca. Site on the southern coast of PERU known for huge geometric designs drawn on the desert floor that are visible from nearby elevated areas and from aircraft. The Nazca lines are thought to have been drawn between 900 BCE. and 600 CE. Some animals represented are a lizard, monkey, and condor. The designs might make up a vast pre-Inca calendar.

Negro River. Descending from the ANDES MOUNTAINS in COLOMBIA, where the river is named the Guainía, it becomes the Negro River on entering BRAZIL, coursing northwest/southeast in the state of AMAZONAS. It empties into the Solimões River below MANAUS, forming the Brazilian AMAZON RIVER. It is more than 1,400 miles (2,253 km.) long and has numerous major tributaries.

Negro River Reservoir. Most important hydroelectric and irrigation project in URUGUAY. Hydroelectricity helps industrialization where no fossil fuels exist, and irrigation of the central plains has converted unfenced pastures into farms that raise animals and grow grains, grapes, planted pastures, and vegetables, all important in an agricultural, pastoral economy.

Netherlands Guiana. See SURINAME.

Neuquén. Most important oasis city in northern PATAGONIA region of ARGENTINA. Population in 2010 was 225,000. The center of extensive irrigation and hydroelectric works on the Limay and Neuquén Rivers. Became important as the center of a vast fruit orchard in 1886 after the Amerindian wars ended. Industries include fruit processing, many wineries, and heavy machinery and construction materials for the nearby oil fields. National parks and beautiful Andean scenery are the backdrop for the city.

New World. Term first applied to the WESTERN HEMISPHERE by early sixteenth century geographers. "New World" differentiates the AMERICAS from the "Old World," which was understood as consisting of Europe, Africa, and Asia.

Nicaragua. Largest country in CENTRAL AMERICA, covering 50,336 square miles (130,370 sq. km.)—an area close to that of the U.S. state of Alabama. It is bordered by HONDURAS on its north, the CARIBBEAN SEA on its east, COSTA RICA on its south, and the Pacific Ocean on its west. Its climate is tropical and somewhat consistent throughout the country. Its capital is MANAGUA. Nicaragua's maximum length from north to south is about 275 miles (440 km.), and its maximum width from east to west is about 280 miles (450 km.). Its population was 6,085,000 in 2018.

Nicaragua, Lake. Lake Nicaragua and its sister lake, Managua, are the largest natural bodies of fresh water in CENTRAL AMERICA. Located in the southern part of NICARAGUA, both drain into the CARIBBEAN SEA by way of the San Juan River. Originally an inlet of the Pacific Ocean, the lake gradually became separated from the sea and evolved into a freshwater lake, transforming saltwater animal life-forms—including sharks—into freshwater forms. Lake Nicaragua has more than 350 islands, generally referred to as Las Isletas. The islands' inhabitants fish and engage in small-scale agriculture. Zapatera Island is a national park developed to preserve archaeological remains; the island of San Pablo has a small Spanish fort that was used to defend the area from the British in the eighteenth century. Solentiname Island, in the southern part of Lake Nicaragua, noted for its artist colony, was established by Nicaraguan poet Ernesto Cardenal.

North Brazil. Largest region of BRAZIL, comprising the states of ACRE, AMAPÁ, AMAZONAS, PARÁ, RONDÔNIA, RORAIMA, and TOCANTINS. Covers 1.5 million square miles (3.8 million sq. km.), with a 2016 population of over 17 million, including many Amerindians. Dominated by the AMAZON RIVER and its tributaries and equatorial rain forest.

Northeast Brazil. Region of BRAZIL, comprising the states of ALAGOAS, BAHIA, CEARÁ, MARANHÃO, PARAÍBA, PERNAMBUCO, PIAUÍ, RIO GRANDE DO NORTE, and SERGIPE. Covers 580,000 square miles (more than 1.5 million sq. km.), with a 2005 population of more than 53 million. Region thrived in colonial times but is known for poverty, and for aridity and droughts in the upland interior. Divided into three zones: *zona da mata*, along the coast; *agreste*, between *zona da mata* and uplands; and *sertão*, in the upland interior.

Orinoco River. Largest river in VENEZUELA and the world's fourth-largest river by volume of water. Beginning at its headwaters in the GUIANA HIGHLANDS, it flows 1,615 miles (2,600 km.) eastward to the Atlantic Ocean. When Christopher Columbus first sighted the delta of the Orinoco in 1498 on his third trip to the AMERICAS, he knew that he had discovered a great continent, not just an island.

Oruro. Major mining center in BOLIVIA, also known for its colorful carnival parade. Located in the ANDES MOUNTAINS at an elevation of 12,150 feet (3,700 meters), with a 2012 population of 265,000. Oruro's mines produce tin, silver, and tungsten. A revolution in 1952 resulted in a government takeover of the mines and the freeing of slave laborers.

Otavalo. Market town in the northern highlands of ECUADOR named for the local Amerindian peoples. The Otavalans, known for their distinctive blue-and-white dress, sell their traditional woven goods throughout Ecuador, as well as in the United States and Europe. The population was nearly 40,000 in 2010.

Palmas. Capital of TOCANTINS, newest state of BRAZIL. A planned city, located on the right bank of the TOCANTINS RIVER. Brazil's smallest state capital, with a 2017 population of 287,000. Founded in 1990 and located in center of the state. Center of an agricultural and cattle-raising area.

Pampas. The grassy plains of ARGENTINA and URUGUAY. Extending from BUENOS AIRES and MONTEVIDEO to the base of the ANDES MOUNTAINS, the plains are humid in the east with 40 inches (100 centimeters) of rainfall annually, becoming dry (7 inches/18 centimeters) at the Andes. One of the world's great food-producing areas: Cattle, maize (corn), and soybeans are grown in the wetter regions; sheep, wheat, and cotton come from the drier parts of the plains.

Panama. Central American nation that meets the South American continent at COLOMBIA—of which Panama was originally a part. Panama's only other neighbor, COSTA RICA, borders it on the west. Panama's area of 29,120 square miles (75,420 sq. km.) makes it somewhat smaller than the U.S. state of South Carolina. It had a population of about 3,800,644 in 2018. Panama's location at the narrowest part of CENTRAL AMERICA made it a logical region through which to cut a canal to join the Atlantic and Pacific Oceans. With U.S. help, Panama seceded from Colombia in 1903 and signed a treaty with the United States calling for construction of a canal and U.S. sovereignty over a strip of land on either side of the structure (the Panama CANAL ZONE). Panama retook possession of the canal on December 31, 1999.

Panama Canal. Ship canal built through the Isthmus of PANAMA to provide a direct route between the Pacific and Atlantic Oceans. Constructed by the United States (1907–1914), it is a lake-and-lock type of canal consisting of sets of water locks that lift ships up to lake level, where they proceed through the canal. Stretches about 50 miles (80 km.) from PANAMA CITY on the Pacific Ocean to Colón on the Atlantic Ocean. Over 14,000 ships go through the canal each year. The United States operated the canal until a series of treaties began turning it over to Panamanian control. On January 1, 2000, Panama assumed full control of the CANAL ZONE. A US$5 billion, nine-year expansion of the canal was completed in 2016

Panama Canal Zone. See CANAL ZONE.

Panama City. Capital of PANAMA. Population was 880,000 in 2013. Sections of the city have historical charm, but most is a modern, bustling, cosmopolitan center. Places of interest include the Interoceanic Canal Museum of Panama, the History Museum of Panama, the sixteenth century ruins of Panamá Viejo, and the tropical rain forest of the Soberanía National Park.

Pan-American Highway. System of highways extending from Alaska through South America. The northern section of the route, beginning in Fairbanks, Alaska, and continuing to the city of Dawson Creek, British Columbia, is called the Alaska Highway. In Mexico and CENTRAL AMERICA, the segment known as the Inter-American Highway runs from Laredo, Texas, to PANAMA

City, Panama. Routes in the United States and Canada connect the Alaska and Inter-American highways. The system is complete except for the 100-mile (160-km.) segment through the DARÍEN GAP in PANAMA and COLOMBIA. The projected length is about 17,000 miles (27,000 km.).

Pantanal. Vast wetlands in western part of Brazilian states of MATO GROSSO and MATO GROSSO DO SUL, extending into BOLIVIA and PARAGUAY, covering almost 60,000 square miles (about 150,000 sq. km.). Water level rises and falls in relation to the summer rainy season and the volume of overflow water from the PARAGUAY RIVER and its tributaries. Used as pastureland during low-water periods.

Pará. State in NORTH region of BRAZIL. Total area of 481,869 square miles (1,248,035 sq. km.), with a 2010 population of 7.6 million. Capital is BELÉM. Land forms the eastern half of AMAZON RIVER plain, with delta occupying northeast corner of state. Mouth of the river holds island of MARAJÓ, larger in size than Ireland. Equatorial climate with agriculture and ranching; mineral reserves of bauxite, iron, manganese, and gold.

Paraguay. Landlocked South American country surrounded by BOLIVIA, BRAZIL, and ARGENTINA. Its total area of 157,048 square miles (406,752 sq. km.) makes it somewhat larger than the U.S. state of Montana. Rivers play an important role in Paraguay's economic life, providing it with shipping access to the Atlantic Ocean and sites for hydroelectric power plants. Indeed, the very name of the country apparently derives from a local Guaraní word meaning "river that gives birth to the sea." Paraguay's populace is mainly mestizo of Guaraní Amerindian ancestry. Its population was 7,026,000 in 2018. Its capital is ASUNCIÓN.

Paraguay River. Major South American waterway that rises in southwestern BRAZIL's MATO GROSSO state and flows generally south, eventually bisecting PARAGUAY, before feeding the PARANÁ RIVER at Paraguay's southwestern tip. The river is navigable through most of its roughly 1,600-mile (2,550-km.) length.

Paraíba. State in NORTHEAST region of BRAZIL. Total area of 20,833 square miles (53,957 sq. km.) with a 2010 population of 3.77 million. Capital is JOÃO PESSOA. Arid, mountainous interior descends to tropical Atlantic coast. Cattle raising and agriculture are the main occupations. Tourism and fishing are also important.

Paraíba do Sul River. Descending from mountains in the state of SÃO PAULO, this river runs from southwest to northeast through a valley in the mountains of the northern part of the state of RIO DE JANEIRO. More than 600 miles (1,000 km.) long, it empties into the Atlantic Ocean near Campos. Because of its altitude, temperature, and soils, Campos became a site for coffee cultivation in BRAZIL in the nineteenth century. The river runs through numerous industrial cities.

Paramaribo. Capital, principal port, and only large city of SURINAME. Located on the Suriname River 7 miles (12 km.) from the Atlantic Ocean, with a 2012 population of 241,000. Established by the British in 1651. Its diverse population is reflected in its churches, temples, mosques, and colorful ethnic markets.

Paraná, Argentina. River port in ARGENTINA. Located on the PARANÁ RIVER, with a 2010 population of 248,000. For ten years in the nineteenth century, during a bitter conflict between BUENOS AIRES and the interior provinces over political control, it was Argentina's national capital. Grain, fish, cattle, and lumber are shipped from there. Cement, furniture, and ceramics are important industries.

Paraná, Brazil. State in SOUTH region of BRAZIL. Total area of 76,959 square miles (199,323 sq. km.) with a 2010 population of 10.4 million. Capital is CURITIBA. Its rich farmland stretches from its western border with ARGENTINA (location of IGUAÇÚ FALLS) and PARAGUAY to Atlantic coastal mountains. Site of giant hydroelectric project at Itaipú Dam on border with Paraguay. It has subtropical agriculture; agribusinesses for coffee, sugarcane, and soybeans; and forestry and paper processing.

Paraná-La Plata River System. South American river system draining a huge basin that includes parts of BRAZIL, BOLIVIA, PARAGUAY, ARGENTINA, and URUGUAY. Waterfalls, sandbars, and great seasonal variation in water levels make the system poor for navigation. Only small river craft can navigate inland from ROSARIO, Argentina, and FRAY BENTOS, Uruguay.

Paraná Plateau. One of the world's largest lava plateaus, covering large parts of southern BRAZIL, northeast ARGENTINA, URUGUAY, and PARAGUAY. Centered at 26°0′ south latitude, longitude 53°0′ west. Dark reddish soils (*terra roxa*) result from the weathering of the diabase lava, providing excellent conditions for raising coffee, cattle, and grains.

Paraná River. Second-longest river in BRAZIL, more than 2,500 miles (4,000 km.) long. Forms border between Brazil and PARAGUAY. Part of the massive Paraguay-Paraná-Plata River system. The Paraná River drains from the southwest interior of Brazil, entering the PARAGUAY RIVER, which then empties into the Plata River and the Atlantic Ocean.

Patagonia. Southern plateau region of ARGENTINA. Averages 2,000 feet (610 meters) in elevation at the ANDES MOUNTAINS and ends in a low sea cliff at the south Atlantic shore. Dry, cold winds blowing from the Andes make Patagonia a semiarid, scrub region, which constitutes the largest desert in the WESTERN HEMISPHERE. Its area of about 260,000 square miles (673,400 sq. km.) is greater than that of the entire U.S. state of Texas. People, sheep, and agriculture are confined to the flat valleys cut into the plateau by the eastward flowing streams.

Paysandú. Port on the URUGUAY RIVER and third-largest city in URUGUAY. Population in 2011 of 76,000. Has many meatpacking and meat-freezing plants. Other food-processing and transportation facilities link Uruguay with ARGENTINA and interior parts of the continent.

Pernambuco. State in NORTHEAST region of BRAZIL. Total area of 38,000 square miles (98,000 sq. km.) with a 2010 population of 8.8 million. Capital is RECIFE. Arid, impoverished highland interior gradually descends to tropical Atlantic coast. A historic center of sugar cultivation, it has tropical agriculture along the coast and cattle raising in arid uplands. Small tourism industry.

Peru. Country on the west coast of South America bordered by ECUADOR, COLOMBIA, BRAZIL, BOLIVIA, and CHILE. Total area is 496,224 square miles (1,285,216 sq. km.), with a 2018 population of 31.3 million, many of whom are Amerindian. Spanish and Quechua are the official languages. The center of the Inca Empire, it is unequaled in South America for its archaeological sites and colorful local culture. Peru has three distinct geographic regions. A narrow desert region runs along the Pacific coast; LIMA, the capital, is located there. The Andean highlands have long had the highest population densities and concentration of indigenous peoples. The AMAZON BASIN region has been the focus of recent colonization efforts and oil development. Peru has substantial exports of copper and petroleum and is the largest producer of coca leaf, from which cocaine is made.

Peru Current. Cold-water current flowing south to north along the Pacific coast of PERU. The cold waters collide with warm equatorial waters, resulting in upwelling, a nutrient-rich environment, and one of the world's richest fishing grounds, yielding large harvests of anchovies, sardines, and mackerel. Periodic El Niño events disrupt the economy and weather of the region. Also known as the Humboldt Current.

Piauí. State in NORTHEAST region of BRAZIL. Total area of 97,017 square miles (251,274 sq. km.) with a 2010 population of 3.1 million. Capital is TERESINA. Land gradually descends from arid southern interior to narrow northern tropical coast on the Atlantic Ocean. Economy is one of the least developed in country. Has cattle raising in the interior, and some agriculture, particularly cultivation of *babaçu*.

Pico de Neblina. Situated in northwest AMAZONAS near the border with VENEZUELA in a transitional zone between northeastern terminus of the ANDES MOUNTAINS and beginning of GUIANA

Highlands. The highest peak in BRAZIL at nearly 10,000 feet (3,000 meters).

Pilcomayo River. Major tributary of the PARANÁ RIVER in South America. Rises in the Bolivian ANDES, forms the border between ARGENTINA and PARAGUAY, and joins the PARAGUAY RIVER at ASUNCIÓN, where the Paraná River is formed. Subject to flooding and drought, its shifting channel and sandbars are transportation hazards.

Popoyán. Beautifully preserved colonial town in the Andean highlands of southwest COLOMBIA. Its population in 2017 was 253,000. Founded in 1537 by the Spanish conquistador Sebastián de Benalcázar, it retains much of its classic architecture in the style of Andalucia, a region in southern Spain. Many important pre-Columbian ruins, including San Agustín, are nearby.

Porto Alegre. Capital of RIO GRANDE DO SUL state in BRAZIL. Population in 2017 was more than 1.5 million. Lies on north bank of Guaíba River, where it enters into a large north-south axial lagoon bordering the Atlantic Ocean. Industrial and regional services center.

Porto Velho. Capital of RONDÔNIA state in BRAZIL. Located on right bank of MADEIRA RIVER toward northern tip of state, with a 2018 population of 520,000. City originated during railway trade with BOLIVIA. A center of rubber and wood processing activities.

Potosí. A mining town in the southern highlands of BOLIVIA situated 13,350 feet (4,070 meters) above sea level. Population was 175,000 in 2012. During the second half of the sixteenth century, half of the world's silver output came from Potosí mines, greatly enriching the Spanish Empire. Tin and zinc are the most important minerals produced.

Puerto Madryn. City in ARGENTINA. Located on the northern coast of PATAGONIA with a 2010 population of 94,000. Founded by a group of Welsh immigrants in 1865; Welsh culture is still visible in the architecture of the small village and irrigated farms in the Chubut River Valley. A popular cruise ship stop.

Punta Arenas. Chilean city located on the Strait of MAGELLAN near the southern tip of South America; population was 124,000 in 2012. To reach Punta Arenas, one must pass by boat through a fjordlike landscape that resembles the inland passage of southern Alaska. Petroleum and natural gas development is important to the region.

Punta del Este. Elegant South American seaside resort, with a 2011 population of 9,000, 70 miles (110 km.) east of MONTEVIDEO, Uruguay. Miles of white sand and clear ocean combine with a mild climate to draw vacationers year-round. Luxury hotels, private villas, and picturesque residential areas make the city a destination for conventions and upscale shopping.

Purus River. Major tributary of the AMAZON RIVER in BRAZIL. It drains from the Peruvian ANDES and crosses the Brazilian states of ACRE and AMAZONAS. Coursing southwest-northeast, it empties into the Amazon River at MANAUS. About 1,900 miles (3,000 km.) long, it has numerous tributaries of its own.

Putumayo River. Major river in the AMAZON BASIN portion of COLOMBIA. Located in the extreme southeast of the country, it marks Colombia's border with PERU and is the main transport route in the area.

Quebracho Forests. Hardwood forests in the GRAN CHACO region of PARAGUAY. These trees have helped to make Paraguay a world leader in forest exports. The red quebracho is an excellent source of tannin, much of which is shipped to ARGENTINA and used to tan hides. Quebracho means "break ax," which reflects the tree's tough wood.

Quito. Capital city of ECUADOR. Located 14 miles (22 km.) south of the equator, at an elevation of 9,350 feet (2,850 meters). Founded in 1534 by the Spanish conquistador Sebastián de Benalcázar, it is the former site of the northern capital of the Inca Empire. It has a pleasant, springlike climate and picturesque views of several snow-capped volcanoes. In 1978 the old colonial center was declared a World Heritage Site by the United Nations. Quito's 2010 population was 1.6 million and growing rapidly.

Recife. Capital of PERNAMBUCO state in BRAZIL. Its 2010 population was 1.5 million. A commercial, industrial, and regional services center and major coastal port. Ocean beaches attract growing European tourism.

Recôncavo. Area in BRAZIL, near SALVADOR around the coast of the Bay of Todos os Santos. A traditional producer of tobacco and, more recently, petroleum.

Rio Branco. Capital of ACRE state in BRAZIL. Located in far eastern section of the state on the west bank of Acre River, with a 2011 population of 320,000. Originally, the location of rubber processing enterprise and named Empresa. Name changed to that of Brazilian foreign minister who negotiated acquisition of state from BOLIVIA.

Rio de Janeiro. Capital of RIO DE JANEIRO state in BRAZIL. Brazil's second-largest city, with a 2010 population of 6.3 million; metropolitan area, more than 12 million in 2016. Was the capital of the colony of Brazil (1765–1808), kingdom of Brazil (1808–1822; the only city in the AMERICAS ever to be the seat of a European monarchy), the Brazilian Empire (1822–1889); and the Republic of Brazil (1889–1960). Of unparalleled physical beauty, the city is profiled by mountain peaks wreathed in tropical foliage, laced by beaches bordering waters of the Atlantic Ocean and Bay of GUANABARA. A major air and sea port and world tourist mecca, its economy declined after the transfer of the federal government offices to BRASÍLIA in the 1960s, the loss of shipbuilding work to other developing countries, and the movement of banking services to SÃO PAULO. City declared a UNESCO World Heritage Site in 2012 and hosted the 2016 Summer Olympic Games.

Rio de Janeiro (state). State in SOUTHEAST region of BRAZIL. Total area of 17,092 square miles (44,268 sq. km.) with a 2010 population of more than 16 million. Capital is RIO DE JANEIRO. Dominated by mountains of SERRA DO MAR from west to east, which hold numerous small plains and river valleys. Narrow coastal area; tropical climate with considerable rainfall. Declining economy of agriculture and industry. Offshore oil drilling.

Rio Grande do Norte. State in NORTHEAST region of BRAZIL. Total area of 20,528 square miles (53,168 sq. km.) with a 2010 population of 3.2 million. Capital is NATAL. Arid highland interior gradually descends to tropical coast on Atlantic Ocean. Cattle raising and agriculture, especially cotton and sisal; growing tourism.

Rio Grande do Sul. State in SOUTH region of BRAZIL. Brazil's most southern state, it borders URUGUAY and ARGENTINA. Total area of 108,951 square miles (282,183 sq. km.) with a 2010 population of 10.7 million. Capital is PORTO ALEGRE. Lowland hills and valleys of interior descend to Atlantic coast. Extensive cattle raising, defining dominant *gaúcho* culture; agriculture includes wheat, rice, grape, and soy cultivation; some manufacturing.

Rivera. City on the URUGUAY/BRAZIL border in South America, forming an important link between the countries. Population in 2011 was 64,000. "Portunol," the hybrid Portuguese/Spanish dialect spoken here, reflects an unusual cultural blend in the hinterlands of both countries. Good road and rail transportation connects Rivera with Brazil's productive RIO GRANDE DO SUL state.

Rondônia. State in southern segment of NORTH region of BRAZIL. Total area of 93,839 square miles (243,043 sq. km.) with a 2010 population of 1.6 million. Capital is RIO BRANCO. Gradually declining terrain to AMAZON BASIN. Rubber extraction and agricultural activities, especially cattle raising.

Roraima. State in NORTH region of BRAZIL. Most northern area of the country, with its highest peaks. Total area of 88,843 square miles (230,181 sq. km.); Brazil's least populated state, with a 2010 population of 451,000. Capital is BOA VISTA. Land descends from north to south in equatorial climate. Cattle raising is the dominant activity.

Rosario. Third-largest city in ARGENTINA. Population in 2010 was 1.2 million people. Rail, water, and road transportation link Rosario with BUEnos Aires and the nation's important northwest regions. A national university makes the city an educational center; it is also a major processor of food products and a manufacturing center. Steel is a regional specialty.

Salar de Uyuni. A vast salt lake in the arid southern highlands of BOLIVIA, located at 20°58′ south latitude, longitude 67°9′ west. It is approximately twice as large as the Great Salt Lake in the U.S. state of Utah.

Salta. City in ARGENTINA, founded in 1582. Population in 2010 was 535,000. An oasis on the main route from BUENOS AIRES to the Bolivian silver mines. Mules from the Argentina plains were sent to BOLIVIA to bring out the valuable silver ores. The center of important oil and natural gas fields, and has many food-processing industries. Colonial architecture of the old city is a draw for many tourists interested in the city's rich colonial past.

Salto. Second-largest city in URUGUAY. Located on the URUGUAY RIVER, with a 2011 population of 104,000. Salto Falls is just north of the city. An important shipping center, and has meat-salting and meatpacking plants. Hydroelectricity generated at the falls supplies the city needs, and some is exported to other Uruguayan cities as well.

Salto Grande Lake. Lake on the URUGUAY RIVER. Formed when a dam was built at Salto Falls; full power generation was achieved in 1981. Provides recreational resources for fishing, boating, and swimming. URUGUAY has no fossil fuels, and its mostly flat terrain offers few good sites for hydroelectric development. Rapids on the Uruguay River just north of SALTO created the opportunity for Uruguay to solve its power shortages.

Salvador. Capital of BAHIA state in BRAZIL; Brazil's first colonial capital. Located on Atlantic Ocean and Bay of Todos os Santos, with a 2010 population of 2.7 million. Some industry; regional business and services center. Historic core of Afro-Brazilian culture. Historic buildings in center of city, together with ocean and bay beaches, attract extensive tourism.

San Andrés. Island province and archipelago of COLOMBIA; a popular tourist destination. Located in the CARIBBEAN SEA 435 miles (700 km.) northeast of Colombia and 140 miles (225 km.) east of NICARAGUA. A former British colony, San Andrés has a large population of African descent, and English is still spoken. The main island has an area of 17 square miles (44 sq. km.); the total population of the province was about 68,000 in 2007.

San Blas Archipelago. Caribbean string of islands extending from the Gulf of San Blas to the Colombian border. Comprises almost 400 islands, primarily occupied by the Guna tribe, who administer the islands as an autonomous province. They are allowed and encouraged by the Panamanian government to maintain their customs, language, dress, music, and dance. Because of this survival of Amerindian culture, the tourism sector of the island's economy is growing.

San Carlos de Bariloche. Resort city in the Andean lake region of ARGENTINA, at NAHUEL HUAPÍ LAKE. Mountains, lakes, and national parks provide year-round recreational activities, including skiing, fishing, boating, and hiking. Swiss immigrants built chalet-style hotels and villas that replicated the look of their homeland. International conference attendees, filmmakers, and vacationers come to enjoy the alpine environment. Also known as Bariloche. Population was 108,000 in 2010.

San Jorge Gulf. Largest bay in the PATAGONIA region of ARGENTINA. The gulf is an important oil-producing district; COMODORO RIVADAVIA is its most important urban center. The sea cliffs and great tidal ranges in Argentina's south make navigation difficult. Improved land and air transportation and consequent economic development have made Comodoro Rivadavia a city of more than 180,000 people. Oil and food are shipped 1,000 miles (1,610 km.) to BUENOS AIRES and LA PLATA.

San José. Capital of COSTA RICA, since 1823. Located in the interior highlands of the country, with 1.2 million people in 2011. A growing, cosmopolitan area and the country's transportation and economic hub. Cultural attractions include the National Museum, the Pre-Columbian Museum of Gold, the Museum of Jade, and the city's vibrant central market district.

San Juan. Oasis at the foot of the ANDES MOUNTAINS in ARGENTINA. Irrigation has made the province of San Juan a region of intense agriculture. Wine grapes, cattle, grain, fruits, and vegetables grow prolifically. Industries in San Juan City (population 112,000 in 2001) are based on food processing. The region suffers from numerous earthquakes.

San Nicholás. Planned industrial city in ARGENTINA. Located on the PARANÁ RIVER between BUENOS AIRES and ROSARIO, with a 2010 population of 133,000. In 1960 the government steel plant started production; this project was part of a regional development package, including oil refining, power generation, steel, and fabricated metals manufacturing, which the United Nations helped to plan and finance. Argentina has coal and iron ore, but it is cheaper to make steel out of foreign raw materials brought in to San Nicholás port.

San Rafael Mountain. The highest point in PARAGUAY, reaching 2,789 feet (850 meters). Located 47 miles (75 km.) north of ENCARNACIÓN. This eastern section of the country has Paraguay's best soils, mildest climate, most dependable rainfall, and best drainage. People living there raise an abundance of peanuts, tobacco, wheat, cotton, soybeans, fruits, and vegetables.

San Salvador. Capital and largest city of EL SALVADOR. Located at the foot of San Salvador volcano, with a 2018 population of 1.1 million. Has been the capital since 1839, but has few colonial buildings, since it has fallen victim to earthquakes, floods, and volcanic eruptions over the years. Cultural sites include the National Theater; the David Guzman National Museum, renowned for its archaeological holdings; and the Metropolitan Cathedral, where the famous archbishop Oscar Romero is buried.

San Salvador de Jujuy. City in ARGENTINA, with a 2010 population of 258,000. Founded in 1593 as a waystation between the silver mines of BOLIVIA and the city of BUENOS AIRES, it straddles the PAN-AMERICAN HIGHWAY. With lead, zinc, iron ore, and silver mines in the vicinity, mineral production keeps the city important to the nation despite its remote location.

Santa Catarina. State in SOUTH region of BRAZIL. Total area of 37,060 square miles (95,985 sq. km.) with a 2010 population of 6.2 million. Capital is FLORIANÓPOLIS. Low mountain range extends along eastern half of land; its waters drain east to Atlantic Ocean and west to PARANÁ RIVER. Subtropical climate with wide range of agricultural cultivation; extensive cattle ranching.

Santa Cruz. Largest city in BOLIVIA, with a 2012 population of 1.4 million. A boomtown in Bolivia's resource-rich eastern lowlands, it is a center of production of rice, soybeans, sugarcane, cattle, and timber for export.

Santa Fe. Transportation hub in ARGENTINA, connecting the nation's heartland of BUENOS AIRES Province with the northwest oases and northern humid areas. Population in 2010 was 653,000. It serves a rich agricultural area by processing grain, vegetable oils, and meats. Road, rail, and river traffic on the PARANÁ RIVER make it one of the nation's oldest and most important interior settlements.

Santiago. Capital and largest city in CHILE, with a 2012 population of 4.9 million. Situated in the fertile central valley against the magnificent backdrop of the ANDES MOUNTAINS. A modern, bustling city and the economic and cultural center of Chile, but plagued by air pollution and earthquakes.

Santiago del Estero. Oldest city in ARGENTINA; founded in 1553. Strategically located on the Dulce River, providing a watering and resting place on the road from BOLIVIA and across Argentina's semiarid plains. The city of 252,000 people (2010) is the center of extensive agricultural operations, such as wheat farming and cattle raising on the open range.

São Francisco River. The longest Brazilian river running entirely in the country; 1,800 miles (2,900 km.) long. Descending from the mountains of MINAS GERAIS, it runs from south to north-northeast through the states of BAHIA, PERNAMBUCO, ALAGOAS, and SERGIPE. The Paulo Afonso hydroelectric dam in Bahia is located on this river. Work on building canals to divert water to the arid interior began in 2009; expected to be completed by 2025.

São Luís. Capital of MARANHÃO state in BRAZIL. Located on an Atlantic coastal bay island of same name, with a 2018 population of 1.1 million. Originally founded by French; now a business and services center. Because of its well-preserved colonial architecture, it has been declared a UNESCO World Heritage Site.

São Paulo. Capital of Brazilian state of same name. Largest city in BRAZIL and South America, with a 2010 population of 11 million; metropolitan area, almost 21 million. National industrial, commercial, banking, and services center. Stretches along hilly peaks of SERRA DO MAR. Port is located 50 miles (80 km.) below, in city of Santos, situated on coastal island.

São Paulo (state). State in SOUTHEAST region of BRAZIL. Total area of 95,852 square miles (248,257 sq. km.); largest state in Brazil, with a 2010 population of 41 million. Capital is SÃO PAULO. Southern area that is a mountain range dropping to the Atlantic coast. Stretching north from this area are hills and plains of rich soil. Intensive cultivation of coffee, soybeans, and sugarcane, along with other agricultural products. Extensive manufacturing, including auto, computer, and aviation industries.

Sergipe. State in NORTHEAST region of BRAZIL. Smallest state in Brazil, with a total area of 8,491 square miles (21,862 sq. km.); 2010 population was 2 million. Capital is ARACAJU. Land descends from southwest to lowland plain in north along SÃO FRANCISCO RIVER and to eastern Atlantic coast. Tropical climate; agricultural production in lowlands, cattle raising in interior.

Serra do Mar. Mountain range that extends for more than 1,600 miles (2,600 km.) along the east coast of BRAZIL, from RIO GRANDE DO SUL to BAHIA. The mountains in the city of RIO DE JANEIRO, such as Sugarloaf and Corcovado, are

part of this range. This escarpment at times falls directly into the sea, but most of it is several to 50 miles (80 km.) distant from sea so that there is a coastal ribbon of land between it and the ocean. This mountain range is much older and lower than the ANDES MOUNTAINS and reveals where the eastern plate of the South American continent broke from the western plate of Africa. At their highest, these mountains rarely exceed 7,000 feet (2,100 meters) in altitude in their middle sections, in ESPÍRITO SANTO, and are generally half that altitude.

Aerial view of Sugarloaf. (Image by Rejane Dominguez)

South Brazil. Region of BRAZIL, comprising states of PARANÁ, SANTA CATARINA, and RIO GRANDE DO SUL. Covers about 230,000 square miles (600,000 sq. km.); 2010 population of more than 29 million, with many immigrants of central and southern European descent. The "breadbasket" of Brazil in terms of richness and productivity of land for rice, wheat, soybeans, vegetables, and other food staples.

Southeast Brazil. Region of BRAZIL, including states of ESPÍRITO SANTO, MINAS GERAIS, RIO DE JANEIRO, and SÃO PAULO. Covers about 357,000 square miles (925,000 sq. km.), with a 2010 population of 80 million. Because São Paulo is located there, the region is the economic powerhouse of Brazil, with the highest per capita income, contributing two-thirds of country's gross domestic product.

Sucre. The constitutional capital of BOLIVIA. Founded in 1538, Sucre is a charming colonial town of 239,000 inhabitants (2012). Because of its well-preserved center, with adobe structures, white walls, and red-tiled roofs, Sucre was named a World Heritage Site by the United Nations in 1991.

Suriname. Country on the north coast of South America, bordered by GUYANA to the west, FRENCH GUIANA to the east, and BRAZIL to the south. Total area of 63,251 square miles (163,820 sq. km.) with a 2018 population of 598,000, mostly settled on the coast. Capital is PARAMARIBO. First permanent European settlement established in 1651 by the English. Ceded to the Dutch in 1667 (then known as Dutch Guiana) in exchange for Manhattan Island (now part of New York City). Gained independence in 1975. Diverse population, composed of Creoles (people of African ancestry) and descendants of laborers from India, China, and Java. Dutch is the official language, but Hindi and Javanese are widely spoken. Christianity, Hinduism, and Islam are the principal religions. Leading exports are bauxite and aluminum products; rice, shrimp, bananas, and some petroleum are also produced. (Formerly spelled "Surinam.")

Tajumulco, Volcán. Guatemalan volcano that is the highest peak in CENTRAL AMERICA, at 13,845 feet (4,220 meters).

Tapajós River. Major tributary of AMAZON RIVER in South America, that drains from MATO GROSSO and forms the border between Brazilian states of AMAZONAS and PARÁ. More than 1,100 miles (1,800 km.), it flows southwest-northeast and empties into the Amazon River at the port city of Santarém.

Tarabuco. Most colorful traditional market for Amerindian weavings in BOLIVIA. The intricate designs are handmade, often taking months to complete one item, and artisans can be watched at work. Located in the southern highlands approximately 40 miles (64 km.) from SUCRE. Population was 2,442 in 2001.

Tazumal Ruins. Best-known, best-preserved Mayan ruins in EL SALVADOR. Located in the town of Chalchuapa, about 47 miles (75 km.) northwest of SAN SALVADOR. First occupied about 7,000 years ago but best known for the Maya who lived there about 1,000 years ago. Archaeologists believe the site was a trade center, based on artifacts from as far away as Mexico and PANAMA. Name is Mayan for "pyramid where the victims were burned."

Tazumal, El Salvador. (Image by otrarove)

Tegucigalpa. Capital of HONDURAS. Had a 2009 population of 990,000. Founded as a mining center; became the national capital in 1880. Its colonial architecture includes the central park, which is the hub of the city; the legislative plaza; the president's house; and the church of San Francisco, built during the sixteenth century. La Tigra National Park, just northeast of the city, has a well-preserved cloud forest and associated wildlife. Name means "silver hill" in the local dialect, referring to its founding by the Spanish in 1578.

Teresina. Capital of PIAUÍ state in BRAZIL; 2010 population of about 767,000. A nineteenth century planned city, it lies on right bank of Parnaíba River in the central west portion of state near the border with MARANHÃO. The center of an impoverished area.

Tierra del Fuego. Archipelago at the southern extremity of South America. In shape, the main island separated from the mainland by the Strait of MAGELLAN is a triangle with its base on BEAGLE CHANNEL. The total area is 28,473 square miles (73,746 sq. km.), about two-thirds of which is part of CHILE and the rest part of ARGENTINA. Given its latitude and proximity to

Antarctica, the climate is cold. The discovery of oil at Manantiales in 1945 converted the northern part of Tierra del Fuego into Chile's only oil field. The archipelago takes its name from Tierra del Fuego Island, at its southern tip. Located at 54°0′ south latitude, longitude 70°0′ west, the island is shared by Chile and Argentina. Sheep, forestry, fishing, and oil are mainstays of the economy. USHUAIA, Argentina, on Beagle Channel, is the world's southernmost city.

Tikal. One of the most impressive Mayan sites in CENTRAL AMERICA. Located in the extreme north of GUATEMALA in the Peten, which is likened to the Mexican Yucatán. Tikal National Park contains thousands of structures and artifacts, including massive temples, such as the Temple of the Giant Jaguar; ball courts; house sites; ceremonial platforms; palaces; reservoirs; and stelae—massive stone sculptures documenting important Mayan events. Visitors can often watch archaeologists studying the site, a long-term, continuous process.

Tikal, Guatemala. (Image by Maurice Mar cellin)

Titicaca, Lake. Located in the ANDES MOUNTAINS region of South America on the border between PERU and BOLIVIA. At 12,500 feet (3,810 meters) above sea level, it is the world's highest large freshwater lake. The lands around the lake are populated by farmers who raise potatoes, sheep, and llamas. Reed boats once were commonly used for transportation. Landlocked Bolivia's navy is based on Lake Titicaca.

Tocantins. State in NORTH region of BRAZIL. Total area of 107,190 square miles (277,622 sq. km.) with a 2010 population of nearly 1.4 million. Capital is PALMAS. From north-south axis of the central highlands, land descends to east and west in transitional equatorial climate. Cattle raising in north; agriculture in south.

Tocantins River. Major tributary of AMAZON RIVER in BRAZIL. Almost 1,700 miles (2,700 km.) long, it flows south and north from GOIÁS across TOCANTINS, forming the border with MARANHÃO. Joined by the Araguaia River, it crosses PARÁ, emptying into the southeast part of the Amazon delta below BELÉM and MARAJÓ ISLAND. The Tucuruí hydroelectric dam is in Pará, below where the Araguaia and Tocantins Rivers join.

Torres del Paine National Park. South American park in the Chilean PATAGONIA, known for its spectacular scenery and wildlife. Located 70 miles (112 km.) north of Puerto Natales in the far south of CHILE. Adventurous travelers can view glaciers, towering granite pillars, and such wildlife as the Andean condor, flamingo, emu (a relative of the ostrich), and guanaco (a relative of the llama).

Trujillo. Third-largest city in PERU (2007 population, 683,000) and the main settlement on its north coast. An attractive colonial city, founded in 1536 and named after Spanish conqueror Francisco Pizarro's hometown in Spain. Nearby are the ruins of the monumental Moche pyramids and the adobe Chimú ruin city of CHAN CHÁN.

Tucumán. Second-largest oasis in the arid west of ARGENTINA. Population in 2010 was 549,000. Irrigation from Andean streams and warm weather combine to give the region a unique agricultural prominence. Rice, tobacco, sugarcane, and fruit provide the raw materials for many factories. With two universities and the provincial capital, the city is important politically, culturally, and economically. A colonial city dating from 1565; Argentine independence was declared from its municipal assembly building.

Uruguay. Country located on the southeastern coast of South America and bounded by BRAZIL and ARGENTINA. After SURINAME, Uruguay is the smallest country in South America, with an area of about 68,000 square miles (176,120 sq. km.)—very similar to the area of the U.S. state of Missouri. Uruguay's economy is based largely on such agricultural exports as beef. Its capital is MONTEVIDEO, which has almost half of the country's population in its metropolitan area. Its population—mostly of European descent as in Argentina—was 3,369,000 in 2018.

Uruguay River. River flowing from southern BRAZIL, forming the northeast border of ARGENTINA and the western border of URUGUAY. About 1,000 miles (1,610 km.) long, it merges with the LA PLATA RIVER ESTUARY a little north of BUENOS AIRES. The river floodplain is some of Uruguay's best farmland, and there are important hydroelectric works along the river.

Ushuaia. Southernmost city in the world. Located in ARGENTINA. English naturalist Charles Darwin visited this site on the BEAGLE CHANNEL in 1832; the glaciers, ice-sculptured mountains, and dense arctic woodlands sheltered the wildlife that helped form his evolutionary theories. The city of 45,000 (2003) people is a center for fishing, lumbering, and sheep ranching; today, Ushuaia dominates Argentina's production of computers, cell phones, and related products. Argentina has a navy base here. An estimated 120,000 tourists visit annually.

Uspallata Pass. Easiest pass through the ANDES MOUNTAINS between MENDOZA, ARGENTINA, and SANTIAGO, Chile. The climb to 12,572 feet (3,832 meters) is direct, but heavy winter snows make passage difficult. European settlement of Argentina was started by Spaniards from CHILE and BOLIVIA filtering through the Andean passes. A statue to Christ the Redeemer in the Uspallata Pass marks the peaceful settlement of the international boundary.

Valencia. Third-largest city in VENEZUELA, with a 2011 population of 830,000. Founded in 1555 on Lake Valencia; named after a city in southern Spain. One of Venezuela's most important commercial and industrial centers; products include agricultural products, pharmaceuticals, and automobiles. The surrounding area is known for sugarcane production.

Valparaíso. Principal port in CHILE and the location of the Chilean parliament, which was moved there in the 1980s as part of a movement to decentralize government functions. Located 75 miles (120 km.) northwest of SANTIAGO, with a 2012 population of 270,000. It was an important trading city and a British Naval center until the PANAMA CANAL opened in 1914.

Venezuela. Country on the north coast of South America, bordered by COLOMBIA, BRAZIL, and GUYANA. Total area of 352,144 square miles (912,050 sq. km.), with a 2018 population of 31.7 million. Capital is CARACAS. Venezuela has a diverse physical geography. ANDES MOUNTAINS in the west rise to 16,000 feet (5,000 meters). Vast plains of the LLANOS and ORINOCO RIVER dissect the country. The largely undeveloped GUIANA HIGHLANDS and AMAZON territory encompass the east and south. The first colony to revolt against Spanish rule, Venezuela achieved independence in 1830 under Simon Bolívar. Venezuela was once one of LATIN AMERICA's richest countries, based on the oil wealth of the Lake MARACAIBO area. In recent years, Venezuela has suffered from corruption, government mismanagement, and other problems, with catastrophic impact on the economy. Iron ore, steel, and aluminum are also important exports. The name "Venezuela" means "Little Venice," so named because explorers observed Amerindians living in houses on stilts on Lake Maracaibo.

Viña del Mar. One of South America's premier seaside resorts, located in central CHILE near SANTIAGO. The site of an internationally recognized music festival, a casino, and Chile's presidential summer palace. The population in 2012 was 324,800.

Vitória. Capital of Espírito Santo state in Brazil. Located on an Atlantic coastal island of same name; 2015 population of 359,000. Noted for its fish cuisine; growing tourism. Port of Tubarão is a major mineral exporter.

Volcán Tajumulco. See Tajumulco, Volcán.

Western Hemisphere. Portion of Earth that contains North and South America; generally demarcated as within longitude 20° west and 160° east. Historically known as the New World.

Xingú River. Major tributary of Amazon River in Brazil. It flows south and north from Mato Grosso across Pará for almost 1,300 miles (about 2,100 km.). It enters the Amazon where the river's delta begins to form. Upper reaches of river house Xingú National Park, a major indigenous homeland.

Xunantunich. Most impressive archaeological park in Belize. near the Guatemalan border. Was a large ceremonial center controlling the area from the present-day Belize-Guatemala border to the Caribbean Sea. Archaeologists believe the site was abandoned about 1,000 years ago when a major earthquake devastated the area. El Castillo, the largest and tallest structure there, rises about 130 feet (40 meters) above the tropical forest. Name is Mayan for "stone maiden."

Ypacaraí Lake. Popular vacation spot in Paraguay. Located about 18 miles (29 km.) east of Asunción, at the edge of the Paraná Plateau. Part of Paraguay's best farmland. Early German immigrants helped make this region prosperous through intensive cultivation of grains and fruits, and the raising of meat animals. San Bernardino is the resort city on the lake.

Ypoa Lake. Lake in Paraguay. Located 40 miles (65 km.) south of Asunción. Occupies a lowland at the southern edge of the Paraná Plateau; its borders merge into extensive swamps. An estimated 400 species of birds make homes in the swamp; large mammals found here include jaguar, wild boar, deer, and capybara (water hog).

Yungas. A frontier and colonization zone in Bolivia. Located north of La Paz in the foothills of the Amazon Basin, the Yungas provide a welcome retreat from the high altitude of the capital. The area produces citrus, bananas, coffee, and coca—sometimes for use as a traditional herb, and sometimes converted into cocaine.

Douglas Heffington; James R. Keese;
Dana P. McDermott; Judith C. Mimbs;
Edward A. Riedinger; Robert J. Tata

APPENDIX

THE EARTH IN SPACE

THE SOLAR SYSTEM

Earth's solar system comprises the Sun and its planets, as well as all the natural satellites, asteroids, meteors, and comets that are captive around it. The solar system formed from an interstellar cloud of dust and gas, or nebula, about 4.6 billion years ago. Gravity drew most of the dust and gas together to make the Sun, a medium-size star with an estimated life span of 10 billion years. Its system is located in the Orion arm of the Milky Way galaxy, about two-thirds of the way out from the center.

During the Sun's first 100 million years, the remaining rock and ice smashed together into increasingly larger chunks, or planetesimals, until the planets, moons, asteroids, and comets reached their present state. The resulting disk-shaped solar system can be divided into four regions—terrestrial planets, giant planets, the Kuiper Belt, and the Oort Cloud—each containing its own types of bodies.

Terrestrial Planets

In the first region are the terrestrial (Earth-like) planets Mercury, Venus, Earth, and Mars. Mercury, the nearest to the Sun, orbits at an average distance of 36 million miles (58 million km.) and Mars, the farthest, at 142 million miles (228 million km.). Astronomers call the distance from the Sun to Earth (93 million miles/150 million km.) an astronomical unit (AU) and use it to measure planetary distances.

Terrestrial planets are rocky and warm and have cores of dense metal. All four planets have volcanoes, which long ago spewed out gases that created atmospheres on all but Mercury, which is too close to the Sun to hold onto an atmosphere. Mercury is heavily cratered, like the earth's moon. Venus has a permanent thick cloud cover and a surface temperature hot

enough to melt lead. The air on Mars is very thin and usually cold, made mostly of carbon dioxide. Its dry, rock-strewn surface has many craters. It also has the largest known volcano in the solar system, Olympus Mons, which is 16 miles (25 km.) high.

Average temperatures and air pressures on Earth allow liquid water to collect on the surface, a unique feature among planets within the solar system. Meanwhile, Earth's atmosphere—mostly nitrogen and oxygen—and a strong magnetic field protect the surface from harmful solar radiation. These are the conditions that nurture life, according to scientists. It is widely accepted that Mars had abundant water very early in its history, but all large areas of liquid water have since disappeared. A fraction of this water is retained on modern Mars as both ice and locked into the structure of abundant water-rich materials, including clay minerals (phyllosilicates) and sulfates. Studies of hydrogen isotopic ratios indicate that asteroids and comets from beyond 2.5 AU provide the source of Mars' water. Like Earth, Mars has polar ice caps, although those on Mars are made up mostly of carbon dioxide ice (dry ice), while those on Earth are made up of water ice.

A single natural satellite, the Moon, orbits Earth, probably created by a collision with a huge planetesimal more than 4 billion years ago. Mars has two tiny moons that may have drifted to it from the asteroid belt. A broad ring from 2 to 3.3 AU from the Sun, this belt is composed of space rocks as small as dust grains and as large as 600 miles (1,000 km.) in diameter. Asteroids are made of mineral compounds, especially those containing iron, carbon, and silicon. Although the asteroid belt contains

FORMATION OF THE SOLAR SYSTEM

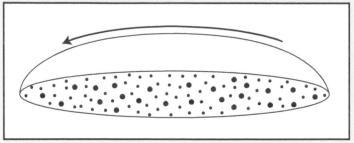

1. The solar system began as a cloud of rotating interstellar gas and dust.

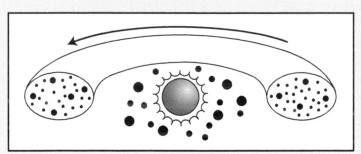

2. Gravity pulled some gases toward the center.

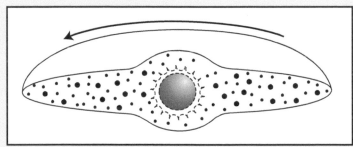

3. Rotation accelerated, and centrifugal force pushed icy, rocky material away from the proto-Sun. Small planetesimals rotate around the Sun in interior orbits.

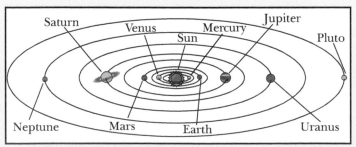

4. The interior, rocky material formed Mercury, Venus, Earth, and Mars. The outer, gaseous material formed Jupiter, Saturn, Uranus, Neptune, and Pluto.

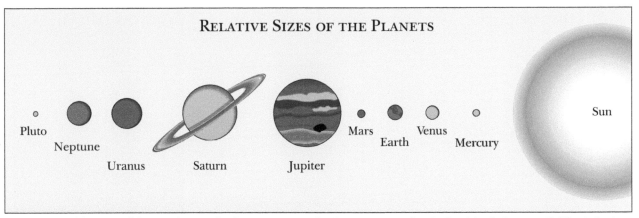

RELATIVE SIZES OF THE PLANETS

Note: The size of the Sun and distances between the planets are not to scale.; Source: Data are from Jet Proulsion Laboratory, California Institute of Technology. The Deep Space Network. Pasadena, Calif.:JPL, 1988, p. 17.

enough material for a planet, one did not form there because Jupiter's gravity prevented the asteroids from crashing together. The belt separates the first region of the solar system from the second.

The Giant Planets

The second region belongs to the gas giants Jupiter, Saturn, Uranus, and Neptune. The closest, Jupiter, is 5.2 AU from the Sun, and the most distant, Neptune, is 30.11 AU. Jupiter is the largest planet in the solar system, its diameter 109 times larger than Earth's. The giant planets have solid cores, but most of their immense size is taken up by hydrogen, helium, and methane gases that grow thicker and thicker until they are like sludge near the core. On Jupiter, Saturn, and Uranus, the gases form wide bands over the surface. The bands sometimes have immense circular storms like hurricanes, but hundreds of times larger. The Great Red Spot of Jupiter is an example. It has winds of up to 250 miles (400 km.) per hour, and is at least a century old.

These planets have such strong gravity that each has attracted many moons to orbit it. In fact, they are like miniature solar systems. Jupiter has the most moons—eighteen—and Neptune has the fewest—eight—but Neptune's moon Triton is the largest of all. Most moons are balls of ice and rock, but Jupiter's Europa and Saturn's Titan may have liquid water below ice-bound surfaces. Several moons appear to have volcanoes, and a wispy atmosphere covers Titan. Additionally, the giant planets have rings of broken rock and ice around them, no more than 330 feet (100 meters) thick. Saturn's hundreds of rings are the brightest and most famous.

The Kuiper Belt

The third region of the solar system, the Kuiper Belt, contains the dwarf planet, Pluto. Pluto has a single moon, Charon. It does not orbit on the same plane, called the ecliptic, as the rest of the planets do. Instead, its orbit diverges more than seventeen degrees above and below the ecliptic. Its orbit's oval shape brings Pluto within the orbit of Neptune for a large percentage of its long year, which is equal to 248 Earth years. Two-thirds the size of the earth's moon, Pluto has a thin, frigid methane atmosphere. Charon is half Pluto's size and orbits less than 32,000 miles (20,000 km.) from Pluto's surface. Some astronomers consider Pluto and Charon to be a double planet.

The Kuiper Belt holds asteroids and the "short-period" comets that pass by Earth in orbits of 20 to 200 years. These bodies are the remains of

OTHER EARTHS

By the year 2000 astronomers had detected twenty-eight planets circling stars in the Sun's neighborhood of the galaxy. Planets, they think, are common. Those found were all gas giants the size of Saturn or larger. Earth-size planets are much too small to spot at such great distances. Where there are gas giants, there also may be terrestrial dwarfs, as in Earth's solar system. Where there are terrestrial planets, there may be liquid water and, possibly, life.

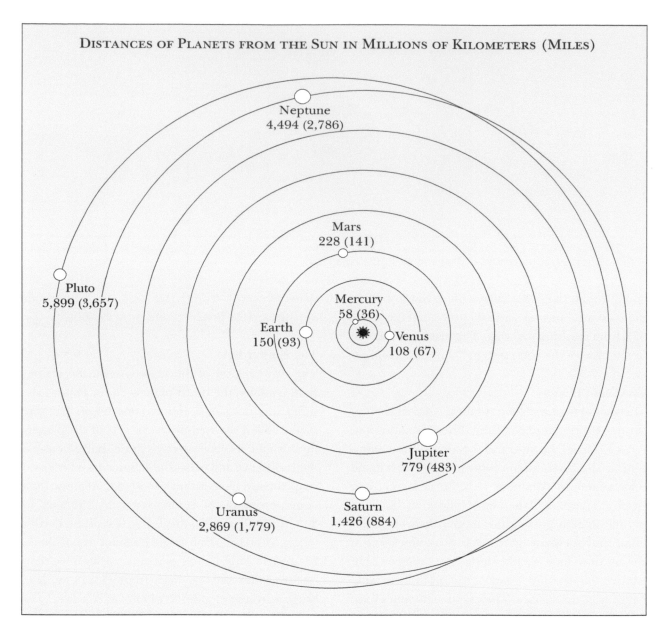

DISTANCES OF PLANETS FROM THE SUN IN MILLIONS OF KILOMETERS (MILES)

Neptune
4,494 (2,786)

Mars
228 (141)

Pluto
5,899 (3,657)

Mercury
58 (36)

Earth
150 (93)

Venus
108 (67)

Jupiter
779 (483)

Uranus
2,869 (1,779)

Saturn
1,426 (884)

planet formation and did not collect into planets because distances between them are too great for many collisions to occur. Most of them are loosely compacted bodies of ice and mineral—"dirty snow-balls," as they were termed by the famous astronomer Fred Lawrence Whipple (November 5, 1906–August 30, 2004). An estimated 200 million Kuiper Belt objects orbit within a band of space from 30 to 50 AU from the Sun.

The Oort Cloud

In contrast to the other regions of the solar system, the Oort Cloud is a spherical shell surrounding the entire solar system. It is also a collection of com-

ets—as many as two trillion, scientists calculate. The inner edge of the cloud forms at a distance of about 20,000 AU from the Sun and extends as far out as 100,000 AU. The Oort Cloud thus gives the solar system a theoretical diameter of 200,000 AU—a distance so vast that light needs more than three years to cross it. No astronomer has yet detected an Oort Cloud object, because the cloud is so far away. Occasionally, however, gravity from a nearby star dislodges an object in the cloud, causing it to fall toward the Sun. When observers on Earth see such an object sweep by in a long, cigar-shaped orbit, they call it a long-period comet.

The outer edge of the Oort Cloud marks the farthest reach of the Sun's gravitational power to bind bodies to it. In one respect, the Oort Cloud is part of interstellar space.

In addition to light, the Sun sends out a constant stream of charged particles—atoms and subatomic particles—called the solar wind. The solar wind shields the solar system from the interstellar medium, but it only does so out to about 100 AU, a boundary called the heliopause. That is a small fraction of the distance to the Oort Cloud.

Roger Smith

EARTH'S MOON

The fourth-largest natural satellite in the solar system, Earth's moon has a diameter of 2,159.2 miles (3,475 km)—less than one-third the diameter of Earth. The Moon's mass is less than one-eightieth that of Earth.

The Moon orbits Earth in an elliptical path. When it is at perigee (when it is closest to Earth), it is 221,473 miles (356,410 km.) distant. When it is at apogee (farthest from Earth), it is 252,722 miles (406,697 km.) distant.

The Moon completes one orbit around Earth every 27.3 Earth days. Because it rotates at about the same rate that it orbits the earth, observers on Earth only see one side of the Moon. The changing angles between Earth, the Sun, and the Moon determine how much of the Moon's illuminated surface can be seen from Earth and cause the Moon's changing phases.

Volcanism

Naked-eye observations of the Moon from Earth reveal dark areas called *maria*, the plural form of the Latin word *mare* for sea. The maria are the remains of ancient lava flows from inside gigantic impact craters; the last eruptions were more than 3 billion years ago. The lava consists of basalt, similar in composition to Earth's oceanic crust and many volcanoes. The maria have names such as Mare Serenitatis (15° to 40°N, 5° to 20°E) and Mare Tranquillitatis (0° to 20°N, 15° to 45°E). Some of the smaller dark areas on the Moon also have names that are water-related: lacus (lake), sinus (bay), and palus (marsh).

Impact Craters

Observing the Moon with an optical aid, such as a telescope or a pair of binoculars, provides a closer view of impact craters. Impact craters of various sizes cover 83 percent of the Moon's surface. More than 33,000 craters have been counted on the Moon.

One of the easiest craters to observe from the Earth is Tycho. Located at 43.3°S, longitude 11.2 degrees west, it is about 50 miles (85 km.) wide. Surrounding Tycho are rays of dusty material, known as

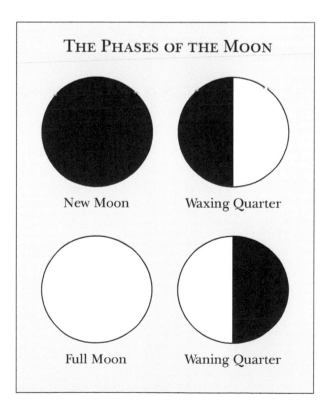

THE PHASES OF THE MOON

New Moon

Waxing Quarter

Full Moon

Waning Quarter

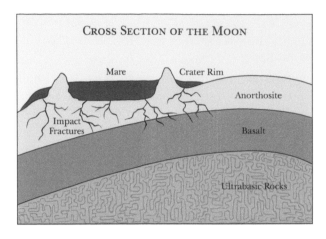

ejecta, that appear to radiate from the crater. When an object from space, such as a meteoroid, slams into the Moon's surface, it is vaporized upon impact. The dust and debris from the interior of the crater fall back onto the lunar surface in a pattern of rays. Because the ejecta is disrupted by subsequent impacts, only the youngest craters still have rays. Sometimes, pieces of the ejecta fall back and create smaller craters called secondary craters. The ejecta rays of Tycho extend to almost 1,865 miles (3,000 km.) beyond the crater's edge.

Other Lunar Features

Near the crater called Archimedes is the Apennines mountain range, which has peaks nearly 20,000 feet (60,000 meters) high—altitudes comparable to South America's Andes.

The Moon also has valleys. Two of the most well known are the Alpine Valley, which is about 115 miles (185 km.) long; and the Rheita Valley, located about 155 miles (250 km.) from the Stevinus crater, which is 238 miles (383 km.) long, 15.5 miles (25 km.) wide, and 2,000 feet (609 meters) deep.

Smaller than valleys and resembling cracks in the lunar surface are features called rilles, which are thought to be places of ancient lava flow. Many rilles can be seen near the Aristarchus crater. Rilles are often up to 3 miles (5 km.) wide and can stretch for more than 104 miles (167 km.).

A wrinkle in the lunar surface is called a ridge. Many ridges are found around the boundaries of the maria. The Serpentine Ridge cuts through Mare Serenitatis.

Exploration of the Moon

Robotic spacecraft were the first visitors to explore the Moon. The Russian spacecraft Luna 1 made the first flyby of the Moon in January, 1959. Eight months later, Luna 2 made the first impact on the Moon's surface. In October, 1959, Luna 3 was the first spacecraft to photograph the side of the Moon not visible from Earth. In 1994 the United States' *Clementine* spacecraft was the first probe to map the Moon's composition and topography globally.

The first humans to land on the Moon were the U.S. astronauts Neil Armstrong and Edwin "Buzz" Aldrin. On July 20, 1969, they landed in the *Eagle* lunar module, during the Apollo 11 mission. Armstrong's famous statement, "That's one small step for man, one giant leap for mankind," was heard around the world by millions of people who watched the first humans set foot on the lunar surface, at the Sea of Tranquillity. The last twentieth century human mission to reach the lunar surface, Apollo 17, landed there in December, 1972. Astronauts Gene Cernan and geologist Jack Schmitt landed in the Taurus-Littrow Valley (20°N, 31°E).

Noreen A. Grice

THE SUN AND THE EARTH

Of all the astronomical phenomena that one can consider, few are more important to the survival of life on Earth than the relationship between Earth and the Sun. With the exception of small amounts of residual (endogenic) energy that have remained inside the earth from the time of its formation some 4.5 billion years ago and which sustain some specialized forms of life along some oceanic rift systems, almost all other forms of life, including human, depend on the exogenic light and energy that the earth receives directly from the Sun.

The enormous variety of ecosystems on Earth are highly dependent on the angles at which the Sun's rays strike Earth's spherical surface. These angles, which vary greatly with latitude and time of year, determine many commonly observed phenomena, such as the height of the Sun above the horizon, the changing lengths of day and night throughout the year, and the rhythm of the seasons. Daily and seasonal changes have profound effects on the many climatic regions and life cycles found on earth.

The Sun

The center of Earth's solar system, the Sun is but one ordinary star among some 100 billion stars in an ordinary cluster of stars called the Milky Way gal-

axy. There are at least 10 billion galaxies in the universe, each with billions of stars. Statistically, the chances are good that many of these stars have their own solar systems. Late twentieth century astronomical observations discovered the presence of what appear to be planets, large ones similar in size to Jupiter, orbiting other stars.

Earth's Sun is an average star in terms of its physical characteristics. It is a large sphere of incandescent gas that has a diameter more than 100 times that of Earth, a mass more than 300,000 times that of Earth, and a volume 1.3 million times that of Earth. The Sun's surface gravity is thirty-four times that of Earth.

The conversion of hydrogen into helium in the Sun's interior, a process known as nuclear fusion, is the source of the Sun's energy. The amount of mass that is lost in the fusion process is miniscule, as evidenced by the fact that it will take perhaps 15 million years for the Sun to lose one-millionth of its total mass. The Sun is expected to continue shining through another several billion years.

Earth Revolution

The earth moves about the Sun in a slightly elliptical orbit called a revolution. It takes one year for the earth

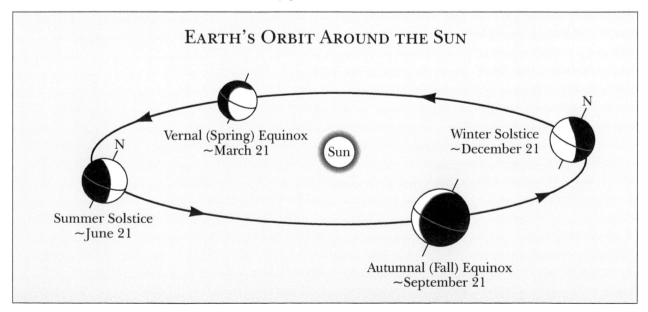

EARTH'S ORBIT AROUND THE SUN

Vernal (Spring) Equinox
~March 21

Sun

Winter Solstice
~December 21

N

Summer Solstice
~June 21

Autumnal (Fall) Equinox
~September 21

to make one revolution at an average orbital velocity of about 29.6 kilometers per second (18.5 miles per second). Earth-sun relationships are described by a tropical year, which is defined as the period of time (365.25 average solar days) from one vernal equinox to another. To balance the tropical year with the calendar year, a whole day (February 29) is added every fourth year (leap year). Other minor adjustments are necessary so as to balance the system.

Perihelion and Aphelion

The average distance between Earth and the Sun is approximately 93 million miles (150 million km.). At that distance, sunlight, which travels at the speed of light (186,000 miles/300,000 kilometers per second), takes about 8.3 minutes to reach the earth. Since the earth's orbit is an ellipse rather than a circle, the earth is closest to the Sun on about January 3—a distance of 91.5 million miles (147 million km.). This position in space is called perihelion, which comes from the Greek *peri*, meaning "around" or "near," and *helios*, meaning the Sun. Earth is farthest from the Sun on about July 4 at aphelion (Greek *ap*, "away from," and *helios*), with a distance of 152 million kilometers (94.5 million miles).

Axial Inclination

Astronomers call the imaginary surface on which Earth orbits around the Sun the plane of the ecliptic. The earth's axis is inclined 66.5 degrees to the plane of the ecliptic (or 23.5 degrees from the perpendicular to the plane of the ecliptic), and it maintains this orientation with respect to the stars. Thus, the North Pole points in the same direction to Polaris, the North Star, as it revolves about the Sun. Consequently, the Northern Hemisphere tilts away from the Sun during one-half of Earth's orbit and toward the Sun through the other half.

Winter solstice occurs on December 21 or 22, when the tilt of the Northern Hemisphere away from the Sun is at its maximum. The opposite condition occurs during summer solstice on June 21 or 22, when the Northern Hemisphere reaches its maximum tilt toward the Sun. The equinoxes occur midway between the solstices when neither the Southern nor the Northern Hemisphere is tilted toward the Sun. The

---

ECLIPSES

The Sun's diameter is 400 times larger than the moon's; however, the moon is 400 times closer to Earth than the Sun, making the two objects appear nearly the same size in the sky to observers on Earth. As the moon orbits Earth, it crosses the plane of the Earth-Sun orbit twice each month. If one of the orbit-crossing points (called nodes) occurs during a new or full moon phase, a solar or lunar eclipse can occur.

A solar eclipse occurs when the moon and the Sun appear to be in the exact same place in the sky during a new moon phase. When that happens, the moon blocks the light of the Sun for up to seven minutes. Because solar eclipses can be seen only from certain places on Earth, some people travel around the world—sometimes to remote places—to view them.

A lunar eclipse occurs when Earth is positioned between the Sun and the moon and casts its shadow on the moon. In contrast to solar eclipses, lunar eclipses are visible from every place on Earth from which the moon can be seen.

---

vernal and autumnal equinoxes occur on March 20 or 21 and September 22 or 23, respectively.

The axial inclination of 66.6 degrees (or 23.5 degrees from the perpendicular) explains the significance of certain parallels on the earth. The noon sun shines directly overhead on the earth at varying latitudes on different days—between 23.5°S and 23.5°N. The parallels at 23.5°S and 23.5°N are called the Tropics of Capricorn and Cancer, respectively.

During the winter and summer solstices, the area on the earth between the Arctic Circle (at 66.5°N) and the North Pole has twenty-four hours of darkness and daylight, respectively. The same phenomena occurs for the area between the Antarctic Circle (at 66.5°S) and the South Pole, except that the seasons are reversed in the Southern Hemisphere. At the poles, the Sun is below the horizon for six months of the year.

For those living outside the tropics (poleward of 23.5 degrees north and south latitude), the noon sun will never shine directly overhead. Hours of daylight will also vary greatly during the year. For example, daylight will range from approximately nine hours during the winter solstice to fifteen hours during the summer solstice for persons living near 40°N, such as in Philadelphia, Denver, Madrid, and Beijing.

Solar Radiation

Given the size of the earth and its distance from the Sun, it is estimated that this planet receives only about one two-billionth part of the total energy released by the Sun. However, this seemingly small amount is enough to drive the massive oceanic and atmospheric circulation systems and to support all life processes on Earth.

Solar energy is not evenly distributed on Earth. The higher the angle of the Sun in the sky, the greater the duration and intensity of the insolation.

To illustrate this, note how easy it is look at the Sun when it is very low on the horizon—near dawn and sunset. At those times, the Sun's rays have to penetrate much more of the atmosphere, so more of the sunlight is absorbed. When the Sun's rays are coming in at a low angle, the same solar energy is spread over a larger area, thereby leading to less insolation per unit of area. Thus, the equatorial region receives much more solar energy than the polar region. This radiation imbalance would make the earth decidedly less habitable were it not for the atmospheric and oceanic circulation systems (such as the warm Gulf Stream) that move the excess heat from the Tropics to the middle and high latitudes.

Robert M. Hordon

THE SEASONS

Earth's 365-day year is divided into seasons. In most parts of the world, there are four seasons—winter, spring, summer, and fall (also called autumn). In some tropical regions—those close to the equator—there are only two seasons. In areas close to the equator, temperatures change little throughout the year; however, amounts of rainfall vary greatly, resulting in distinct wet and dry seasons. The polar regions of the Arctic and Antarctic also have little variation in temperature, remaining cold throughout the year. Their seasons are light and dark, because the Sun shines almost constantly in the summer and hardly at all in the winter.

The four seasons that occur throughout the northern and southern temperate zones—between the tropics and the polar regions—are climatic seasons, based on temperature and weather changes. Winter is the coldest season; it is the time when days are short and few crops can be grown. It is followed by spring, when the days lengthen and the earth warms; this is the time when planting typically be-

gins, and animals that hibernate (from the French word for winter) during the winter leave their dens.

Summer is the hottest time of the year. In many areas, summer is marked by drought, but other regions experience frequent thunderstorms and humid air. In the fall, the days again become shorter and cooler. This is the time when many crops are harvested. In ancient cultures, the turning of the seasons was marked by festivals, acknowledging the importance of seasonal changes to the community's survival.

Each season is defined as lasting three months. Winter begins at the winter solstice, which is the time when the Sun is farthest from the equator. In the Northern Hemisphere, this occurs on December 21 or 22, when the Sun is directly over the tropic of Capricorn. Summer begins at the other solstice, June 20 or 21 in the Northern Hemisphere, when the Sun is directly over the tropic of Cancer. The winter solstice is the shortest day of the year; the summer solstice is the longest.

Spring and fall begin on the two equinoxes. At an equinox, the Sun is directly above the earth's equator and the lengths of day and night are approximately equal everywhere on Earth. In the Northern Hemisphere, the vernal (spring) equinox occurs on March 21 or 22; in the Southern Hemisphere, it is the autumnal (fall) equinox. The Northern Hemisphere's autumnal equinox (and the Southern Hemisphere's vernal equinox) occurs September 22 or 23.

Seasons and the Hemispheres

The relationship of the seasons to the calendar is opposite in the Northern and Southern Hemispheres. On the day that a summer solstice occurs in the Northern Hemisphere, the winter solstice occurs in the Southern Hemisphere. Thus, when it is summer in the Southern Hemisphere, it is winter in the Northern Hemisphere, and vice versa.

The Sun and the Seasons

The reason why summers and winters differ in the temperate zones is often misunderstood. Many people think that winter happens when the Sun is more distant from Earth than it is in summer. What causes Earth's seasons is not the changing distances between the earth and the Sun, but the tilt of the earth's axis. A line drawn from the North Pole to the South Pole through the center of the earth (the earth's axis) is not perpendicular to the plane of the earth's orbit (the ecliptic). The earth's axis and the perpendicular to the ecliptic make an angle of 23.5 degrees. This tilts the Northern Hemisphere toward the Sun when the earth is on one side of its orbit around the Sun, and tilts the Southern Hemisphere toward the Sun when the earth moves around to the Sun's opposite side. When the Sun appears to be at its highest in the sky, and its rays are most direct, summer occurs. When the Sun appears to be at its lowest, and its rays are indirect, there is winter.

Local Phenomena

Local conditions can have important effects on seasonal weather. At locations near oceans, sea breezes develop during the day, and evenings are characterized by land breezes. Sea breezes bring cooler ocean air in toward land. This results in temperatures at the shore often being 5 to 11 degrees Fahrenheit (3 to 6 degrees Celsius) lower than temperatures a few miles inland.

At night, when land temperatures are lower than ocean temperatures, land breezes move air from the land toward the water. As a result, coastal regions have less seasonal temperature variations than inland areas do. For example, coastal areas seldom become cold enough to have snow in the winter, even though inland areas at the same latitude do.

Hailstorms

Hail usually occurs during the summer, and is associated with towering thunderstorm clouds, called cumulonimbus. Hail is occasionally confused with sleet. Sleet is a wintertime event, and occurs when warmer layers of air sit above freezing layers near the ground. Rain that forms in the warmer, upper layer solidifies into tiny ice pellets in the lower, subfreezing layer before hitting the ground.

Hail is an entirely different phenomenon. When cold air plows into warmer, moist air—called a cold front boundary—powerful updrafts of rising air can be created. The warm, moist air propelled upward by the heavier cold air can reach velocities approaching 100 miles (160 kilometers) per hour. Ice crystals form above the freezing level in the cumulonimbus clouds and fall into lower, warmer parts of the clouds, where they become coated with water. Picked up by an updraft, the coated ice crystals are carried back to a higher, colder levels where their water coatings freeze. This cycle can repeat many times, producing hailstones that have multiple, concentric layers of ice.

Hailstorms can be very damaging. Hail can ruin crops, dent car bodies, crack windshields, and injure people. The Midwest of the United States is particularly susceptible to hailstorms. There, warm, moist air from the Gulf of Mexico often meets much colder, drier air originating in Canada. This combination produces the extreme atmospheric instability necessary for that kind of weather.

Alvin S. Konigsberg

EARTH'S INTERIOR

EARTH'S INTERNAL STRUCTURE

Earth is one of the nine known planets in the Sun's solar system that formed from a giant cloud of cosmic dust called a nebula. This event is thought to have happened between 4.44 billion years ago (based on the age of the oldest-known Moon rock) and 4.56 billion years ago (the age of meteorite bombardment). After Earth's formation, heat released by colliding particles combined with the heat energy released by the decay of radioactive elements to cause some or all of Earth's interior to melt. This melting began the process of differentiation, which allowed the heavier elements, mainly iron and nickel, to sink toward Earth's center while the lighter, rocky components moved upward, as a result of the contrast in density of the earth's forming elements.

This process of differentiation was probably the most important event of Earth's early history. It changed the planet from a homogeneous mixture with neither continents nor oceans to a planet with three layers: a dense core beginning at 1,800 miles (2,900 km.) deep and ending at Earth's center, 3,977 miles (6,400 km.) below the surface; a mantle beginning between 3 and 44 miles (5-70 km.) deep and ending at Earth's core; and a crust going from Earth's surface to about 3-6 miles (5-10 km.) deep for oceanic crust and 22-44 miles (35-70 km.) deep for continental crust.

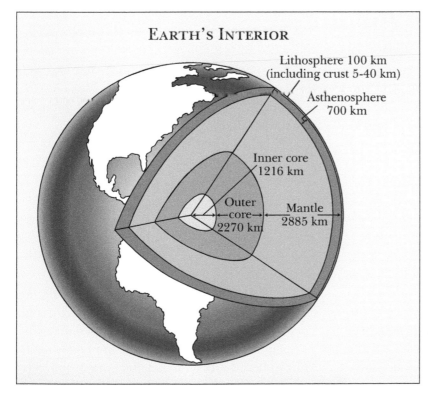

EARTH'S INTERIOR

Lithosphere 100 km
(including crust 5-40 km)

Asthenosphere
700 km

Inner core
1216 km

Outer
core
2270 km

Mantle
2885 km

Layering of the Earth

Earth's layers can be classified either by their composition (the traditional method) or by their mechanical behavior (strength). Compositional classification identifies several distinct concentric layers, each with its own properties. The outermost layer of Earth is the crust or skin. This is divided into continental and oceanic crusts. The continental crust varies in thickness between 22 and 25 miles (35 and 40 km.) under flat continental regions and up to 44 miles (70 km.) under high mountains. The oceanic crust is made up of igneous rocks rich in iron and magnesium, such as basalt and peridotite. The upper continental crust is composed mainly of alumino-silicates. The old-

PROPERTIES OF SEISMIC WAVES

Seismologists use two types of body waves—primary (P-waves) and secondary (S-waves) waves—to estimate seismic velocities of the different layers within the earth. In most rock types P-waves travel between 1.7 and 1.8 times more quickly than S-waves; therefore, P-waves always arrive first at seismographic stations. P-waves travel by a series of compressions and expansions of the material through which they travel. P-waves can travel through solids, liquids, or gases. When P-waves travel in air, they are called sound waves.

The slower S-waves, also called shear waves, move like a wave in a rope. This movement makes the S-wave more destructive to structures like buildings and highway overpasses during earthquakes. Because S-waves can travel only through solids and cannot travel through Earth's outer core, seismologists concluded that Earth's outer core must be liquid or at least must have the properties of a fluid.

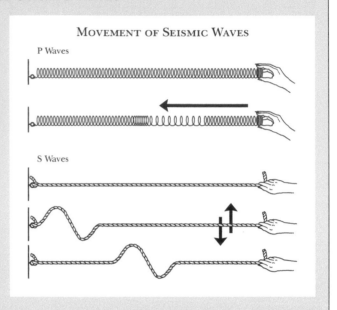

MOVEMENT OF SEISMIC WAVES

P Waves

S Waves

est continental crustal rock exceeds 3.8 billion years, while oceanic crustal rocks are not older than 180 million years. The oceanic crust is heavier than the continental crust.

Earth's next layer is the mantle, which is made up mostly of ferro-magnesium silicates. It is about 1,800 miles (2,900 km.) thick and is separated into the upper and lower mantle. Most of Earth's internal heat is contained within the mantle. Large convective cells in the mantle circulate heat and may drive plate-tectonic processes.

The last layer is the core, which is separated into the liquid outer core and the solid inner core. The outer core is 1,429 miles (2,300 km.) thick, twice as thick as the inner core. The outer core is mainly composed of a nickel-iron alloy, while the inner core is almost entirely composed of iron. Earth's magnetic field is believed to be controlled by the liquid outer core.

In the mechanical layering classification of the earth's interior, the layers are separated based on mechanical properties or strength (resistance to flowing or deformation) in addition to composition. The uppermost layer is the lithosphere (sphere of rock), which comprises the crust and a solid portion of the upper mantle. The lithosphere

is divided into many plates that move in relation to each other due to tectonic forces. The solid lithosphere floats atop a semiliquid layer known as the asthenosphere (weak sphere), which enables the lithosphere to move around.

Exploring Earth's Interior

Volcanic activity provides natural samples of the outer 124 miles (200 km.) of Earth's interior. Meteorites—samples of the solar system that have collided with Earth—also provide clues about Earth's composition and early history. The most ambitious human effort to penetrate Earth's interior was made by the former Soviet Union, which drilled a super-deep research well, named the Kola Well, near Murmansk, Russia. This was an attempt to penetrate the crust and reach the upper mantle. The reported depth of the Kola Well is a little more than 7.5 miles (12 km.). Although impressive, the drilled depth represents less than 0.2 percent of the distance from the earth's surface to its center.

A great deal of knowledge about Earth's composition and structure has been obtained through computer modeling, high-pressure laboratory experiments, and meteorites, but most of what is known about Earth's interior has been acquired by studying seismic waves generated by earthquakes and nuclear explosions. As seismic waves are transmitted, reflected, and refracted through the earth, they carry information to the surface about the materials through which they have traveled. Seismic waves are recorded at receiver stations (seismographic stations) and processed to provide a picturelike image of Earth's interior.

Changes in P- and S-wave velocities within Earth reveal the sequence of layers that make up Earth's interior. P-wave velocity depends on the elasticity, rigidity, and density of the material. By contrast, S-wave velocity depends only on the rigidity and density of the material. There are sharp variations in velocity at different depths, which correspond to boundaries between the different layers of Earth. P-wave velocity within crustal rocks ranges from 3.6-4.2 miles (6-7 km.) per second.

The boundary between the crust and the mantle is called the Mohorovičić discontinuity or Moho. At Moho, P-wave velocity increases from 4.2-4.8 miles (7-8 km.) per second. Beyond the crust-mantle boundary, P-wave velocity increases gradually up to about 8.1 miles (13.5 km.) per second at the core-mantle boundary. At this depth, S-waves are not transmitted and P-wave velocity, decreases from 8.1 to 4.8 miles (13.5 to 8 km.) per second, which strongly supports the concept that the outer core is liquid, since S-waves cannot travel through liquids. As P-waves enter the inner core, their velocity again increases, to about 6.8 miles (11.3 km.) per second.

Earth's interior seems to be characterized by a gradual increase with depth in temperature, pressure, and density. Extensive experimental and modeling work indicates that the temperature at 62 miles (100 km.) is between 1,200 and 1,400 degrees Celsius (2,192 to 2,552 degrees Fahrenheit). The temperature at the core-mantle boundary—about 1,802 miles (2,900 km.) deep—is calculated to be about 8,130 degrees Fahrenheit (4,500 degrees Celsius). At Earth's center the temperature may exceed 12,092 degrees Fahrenheit (6,700 degrees Celsius). Although at Earth's surface, heat energy is slowly but continuously lost as a result of outgassing, such as from volcanic eruptions, its interior remains hot.

Seismic Tomography and Future Exploration
Seismic tomography is one of the newest tools that earth scientists are using to develop three-dimensional velocity images of Earth's interior. In seismic tomography, several crossing seismic waves from different sources (earthquakes and nuclear explosions) are analyzed in much the same way that computerized axial tomography (CAT) scanners are used in medicine to obtain images of human organs. Seismic tomography is providing two- and three-dimensional images from the crust to the core-mantle boundary. Fast P-wave velocities have been correlated to cool material—for example, a piece of sinking lithosphere (cool rigid layer) such as in regions underneath the Andes Mountains (subduction zone); slow P-wave velocities have been correlated with hot materials—for example, rising mantle plumes of hot spots such as the one responsible for volcanic activity in the Hawaiian Islands.

Rubén A. Mazariegos-Alfaro

PLATE TECTONICS

The theory of plate tectonics provides an explanation for the present-day structure of the large landforms that constitute the outer part of the earth. The theory accounts for the global distribution of continents, mountains, hills, valleys, plains, earthquake activity, and volcanism, as well as various associations of igneous, metamorphic, and sedimentary rocks, the formation and location of min-

MAJOR TECTONIC PLATES AND MID-OCEAN RIDGES

Types of Boundaries: Divergent ⫽   Convergent ⤙   Transform ⟋

eral resources, and the geology of ocean basins. Everything about the earth is related either directly or indirectly to plate tectonics.

Basic Theory

Plate-tectonic theory is based on an Earth model in which a rigid, outer shell—the lithosphere—lies above a hotter, weaker, partially molten part of the mantle called the asthenosphere. The lithosphere varies in thickness between 6 and 90 miles (10 and 150 km.), and comprises the crust and the underlying, upper mantle. The asthenosphere extends from the base of the lithosphere to a depth of about 420 miles (700 km.). The brittle lithosphere is broken into a pattern of internally rigid plates that move horizontally relative to each other across the earth's surface.

More than a dozen plates have been distinguished, some extending more than 2,500 miles (4,000 km.) across. Exhibiting independent motion, the plates grind and scrape against each other, similar to chunks of ice in water, or like giant rafts cruising slowly on the asthenosphere. Most of the earth's dynamic activity, including earthquakes and volcanism, occurs along plate boundaries. The global distribution of these tectonic phenomena delineates the boundaries of the plates.

Geological observations, geophysical data, and theoretical models support the existence of three types of plate boundaries. Divergent boundaries occur where adjacent plates move away from each other. Convergent boundaries occur where adjacent plates move toward each other. Transform boundaries occur where plates slip past one another in directions parallel to their common boundaries.

The continents were formed by the movement at plate boundaries, and continental landforms were generated by volcanic eruptions and continental plates colliding with each other. The velocity of plate movement varies from plate to plate and even within portions of the same plate, ranging from 0.8 to 8 inches (2 to 20 centimeters) per year. The rates

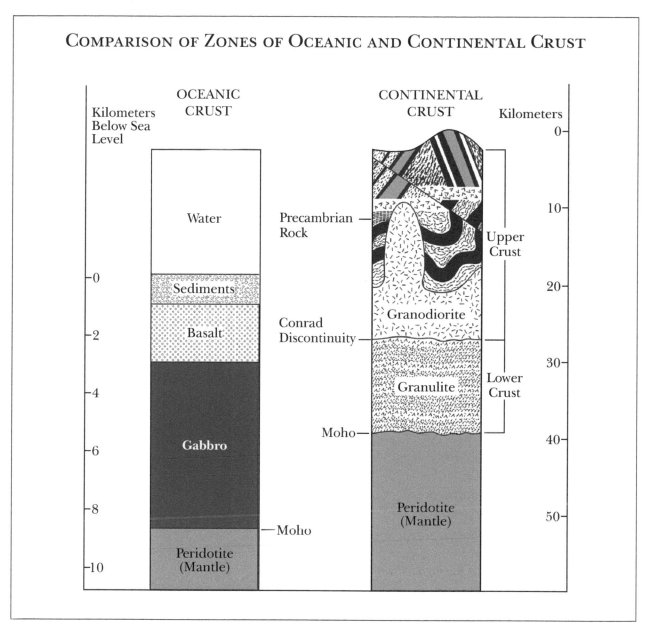

COMPARISON OF ZONES OF OCEANIC AND CONTINENTAL CRUST

are calculated from the distance to the midoceanic ridge crests, along with the age of the sea floor as determined by radioactive dating methods.

Convection currents that are driven by heat from radioactive decay in the mantle are important mechanisms involved in moving the huge plates. Convection currents in the earth's mantle carry magma (molten rock) up from the asthenosphere. Some of this magma escapes to form new lithosphere, but the rest spreads out sideways beneath the lithosphere, slowly cooling in the process. Assisted by gravity, the magma flows outward, dragging the overlying lithosphere with it, thus continu-

ing to open the ridges. When the flowing hot rock cools, it becomes dense enough to sink back into the mantle at convergent boundaries.

A second plate-driving mechanism is the pull of dense, cold, down-flowing lithosphere in a subduction zone on the rest of the trailing plate, further opening up the spreading centers so magma can move upward.

Divergent Plate Boundaries
During the 1950s and 1960s, oceanographic studies revealed that Earth's seafloors were marked by a nearly continuous system of submarine ridges,

THE SUPERCONTINENTS

The theory of plate tectonics explains the present-day distribution of major landforms, seismic and volcanic activity, and physiographic features of ocean basins. Many scientists also use the theory to explain the history of Earth's surface. Evidence indicates that the modern continents once formed a single landmass called Pangaea, meaning "all lands." According to the theory of plate tectonics, approximately 200 million years ago Pangaea began to split into two supercontinents, Laurasia and Gondwanaland. Eventually, as a result of tectonic forces, Laurasia split into North America, Europe, and most of Asia. Gondwanaland broke up into India, South America, Africa, Australia, and Antarctica.

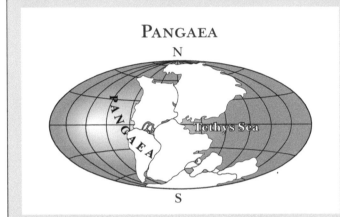

PANGAEA

LAURASIA AND GONDWANALAND

more than 40,000 miles (64,000 km.) in length. Detailed investigations revealed that the midoceanic ridge system has a central rift valley that runs along its length and that the ridge system is associated with volcanic and earthquake activity. The earthquakes are frequent, shallow, and mild.

Magnetic studies of the seafloor indicate that the oceanic lithosphere has been segmented into a series of long magnetic strips that run parallel to the axis of the midoceanic ridges. On either side of the ridge, the ocean floor consists of alternating bands of rock, magnetized either parallel to or exactly opposite of the present-day direction of the earth's magnetic field.

Midoceanic ridges, or divergent plate boundaries, are tensional features representing zones of weakness within the earth's crust, where new seafloor is created by the welling up of mantle material from the asthenosphere into cracks along the ridges. As rifting proceeds, magma ascends to fill in the fissures, creating new oceanic crust. Iron minerals within the magma become aligned to the existing Earth polarity as the rock cools and crystallizes. The oceanic floor slowly moves away from the oce-

anic ridge toward deep ocean trenches, where it descends into the mantle to be melted and recycled to the earth's surface to generate new rocks and landforms.

As the seafloor spreads outward from the rift center, about half of the material is carried to either side of the rift, which is later filled by another influx of molten basalt. When the polarity of the earth changes, the subsequent molten basalt is magnetized in the opposite polarity. The continuation of this process over geologic time leads to the young geologic age of the seafloor and the magnetic symmetry around the midoceanic ridges.

Not all spreading centers are underneath the oceans. An example of continental rifting in its embryonic stage can be observed in the Red Sea, where the Arabian plate has separated from the African plate, creating a new oceanic ridge. Another modern-day example of continental divergent activity is East Africa's Great Rift Valley system. If this rifting continues, it will eventually fragment Africa, producing an ocean that will separate the resulting pieces. Through divergence, large plates are made into smaller ones.

Convergent Plate Boundaries

Because Earth's volume is not changing, the increase in lithosphere created along divergent boundaries must be compensated for by the destruction of lithosphere elsewhere. Otherwise, the radius of Earth would change. The compensation occurs at convergent plate boundaries, where plates are moving together. Three scenarios are possible along convergent boundaries, depending on whether the crust involved is oceanic or continental.

If both converging plates are made of oceanic crust, one will inevitably be older, cooler, and denser than the other. The denser plate eventually subducts beneath the less-dense plate and descends into the asthenosphere. The boundary along the two interacting plates, called a subduction zone, forms a trench. Some trenches are more than 620 miles (1,000 km.) long, 62 miles (100 km.) wide, and 6.8 miles (11 km.) deep. Heated by the hot asthenosphere beneath, the subducted plate becomes hot enough to melt.

Because of buoyancy, some of the melted material rises through fissures and cracks to generate volcanoes along the overlying plate. Over time, other parts of the melted material eventually migrate to a divergent boundary and rise again in cyclic fashion to generate new seafloor. The volcanoes generated along the overriding plate often form a string of islands called island arcs. Japan, the Philippines, the Aleutians, and the Mariannas are good examples of island arcs resulting from subduction of two plates consisting of oceanic lithosphere. Intense earthquakes often occur along subduction zones.

If the leading edge of one of the two convergent plates is oceanic crust and the other is continental crust, the oceanic plate is always the one subducted, because it is always denser. A classic example of this case is the western boundary of South America. On the oceanic side of the boundary, a trench was formed where the oceanic plate plunged underneath the continental plate. On the continental side, a fold mountain belt—the Andes—was formed as the oceanic lithosphere pushed against the continental lithosphere.

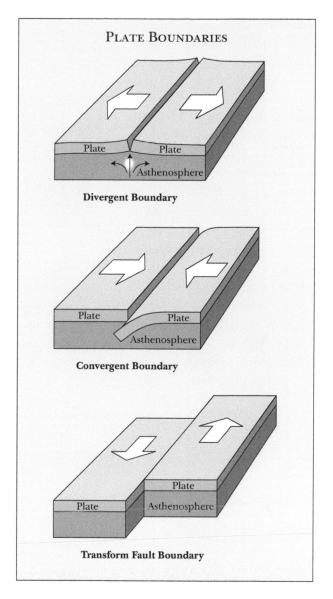

PLATE BOUNDARIES

Divergent Boundary

Convergent Boundary

Transform Fault Boundary

When the oceanic plate descends into the mantle, some of the material melts and works its way up through the mountain belt to produce rather violent volcanoes. The boundary between the plates is a region of earthquake activity. The earthquakes range from shallow to relatively deep, and some are quite severe.

The last type of convergent plate boundary involves the collision of two continental masses of lithosphere, which can result in folding, faulting, metamorphism, and volcanic activity. When the plates collide, neither is dense enough to be forced into the asthenosphere. The collision compresses and thickens the continental edges, twisting and deforming the rocks and uplifting the land to form unusually high fold mountain belts. The prototype example is the collision of India with Asia, resulting in the formation of the Himalayas. In this case, the earthquakes are typically shallow, but frequent and severe.

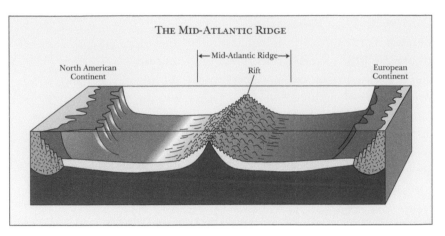

THE MID-ATLANTIC RIDGE

Transform Plate Boundaries

The actual structure of a seafloor spreading ridge is more complex than a single, straight crack. Instead, ridges comprise many short segments slightly offset from one another. The offsets are a special kind of fault, or break in the lithosphere, known as a transform fault, and their function is to connect segments of a spreading ridge. The opposite sides of a transform fault belong to two different plates that are grinding against each other in opposite directions.

CALIFORNIA'S SAN ANDREAS FAULT

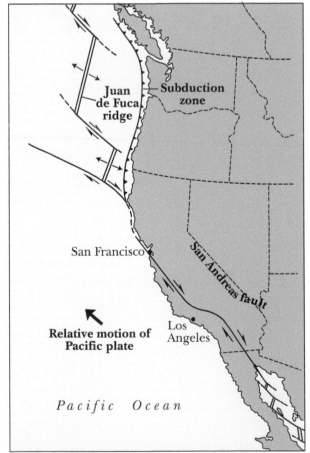

Transform faults form the boundaries that allow the plates to move relative to each another. The classic case of a transform boundary is the San Andreas Fault. It slices off a small piece of western California, which rides on the Pacific plate, from the rest of the state, which resides on the North American plate. As the two plates scrape past each other, stress builds up, eventually being released in earthquakes that can be quite violent.

Mantle Plumes and Hot Spots

Most plate tectonic features are near plate boundaries, but the Hawaiian Islands are not. In the late twentieth century, the only active volcanoes in the Hawaiian Islands were on the island of Hawaii, at the southeast end of the chain. Radiometric dating and examination of states of erosion show that, when proceeding along the chain to the northwest, successive islands are progressively older.

Evidently, the same heat source produced all the volcanoes in the Hawaiian chain. Known as a mantle plume, it has remained stationary while the Pacific plate rides over it, producing a volcanic trail from which absolute motion of the plate can be determined. Since mantle plumes do not move with the plates, the plumes must originate beneath the lithosphere, probably far below it. Resulting volcanoes are called hot spots to distinguish them from subduction-zone volcanoes. Iceland is a good example of a hot spot, as is Yellowstone. At least 100 hot spots are distributed around Earth.

Alvin K. Benson

VOLCANOES

Volcanoes form mountains both on land and in the sea and either do it on a grand scale or merely create minute bumps on the seafloor. Volcanoes do not occur in a random pattern, but are found in distinct zones that are related to plate dynamics. Each of the three types of volcanism on Earth is characterized by specific types of eruptions and magma compositions. Molten magma is the rock material below the earth's crust that forms igneous rock as it cools.

Types of Volcanoes

Geologists generally group volcanoes into four main kinds—cinder cones, composite volcanoes, shield volcanoes, and lava domes.

Cinder cones are built from congealed lava ejected from a single vent. As the gas-charged lava is blown into the air, it breaks into small fragments that solidify and fall as *cinders* around the vent to form a circular or oval cone. Most cinder cones have a bowl-shaped *crater* at the summit and rarely rise more than a thousand feet or so above their surroundings. Cinder cones are numerous in western North America and in other volcanic terrains of the world.

Composite volcanoes —sometimes called stratovolcanoes—include some of the Earth's grandest mountains, including Mount Fuji in Japan, Mount Cotopaxi in Ecuador, Mount Shasta in California, Mount Hood in Oregon, and Mount St. Helens and Mount Rainier in Washington. The essential feature of a composite volcano is a conduit system through which magma deep in the Earth's crust rises to the surface. They are typically steep-sided, symmetrical cones of large dimension built of alternating layers of lava flows, volcanic ash, cinders, blocks, and bombs. They may rise as much as 8,000 feet above their bases. Most have a crater at the summit that contains a central vent or a clustered group of vents. Lavas either flow through breaks in the crater wall or fissures on the flanks of the cone.

Shield volcanoes, the third type of volcano, are built almost entirely of fluid lava flows that pour out in all directions from a central vent, or group of vents, building a broad, gently sloping cone. They are built up slowly as thousands of highly fluid lava flows—basalt lava—spread over great distances, and then cool into thin sheets. Some of the largest volcanoes in the world are shield volcanoes. The Hawaiian Islands are composed of linear chains of these volcanoes including Kilauea and Mauna Loa on the island of Hawaii—two of the world's most active volcanoes. The floor of the ocean is more than 15,000 feet deep at the bases of the islands. As Mauna Loa, the largest of the shield volcanoes (and also the world's largest active volcano), projects 13,679 feet above sea level, its top is over 28,000 feet above the deep ocean floor.

Volcanic Composition

Volcanoes in the midocean ridges and plume environments draw most of their magmas from the earth's mantle and produce mainly dark, magnesium-rich basaltic magmas. When basaltic magmas accumulate in the continental crust (for example, at Yellowstone), the large-scale crustal melting leads to rhyolitic volcanism, the volcanic equivalent of granites. Arc magmas cover a wider range of magmatic compositions, ranging from arc basalt to light-colored, silica-rich rhyolites; the latter are commonly erupted in the form of the silica-rich volcanic rock known as pumice, or the black volcanic glass known as obsidian. Andesites, named after the Andes Mountains, are a common volcanic rock in stratovolcanoes, intermediate in composition between basalt and rhyolite.

Magmas form from several processes that lead to partial melting of a solid rock. The simplest is adding heat—for example, plumes carrying heat from deep levels in the mantle to shallower levels, where melting occurs. Decompressional (lowering the pressure) melting of the mantle occurs where the

SOME VOLCANIC HOT SPOTS AROUND THE WORLD

ocean floor is thinned or carried away by seafloor spreading in midocean ridge environments.

Genesis of Magma

Adding a "flux" to a solid mineral mixture may lower the substance's melting point. The most common theory about arc magma genesis invokes the addition of a low-melting-point substance to the arc mantle, a layer of mantle material at about 60 to 90 miles (100 to 150 km.) below the volcanic arc. The relatively dry arc mantle would usually start to melt at about 2,100 to 2,300 degrees Fahrenheit (1,200 to 1,300 degrees Celsius). However, the addition of water and other gases can lower the melting point of the mixture. The water and its dissolved chemicals are supposedly derived from the subducted slab, the former ocean floor that is pushed back into the earth.

The sequence of events is as follows: New basaltic ocean floor forms at midocean ridge volcanoes. The new hot magma interacts with seawater, leading to vents at the seafloor with their mineralized

deposits. The seafloor becomes hydrated, and sulfur and chlorine from seawater are locked up in newly formed minerals. During subduction, this altered seafloor with slivers of sediment, including limestone, is gradually warmed up and starts to decompose, adding a flux to the surrounding mantle rocks. The mantle rocks then start to melt, and these magmas with minor inherited oceanic materials start to rise and pond at the bottom of the crust. There the magmas sit and wait for an opportunity to erupt, while cooling and crystallizing. Thus, arc magmas bear a chemical signature of subducted oceanic components while their chemical compositions range from basalt to rhyolite.

Volcanic Eruptions

Volcanic eruptions occur as a result of the rise of magma into the volcano (from depths as great as several miles) and then into the throat of the volcano. In basaltic volcanoes, the magmas have relatively little gas, and the magma simply overflows and forms large lava flows, sometimes associated

VOLCANIC ERUPTION AND CALDERA FORMATION

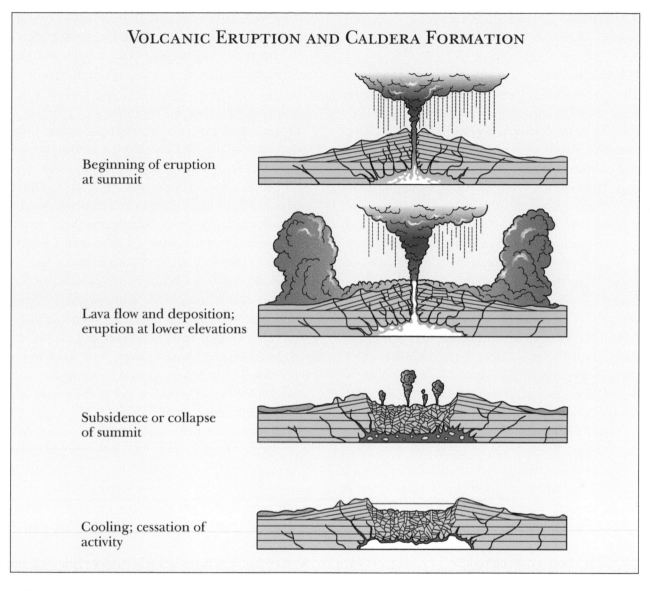

Beginning of eruption
at summit

Lava flow and deposition;
eruption at lower elevations

Subsidence or collapse
of summit

Cooling; cessation of
activity

with fire fountains. Stratovolcanoes can erupt regularly with small explosions or catastrophically after long periods of dormancy. Mount Stromboli, a volcano in Italy, erupts every twenty minutes, with an explosion that creates a column 650 to 980 feet (200 to 300 meters) high. Mount St. Helens in the U.S. state of Washington had a catastrophic eruption in 1980 after about 200 years of dormancy. It emitted an ash plume that reached more than 12 miles (20 km.) into the atmosphere.

After long magma storage periods in the crust, crystallization and melting of crustal material can lead to silica-rich magmas. These are viscous and can have high dissolved water contents—up to 4 to 6 percent by weight. When these magmas break out,

the eruption can be violent and form an eruption column 12 to 35 miles (20 to 55 km.) high. Many cubic miles of magma can be ejected. This leads to so-called plinian ash falls, with showers of pumice and ash over thousands of square miles, with the ash commonly carried around the globe by the high-level winds known as jet streams.

If the volume of ejected magma is large, the volcano empties itself and collapses into the hole, leading to a caldera—a volcanic collapse structure. The caldera at Crater Lake in Oregon is related to a large pumice eruption about 76,000 years ago. Basaltic volcanoes can also form collapse calderas when large volumes of lava have been extruded in a short time. Examples of famous basaltic calderas can be found in Hawaii's Mount Kilauea and the Galapagos Islands.

Volcanic Plumes

The dynamics of volcanic plumes has been studied from eruption photographs, experiments, and theoretical work. The rapidly expanding hot gases force the viscous magma out of the throat of the volcano, where it freezes into pumice. The kinetic energy of the ejected mass carries it 2 to 2.5 miles (3-4 km.) above the volcano. During this phase, air is entrained in the column, diluting the concentration of ash and pumice particles. The hot particles heat the entrained air, the mixture of hot air and solids becomes less dense than the surrounding atmosphere, and a buoyant column rises high into the sky.

The height of an eruption column is not directly proportional to the force of the eruption but is strongly dependent on the rate of heat release of the volcano. If little of the entrained air is heated up, the column will collapse back to the ground and an ash flow forms, which may deposit ash around the volcano. These types of eruptions are among the most devastating, creating glowing ash clouds traveling at speeds up to 60 miles (100 km.) per hour, burning everything in their path. The 1902 eruption of Mount Pelée on Martinique in the Caribbean was such an eruption and killed nearly 30,000 people in a few minutes.

Many volcanoes that are high in elevation are glaciated, and their eruptions lead to large-scale ice melting and possibly mixing of water, magma, and volcanic debris. Massive hot mudflows can race down from the volcano, following river valleys and filling up low areas. The 1980 Mount St. Helens eruption created many mudflows, some of which reached the Pacific Ocean, ninety miles to the west. A catastrophic mudflow event occurred in 1984 at Nevado del Ruiz, a volcano in Colombia, where 20,000 people were buried in mud and perished. When magma intrudes under the ice, meltwater can accumulate and then escape catastrophically, but such meltwater bursts are rare outside Iceland.

Minerals and Gases in Eruptions

The gas-rich character of arc magmas leads to fluid escape at various levels in the volcanoes, and these fluids tend to be rich in chlorine. They can transport metals such as copper, lead, zinc, and gold at high concentrations, and lead to the enrichment of these metals in the fractured volcanic rocks. Many of the world's largest copper ore deposits are associated with older arc volcanism, where erosion has removed most of the volcanic structure and laid the volcano innards bare. Many active volcanoes have modern hydrothermal (hot-water) systems, leading to acid hot springs and crater lakes and the potential to harness geothermal energy. Some areas in Japan, New Zealand, and Central America have an abundance of geothermal energy resources, which are gradually being developed.

Apart from the dangers of eruptions, continuous emissions of large amounts of sulfur dioxide, hydrochloric acid, and hydrofluoric acid present a danger of air pollution and acid rain. Incidences of emphysema and other irritations of the respiratory system are common in people living on the slopes of active volcanoes. The large lava emissions in Iceland in the eighteenth century led to acid fogs all over Europe. Many cattle died in Iceland during this period from the hydrofluoric acid vapors. High levels of fluorine in drinking water can lead to fluorosis, a disease that attacks the bone structure. The discharge of highly acidic fluids from hot springs and crater lakes can cause widespread environmental contamination, which can present a danger for crops gathered from fields irrigated with these waters and for local ecosystems in general.

Johan C. Varekamp

GEOLOGIC TIME SCALE

A major difference between the geosciences (earth sciences) and other sciences is the great enormity of their time scale. One might compare the magnitude of geologic time for geoscientists to the vastness of space for astronomers. Every geological process, such as the movement of crustal plates (plate tectonics), the formation of mountains, and the advance and retreat of glaciers, must be considered within the context of time.

Although certain geologic events, such as floods and earthquakes, seem to occur over short periods of time, the vast majority of observed geological features formed over a great span of time. Consequently, modern geoscientists consider Earth to be exceedingly old. Using radiometric age-dating techniques, they calculate the age of Earth as 4.6 billion years old.

Early miners were probably the first to recognize the need for a scale by which rock and mineral units could be compared over large geographic areas. However, before a time scale—and even geology as a science—could develop, certain principles had to be established. This did not occur until the late eighteenth century when James Hutton, a Scottish naturalist, began his extensive examinations of rock relationships and natural processes at work on the earth. His work was amplified by Charles Lyell in his textbook *Principles of Geology* (1830-1833). After careful observation, Hutton concluded that the natural processes and functions he observed had operated in the same basic manner in the past, and that, in general, natural laws were invari-

able. That idea became known as the principle of uniformitarianism.

The Birth of Stratigraphy

In 1669 Nicholas Steno, a Danish physician working in Italy, recognized that horizontal rock layers contained a chronological record of Earth history and formulated three important principles for interpreting that history. The principle of superposition states that in a succession of undeformed strata, the oldest stratum lies at the bottom, with successively younger ones above. The principle of original horizontality states that because sedimentary particles settle from fluids under graviational influence, sedimentary rock layers must be horizon-

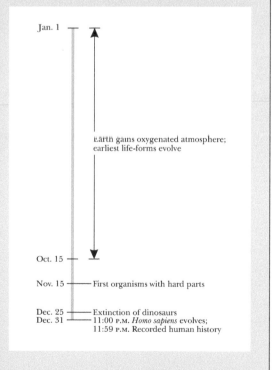

EARTH'S HISTORY COMPRESSED INTO ONE CALENDAR YEAR

One way to visualize events in Earth's history is to compress geologic events into a single calendar year. Earth's birth, 4.6 billion years ago, would occur during the first minute of January 1. The first three-quarters of Earth's history is obscure and would take place from January to mid-October. During this time, Earth gained an oxygenated atmosphere, and the earliest life-forms evolved. The first organisms with hard parts preserved in the fossil record (approximately 570 million years ago) would appear around November 15. The extinction of the dinosaurs (65 million years ago) would occur on Christmas Day. Homo sapiens would first appear at approximately 11 P.M. on December 31, and all of recorded human history would occur in the last few seconds of New Year's Eve.

tal; if not, they have suffered from subsequent disturbance. The principle of original lateral continuity states that strata originally extended in all directions until they thinned to zero or terminated against the edges of the original area of deposition.

In the late eighteenth century, the English surveyor William Smith recognized the wide geographic uniformity of rock layers and discovered the utility of fossils in correlating these layers. By 1815, Smith had completed a geologic map of England and was able to correlate English rock layers with layers exposed across the English Channel in France.

From the need to classify and organize rock layers into an orderly form arose a subdiscipline of modern geology—stratigraphy, the study of rock layers and their age relationships. In 1835 two British geologists, Adam Sedgwick and Roderick Murchison, began organizing rock units into a formal stratigraphic classification. Large divisions, called eras, were based upon well known and characteristic fossils, and included a number of smaller subdivisions, called periods.

The periods are often subdivided into smaller units called epochs. Each period is defined by a representative sequence of rock strata and fossils. For instance, the Devonian period is named for exposures of rock in Devonshire in southern England, while the Jurassic period is defined by strata exposed in the Jura Mountains in northern Switzerland.

Approximately 80 percent of Earth's history is included in the Crypotozoic era (meaning obscure life). Fossils from the Crypotozoic era are rare, and the rock record is very incomplete. After the Crypotozoic era came the Paleozoic (ancient life), Mesozoic (middle life), and Cenozoic (recent life) eras. Most of the life forms that evolved during the Paleozoic and Mesozoic eras are now extinct, whereas 90 percent of the life-

forms that evolved up to the middle Cenozoic era still exist.

The Geologic Time Scale

The geologic time scale is continually in revision as new rock formations are discovered and dated. The ages shown in the table below are in millions of years ago (MYA) before the present and represent the beginning of that particular period. It would be impossible to list all the significant events in Earth's history, but one or two are provided for each period. Note that in the United States, the Carboniferous period has been subdivided into the Mississippian period (older) and the Pennsylvanian period (younger).

The Fossil Record

The word "fossil" comes from the Latin *fossilium*, meaning "dug from beneath the surface of the ground." Fossils are defined as any physical evidence of past life. Fossils can include not only shells, bones, and teeth, but also tracks, trails, and bur-

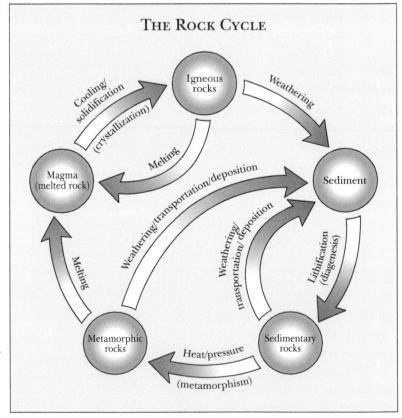

rows. The latter group are referred to as trace fossils. Fossils demonstrate two important truths about life on Earth: First, thousands of species of plants and animals have existed and later became extinct. Second, plants and animals have evolved through time, and the communities of life that have existed on Earth have changed.

Some organisms are slow to evolve and may exist in several geologic time periods, while others evolve quickly and are restricted to small intervals of time within a particular period. The latter, referred to as index fossils, are the most useful to geoscientists for correlating rock layers over wide geographic areas and for recognizing geologic time.

The fossil record is incomplete, because the process of preservation favors organisms with hard parts that are rapidly buried by sediments soon after death. For this reason, the vast majority of fossils are represented by marine invertebrates with exoskeletons, such as clams and snails. Under special circumstances, soft-bodied organism can be preserved, for instance the preservation of insects in amber, made famous by the feature film *Jurassic Park* (1993).

The Rock Cycle

A rock is a naturally formed aggregate of one or more minerals. Three types of rocks exist in the earth's crust, each reflecting a different origin. Igneous rocks have cooled and solidified from molten material either at or beneath Earth's surface. Sedimentary rocks form when preexisting rocks are weathered and broken down into fragments that accumulate and become compacted or cemented together. Fossils are most commonly found in sedimentary rocks. Metamorphic rocks form when heat, pressure, or chemical reactions in Earth's interior change the mineral or chemical composition and structure of any type of preexisting rock.

Over the huge span of geologic time, rocks of any one of these basic types can change into either of the other types or into a different form of the same type. For this reason, older rocks become increasingly more rare. The processes by which the various rock types change over time are illustrated in the rock cycle.

Larry E. Davis

EARTH'S SURFACE

INTERNAL GEOLOGICAL PROCESSES

The earth is layered into a core, a mantle, and a crust. The topmost mantle and the crust make up the lithosphere. Beneath this is a layer called the asthenosphere, which is composed of moldable and partly liquid materials. Heat transference within the asthenosphere sets up convection cells that diverge from hot regions and converge to cold regions. Consequently, the overlying lithosphere is segmented into ridged plates that are moved by the convection process. The hot asthenosphere does not rise along a line. This causes the development of a structure called a transform plate boundary, which is perpendicular to and offsetting the divergent boundary.

The topographic features at Earth's surface, such as mountains, rift valleys, oceans, islands, and ocean trenches, are produced by extension or compression forces that act along divergent, convergent, or transform plate boundaries. The extension and compression forces at Earth's surface are powered by convection within the asthenosphere.

Mountains and Depressions in Zones of Compression

Compression along convergent plate boundaries yields three types of mountain: island arcs that are partly under water; mountains along a continental edge, such as the Andes; and mountains at continental interiors, such as the Alps. At convergent plate boundaries, the denser of the two colliding plates slides down into the asthenosphere and causes volcanic activity to form on the leading edge of the upper plate. Island arcs such as the Aleutians and the Caribbean are formed when an oceanic plate descends beneath another oceanic plate.

Volcanic mountain chains such as the Andes of South America are formed when an oceanic plate descends beneath a continental plate. In both the island arc type and Andean type collisions, a deep depression in the oceans, called a trench, marks the place where neighboring plates are colliding and where the denser plates are pulled downward into the asthensophere. If the colliding plates are of similar density, neither plate will go into the asthenosphere. Instead, the edges of the neighboring plates will be folded and faulted and excess material will be pushed upward to form a block mountain, such as the mountain chain that stretches from the Alps through to the Himalayas. This type of mountain chain is not associated with a trench.

The Appalachians of the eastern United States are an example of the alpine type of mountain belt. When the Appalachians were forming 300 million years ago, rock layers were deformed. The deformation included folding to form ridges and valleys; fracturing along joint sets, with one joint set being parallel to ridges, while the other set is perpendicular; and thrust faulting, in which rock blocks were detached and shoved upward and northwestward.

Millions of years of erosion have reduced the height of the mountains and have produced topographic inversion in the foothills. Topographic inversion occurs because joints create wider fractures at upfolded ridges and narrower fractures at downfolded valleys. Erosion is then accelerated at upfolded ridges, converting ancient ridges into valleys, while ancient valleys stand as ridges. The Valley and Ridge Province of the Appalachians is noted for such topographic inversion.

West of the Valley and Ridge Province of the Appalachians is the Allegheny Plateau, which is bounded by a cliff on its eastern side. In general, plateaus are flat topped because the rock layer that covers the surface is resistant to weathering. The cliff side is formed by erosion along joint or fault surfaces.

The Sierra Nevada range, which formed 70 million years ago, is an example of an Andean type of mountain belt. Millions of years of erosion there has exposed igneous rocks that formed at depth. Over the years, the force of compression that formed the Sierras has evolved to form a zone of extension between the Sierras and the Colorado Plateau.

Mountains and Depressions in Zones of Extension

Extension is a strain that involves an increase in length and causes crustal thinning and faulting. Extension is associated with convergent boundaries, divergent boundaries, and transform boundaries.

Extension Associated with a Convergent Boundary

During the formation of the Sierra Nevada, an oceanic plate that was subducted beneath California declined at a shallow angle eastward toward the Colorado Plateau. Later, the subducted plate peeled off and molten asthenosphere took its place. From the asthenosphere, lava ascended through fractures to form volcanic mountains in Arizona and Utah, and lava flowed and volcanic ash fell as far west as California. The lithosphere has been heated up and has become buoyant, so the Colorado Plateau rises to higher elevations, and rock layers slide westward from it in a zone of extension that characterizes the Basin and Range Province.

In the extension zone, the top rock layers move westward on curved displacement planes that are steep at the surface and nearly horizontal at depth. When rock layers move westward over a curved detachment surface, the trailing edge of the rock layers roll over and are tilted toward the east so they do not leave space in buried rocks. On the other hand, a west-facing slope is left behind on a mountain from which the rock layers were detached. There-

fore, movement along one curved detachment surface creates a valley, and movement along several such detachment surfaces forms a series of valleys separated by ridges, as in the Basin and Range Province. The amount of the displacement along the curved surfaces is not uniform. For example, more displacement has created wide zones of valleys such as the Las Vegas valley in Nevada, and Death Valley in California.

Extension Associated with a Divergent Boundary

The longest mountain chain on Earth lies under the Pacific Ocean. It is about 37,500 miles (60,000 km.) long, 31.3 miles (50 km.) wide, and 2 miles (3 km.) high. The central part of this midoceanic ridge is marked by a depression, about 3,000 feet (1,000 meters) deep, and is called a rift valley. A part of the submarine ridge, called the East Pacific Rise, forms the seafloor sector in the Gulf of California and reappears off the coast of northern California, Oregon, and Washington as the Juan de Fuca Ridge. Another part forms the seafloor sector in the Gulf of Aden and Red Sea seafloor, part of which is exposed in the Afar of Ethiopia. From the Afar southward to the southern part of Mozambique is the longest exposed rift valley on land, the East African Great Rift Valley.

A rift valley is the place where old rocks are pushed aside and new rocks are created. Blocks of rock that are detached from the rift walls slide down by a series of normal fault displacements. The ridge adjacent to the central rift is present because hot rocks are less dense and buoyant. If the process of divergences continues from the rifting stage to a drifting stage, as the rocks move farther away from the central rift, the rocks become older, colder, and denser, and push on the underlying asthenosphere to create basins. These basins will be flooded by oceanic water as neighboring continents drift away. However, not all processes of divergence advance from the rifting to the drifting stage.

Extension Associated with Transform Boundary

The best-known example of a transform boundary is the San Andreas Fault that offsets the East Pacific

Rise from the Juan de Fuca Ridge, and is exposed on land from the Gulf of California to San Francisco. Along transform boundaries, there are pull-apart basins that may be filled to form lakes, such as the Salton Sea in Southern California. Another example is the Aqaba transform of the Middle East, along which the Sea of Galilee and the Dead Sea are located.

H. G. Churnet

EXTERNAL PROCESSES

Continuous processes are at work shaping the earth's surface. These include breaking down rocks, moving the pieces, and depositing the pieces in new locations. Weathering breaks down rocks through atmospheric agents. The process of moving weathered pieces of rock by wind, water, ice, or gravity is called erosion. The materials that are deposited by erosion are called sediment.

Mechanical weathering occurs when a rock is broken into smaller pieces but its chemical makeup is not changed. If the rock is broken down by a change in its chemical composition, the process is called chemical weathering.

Mechanical Weathering

Different types of mechanical weathering occur, depending on climatic conditions. In areas with moist climates and fluctuating temperatures, rocks can be broken apart by frost wedging. Water fills in cracks in rocks, then freezes during cold nights. As the ice expands and pushes out on the crack walls, the crack enlarges. During the warm days, the water thaws and flows deeper into the enlarged crack. Over time, the crack grows until the rock is broken apart. This process is active in mountains, producing a pile of rock pieces at the mountain base called talus.

Salt weathering occurs in areas where much salt is available or there is a high evaporation rate, such as along the seashore. Salt crystals form when salty moisture enters rock cracks. Growing crystals settle in the bottom of the crack and apply pressure on the crack walls, enlarging the crack.

Thermal expansion and contraction occur in climates with fluctuating temperatures, such as deserts. All minerals expand during hot days and contract during cold nights, and some minerals expand and contract more than others. This process continues until the rock loosens up and breaks into pieces.

Mechanical exfoliation can happen to a rock body overlain by a thick rock or sediment layer. If the heavy overlying layer over a portion of the rock body is removed, pressure is relieved and the exposed rock surface will expand in response. This expanding surface will break off into sheets parallel to the surface, but the remaining rock body remains under pressure and unchanged.

When plant roots grow into cracks in rocks, they enlarge the cracks and break up the rocks. Finally, abrasion can occur to rock fragments during transport. Either the fragments collide, breaking apart, or fragments are scraped against rocks, breaking off pieces.

Chemical Weathering

Water and oxygen create two common causes of chemical weathering. For example, dissolution occurs when water or another solution dissolves minerals within a rock and carries them away. Hydrolysis can occur when water flows through earth materials. The hydrogen ions or the hydroxide ions of the water may react with minerals in the

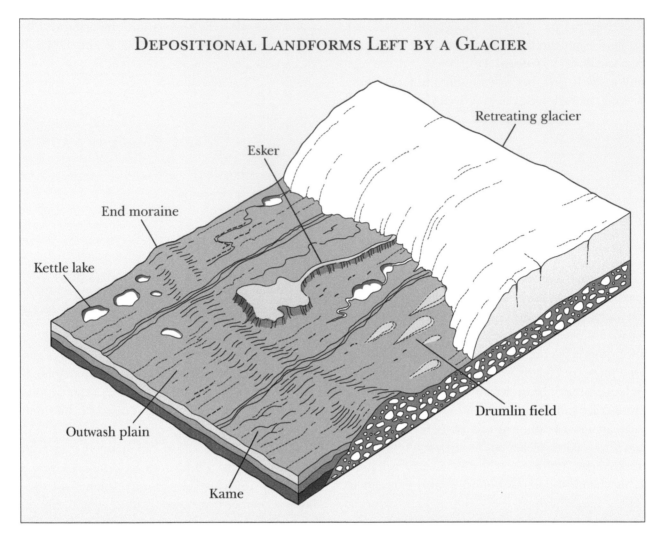

DEPOSITIONAL LANDFORMS LEFT BY A GLACIER

Retreating glacier

Esker

End moraine

Kettle lake

Drumlin field

Outwash plain

Kame

rocks. When this occurs, the chemical composition of the mineral is changed, and a new mineral is formed. Hydrolysis often produces clay minerals.

Some elements in minerals combine with oxygen from the atmosphere, creating a new mineral. This process is called oxidation. Some of these oxidation minerals are commonly referred to as rust.

Mass Movement

Weathered rock pieces (sediments) are transported (eroded) by one or more of four transport processes: water (streams and oceans), wind, ice (glaciers), or gravity. Mass movement transports earth materials down slopes by the pull of gravity. Gravity, constantly working to pull surface materials down, parallel to the slope, is the most important factor affecting mass movement. There is also a force in-

volved perpendicular to the slope that contributes to the effects of friction.

Friction, the second factor, is determined by the earth material type involved. For example, weathering may create cracks in rocks, which form planes of weakness on which the mass movement can occur. Loose sediments always tend to roll downhill.

The third factor is the slope angle. Each earth material has its own angle of repose, which is the steepest slope angle on which the materials remain stable. Beyond this slope angle, earth materials will move downslope.

Water, the fourth factor, affects the stability of the earth material in the slope. Friction is weakened by water between the mineral grains in the rock. For example, water can make clay quite slippery, causing the mass movement.

HOW HYDROLOGY SHAPES GEOGRAPHY

Water and ice sculpt the landscape over time. Fast-flowing rivers erode the soil and rock through which they flow. When rivers slow down in flatter areas, they deposit eroded sediments, creating areas of rich soils and deltas at the mouths of the rivers. Over time this process wears down mountain ranges. The Appalachian Mountain range on the eastern side of the North American continent is hundreds of millions of years older than the Rocky Mountain range on the continent's western side. Although the Appalachians once rivaled the Rockies in size, they have been made smaller by time and erosion.

Canyons are carved by rivers, as the Grand Canyon was carved by the Colorado River, which exposed rocks billions of years old. Ice also changes the landscape. Large ice sheets from past ice ages could have been well over 1 mile (1,600 meters) thick, and they scoured enormous amounts of soil and rock as they slowly moved over the land surface. Terminal moraines are the enormous mounds of soil pushed directly in front of the ice sheets. Long Island, New York, and Cape Cod, Massachusetts, are two examples of enormous terminal moraines that were left behind when the ice sheets retreated.

The rooting system of vegetation, the fifth factor, helps make the surficial materials of the slope stable by binding the loose materials together.

Mass movements can be classified by their speed of movement. Creep and solifluction are the two types of slow mass movement, which are measured in fractions of inches per year. Creep is the slowest mass movement process, where unconsolidated materials at the surface of a slope move slowly downslope. The materials move slightly faster at the surface than below, so evidence of creep appears in the form of slanted telephone poles. During solifluction, the warm sun of the brief summer season in cold regions thaws the upper few feet of the earth. This waterlogged soil flows downslope over the underlying permafrost.

Rapid mass movement processes occur at feet per second or miles per hour. Falls occur when loose rock or sediment is dislodged and drops from a steep slope, such as along sea cliffs where waves erode the cliff base. Topples occur when there is an overturning movement of the mass. A topple can turn into a fall or a slide. A slide is a mass of rock or sediment that becomes dislodged and moves along a plane of weakness, such as a fracture. A slump is a slide that separates along a concave surface. Lateral spreads occur when a fractured earth mass spreads out at the sides.

A flow occurs when a mass of wet or dry rock fragments or sediment moves downslope as a highly viscous fluid. There are several different flow types. A debris flow is a mass of relatively dry, broken pieces of earth material that suddenly has water added. The debris flow occurs on steeper slopes and moves at speeds of 1-25 miles (2-40 km.) per hour. A debris avalanche occurs when an entire area of soil and underlying weathered bedrock becomes detached from the underlying bedrock and moves quickly down the slope. This flow type is often triggered by heavy rains in areas where vegetation has been removed. An earthflow is a dry mass of clayey or silty material that moves relatively slowly down the slope. A mudflow is a mass of earth material mixed with water that moves quickly down the slope.

A quick clay can occur when partially saturated, solid, clayey sediments are subjected to an earthquake, explosion, or loud noise and become liquid instantly.

Sherry L. Eaton

FLUVIAL AND KARST PROCESSES

Earth's landscape has been sculptured into an almost infinite variety of forms. The earth's surface has been modified by various processes for thousands, even hundreds of millions, of years to arrive at the modern configuration of landscapes.

Each process that transforms the surface is classified as either endogenic or exogenic. Endogenic processes are driven by the earth's internal heat and energy and are responsible for major crustal deformation. Endogenic processes are considered constructional, because they build up the earth's surface and create new landforms, such as mountain systems. Conversely, exogenic processes are considered destructional because they result in the wearing away of landforms created by endogenic processes. Exogenic processes are driven by solar energy putting into motion the earth's atmosphere and water, resulting in the lowering of features originally created by endogenic processes.

The most effective exogenic processes for wearing away the landscape are those that involve the action of flowing water, commonly referred to as fluvial processes. Water flows over the surface as runoff, after it evaporates into the atmosphere and infiltrates into the soil. The water that is left over flows down under the influence of gravity and has tremendous energy for sculpturing the earth's surface. Although flowing water is the most effective agent for modifying the landscape, it represents less than 0.01 percent of all the water on Earth's surface. By comparison, nearly 75 percent of the earth's surface water is stored within glaciers.

Drainage Basins

Fluvial processes can be considered from a variety of spatial scales. The largest scale is the drainage basin. A drainage basin is the area defined by topographic divides that diverts all water and material within the basin to a single outlet. Every stream of any size has its own drainage basin, and every portion of the earth's land surfaces are located within a drainage basin. Drainage basins vary tremendously in size, de-

pending on the size of the river considered. For example, the largest drainage basin on earth is the Amazon, which drains about 2.25 million square miles (5.83 million sq. km.) of South America.

The Amazon Basin is so large that it could contain nearly the entire continent of Australia. By comparison, the Mississippi River drainage basin, the largest in North America, drains an area of about 1,235,000 square miles (3,200,000 sq. km.). Smaller rivers have much smaller basins, with many draining only an area roughly the size of a football field. While basins vary tremendously in size, they are spatially organized, with larger basins receiving the drainage from smaller basins, and eventually draining into the ocean. Because drainage basins receive water and material from the landscape within the basin, they are sensitive to environmental change that occurs within the basin. For example, during the twentieth century, the Mississippi River was influenced by many human-imposed changes that occurred either within the basin or directly within the channel, such as agriculture, dams and reservoirs, and levees.

Drainage Networks and Surface Erosion

Drainage basins can be subdivided into drainage networks by the arrangement of their valleys and interfluves. Interfluves are the ridges of higher elevation that separate adjacent valleys. Where an interfluve represents a natural boundary between two or more basins, it is referred to as a drainage divide. Valleys contain the larger rivers and are easily distinguished from interfluves by their relatively low, flat surfaces. Interfluves have relatively steep slopes and, for this reason, are eroded by runoff. The term erosion refers to the transport of material, in this case sediment that is dislodged from the surface.

Runoff starts as a broad sheet of slow-moving water that is not very erosive. As it continues to flow downslope, it speeds up and concentrates into rills, which are narrow, fast-moving lines of water. Because the runoff is concentrated within rills, the water travels faster and has more energy for erosion. Thus, rills are responsible for transporting sediment from higher points of elevation within the basin to the valleys, which are at a lower elevation. Rills can become powerful enough to scour deeply into the surface, developing into permanent channels called gullies.

The presence of many gullies indicates significant erosion on the landscape and represents an expensive and long-lasting problem if it is not remedied after initial development. The formations of gullies is often associated with human manipulation of the earth. For example, gullies can develop after improper land management, particularly intensive agricultural and grazing practices. A change in land use from natural vegetation, such as forests or prairie, can result in a type of land cover that is not suited for preventing erosion. Such land surfaces become susceptible to the formation of gullies during heavy, prolonged rains.

At a smaller scale, fluvial processes can be considered from the perspective of the river channel. River channels are located within the valleys of basins, offering a permanent conduit for drainage. Higher in the basin, river channels and valleys are relatively narrow, but grow larger toward the mouth of the basin as they receive drainage from smaller rivers within the basin. River channels may be categorized by their planform pattern, which refers to their overhead appearance, such as would be viewed from the window of an airplane.

The two major types of rivers are meandering and braided. Meandering rivers have a single channel that is sinuous and winding. These rivers are characterized as having orderly and symmetrical bends, causing the river to alternate directions as it flows across its valley. In contrast, braided rivers contain numerous channels divided by small islands, which results in a disorganized pattern. The islands within a braided river channel are not permanent. Instead, they erode and form over the course of a few years, or even during large flood events. Meandering channels usually have narrow and deep channels, but braided river channels are shallow and wide.

Sediment and Floodplains

Another distinction between braided and meandering river channels is the types of sediment they transport. Braided rivers transport a great amount of sediment that is deposited into midchannel islands within the river. Also, because braided rivers are frequently located higher in the drainage basin, they may have larger sediments from the erosion of adjacent slopes. In contrast, meandering river channels are located closer to the mouth of the basin and transport fine-grained sediment that is easily stored within point bars, which results in symmetrical bends within the river.

The sediments of both meandering and braided rivers are deposited within the valleys onto floodplains. Floodplains are wide, flat surfaces formed from the accumulation of alluvium, which is a term for sediment that is deposited by water. Floodplain sediments are deposited with seasonal flooding. When a river floods, it transports a large amount of sediment from the channel to the adjacent floodplain. After the water escapes the channel, it loses energy and can no longer transport the sediment. As a result, the sediment falls out of suspension and is deposited onto the floodplain. Because flooding occurs seasonally, floodplain deposits are layered and may accumulate into very thick alluvial deposits over thousands of years.

Karst Processes and Landforms

A specialized type of exogenic process that is also related to the presence of water is karst. Karst processes and topography are characterized by the solution of limestone by acidic groundwater into a number of distinctive landforms. While fluvial processes lower the landscape from the surface, karst processes lower the landscape from beneath the surface. Because limestone is a very permeable sedimentary rock, it allows for a large amount of groundwater flow. The primary areas for solution of the limestone occur along bedding planes and joints. This creates a positive feedback by increasing the amount of water flowing through the rock, thereby further increasing solution of the limestone. The result is a complex maze of underground conduits and caverns, and a surface with few rivers because of the high degree of infiltration.

The surface topography of karst regions often is characterized as undulating. A closer inspection reveals numerous depressions that lack surface outlets. Where this is best developed, it is referred to as cockpit karst. It occurs in areas underlain by extensive limestone and receiving high amounts of precipitation, for example, southern Illinois and Indiana in the midwestern United States, and in Puerto Rico and Jamaica.

Sinkholes are also common to karstic regions. Sinkholes are circular depressions having steep-sided vertical walls. Sinkholes can form either from the sudden collapse of the ceiling of an underground cavern or as a result of the gradual solution and lowering of the surface. Sinkholes can fill with sediments washed in from surface runoff. This reduces infiltration and results in the development of small circular lakes, particularly common in central Florida. Over time, erosion causes the vertical walls to retreat, resulting in uvalas, which are much larger flat-floored depressions.

Where there are numerous adjacent sinkholes, the retreat and expansion of the depressions causes them to coalesce, resulting in the formation of poljes. Unlike uvalas, poljes have an irregular shape, and the floor of the basin is not flat because of differences between the coalescing sinkholes.

Caves are among the most characteristic features of karst regions, but can only be seen beneath the surface. Caves can traverse the subsurface for miles, developing into a complex network of interconnected passages. Some caves develop spectacular formations as a result of the high amount of dissolved limestone transported by the groundwater. The evaporation of water results in the accumulation of carbonate deposits, which may grow for thousands of years. Some of the most common deposits are stalactites, which grow downward from the ceiling of the cave, and stalagmites, which grow upward and occasionally connect with stalactites to form large vertical columns.

Paul F. Hudson

GLACIATION

In areas where more snow accumulates each winter than can thaw in summer, glaciers form. Glacier ice, called firn, looks like rock but is not as strong as most rocks and is subject to intermittent thawing and freezing. Glacier ice can be brittle and fracture readily into crevasses, while other ice behaves as a plastic substance. A glacier is thickest in the area receiving the most snow, called the zone of accumulation. As the thickness piles up, it settles down and squeezes the limit of the ice outward in all directions. Eventually, the ice reaches a climate where the ice begins to melt and evaporate. This is called the zone of ablation.

Alpine Glaciation

Varied topographic evidence throughout the alpine environment attests to the sculpturing ability of glacial ice. The world's most spectacular mountain scenery has been produced by alpine glaciation, including the Matterhorn, Yosemite Valley, Glacier National Park, Mount Blanc, the Tetons, and Rocky

A FUTURE ICE AGE

If past history is an indicator, some time in the future conditions again will become favorable for the growth of glaciers. As recently as 1300 to 1600 CE, a cold period known at the Little Ice Age settled over Northern Europe and Eastern North America. Viking colonies perished as agriculture became unfeasible, and previously ice-free rivers in Europe froze over.

Another ice age would probably develop rapidly and be impossible to stop. Active mountain glaciers would bury living forests. Great ice caps would again cover Europe and North America, and move at a rate of 100 feet (30 meters) per day. Major cities and populations would shift to the subtropics and the topics.

Mountain National Park, all of which are visited by large numbers of people annually. Although alpine glaciation is still an active process of land sculpture in the high mountain ranges of the world, it is much less active than it was in the Ice Age of the Pleistocene epoch.

The prerequisites for alpine, or mountain, glaciation to become active are a mountainous terrain with Arctic climatic conditions in the higher elevations, and sufficient moisture to help snow and ice develop into glacial ice. As glaciers move out from their points of origin, they erode into the sides of mountains and increase the local relief in the higher elevations. The erosional features produced by alpine glaciation dominate mountain topography and usually are the most visible features on topographic maps. The eroded material is transported downvalley and deposited in a variety of landforms.

One kind of an erosional feature is a cirque, a hollow bowl-shaped depression. The bowl of the cirque commonly contains a small round lake or tarn. A steep-walled mountain ridge called an arête forms between two cirques. A high pyramidal peak, called horn, is formed by the intersecting walls of three or more cirques.

Erosion is particularly rapid at the head of a glacier. In valleys, moving glaciers press rock fragments against the sides, widening and deepening them by abrasion and forming broad U-shaped valleys. When glaciers recede, tributary streams become higher than the floor of the U-shaped valley and waterfalls occur over these hanging valleys. As the ice continues to melt, residual sediments called moraines may be deposited. Moraines are made up of glacier till, a collection of sediment of all sizes. Bands of sediment along the side of a valley glacier are lateral moraines; those crossing the valley are end or recessional moraines; where two glaciers join, a medial moraine is formed. Meltwater may also sort out the finer materials, transport them downvalley, and deposit them in beds as outwash.

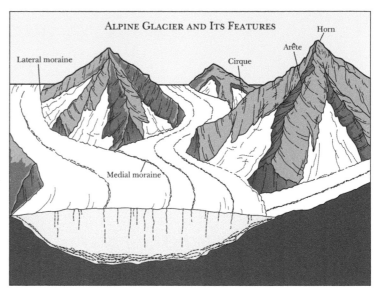

ALPINE GLACIER AND ITS FEATURES

Lateral moraine • Cirque • Arête • Horn • Medial moraine

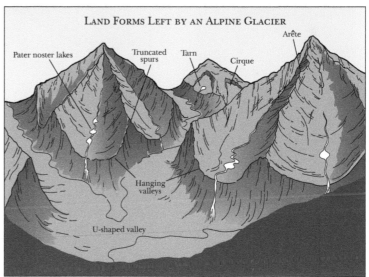

LAND FORMS LEFT BY AN ALPINE GLACIER

Pater noster lakes • Truncated spurs • Tarn • Cirque • Arête • Hanging valleys • U-shaped valley

Continental Glaciation

In the modern world, continental glaciation operates on a large scale only in Greenland and Antarctica. However, its existence in previous geologic ages is evidenced by strata of tillite (a compacted rock formed of glacial deposits) or, more frequently, by surficial deposits of glacial materials.

Much of the geomorphology of the northeastern quadrant of North America and the northwestern portion of Europe was formed during the Ice Age. During that time, great masses of ice accumulated on the continents and moved out from centers near the Hudson Bay and the Fenno-Scandian Shield, extending over the continents in great advancing and retreating lobes. In North America, the four

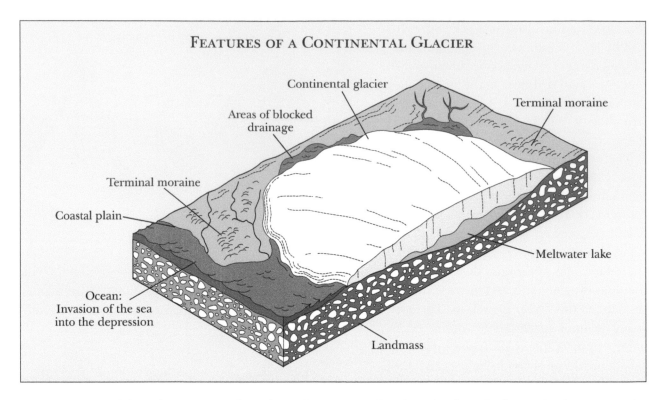

FEATURES OF A CONTINENTAL GLACIER

Continental glacier

Areas of blocked drainage

Terminal moraine

Terminal moraine

Coastal plain

Meltwater lake

Ocean: Invasion of the sea into the depression

Landmass

major stages of lobe advance were the Wisconsin (the most recent), the Illinoian, the Kansan, and the Nebraskan (the oldest). Between each of these major advances were pluvial periods in which the ice melted and great quantities of water rushed over or stood on the continents, creating distinctive features which can still be detected today.

The two major functions of gradation are accomplished by the processes of scour (degradation) in the areas close to the centers and deposition (aggradation) adjacent to the terminal or peripheral areas of the lobes. Thus, the overall effect of continental glaciation is to reduce relief—to scour high areas and fill in lower regions—unlike the changes caused by alpine glaciation.

Although continental glaciation usually does not result in the spectacular scenery of alpine glaciation, it was responsible for creating most of the Great Lakes and the lakes of Wisconsin, Michigan, Minnesota, Finland, and Canada; for gravel deposits; and for the rich agricultural lands of the Midwest, to mention just a few of its effects.

While glaciers were leveling hilly sections of North America and Europe by scraping them bare

of soil and cutting into the ice itself, they acquired a tremendous load of material. As a glacier warms and melts, there is a tremendous outflow of water, and the streams thus formed carry with them the debris of the glacier. The material deposited by glaciers is called drift or outwash. Glaciofluvial drift can be recognized by its separation into layers of finer sands and coarser gravels.

Kettles and kames are the most common features of the end moraines found at the outermost edges of a glacier. A kettle is a depression left when a block of ice, partially or completely buried in deposits of drift, melts away. Most of the lakes in the upper Great Lakes of the United States are kettle lakes. A kame is a round, cone-shaped hill. Kames are produced by deposition from glacial meltwater. Sometimes, the outwash material poured into a long and deep crevasse, rather than a hole. These tunnels have had their courses choked by debris, revealed today by long, narrow ridges, generally referred to as eskers.

Ron Janke

DESERT LANDFORMS

Deserts are often striking in color, form, or both. The underlying lack of water in deserts produces unique features not found in humid regions. Arid lands cover approximately 30 percent of the earth's land surface, an area of about 15.4 million square miles (40 million sq. km.). Arid lands include deserts and surrounding steppes, semiarid regions that act as transition zones between arid and humid lands.

Many of the world's largest and driest deserts are found between 20 and 40 degrees north and south latitude. These include the Mojave and Sonoran Deserts of the United States, the Sahara in northern Africa, and the Great Sandy Desert in Australia. In these deserts, the subtropical high prevents cloud formation and precipitation while increasing rates of surface evaporation.

Some arid lands, like China's Gobi Desert, form because they are far from oceans that are the dominant source for atmospheric water vapor and precipitation. Others, like California's Death Valley, are arid because mountain ranges block moisture from coming from the sea. The combination of mountain barriers and very low elevations makes Death Valley the hottest, driest desert in North America.

Sand Dunes

Many people envision deserts as vast expanses of blowing sand. Although wind plays a more important role in deserts than it does elsewhere, only about 25 percent of arid lands are covered by sand. Broad regions that are covered entirely in sand (such as portions of northwestern Africa, Arabia, and Australia) are referred to as sand seas. Why is wind more effective here than elsewhere?

The lack of soil water and vegetation, both of which act to bind grains together, allows enhanced eolian (wind) erosion. Very small particles are picked up and suspended within the moving air mass, while sand grains bounce along the surface. Removal of material often leaves behind depressions called blowouts or deflation hollows. Moving grains abrade cobbles and boulders at the surface, creating uniquely sculpted and smoothed rocks known as ventifacts. Bedrock outcrops can be streamlined as they are blasted by wind-borne grains to form features called yardangs. As these rocks are ground away, they contribute additional sediment to the wind.

Desert sand dunes are not stationary features—instead, they represent accumulations of moving sand. Wind blows sand along the desert floor. Where it collects, it forms dunes. Typically, dunes have relatively shallow windward faces and steeper slip faces. Sand grains bounce up the windward face and then eventually cascade down the slip face, the movement of individual grains driving movement of the entire dune in a downwind direction.

Four major dune types are found within arid regions. Barchan dunes are crescent-shaped features, with arms that point downwind. They may occur as isolated structures or within fields. They form where winds blow in a single direction and where the supply of sand is limited. With a larger supply of sand, barchan dunes can join with one another to form a transverse dune field.

There, ridges are perpendicular to the predominant wind direction. With quartering winds (that is, winds that vary in direction throughout a range of about 45 degrees) dune ridges form that are parallel to the average wind direction. These so-called longitudinal dunes have no clearly defined windward and slip faces. Where winds blow sand from all directions, star dunes form. Sand collects in the middle of the feature to form a peaked center with arms that spiral outward.

Badlands, Mesas, and Buttes

As scarce as it may be, water is still the dominant force in shaping desert landscapes. Annual precipitation may be low, but the amount of precipitation in a single storm may be a large fraction of the yearly total. An arid landscape that is underlain by poorly cemented rock or sediment, such as that found in western South Dakota, may form badlands as a result of the erosive ability of storm-water run-off. Overall aridity prevents vegetation from establishing the interconnected root system that holds soil particles together in more humid regions.

Cloudbursts cause rapid erosion that forms numerous gullies, deeply incised washes, and hoodoos, which are created when rock or sediment that is more resistant protects underlying material from erosion. Over time, protected sections stand as prominent spires while surrounding material is removed. Landscapes like those found in Badlands National Park in South Dakota are devoid of vegetation and erode rapidly during storms.

Arid regions that are underlain by flat-lying rock units can form mesas and buttes. Water follows fractures and other lines of weakness, forming ever-widening canyons. Over time, these grow into broad valleys. In northern Arizona's Monument Valley, remnants of original bedrock stand as isolated, flat-topped structures. Broad mesas are marked by their flat tops (made of a resistant rock like sandstone or basalt) and steep sides. Buttes are much narrower, with a small resistant cap, but are often as tall and steep as neighboring mesas.

Desert Pavement and Desert Varnish

Much of the desert floor is covered by desert pavement, an accumulation of gravel and cobbles that forms a surface fabric that can interconnect tightly. Fine material has been removed by wind and water, leaving behind larger fragments that inhibit further erosion. In many areas, desert pavements have been stable for long periods of time, as evidenced by their surface patina of desert varnish. Desert varnish is a thin outer coating of wind-deposited clay mixed with iron and manganese oxides. Varying in color from light brown to black, these coatings are thought to adhere to rocks by the action of single-celled microorganisms. Under a microscope, desert varnish can be seen to be made up of very fine layers. A thick, dark patina means that a rock has been exposed for a long time.

Playas

Where neither dunes nor rocky pavements cover the desert floor, one may find an accumulation of saline minerals. A playa is a flat surface that is often blindingly white in color. Playas are usually found in the centers of desert valleys and contain material that mineralized during the evaporation of a lake. Dry lake beds are a common feature of the Great Basin in the western United States. During glacial stages, the last of which occurred about 20,000 years ago, lakes grew in what are now arid, closed valleys. As the climate warmed, these lakes shrank, and many dried completely. As a lake evaporates, minerals that were held in solution crystallize, forming salts, including halite (table salt). These salt deposits frequently are mined for useful household and industrial chemicals.

Richard L. Orndorff

DEATH VALLEY PLAYA

California's Death Valley is the driest desert in the United States, with an average rainfall of only 1.5 inches (38 millimeters) per year at the town of Furnace Creek. It is also consistently one of the hottest places on Earth, with a record high of 134 degrees Fahrenheit (57 degrees Celsius). In the distant past, however, Death Valley held lakes that formed in response to global cooling. Over 120,000 years ago, Death Valley hosted a 295-foot-deep (90 meters) body of water called Lake Manley. Evidence of this lake remains in evaporite deposits that make up the playa in the valley's center, in wave-cut shorelines, and in beach bars.

Ocean Margins

Ocean margins are the areas where land borders the sea. Although often referred to as coastlines or beaches, ocean margins cover far greater territory than beaches. An ocean margin extends from the coastal plain—the fertile farming belt of land along the seacoast—to the edge of the gently sloping land submerged in water, called the continental shelf.

Ocean margin constitutes 8 percent of the world's surface. It is rich in minerals, both above and below water, and is home to 25 percent of Earth's people, along with 90 percent of the marine life. This fringe of land at the border of the ocean is ever changing. Tides wash sediment in and leave it behind, just below sea level. This process, called deposition, builds up land in some areas of the coastline. At the same time, ocean waves, winds, and storms wear away or erode parts of the shoreline. As land is worn away or built up, the amount of land above sea level changes. Factors such as climate, erosion, deposition, changes in sea level, and the effects of humans constantly change the shape of the ocean margin on Earth.

Beach Dynamics

The two types of coasts or land formations at the ocean margin are primary coasts and secondary coasts. Primary coasts are formed by systems on land, such as the melting of glaciers, wind or water erosion, and sediment deposited by rivers. Deltas and fjords are examples of primary coasts. Secondary coasts are formed by ocean patterns, such as erosion by waves or currents, sediment deposition by waves or currents, or changes by marine plants or animals. Beaches, coral reefs, salt marshes, and mangrove swamps are examples of secondary coasts.

Sediment carried by rivers to the sea is deposited to form deltas at the mouths of the rivers. Some of the sediment can wash out to sea, causing formations to build up at a distance from the shore. These formations eventually become barrier islands, which are often little more than 10 feet (3 km.) above sea level. As

a consequence, heavy storms, such as hurricanes, can cause great damage to barrier islands. Barrier islands naturally protect the coastline from erosion, however, especially during heavy coastal storms.

Sea level changes also affect the shape of the coastline. As oceans slowly rise, land is slowly consumed by the ocean. Barrier islands, having low sea levels, may slowly be covered with water. The melting of continental glaciers increased the sea level 0.06 inch (0.15 centimeter) per year during the twentieth century. As ocean waters warm, they expand, eating away at sea levels. Global warming caused by carbon dioxide levels in the atmosphere could cause sea levels to rise as much as 0.24 inch (0.6 centimeter) per year as a result of the warming of the water and glacial melting.

Human Influence

The shape of the ocean margin also changes radically as a result of human influence. According to the United States Geological Survey, 39 percent of the people living in the United States live directly on the the coasts. According to UN Atlas of Oceans, about 44 percent of the world's population lives within 93 miles (150 kilometers) of the coast. Pollution from toxins, dredging, recreational boating, and waste disposal kills plants and animals along the ocean margin. This changes the coastal shape, as mangrove forests, coral reefs, and other coastal lifeforms die.

A greater concern along the coastal fringe, however, is human development. Not only are people drawn to the fertile soil along the coastal zone of the continent, but they also develop islands and coves into resort communities. To protect homes and hotels along the coastal zone from coastal erosion, people build breakwalls, jetties, and sand and stone bars called groins.

These human-made barriers disrupt the natural method by which the ocean carries material along the coast. Longshore drift, a zigzag movement, deposits sediment from one area of the beach farther along the shoreline. Breakwalls, jetties, and groins disrupt this flow. As the ocean smashes against a breakwall, the property behind it may be safe for the present, but the coastline neighboring the breakwall takes a greater beating. The silt and sediment from upshore, which would replace that carried downshore, never arrives. Eventually, the breakwall will break down under the impact of the ocean force. Areas with breakwalls and jetties often suffer greater damage in coastal storms than areas that remain naturally open to the changing forces of the ocean.

To compensate for the destructive nature of human-made barriers, many recreational beaches replace lost sand with dredgings or deposit truckloads of sand from inland sources. For example, Virginia Beach in the United States spends between US$2 million and US$3 millon annually to restore beaches for the tourist season in this way.

Despite the changes in the shape of the ocean margin, it continues to provide a stable supply of resources—fish, seafood, minerals, sponges, and other marine plants and animals. Offshore drilling of oil and natural gas often takes place within 200 miles (322 km.) of shorelines.

Lisa A. Wroble

EARTH'S CLIMATES

THE ATMOSPHERE

The thin layer of gases that envelops the earth is the atmosphere. This layer is so thin that if the earth were the size of a desktop globe, more than 99 percent of its atmosphere would be contained within the thickness of an ordinary sheet of paper. Despite its thinness, the atmosphere sustains life on Earth, protecting it from the Sun's searing radiation and regulating the earth's temperature. Storms of the atmosphere carry water to the continents, and weathering by its wind and rain helps shape them.

Composition of the Atmosphere

The earth's atmosphere consists of gases, microscopic particles called aerosol, and clouds consisting of water droplets and ice particles. Its two principal gases are nitrogen and oxygen. In dry air, nitrogen occupies 78 percent, and oxygen 21 percent, of the atmosphere's volume. Argon, neon, xenon, helium, hydrogen, and other trace gases together equal less than 1 percent of the remaining volume.

These gases are distributed homogeneously in a layer called the homosphere, which occurs between the earth's surface and about 50 miles (80 km.) altitude. Above 50 miles altitude, in the heterosphere, the concentration of heavier gases decreases more rapidly than lighter gases.

The atmosphere has no firm top. It simply thins out until the concentration of its gas molecules approaches that of the gases in outer space. The concentration of nitrogen and oxygen remains essentially constant in the atmosphere because a balance exists between the production and removal of these gases at the earth's surface. Decaying organic matter adds nitrogen to the atmosphere, while soil bacteria remove nitrogen. Oxygen enters the atmosphere primarily through photosynthesis and is removed through animal respiration, combustion, and decay of organic material, and by chemical reactions involving the creation of oxides.

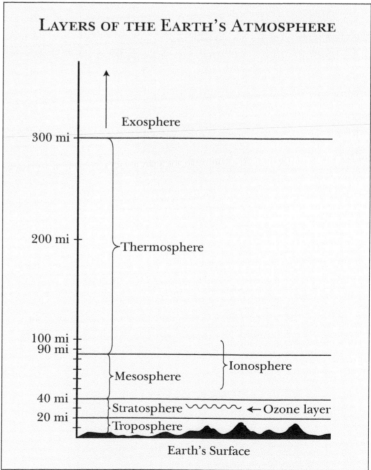

LAYERS OF THE EARTH'S ATMOSPHERE

THE GREENHOUSE EFFECT

Clouds and atmospheric gases such as water vapor, carbon dioxide, methane, and nitrous oxide absorb part of the infrared radiation emitted by the earth's surface and reradiate part of it back to the earth. This process effectively reduces the amount of energy escaping to space and is popularly called the "greenhouse effect" because of its role in warming the lower atmosphere. The greenhouse effect has drawn worldwide attention because increasing concentrations of carbon dioxide from the burning of fossil fuels result in a global warming of the atmosphere.

Scientists know that the greenhouse analogy is incorrect. A greenhouse traps warm air within a glass building where it cannot mix with cooler air outside. In a real greenhouse, the trapping of air is more important in maintaining the temperature than is the trapping of infrared energy. In the atmosphere, air is free to mix and move about.

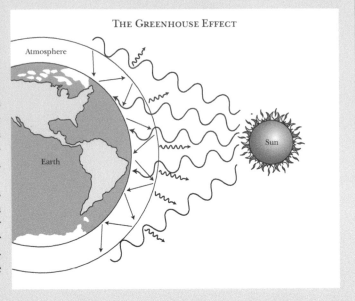

THE GREENHOUSE EFFECT

Atmosphere

Earth

Sun

The atmosphere contains many gases that are present in small, variable concentrations. Three gases—water vapor, carbon dioxide and ozone—are vital to life on Earth. Water vapor enters the atmosphere through evaporation, primarily from the oceans, and through transpiration by plants. It condenses to form clouds, which provide the rain and snow that sustain life outside the oceans. The concentration of water vapor varies from about 4 percent by volume in tropical humid climates to a small fraction of a percent in polar dry climates. Water vapor plays an important role in regulating the temperature of the earth's surface and the atmosphere. Clouds reflect some of the incoming solar radiation, while water vapor and clouds both absorb earth's infrared radiation.

Carbon dioxide also absorbs the earth's infrared radiation. The global average atmospheric carbon dioxide in 2018 was 407.4 parts per million (ppm for short), with a range of uncertainty of plus or minus 0.1 ppm. Carbon dioxide levels today are higher than at any point in at least the past 800,000 years. The annual rate of increase in atmospheric carbon dioxide over the past 60 years is about 100 times faster than previous natural increases, such as those that occurred at the end of the last ice age 11,000–17,000 years ago.

Carbon dioxide enters the atmosphere as the result of decay of organic material, through respiration, during volcanic eruptions, and from the burning of fossil fuels. It is removed during photosynthesis and by dissolving in ocean water, where it is used by organisms and converted to carbonates. The increase in atmospheric carbon dioxide associated with the burning of fossil fuels has raised concerns that the earth's atmosphere may be warming through enhancement of the greenhouse effect.

Ozone, a gas consisting of molecules containing three oxygen atoms, forms in the upper atmosphere when oxygen atoms and oxygen molecules combine. Most ozone exists in the upper atmosphere between 6.2 and 31 miles (10 and 50 km.) in altitude, in concentrations of no more than 0.0015 percent by volume. This small amount of ozone sustains life outside the oceans by absorbing most of the Sun's ultraviolet radiation, thereby shielding the earth's surface from the radiation's harmful effects on living organisms. Paradoxically, ozone is an irritant near the earth's surface and is the major component of photochemical smog. Other gases that contribute to pollution include methane, nitrous oxide, hydrocarbons, and chlorofluorocarbons.

Aerosols represent another component of atmospheric pollution. Aerosols form in the atmosphere during chemical reactions between gases, through mechanical or chemical interactions between the earth, ocean surface, and atmosphere, and during evaporation of droplets containing dissolved or solid material. These microscopic particles are always present in air, with concentrations of about a few hundred per cubic centimeter in clean air to as many as a million per cubic centimeter in polluted air. Aerosols are essential to the formation of rain and snow, because they serve as centers upon which cloud droplets and ice particles form.

Energy Exchange in the Atmosphere

The Sun is the ultimate source of the energy in Earth's atmosphere. Its radiation, called electromagnetic radiation because it propagates as waves with electric and magnetic properties, travels to the surface of the earth's atmosphere at the speed of light. This energy spans many wavelengths, some of which the human eye perceives as colors. Visible wavelengths make up about 44 percent of the Sun's energy. The remainder of the Sun's radiant energy cannot be seen by human eyes. About 7 percent arrives as ultraviolet radiation, and most of the remaining energy is infrared radiation.

The Sun is not the only source of radiation. All objects emit and absorb radiation to some degree. Cooler objects such as the earth emit nearly all their energy at infrared wavelengths. Objects heat when they absorb radiation and cool when they emit radiation. The radiation emitted by the earth and atmosphere is called terrestrial radiation.

The balance between absorption of solar radiation and emission of terrestrial radiation ultimately determines the average temperature of the earth-atmosphere system. The vertical temperature distribution within the atmosphere also depends on the absorption and emission of radiation within the atmosphere, and the transfer of energy by the processes of conduction, convection, and latent heat exchange. Conduction is the direct transfer of heat from molecule to molecule. This process is most important in transferring heat from the earth's surface to the first few centimeters of the at-

---

THE OZONE HOLE

Since the 1970s, balloon-borne and satellite measurements of stratospheric ozone have shown rapidly declining stratospheric ozone concentrations over the continent of Antarctica, termed the "ozone hole." The lowest concentrations occur during the Antarctic spring, in September and October. The decrease in ozone has been associated with an increase in the concentration of chlorine, a gas introduced into the stratosphere through chemical reactions involving sunlight and chlorofluorocarbons— synthetic chemicals used primarily as refrigerants. The ozone hole over Antarctica has raised concern about possible worldwide reduction in the concentration of upper atmospheric ozone.

---

mosphere. Convection, the transfer of heat by rising or sinking air, transports heat energy vertically through the atmosphere.

Latent heat is the energy required to change the state of a substance, for example, from a liquid to a gas. Energy is transferred from the earth's surface to the atmosphere through latent heat exchange when water evaporates from the oceans and condenses to form rain in the atmosphere.

Only 48 percent of the solar energy reaching the top of the earth's atmosphere is absorbed by the earth's surface. The atmosphere absorbs another 23 percent. The remaining 30 percent is scattered back to space by atmospheric gases, clouds and the earth's surface. To understand the importance of terrestrial radiation and the greenhouse effect in the atmosphere's energy balance, consider the solar radiation arriving at the top of the earth to be 100 energy units, with 48 energy units absorbed by the earth's surface and 23 units by the atmosphere.

The earth's surface actually emits 117 units of energy upward as terrestrial radiation, more than twice as much energy as it receives from the Sun. Only 6 of these units are radiated to space—the atmosphere absorbs the remaining energy. Latent heat exchange, conduction, and convection account for another 30 units of energy transferred from the surface to the atmosphere. The atmosphere, in turn, radiates 96 units of energy back to the earth's surface (the greenhouse effect), and 64 units to space. The earth's and atmosphere's energy budget remains in balance, the atmosphere gaining and losing 160 units of energy, and the earth gaining and losing 147 units of energy.

Vertical Structure of the Atmosphere

Temperature decreases rapidly upward away from the earth's surface, to about –60 degrees Farenheit (–51 degrees Celsius) at an altitude of about 7.5 miles (12 km.). Above this altitude, temperature increases with height to about 32 degrees Farenheit (0 degrees Celsius) at an altitude of 31 miles (50 km.). The layer of air in the lower atmosphere where temperature decreases with height is called the troposphere. It contains about 75 percent of the atmosphere's mass. The layer of air above the troposphere, where temperature increases with height, is called the stratosphere. All but 0.1 percent of the remaining mass of the atmosphere resides in the stratosphere.

The stratosphere exists because ozone in the stratosphere absorbs ultraviolet light and converts it to heat. The boundary between the troposphere and stratosphere is called the tropopause. The tropopause is extremely important because it acts as a lid on the earth's weather. Storms can grow vertically in the troposphere, but cannot rise far, if at all, beyond the tropopause. In the polar regions, the tropopause can be as low as 5 miles (8 km.) above the surface, while in the tropics, the tropopause can be as high as 11 miles (18 km.). For this reason, tropical storms can extend to much higher altitudes than storms in cold regions.

The mesosphere extends from the top of the stratosphere, the stratopause, to an altitude of about 56 miles (90 km.). Temperature decreases with height within the mesosphere. The lowest average temperatures in the atmosphere occur at the mesopause, the top of the mesosphere, where the

temperature is about –130 degrees Farenheit (–90 degrees Celsius). Only 0.0005 percent of the atmosphere's mass remains above the mesopause. In this uppermost layer, the thermosphere, there are few atoms and molecules. Oxygen molecules in the thermosphere absorb high-energy solar radiation. In this near vacuum, absorption of even small amounts of energy causes a large increase in temperature. As a result, temperature increases rapidly with height in the lower thermosphere, reaching about 1,300 degrees Farenheit (700 degrees Celsius) above 155 miles (250 km.) altitude.

The upper mesosphere and thermosphere also contain ions, electrically charged atoms or molecules. Ions are created in the atmosphere when air molecules collide with high-energy particles arriving from space or absorb high-energy solar radiation. Ions cannot exist very long in the lower atmosphere, because collisions between newly formed ions quickly restore ions to their uncharged state. However, above about 37 miles (60 km.) collisions are less frequent and ions can exist for longer times. This region of the atmosphere, called the ionosphere, is particularly important for amplitude-modulated (AM) radio communication because it reflects standard AM radio waves. At night, the lower ionosphere disappears as ions recombine, allowing AM radio waves to travel longer distances when reflected. For this reason, AM radio station signals can sometimes travel great distances at night.

The top of the atmosphere occurs at about 310 miles (500 km.). At this altitude, the distance between individual molecules is so great that energetic molecules can move into free space without colliding with neighbor molecules. In this uppermost layer, called the exosphere, the earth's atmosphere merges into space.

Robert M. Rauber

GLOBAL CLIMATES

A region's climate is the sum of its long-term weather conditions. Most descriptions of climate emphasize temperature and precipitation characteristics, because these two climatic elements usually exert more impact on environmental conditions and human activities than do other elements, such as wind, humidity, and cloud cover. Climatic descriptions of a region generally cover both mean conditions and extremes. Climatic means are important because they represent average conditions that are frequently experienced; extreme conditions, such as severe storms, excessive heat and cold, and droughts, are important because of their adverse impact.

Important Climate Controls

A region's climate is largely determined by the interaction of six important natural controls: sun angle, elevation, ocean currents, land and water heating and cooling characteristics, air pressure and wind belts, and orographic influence.

Sun angle—the height of the Sun in degrees above the nearest horizon—largely controls the amount of solar heating that a site on Earth receives. It strongly influences the mean temperatures of most of the earth's surface, because the Sun is the ultimate energy source for nearly all the atmosphere's heat. The higher the angle of the Sun in the sky, the greater the concentration of energy, per unit area, on the earth's surface (assuming clear skies). From a global perspective, the Sun's mean angle is highest, on average, at the equator, and becomes progressively lower poleward. This causes a gradual decrease in mean temperatures with increasing latitude.

Sun angles also vary seasonally and daily. Each hemisphere is inclined toward the Sun during spring and summer, and away from the Sun during fall and winter. This changing inclination causes mean sun angles to be higher, and the length of daylight longer, during the spring and summer. Therefore, most locations, especially those outside the tropics, have warmer temperatures during these two seasons. The earth's rotation causes sun angles to be higher during midday than in the early morning and late afternoon, resulting in warmer temperatures at midday. Heating and cooling lags cause both seasonal and daily maximum and minimum temperatures typically to occur somewhat after the periods of maximum and minimum solar energy receipt.

Variations in elevation—the distance above sea level—can cause locations at similar latitudes to vary greatly in temperature. Temperatures decrease an average of about 3.5 degrees Fahrenheit per thousand feet (6.4 degrees Celsius per thousand meters). Therefore, high mountain and plateau stations are much colder than low-elevation stations at the same latitude.

Surface ocean currents can transport masses of warm or cold water great distances from their source regions, affecting both temperature and moisture conditions. Warm currents facilitate the evaporation of copious amounts of water into the atmosphere and add buoyancy to the air by heating it from below. This results in a general increase in precipitation totals. Cold currents evaporate water relatively slowly and chill the overlying air, thus stabilizing it and reducing its potential for precipitation.

The influence of ocean currents on land areas is greatest in coastal regions and decreases inland. The west coasts of continents (except for Europe) generally are paralleled by relatively cold currents, and the east coasts by relatively warm currents. For example, the warm Gulf Stream flows northward off the eastern United States, while the West Coast is cooled by the southward-flowing California Current.

Land can change temperature much more readily than water. As a result, the air over continents typically experiences larger annual temperature ranges (that is, larger temperature differences between summer and winter) and shorter heating and cooling lags than does the air over oceans. This same effect causes continental interiors and the leeward (downwind) coasts of continents typically to have larger temperature ranges than do windward (upwind) coasts. Climates that are dominated by air from landmasses are often described as continental climates. Conversely, climates dominated by air from oceans are described as maritime climates.

The seasonal heating and cooling of continents can also produce a monsoon influence, which has to do with annual shifts of wind patterns. Areas influenced by a monsoon, such as Southeast Asia, tend to have a predominantly onshore flow of moist maritime air during the summer. This often produces heavy rains. An offshore flow of dry air predominates in winter, producing fair weather.

Earth's atmosphere displays a banded, or beltlike, pattern of air pressure and wind systems. High pressure is associated with descending air and dry weather; low pressure is associated with rising air, which produces cloudiness and often precipitation. Wind is produced by differences in air pressure. The air blows outward from high-pressure systems and into low-pressure systems in a constant attempt to equalize air pressures.

The direction and speed of movement of weather systems, such as weather fronts and storms, are controlled by wind patterns, especially those several kilometers above the surface. The seasonal shift of global temperatures caused by the movement of the Sun's vertical rays between the Tropics of Cancer and Capricorn produces a latitudinal migration of both air pressure and wind belts. This shift affects the annual temperature and precipitation patterns of many regions.

Four air-pressure belts exist in each hemisphere. The intertropical convergence zone (ITCZ) is a broad belt of low pressure centered within a few degrees of latitude of the equator. The subtropical highs are high-pressure belts centered between 20 and 40 degrees north and south latitude, which are responsible for many of the world's deserts. The subpolar lows are low-pressure belts centered about 50 or 70 degrees north and south latitude. Finally, the polar highs are high-pressure centers located near the North and South Poles.

The air pressure gradient between these belts produces the earth's major wind belts. The regions between the ITCZ and the subtropical highs are dominated by the trade winds, a broad belt in each hemisphere of easterly (that is, moving east to west) winds. The middle latitudes are mostly situated between the subtropical highs and the subpolar lows and are within the westerly wind belt. This wind belt causes winds, and weather systems, to travel generally from west to east in the United States and Canada. Finally, the high-latitude zones between the subpolar lows and polar highs are situated within the polar easterlies.

The final factor affecting climate— orographic influence—is the lifting effect of mountain peaks or ranges on winds that pass over them. As air approaches a mountain barrier, it rises, typically producing clouds and precipitation on the windward (upwind) side of the mountains. After it crosses the crest, it descends the leeward (downwind) side of the mountains, generally producing dry weather. Most of the world's wettest locations are found on the windward sides of high mountain ranges; some deserts, such as those of the western interior United States, owe their aridity to their location on the leeward sides of orographic barriers.

World Climate Types
The global distribution of the world climate controls is responsible for the development of fourteen widely recognized climate types. In this section, the major characteristics of each of these climates will be briefly described. The climates are discussed in a rough poleward sequence.

Tropical Wet Climate
Sometimes called the tropical rain forest climate, the tropical wet climate exists chiefly in areas lying within 10 degrees of the equator. It is an almost seasonless climate, characterized by year-round warm, humid, rainy conditions that allow land areas to support a dense broadleaf forest cover. The warm temperatures, which for most locations average near 80 degrees Fahrenheit (27 degrees Celsius) throughout the year, result from the constantly high midday sun angles experienced at this low latitude.

WORLD CLIMATE REGIONS

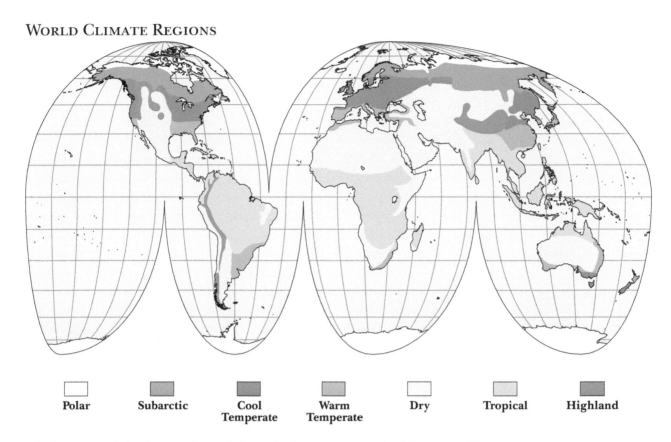

| Polar | Subarctic | Cool Temperate | Warm Temperate | Dry | Tropical | Highland |

The heavy precipitation totals result from the heating and subsequent rising of the warm moist air to form frequent showers and thunderstorms, especially during the afternoon hours. The dominance of the ITCZ enhances precipitation totals, helping make this climate type one of the world's rainiest.

Tropical Monsoonal Climate

The tropical monsoonal climate occurs in low-latitude areas, such as Southeast Asia, that have a warm, rainy climate with a short dry season. Temperatures are similar to those of the tropical wet climate, with the warmest weather often occurring during the drier period, when sunshine is more abundant. The heavy rainfalls result from the nearness of the ITCZ for much of the year, as well as the dominance of warm, moist air masses derived from tropical oceans. During the brief dry season, however, the ITCZ has usually shifted into the opposite hemisphere, and windflow patterns often have changed so as to bring in somewhat drier air derived from continental sources.

Tropical Savanna Climate

The tropical savanna climate, also referred to as the tropical wet-dry climate, occupies a large portion of the tropics between 5 and 20 degrees latitude in both hemispheres. It experiences a distinctive alternation of wet and dry seasons, caused chiefly by the seasonal shift in latitude of the subtropical highs and ITCZ. Summer is typically the rainy season because of the domination of the ITCZ. In many areas, an onshore windflow associated with the summer monsoon increases rainfalls at this time. In winter, however, the ITCZ shifts into the opposite hemisphere and is replaced by drier and more stable air associated with the subtropical high. In addition, the winter monsoon tendency often produces an outflow of continental air. The long dry season inhibits forest growth, so vegetation usually consists of a cover of drought-resistant shrubs or the tall savanna grasses after which the climate is named.

Subtropical Desert Climate

The subtropical desert climate has hot, arid conditions as a result of the year-round dominance of the subtropical highs. Summertime temperatures in this climate soar to the highest readings found anywhere on earth. The world's record high temperature was 134 degrees Fahrenheit (56.7 degrees Celsius), recorded in Furnace Creek Ranch, California (formerly Greenland Ranch) on July 10, 1913. Rainfall totals in this type of climate are generally less than 10 inches (25 centimeters) per year. What rainfall does occur often arrives as brief, sometimes violent, afternoon thunderstorms. Although summer temperatures are extremely hot, the dry air enables rapid cooling during the winter, so that temperatures are cool to mild at this time of year.

Subtropical Steppe Climate

The subtropical steppe climate is a semiarid climate, found mostly on the margins of the subtropical deserts. Precipitation usually ranges from 10 to 30 inches (25 to 75 centimeters), sufficient for a ground cover of shrubs or short steppe grasses. Areas on the equatorward margins of subtropical deserts typically receive their precipitation during a brief showery period in midsummer, associated with the poleward shift of the ITCZ. Areas on the poleward margins of the subtropical highs receive most of their rainfall during the winter, due to the penetration of cyclonic storms associated with the equatorward shift of the westerly wind belt.

Mediterranean Climate

The Mediterranean climate, also sometimes referred to as the dry summer subtropics, has a distinctive pattern of dry summers and more humid, moderately wet winters. This pattern is caused by the seasonal shift in latitude of the subtropical high and the westerlies. During the summer, the subtropical high shifts poleward into the Mediterranean climate regions, blanketing them with dry, warm, stable air. As winter approaches, this pressure center retreats equatorward, allowing the westerlies, with their eastward-traveling weather fronts and cyclonic storms, to overspread this region. The Mediterranean climate is found on the windward sides of continents, particularly the area surrounding the Mediterranean Sea and much of California. This results in the predomi-

nance of maritime air and relatively mild temperatures throughout the year.

Humid Subtropical Climate

The humid subtropical climate is found on the eastern, or leeward, sides of continents in the lower middle latitudes. The most extensive land area with this climate is the southeastern United States, but it is also seen in large areas in South America, Asia, and Australia. Temperature ranges are moderately large, with warm to hot summers and cool to mild winters. Mean temperatures for a given location are dictated largely by latitude, elevation, and proximity to the coast. Precipitation is moderate. Winter precipitation is usually associated with weather fronts and cyclonic storms that travel eastward within the westerly wind belt. During summer, most precipitation is in the form of brief, heavy afternoon and evening thunderstorms. Some coastal areas are subject to destructive hurricanes during the late summer and autumn.

Midlatitude Desert Climate

This type of climate consists of areas within the western United States, southern South America, and Central Asia that have arid conditions resulting from the moisture-blocking influence of mountain barriers. This climate is highly continental, with warm summers and cold winters. When precipitations occurs, it frequently comes in the form of winter snowfalls associated with weather fronts and cyclonic storms. Rainfall in summer typically occurs as afternoon thunderstorms.

Midlatitude Steppe

The midlatitude steppe climate is located in interior portions of continents in the middle latitudes, particularly in Asia and North America. This climate has semiarid conditions caused by a combination of continentality resulting from the large distance from oceanic moisture sources and the presence of mountain barriers. Like the midlatitude desert climate, this climate has large annual temperature ranges, with cold winters and warm summers. It also receives winter rains and snows chiefly from weather fronts and cyclonic storms; summer rains occur largely from afternoon convectional storms. In the Great Plains of the United States, spring can bring very turbulent conditions, with blizzards in early spring and hailstorms and tornadoes in mid to late spring.

Marine West Coast

This type of climate is typically located on the west coasts of continents just poleward of the Mediterranean climate. Its location in the heart of the westerly wind belt on the windward sides of continents produces highly maritime conditions. As a result, cloudy and humid weather is common, along with frequent periods of rainfall from passing weather fronts and cyclonic storms. These storms are often well developed in winter, resulting in extended periods of wet and windy weather. Precipitation amounts are largely controlled by the presence and strength of the orographic effect; mountainous coasts like the northwestern United States and the west coast of Canada are much wetter than are flatter areas like northern Europe. Temperatures are held at moderate levels by the onshore flow of maritime air. As a consequence, winters are relatively mild and summers relatively cool for the latitude.

Humid Continental Climate

The humid continental climate is found in the northern interiors of Eurasia (Europe and Asia) and North America. It does not occur in the Southern Hemisphere because of the absence of large land masses in the upper midlatitudes of that hemisphere. This climate type is characterized by low to moderate precipitation that is largely frontal and cyclonic in nature. Most precipitation occurs in summer, but cold winter temperatures typically cause the surface to be frozen and snow-covered for much of the late fall, winter, and early spring. Temperature ranges in this climate are the largest in the world. A town in Siberia, Verkhoyansk, holds the Guinness world record for the greatest temperature range at a single location is 221 degrees Fahrenheit (105 degrees Celsius), from -90 degrees Fahrenheit (-68 degrees Celsius) to 99 degrees Fahrenheit (37 degrees Celsius). Winter temperatures in parts of both North America and Siberia can fall well below -49 degrees

Fahrenheit (-45 degrees Celsius), making these the coldest permanently settled sites in the world.

Tundra Climate

The tundra climate is a severely cold climate that exists mostly on the coastal margins of the Arctic Ocean in extreme northern North America and Eurasia, and along the coast of Greenland. The high-latitude location and proximity to icy water cause every month to have average temperatures below 50 degrees Fahrenheit (10 degrees Celsius), although a few months in summer have means above freezing. As a result of the cold temperatures, tundra areas are not forested, but instead typically have a sparse ground cover of grasses, sedges, flowers, and lichens. Even this vegetation is buried by a layer of snow during most of the year. Cold temperatures lower the water vapor holding capacity of the air, causing precipitation totals to be generally light. Most precipitation is associated with weather fronts and cyclonic storms and occurs during the summer half of the year.

Ice Cap Climate

The most poleward and coldest of the world's climates is called the ice cap climate. It is found on the continent of Antarctica, interior Greenland, and some high mountain peaks and plateaus. Because monthly mean temperatures are subfreezing throughout the year, areas with this climate are glaciated and have no permanent human inhabitants.

The coldest temperatures of all occur in interior Antarctica, where a Russian research station named Vostok recorded the world's coldest temperature of -128.6 degrees Fahrenheit (-89.2 degrees Celsius) on July 21, 1983. This climate receives little precipitation because the atmosphere can hold very little water vapor. A major moisture surplus exists, however, because of the lack of snowmelt and evaporation. This causes the build up of a surface snow cover that eventually compacts to form the icecaps that bury the surface. Snowstorms are often accompanied by high winds, producing blizzard conditions.

Global Warming

Though warming has not been uniform across the planet, the upward trend in the globally averaged temperature shows that more areas are warming than cooling. According to the National Oceanic and Atmospheric Administration (NOAA) 2018 Global Climate Summary, the combined land and ocean temperature has increased at an average rate of 0.13°F (0.07°C) per decade since 1880; however, the average rate of increase since 1981 (0.31°F/ 0.17°C) is more than twice as great. It is strongly suspected that human activities that increase the accumulation of greenhouse gases (heat-trapping gases) in the atmosphere may play a key role in the temperature rise.

Levels of carbon dioxide (CO_2) in the atmosphere are higher now than they have been at any time in the past 400,000 years. This gas is responsible for nearly two-thirds of the global-warming potential of all human-released gases. Levels surpassed 407 ppm in 2018 for the first time in recorded history. By comparison, during ice ages, CO_2 levels were around 200 parts per million (ppm), and during the warmer interglacial periods, they hovered around 280 ppm. The recent rise in CO_2 shows a remarkably constant relationship with fossil-fuel burning, which is understandable when one considers that about 60 percent of fossil-fuel emissions stay in the air. Atmospheric carbon dioxide concentrations are also increased by deforestation, which is occurring at a rapid rate in several tropical countries. Deforestation causes carbon dioxide levels to rise because trees remove large quantities of this gas from the atmosphere during the process of photosynthesis.

Research indicates that if atmospheric concentrations of greenhouse gases continue to increase at the 1990s pace, global temperatures could rise an additional 1.8 to 6.3 degrees Fahrenheit (1 to 3.5 degrees Celsius) during the twenty-first century. That level of temperature increase would produce major changes in global climates and plant and animal habitats and would cause sea levels to rise substantially.

Ralph C. Scott

CLOUD FORMATION

Clouds are visible manifestations of water in the air. Cloud patterns can provide even a casual observer with much information about air movements and the processes occurring in the atmosphere. The shapes and heights of the clouds and the directions from which they have come are valuable clues in understanding weather.

Importance of Cooling

Clouds are formed when water vapor in the air is transformed into either water droplets or ice crystals. Sometimes large amounts of moisture are added to the air, producing clouds, but clouds generally are formed when a large amount of air is cooled. The amount of water vapor that air can hold varies with temperature: Cold air can hold less water vapor than warmer air. If air is cooled to the point at which it can hold no more water vapor, the water vapor will condense into water droplets. The temperature at which condensation begins is called the dew point. At below freezing temperatures, the water vapor will turn or deposit into ice crystals.

Cloud droplets do not necessarily form even if the air is fully saturated, that is, holding as much water vapor as possible at a given temperature. Once formed, cloud droplets can evaporate again very easily. Two factors hasten the production and growth of cloud droplets. One is the presence of

CLOUD FORMATION

The hydrologic cycle is the continuous circulation of the earth's waters through evaporation, condensation, and precipitation. The cycle also moves water through runoff, infiltration, and transpiration.

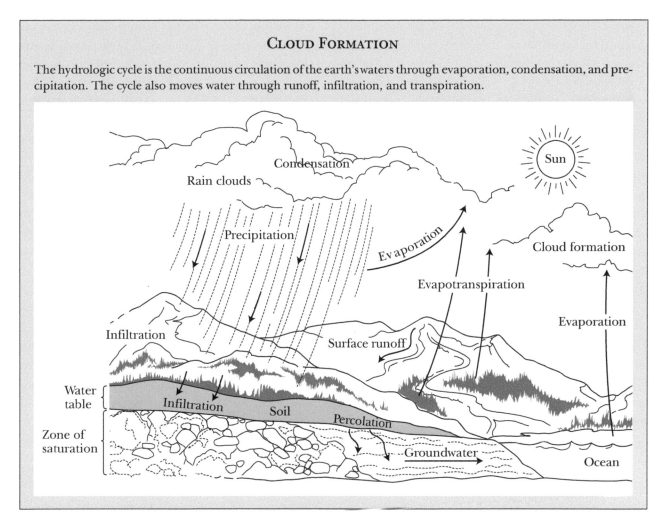

particles in the atmosphere that attract water. These are called hygroscopic particles or condensation nuclei. They include salt, dust, and pollen. Once water vapor condenses on these particles, more condensation can occur. Then the droplets can grow larger and bump into other droplets, growing even larger through this process, called coalescence.

Condensation and cloud droplet growth also is hastened when the air is very cold, at about -40 degrees Farenheit (which is also -40 degrees Celsius). At this temperature ice crystals form, but some water droplets can exist as liquid water. These water droplets are said to be supercooled. The water vapor is more likely to deposit on the ice crystals than on the supercooled water. Thus the ice crystals grow larger and the supercooled water droplets evaporate, resulting in more water vapor to deposit on ice crystals. Whether the cloud droplets start as hygro-

scopic particles or ice crystals, they eventually can grow in size to become a raindrop; around 1 million cloud droplets make one raindrop.

How and Why Rising Air Cools

In order for air to be cooled, it must rise or be lifted. When a volume of air, or an air parcel, is forced to rise through the surrounding air, the parcel expands in size as the pressure of the air around it declines with altitude. Close to the surface, the atmospheric pressure is relatively high because the density of the atmosphere is high. As altitude increases, the atmosphere declines in density, and the still air exerts less pressure. Thus, as an air parcel rises through the atmosphere, the pressure of the surrounding air declines, and the parcel takes up more space as it expands. Since work is done by the parcel as it expands, the parcel cools and its temperature declines.

An alternative explanation of the cooling is that the number of molecules in the air parcel remains the same, but when the volume is larger, the molecules produce less frictional heat because they do not bang into each other as much. The temperature of the air parcel declines, but no heat left the parcel—the change in temperature resulted from internal processes. The process of an air parcel rising, expanding, and cooling is called adiabatic cooling. Adiabatic means that no heat leaves the parcel. If the parcel rises far enough, it will cool sufficiently to reach its dewpoint temperature. With continued cooling, condensation will result—a cloud will be formed. At this height, which is called the lifting condensation level, an invisible parcel of air will turn into a cloud.

Uplift Mechanisms

An initial force is necessary to cause the air parcel to rise and then cool adiabatically. The three major processes are convection, orographic, and frontal or cyclonic.

With certain conditions, convection or vertical movement can cause clouds to form. On a sunny day, usually in the summer, the ground is heated unevenly. Some areas of the ground become warmer and heat the air above, making it warmer and less dense. A stream of air, called a thermal, may rise. As it rises, it cools adiabatically through expansion and may reach its dewpoint temperature. With continued cooling and rising, condensation will occur, forming a cloud. Since the cloud is formed by predominantly vertical motions, the cloud will be cumulus. With continued warming of the surface, the thermals may rise even higher, perhaps producing thunderstorm, or cumulonimbus, clouds. Thus, a sunny summer day can start off without a cloud in the sky, but can be stormy with many thunderstorms by afternoon.

Clouds also can form when air is forced to rise when it meets a mountain or other large vertical barrier. This type of lifting—orographic—is especially prevalent where air moves over the ocean and then is forced to rise up a mountain, as occurs on the west coast of North and South America. As the

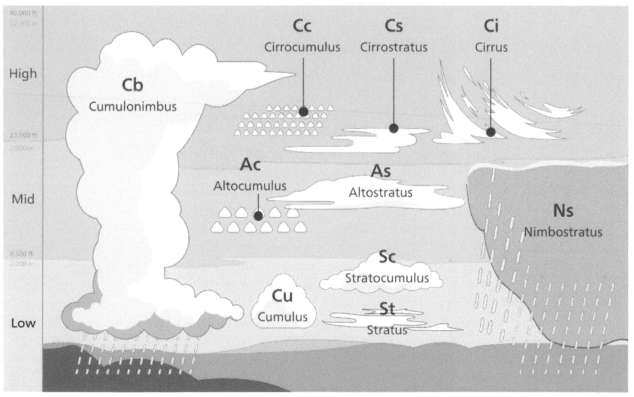

Cloud classification by altitude of occurrence. Multi-level and vertical genus-types not limited to a single altitude level include nimbostratus, cumulonimbus, and some of the larger cumulus species. (Illustration by Valentin de Bruyn)

air rises, it cools adiabatically and eventually becomes so cool that it cannot hold the water vapor. Condensation occurs and clouds form. The air continues to move up the mountain, producing clouds and precipitation on the side of the mountain from which the wind came, the windward side. However, the air eventually must fall down the other side of the mountain, the leeward side. That air is warmed and moisture evaporates, resulting in no clouds.

A third lifting mechanism is frontal, or cyclonic, action. This occurs when a large mass of cold air and a large mass of warm air—often hundreds of miles in area—meet. The warm air mass and the cold air mass will not mix freely, resulting in a border or front between the two air masses. The warm, less dense, air will always rise above the cold, denser, air mass. As the warm air rises, it cools, and when it reaches its dew point, clouds will form. If the warm air displaces the cold air, or a warm front occurs, the warm air will rise gradually, resulting in layered or stratiform clouds. The cloud types will change on an upward diagonal path, with the lowest being stratus, and nimbostratus if rain occurs, followed by altostratus, then cirrostratus, and cirrus.

On the other hand, if the cold air displaces the warm air, the warm air will be forced to rise much more quickly. The clouds formed will be puffy or cumuliform—cumulus at the lowest levels, altocumulus and cirrocumulus at the highest altitudes. Sometimes cumulonimbus clouds will also form.

Sometimes when a cold front meets a warm front, the whole warm air mass is forced off the ground. This forms a cyclone—an area of low pressure—as the warm air rises. As this air rises, it cools. If it reaches its dew point, condensation and clouds will result. In oceanic tropical areas, a cyclone can form within warm, moist air. This air also will cool and, if it reaches its dew point, will condense and form clouds. Sometimes, these tropical cyclones are the precursors of hurricanes. The clouds associated with cyclones are usually cumulus, including cumulonimbus, as they are formed by rapidly rising air.

Margaret F. Boorstein

STORMS

A storm is an atmospheric disturbance that produces wind, is accompanied by some form of precipitation, and sometimes involves thunder and lightning. Storms that meet certain criteria are given specific names, such as hurricanes, blizzards, and tornadoes.

Stormy weather is associated with low atmospheric pressure, while clear, calm, dry weather is associated with high atmospheric pressure. Because of the way atmospheric pressure and wind direction are related, low-pressure areas are characterized by winds moving cyclonically (in a counterclockwise direction in the Northern Hemisphere; clockwise in the Southern Hemisphere) around the center of the low pressure. Storms of all kinds are associated with cyclones, but two classes of cyclones—tropical and extratropical—produce most storms.

Tropical Cyclones
These storms develop during the summer and autumn in every tropical ocean except the South Atlantic and eastern South Pacific Oceans. Tropical cyclones that occur in the North Atlantic and eastern North Pacific Oceans are known as hurricanes; in the western North Pacific Ocean, as typhoons; and in the Indian and South Pacific Oceans, as cyclones.

All tropical cyclones develop in three stages. Arising from the formation of the initial atmospheric disturbance that is characterized by a cluster of thunderstorms, the first stage—tropical depres-

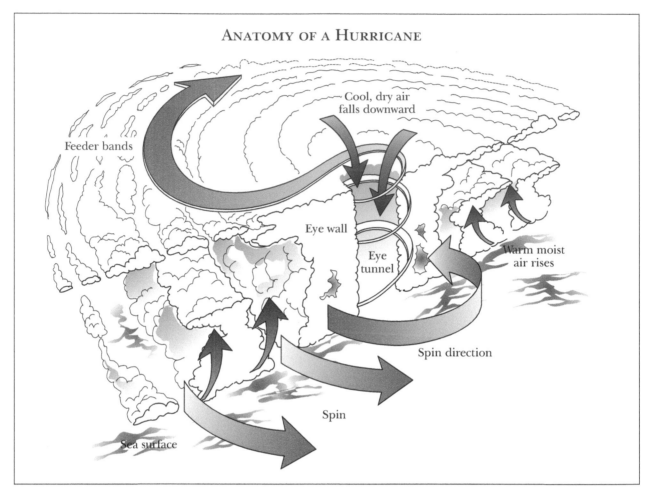

ANATOMY OF A HURRICANE

Feeder bands

Cool, dry air falls downward

Eye wall

Eye tunnel

Warm moist air rises

Spin direction

Spin

Sea surface

sion—occurs when the maximum sustained surface wind speeds (the average speed over one minute) range from 23–39 miles (37–61 km.) per hour. The second stage—tropical storm—occurs when sustained winds range from 40–73 miles (62–119 km.) per hour. At this stage, the storm is given a name. From 80 to 100 tropical storms develop each year across the world, with about half continuing to the final stage—hurricane—at which sustained wind speeds are 74 miles (120 km.) per hour or greater. Moving over land or into colder oceans initiates the end of the hurricane after a week or so by eliminating the hurricane's fuel—warm water.

A mature hurricane is a symmetrical storm, with the "eye" at the center; the eye develops as winds increase and become circular around the central core of low pressure. Within the eye, it is relatively warm, and there are light winds, no precipitation, and few clouds. This is caused by air descending in the center of the storm. Surrounding the eye is the "eye

wall," a ring of intense thunderstorms that can extend high into the atmosphere. Within the eye wall, the strongest winds and heaviest rainfall are found; this is also where warm, moist air, the hurricane's "fuel," flows into the storm. Spiraling bands of clouds, called "rain bands," surround the eye wall. Precipitation and wind speeds decrease from the eye wall out toward the edge of the rain bands, while atmospheric pressure is lowest in the eye and increases outward.

Hurricanes can be the most damaging storms because of their intensity and size. Damage is caused by high winds and the flying debris they carry, flooding from the tremendous amounts of rain a hurricane can produce, and storm surge. A storm surge, which accounts for most of the coastal property loss and 90 percent of hurricane deaths, is a dome of water that is pushed forward as the storm moves. This wall of water is lifted up onto the coast as the eye wall comes in contact with land. For exam-

NAMING HURRICANES

Hurricanes once were identified by their latitudes and longitudes, but this method of naming became confusing when two or more hurricanes developed at the same time in the same ocean. During World War II hurricanes were identified by radio code letters, such as Able and Baker. In 1953 the National Weather Service began using English female names in an alphabetized list. Male names and French and Spanish names were added in 1978. By 2000 six lists of names were used on a rotating basis. When a hurricane causes much death or destruction, as Hurricane Andrew did in August of 1992—its name is retired for at least ten years.

ple, a 25-foot (8-meter) storm surge created by Hurricane Camille in 1969 destroyed the Richelieu Apartments next to the ocean in Pass Christian, Mississippi. Ignoring advice to evacuate, twenty-five people had gathered there for a hurricane party; all but one was killed.

To help predict the damage that an approaching hurricane can cause, the Saffir- Simpson Scale was developed. A hurricane is rated from 1 (weak) to 5 (devastating), according to its central pressure, sustained wind speed, and storm surge height. Michael was a category 5 storm at the time of landfall on October 10, 2018, near Mexico Beach and Tyndall Air Force Base, Florida. Michael was the first hurricane to make landfall in the United States as a category 5 since Hurricane Andrew in 1992, and only the fourth on record. The others are the Labor Day Hurricane in 1935 and Hurricane Camille in 1969.

Extratropical Cyclones

Also known as midlatitude cyclones, these storms are traveling low-pressure systems that are seen on newspaper and television daily weather maps. They are created when a mass of moist, warm air from the south contacts a mass of drier, cool air from the north, causing a front to develop. At the front, the warmer air rides up over the colder air. This causes water vapor to condense and produces clouds and rain during most of the year, and snow in the winter.

Thunderstorms

Thunderstorms also develop in stages. During the cumulus stage, strong updrafts of warm air build the storm clouds. The storm moves into the mature stage when updrafts continue to feed the storm, but cool downdrafts are also occurring in a portion of the cloud where precipitation is falling. When the warm updrafts disappear, the storm's fuel is gone and the dissipating stage begins. Eventually, the cloud rains itself out and evaporates.

Thunderstorms can also form away from a frontal system, usually during summer. This formation is related to a relatively small area of warm, moist air

STORM CLASSIFICATIONS

Tropical Classification	Wind speed
Gale-force winds	>15 meters/second
Tropical depression	20-34 knots and a closed circulation
Tropical storm (named)	35-64 knots
Hurricane	65+ knots (74+ mph)
Saffir-Simpson Scale for Hurricanes	
Category 1	63-83 knots (74-95 mph)
Category 2	83-95 knots (96-110 mph)
Category 3	96-113 knots (111-130 mph)
Category 4	114-135 knots (131-155 mph)
Category 5	>135 knots (>155 mph)

Notes: 1 knot = 1 nautical mile/hour = 1.152 miles/hour = 1.85 kilometers/hour.

Source: National Aeronautics and Space Administration, Office of Space Science, Planetary Data System.

http:/atmos.nmsu.edu/jsdap/encyclopediawork.html

rising and creating a thunderstorm that is usually localized and short lived.

Wind, lightning, hail, and flooding from heavy rain are the main destructive forces of a thunderstorm. Lightning occurs in all mature thunderstorms as the positive and negative electrical charges in a cloud attempt to equal out, creating a giant spark. Most lightning stays within the clouds, but some finds its way to the surface. The lightning heats the air around it to incredible temperatures (54,000 degrees Farenheit/30,000 degrees Celsius), which causes the air to expand explosively, creating the shock wave called thunder. Since lightning travels at the speed of light and thunder at the speed of sound, one can estimate how many miles away the lightning is by counting the seconds between the lightning and thunder and dividing by five. People have been killed by lightning while boating, swimming, biking, golfing, standing under a tree, talking on the telephone, and riding on a lawnmower.

Hail is formed in towering cumulonimbus clouds with strong updrafts. It begins as small ice pellets that grow by collecting water droplets that freeze on contact as the pellets fall through the cloud. The strong updrafts push the pellets back into the cloud, where they continue collecting water droplets until they are too heavy to stay aloft and fall as hailstones. The more an ice pellet is pushed back into the cloud, the larger the hailstone becomes. The largest authenticated hailstone in the United States fell near Vivian, South Dakota, on July 23, 2010. It mea-

sured 8.0 inches (20 cm) in diameter, 18 1/2 inches (47 cm.) in circumference, and weighed in at 1.9375 pounds (879 grams).

Tornadoes

For reasons not well understood, less than 1 percent of all thunderstorms spawn tornadoes. Called funnel clouds until they touch earth, tornadoes contain the highest wind speeds known.

Although tornadoes can occur anywhere in the world, the United States has the most, with an average of 1000 per year. Tornadoes have occurred in every state, but the greatest number hit a portion of the Great Plains from central Texas to Nebraska, known as "Tornado Alley." There cold Canadian air and warm Gulf Coast air often collide over the flat land, creating the wall cloud from which most tornadoes are spawned. May is the peak month for tornado activity, but they have been spotted in every month.

Because tornado winds cannot be measured directly, the tornado is ranked according to its damage, using the Fujita Intensity Scale. The scale ranges from an F0, with wind speeds less than 72 miles (116 km.) per hour, causing light damage, to an F5, with winds greater than 260 miles (419 km.) per hour, causing incredible damage. Most tornadoes are small, but the larger ones cause much damage and death.

Kay R. S. Williams

Earth's Biological Systems

Biomes

The major recognizable life zones of the continents, biomes are characterized by their plant communities. Temperature, precipitation, soil, and length of day affect the survival and distribution of biome species. Species diversity within a biome may increase its stability and capability to deliver natural services, including enhancing the quality of the atmosphere, forming and protecting the soil, controlling pests, and providing clean water, fuel, food, and drugs. Land biomes are the temperate, tropical, and boreal forests; tundra; desert; grasslands; and chaparral.

Temperate Forest
The temperate forest biome occupies the so-called temperate zones in the midlatitudes (from about 30 to 60 degrees north and south of the equator). Temperate forests are found mainly in Europe, eastern North America, and eastern China, and in narrow zones on the coasts of Australia, New Zealand, Tasmania, and the Pacific coasts of North and South America. Their climates are characterized by high rainfall and temperatures that vary from cold to mild.

Temperate forests contain primarily deciduous trees—including maple, oak, hickory, and beechwood—and, secondarily, evergreen trees—including pine, spruce, fir, and hemlock. Evergreen forests in some parts of the Southern Hemisphere contain eucalyptus trees.

The root systems of forest trees help keep the soil rich. The soil quality and color is due to the action of earthworms. Where these forests are frequently cut, soil runoff pollutes streams, which reduces fisheries because of the loss of spawning habitat.

Racoons, opposums, bats, and squirrels are found in the trees. Deer and black bear roam forest floors. During winter, small animals such as groundhogs and squirrels burrow in the ground.

Tropical Forest
Tropical forests are in frost-free areas between the Tropic of Cancer and the Tropic of Capricorn. Temperatures range from warm to hot year-round, because the Sun's rays shine nearly straight down around midday. These forests are found in northern Australia, the East Indies, southeastern Asia, equatorial Africa, and parts of Central America and northern South America.

Tropical forests have high biological diversity and contain about 15 percent of the world's plant species. Animal life lives at different layers of tropical forests. Nuts and fruits on the trees provide food for birds, monkeys, squirrels, and bats. Monkeys and sloths feed on tree leaves. Roots, seeds, leaves, and fruit on the forest floor feed deer, hogs, tapirs, antelopes, and rodents. The tropical forests produce rubber trees, mahogany, and rosewood. Large animals in these forests include the Asian tiger, the African bongo, the South American tapir, the Central and South American jaguar, the Asian and African leopard, and the Asian axis deer. Deforestation for agriculture and pastures has caused reduction in plant and animal diversity.

Boreal Forest
The boreal forest is a circumpolar Northern Hemisphere biome spread across Russia, Scandinavia, Canada, and Alaska. The region is very cold. Evergreen trees such as white spruce and black spruce

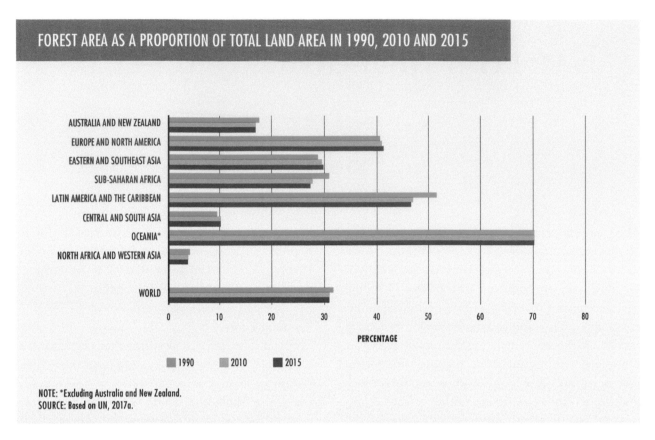

FOREST AREA AS A PROPORTION OF TOTAL LAND AREA IN 1990, 2010 AND 2015

NOTE: *Excluding Australia and New Zealand.
SOURCE: Based on UN, 2017a.

dominate this zone, which also contains larch, balsam, pine, and fir, and some deciduous hardwoods such as birch and aspen. The acidic needles from the evergreens make the leaf litter that is changed into soil humus. The acidic soil limits the plants that develop.

Animals in boreal forests include deer, caribou, bear, and wolves. Birds in this zone include goshawks, red-tailed hawks, sapsuckers, grouse, and nuthatches. Relatively few animals emigrate from this habitat during winter. Conifer seeds are the basic winter food. The disappearing aspen habitat of the beaver has decreased their numbers and has reduced the size of wetlands.

Tundra

About 5 percent of the earth's surface is covered with Arctic tundra, and 3 percent with alpine tundra. The Arctic tundra is the area of Europe, Asia, and North America north of the boreal coniferous forest zone, where the soils remain frozen most of the year. Arctic tundra has a permanent frozen subsoil, called permafrost. Deep snow and low temper-

atures slow the soil-forming process. The area is bounded by a 50 degrees Fahrenheit circumpolar isotherm, known as the summer isotherm. The cold temperature north of this line prevents normal tree growth.

The tundra landscape is covered by mosses, lichens, and low shrubs, which are eaten by caribou, reindeer, and musk oxen. Wolves eat these herbivores. Bear, fox, and lemming also live here. The larger mammals, including marine mammals and the overwintering birds, have large fat layers beneath the skin and long dense fur or dense feathers that provide protection. The small mammals burrow beneath the ground to avoid the harsh winter climate. The most common Arctic bird is the old squaw duck. Ptarmigans and eider ducks are also very common. Geese, falcons, and loons are some of the nesting birds of the area.

The alpine tundra, which exists at high altitude in all latitudes, is acted upon by winds, cold temperatures, and snow. The plant growth is mostly cushion and mat-forming plants.

BIOMES OF THE WORLD

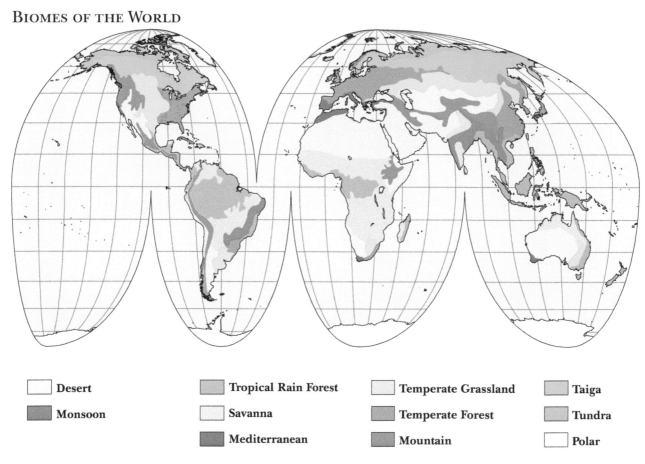

☐ **Desert**	▦ **Tropical Rain Forest**	☐ **Temperate Grassland**	▦ **Taiga**
▦ **Monsoon**	☐ **Savanna**	▦ **Temperate Forest**	▦ **Tundra**
▦ **Mediterranean**	▦ **Mountain**	☐ **Polar**	

Desert

The desert biome covers about one-seventh of the earth's surface. Deserts typically receive no more than 10 inches (25 centimeters) of rainfall a year, but evaporation generally exceeds rainfall. Deserts are found around the Tropic of Cancer and the Tropic of Capricorn. As the warm air rises over the equator, it cools and loses its water content. This dry air descends in the two subtropical zones on each side of the equator; as it warms, it picks up moisture, resulting in drying the land.

Rainfall is a key agent in shaping the desert. The lack of sufficient plant cover removes the natural protection that prevents soil erosion during storms. High winds also cut away the ground.

Some desert plants obtain water from deep below the surface, for example, the mesquite tree, which has roots that are 40 feet (13 meters) deep. Other plants, such as the barrel cactus, store large amounts of water in their leaves, roots, or stems. Other plants slow the loss of water by having tiny leaves or shedding their leaves. Desert plants have very short growth periods, because they cannot grow during the long drought periods.

Desert animals protect themselves from the Sun's heat by eating at night, staying in the shade during the day, and digging burrows in the ground. Among the world's large desert animals are the camel, coyote, mule deer, Australian dingo, and Asian saiga. The digestive process of some desert animals produces water. A method used by some animals to conserve water is the reabsorption of water from their feces and urine.

Grassland

Grasslands cover about a quarter of the earth's surface, and can be found between forests and deserts. Treeless grasslands grow in parts of central North America, Central America, and eastern South America that have between 10 and 40 inches (250-1,000 millimeters) of erratic rainfall. The climate has a high rate of evaporation and periodic major droughts. The biome is also subject to fire.

337

Some grassland plants survive droughts by growing deep roots, while others survive by being dormant. Grass seeds feed the lizards and rodents that become the food for hawks and eagles. Large animals include bison, coyotes, mule deer, and wolves. The grasslands produce more food than any other biome. Poor grazing and agricultural practices and mining destroy the natural stability and fertility of these lands. The reduced carrying capacity of these lands causes an increase in water pollution and erosion of the soil. Diverse natural grasslands appear to be more capable of surviving drought than are simplified manipulated grass systems. This may be due to slower soil mineralization and nitrogen turnover of plant residues in the simplified system.

Savannas are open grasslands containing deciduous trees and shrubs. They are near the equator and are associated with deserts. Grasses grow in clumps and do not form a continuous layer. The northern savanna bushlands are inhabited by oryx and gazelles. The southern savanna supports springbuck and eland. Elephants, antelope, giraffe, zebras, and black rhinoceros are found on the savannas. Lions, leopards, cheetah, and hunting dogs are the primary predators here. Kangaroos are found in the savannas of Australia. Savannas cover South America north and south of the Amazon rain forest, where jaguar and deer can be found.

Chaparral

The chaparral or Mediterranean biome is found in the Mediterranean Basin, California, southern Australia, middle Chile, and Cape Province of South America. This region has a climate of wet winters and summer drought. The plants have tough leathery leaves and may contain thorns. Regional fires clear the area of dense and dead vegetation. Fire, heat, and drought shape the region. The vegetation dwarfing is due to the severe drought and extreme climate changes. The seeds from some plants, such as the California manzanita and South African fire lily, are protected by the soil during a fire and later germinate and rapidly grow to form new plants.

Ocean

The ocean biome covers more than 70 percent of the earth's surface and includes 90 percent of its volume. The ocean has four zones. The intertidal zone is shallow and lies at the land's edge. The continental shelf, which begins where the intertidal zone ends, is a plain that slopes gently seaward. The neritic zone (continental slope) begins at a depth of about 600 feet (180 meters), where the gradual slant of the continental shelf becomes a sharp tilt toward the ocean floor, plunging about 12,000 feet (3,660 meters) to the ocean bottom, which is known as the abyss. The abyssal zone is so deep that it does not have light.

Plankton are animals that float in the ocean. They include algae and copepods, which are microscopic crustaceans. Jellyfish and animal larva are also considered plankton. The nekton are animals that move freely through the water by means of their muscles. These include fish, whales, and squid. The benthos are animals that are attached to or crawl along the ocean's floor. Clams are examples of benthos. Bacteria decompose the dead organic materials on the ocean floor.

The circulation of materials from the ocean's floor to the surface is caused by winds and water temperature. Runoff from the land contains polluting chemicals such as pesticides, nitrogen fertilizers, and animal wastes. Rivers carry loose soil to the ocean, where it builds up the bottom areas. Overfishing has caused fisheries to collapse in every world sector. In some parts of the northwestern Altantic Ocean, there has been a shift from bony fish to cartilaginous fish dominating the fisheries.

Human Impact on Biomes

Human interaction with biomes has increased biotic invasions, reduced the numbers of species, changed the quality of land and water resources, and caused the proliferation of toxic compounds. Managed care of biomes may not be capable of undoing these problems.

Ronald J. Raven

NATURAL RESOURCES

SOILS

Soils are the loose masses of broken and chemically weathered rock mixed with organic matter that cover much of the world's land surface, except in polar regions and most deserts. The two major solid components of soil—minerals and organic matter—occupy about half the volume of a soil. Pore spaces filled with air and water account for the other half. A soil's organic material comes from the remains of dead plants and animals, its minerals from weathered fragments of bedrock. Soil is also an active, dynamic, ever-changing environment. Tiny pores in soil fill with air, water, bacteria, algae, and fungi working to alter the soil's chemistry and speed up the decay of organic material, making the soil a better living environment for larger plants and animals.

Soil Formation

The natural process of forming new soil is slow. Exactly how long it takes depends on how fast the bedrock below is weathered. This weathering process is a direct result of a region's climate and topography, because these factors influence the rate at which exposed bedrock erodes and vegetation is distributed. Global variations in these factors account for the worldwide differences in soil types.

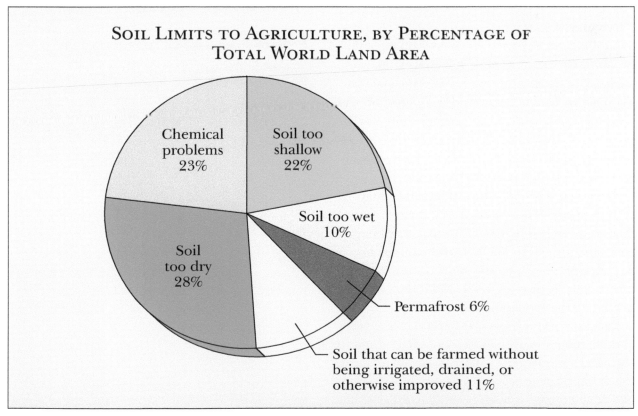

SOIL LIMITS TO AGRICULTURE, BY PERCENTAGE OF TOTAL WORLD LAND AREA

Chemical problems 23%

Soil too shallow 22%

Soil too wet 10%

Soil too dry 28%

Permafrost 6%

Soil that can be farmed without being irrigated, drained, or otherwise improved 11%

Climate is the principal factor in determining the type and rate of soil formation. Temperature and precipitation are the two main climatic factors that influence soil formation, and they vary with elevation and latitude. Water is the main agent of weathering, and the amount of water available depends on how much falls and how much runs off. The amount of precipitation and its distribution during the year influence the kind of soil formed and the rate at which it is formed. Increased precipitation usually results in increased rates of soil formation and deep soils. Temperature and precipitation also determine the kind and amount of vegetation in a region, which determines the amount of available organics.

Topography is a characteristic of the landscape involving slope angle and slope length. Topographic relief governs the amount of water that runs off or enters a soil. On flat or gently sloping land, soil tends to stay in place and may become thick, but as the slope increases so does the potential for erosion. On steep slopes, soil cover may be very thin, possibly only a few inches, because precipitation washes it downhill; on level plains, soil profiles may be several feet thick.

Types of Soil

Typically, bedrock first weathers to form regolith, a protosoil devoid of organic material. Rain, wind, snow, roots growing into cracks, freezing and thawing, uneven heating, abrasion, and shrinking and swelling break large rock particles into smaller ones. Weathered rock particles may range in size from clay to silt, sand, and gravel, with the texture and particle size depending largely on the type of bedrock. For example, shale yields finer-textured soils than sandstone. Soils formed from eroded limestone are rich in base minerals; others tend to be acidic. Generally, rates of soil formation are largely determined by the rates at which silicate minerals in the bedrock weather: the more silicates, the longer the formation time.

In regions where organic materials, such as plant and animal remains, may be deposited on top of regolith, rudimentary soils can begin to form. When waste material is excreted, or a plant or animal dies, the material usually ends up on the earth's surface. Organisms that cause decomposition, such as bacteria and fungi, begin breaking down the remains into a beneficial substance known as humus. Humus restores minerals and nutrients to the soil. It also improves the soil's structure, helping it to retain water. Over time, a skeletal soil of coarse, sandy material with trace amounts of organics gradually forms. Even in a region with good weathering rates and adequate organic material, it can take as long as fifty years to form 12 inches (30 centimeters) of soil. When new soil is formed from weathering bedrock, it can take from 100 to 1,000 years for less than an inch of soil to accumulate.

Water moves continually through most soils, transporting minerals and organics downward by a process called leaching. As these materials travel downward, they are filtered and deposited to form distinct soil horizons. Each soil horizon has its own color, texture, and mineral and humus content. The O-horizon is a thin layer of rotting organics covering the soil. The A-horizon, commonly called topsoil, is rich in humus and minerals. The B-horizon is a subsoil rich in minerals but poor in humus. The C-horizon consists of weathered bedrock; the D-horizon is the bedrock itself.

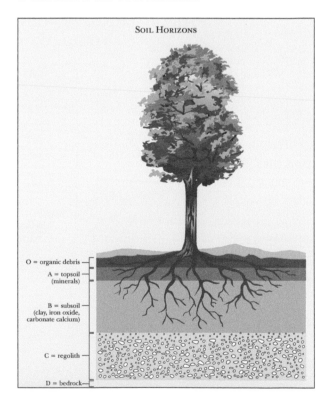

SOIL HORIZONS

O = organic debris

A = topsoil (minerals)

B = subsoil (clay, iron oxide, carbonate calcium)

C = regolith

D = bedrock

Because Earth's surface is made of many different rock types exposed at differing amounts and weathering at different rates at different locations, and because the availability of organic matter varies greatly around the planet due to climatic and seasonal conditions, soil is very diverse and fertile soil is unevenly distributed. Structure and composition are key factors in determining soil fertility. In a fertile soil, plant roots are able to penetrate easily to obtain water and dissolved nutrients. A loam is a naturally fertile soil, consisting of masses of particles from clays (less than 0.002 mm across), through silts (ten times larger) to sands (100 times larger), interspersed with pores, cracks, and crevices.

The Roles of Soil

In any ecosystem, soils play six key roles. First, soil serves as a medium for plant growth by mechanically supporting plant roots and supplying the eighteen nutrients essential for plants to survive. Different types of soil contain differing amounts of these eighteen nutrients; their combination often determines the types of vegetation present in a region, and as a result, influences the number and types of animals the vegetation can support, including humans. Humans rely on soil for crops necessary for food and fiber.

Second, the property of a particular soil is the controlling factor in how the hydrologic system in a region retains and transports water, how contaminants are stored or flushed, and at what rate water is naturally purified. Water enters the soil in the form of precipitation, irrigation, or snowmelt that falls or runs off soil. When it reaches the soil, it will either be surface water, which evaporates or runs into streams, or subsurface water, which soaks into the soil where it is either taken up by plant roots or percolates downward to enter the groundwater system. Passing through soil, organic and inorganic pollutants are filtered out, producing pure groundwater.

Soil also functions as an air-storage facility. Air is pushed into and drawn out of the soil by changes in barometric pressure, high winds, percolating water, and diffusion. Pore spaces within soil provide access to oxygen to organisms living underground as well as to plant roots. Soil pore spaces also contain carbon dioxide, which many bacteria use as a source of carbon.

Soil is nature's recycling system, through which organic waste products and decaying plants and animals are assimilated and their elements made available for reuse. The production and assimilation of humus within soil converts mineral nutrients into forms that can be used by plants and animals, who return carbon to the atmosphere as carbon dioxide. While dead organic matter amounts to only about 1 percent of the soil by weight, it is a vital component as a source of minerals.

Soil provides a habitat for many living things, from insects to burrowing animals, from single microscopic organisms to massive colonies of subterranean fungi. Soils contain much of the earth's genetic diversity, and a handful of soil may contain billions of organisms, belonging to thousands of species. Although living organisms only account for about 0.1 percent of soil by weight, 2.5 acres (one hectare) of good-quality soil can contain at least 300 million small invertebrates—mites, millipedes, insects, and worms. Just 1 ounce (30 grams) of fertile soil can contain 1 million bacteria of a single type, 100 million yeast cells, and 50,000 fungus mycelium. Without these, soil could not convert nitrogen, phosphorus, and sulphur to forms available to plants.

Finally, soil is an important factor in human culture and civilization. Soil is a building material used to make bricks, adobe, plaster, and pottery, and often provides the foundation for roads and buildings. Most important, soil resources are the basis for agriculture, providing people with their dietary needs.

Because the human use of soils has been haphazard and unchecked for millennia, soil resources in many parts of the world have been harmed severely. Human activities, such as overcultivation, inexpert irrigation, overgrazing of livestock, elimination of tree cover, and cultivating steep slopes, have caused natural erosion rates to increase many times over. As a result of mismanaged farm and forest lands, escalated erosional processes wash off or blow away an estimated 75 billion tons of soil annually, eroding away one of civilization's crucial resources.

Randall L. Milstein

WATER

Life on Earth requires water—without it, life on Earth would cease. As human populations grow, the freshwater resources of the world become scarcer and more polluted, while the need for clean water increases. Although nearly three-quarters of Earth's surface is covered with water, only about 0.3 percent of that water is freshwater suitable for consumption and irrigation. This is because more than 97 percent of Earth's water is ocean salt water, and most of the remaining freshwater is frozen in the Antarctic ice cap. Only the small amounts that remain in lakes, rivers, and groundwater is available for human use.

All of earth's water cycles between the ocean, land, atmosphere, plants, and animals over and over. On average, a molecule of surface water cycles from the ocean, to the atmosphere, to the land and back again in less than two weeks. Water consumed by plants or animals takes longer to return to the oceans, but eventually the cycle is completed.

Water's Uses

Water supports the lives of all living creatures. People use for drinking, cooking, cleaning, and bathing. Water also plays a key role in society since humans can travel on it, make electricity with it, fish in it, irrigate crops with it, and use it for recreation. Globally, more than 4 trillion cubic meters of freshwater is used each day. Agriculture accounts for about 70 percent, industry uses 20 percent, and domestic and municipal activities use 10 percent. To produce beef requires between 1,320 and 5,283 gallons (5,000 and 20,000 liters) of water for every 35 ounces (1 kg). A similar amount of wheat requires between 660 and 1056 gallons (2,500 and 4,000 liters) of water. Manufactured goods also require significant amounts of freshwater; a car consumes between 13,000 and 20,000 gallons (49,210 to 75,708 liters), a smartphone has a water footprint of nearly 3,100 gallons (11,734 liters) and a teeshirt uses around 660 gallons (2,498 liters).

The average American family uses more than 300 gallons of drinking quality water per day at home. Roughly 70 percent of this use occurs indoors for drinking, bathing and showering, flushing the toilet, and washing dishes. Outdoor water use for landscape watering, washing cars, and cleaning windows, etc., accounts for 30 percent of household use, although it can be much higher in drier parts of the country. For example, the arid West has some of the highest per capita residential water use because of landscape irrigation.

As the world's population grows, the demand for fresh water will also increase. A study by the World Bank concluded that approximately 80 percent of human illness results from insufficient water supplies and poor water quality caused by lack of sanitation, so careful management of water resources is essential for improving the health of people in the twenty-first century.

Groundwater Supply and Quality

The amount of groundwater in the Earth is seventy times greater than all of the freshwater lakes combined. Groundwater is held within the rocks below the ground surface and is the primary source of water in many parts of the world. In the United States, approximately 50 percent of the population uses some groundwater. However, problems with both groundwater supplies and its quality threaten its future use.

The U.S. Environmental Protection Agency (EPA) found that 45 percent of the large public water systems in the United States that use groundwater were contaminated with synthetic organic chemicals that posed potential health threats. Another major problem occurs when groundwater is used faster than it is replaced by precipitation infiltrating through the ground surface. Many of the arid regions of earth are already suffering from this problem. For example, one-third of the wells in Beijing, China, have gone dry due to overuse. In the United States, the Ogallala Aquifer of the Great Plains, the

THE WORLD AND NORTH AMERICA'S GREATEST RIVERS AND LAKES

Longest river	
Nile (North Africa)	4,130 miles (6,600 km.)
Missouri-Mississippi (United States)	3,740 miles ((6,000 km.)
Largest river by average discharge	
Amazon (South America)	6,181,000 cubic feet/second (175,000 cubic meters/second)
Missouri-Mississippi (United States)	600,440 cubic feet/second (17,000 cubic meters/second)
Largest freshwater lake by volume	
Lake Baikal (Russia)	5,280 cubic miles (22,000 cubic km.)
Lake Superior (United States)	3,000 cubic miles (12,500 cubic km.)

largest in North America, is being severely overused. This aquifer irrigates 30 percent of U.S. farmland, but some areas of the aquifer have declined by up to 60 percent. In one part of Texas, so much water has been pumped out that the aquifer has essentially dried up there. Once depleted, the aquifer will take over 6,000 years to replenish naturally through rainfall.

Surface Water Supply and Quality

Surface water is used for transportation, recreation, electrical generation, and consumption. Ships use rivers and lakes as transport routes, people fish and boat on rivers and lakes, and dams on rivers often are used to generate electricity. The largest river on earth is the Amazon in South America, which has an average flow of 212,500 cubic meters per second, more than twelve times greater than North America's Mississippi River. Earth's largest lake by volume—Lake Baikal in Russia—holds 5,521 cubic miles of water (23,013 cubic kilometers), or approximately 20 percent of Earth's fresh surface water. This is a volume of water approximately equivalent to all five of the North American Great Lakes combined.

Although surface water has more uses, it is more prone to pollution than groundwater. Almost every human activity affects surface water quality. For ex-

ample, water is used to create paper for books, and some of the chemicals used in the paper process are discharged into surface water sources. Most foods are grown with agricultural chemicals, which can contaminate water sources. According to 2018 surveys on national water quality from the U.S. Environmental Protection Agency, nearly half of U.S. rivers and streams and more than one-third of lakes are polluted and unfit for swimming, fishing, and drinking.

Earth's Future Water Supply

Inadequate water supplies and water quality problems threaten the lives of more than 1.5 million people worldwide. The World Health Organization estimates that polluted water causes the death of 361,000 children under five years of age each year and affects the health of 20 percent of Earth's population. As the world's population grows, these problems are likely to worsen.

The United Nations estimates that if current consumption patterns continue, 52 percent of the world's people will live in water-stressed conditions by 2050. Since access to clean freshwater is essential to health and a decent standard of living, efforts must be made to clean up and conserve the planet's freshwater.

Mark M. Van Steeter

EXPLORATION AND TRANSPORTATION

EXPLORATION AND HISTORICAL TRADE ROUTES

The world's exploration was shaped and influenced substantially by economic needs. Lacking certain resources and outlets for trade, many societies built ships, organized caravans, and conducted military expeditions to protect their frontiers and obtain new markets.

Over the last 5,000 years, the world evolved from a cluster of isolated communities into a firmly integrated global community and capitalist world system. By the beginning of the twentieth century, explorers had successfully navigated the oceans, seas, and landmasses and gathered many regional economies into the beginnings of a global economy.

Early Trade Systems

Trade and exploration accompanied the rise of civilization in the Middle East. Egyptian pharaohs, looking for timber for shipbuilding, established trade relations with Mediterranean merchants. Phoenicians probed for new markets off the coast of North Africa and built a permanent settlement at Carthage. By 513 BCE, the Persian Empire stretched from the Indus River in India to the Libyan coast, and it controlled the pivotal trade routes in Iran and Anatolia. A regional economy was taking shape, linking Africa, Asia, and Europe into a blended economic system.

Alexander the Great's victory against the Persian Empire in 330 BCE thrust Greece into a dominant position in the Middle Eastern economy. Trade between the Mediterranean and the Middle East increased, new roads and harbors were constructed,

and merchants expanded into sub-Saharan Africa, Arabia, and India. The Romans later benefited from the Greek foundation. Through military and political conquest, Rome consolidated its control over such diverse areas as Arabia and Britain and built a system of roads and highways that facilitated the growth of an expanding world economy. At the apex of Roman power in 200 CE, trade routes provided the empire with Greek marble, Egyptian cloth, seafood from Black Sea fisheries, African slaves, and Chinese silk.

The emergence of a profitable Eurasian trade route linked people, customs, and economies from the South China Sea to the Roman Empire. Although some limited activity occurred during the Hellenistic period, East-West trade flourished following the rise of the Han Dynasty in China. With the opening of the Great Silk Road from 139 BCE to 200 CE, goods and services were exchanged between people from three different continents.

The Great Silk Road was an intricate network of middlemen stretching from China to the Mediterranean Sea. Eastern merchants sold their products at markets in Afghanistan, Iran, and even Syria, and exchanged a variety of commodities through the use of camel caravans. Chinese spices, perfumes, metals, and especially silk were in high demand. The Parthians from central Asia added their own sprinkling of merchandise, introducing both the East and the West to various exotic fruits, rare birds, and ostrich eggs.

Romans peddled glassware, statuettes, and acrobatic performing slaves. Since communication lines were virtually nonexistent during this period, trade routes were the only means by which ideas regarding art, religion, and culture could mix. The contacts and exchanges enacted along the Great Silk Road initiated a process of cultural diffusion among a diversity of cultures and increased each culture's knowledge of the vast frontiers of world geography.

The Atlantic Slave Trade
Beginning in the fifteenth century, European navigators explored the West African coastline seeking gold. Supplies were difficult to procure, because most of the gold mines were located in the interior along the Senegal River and in the Ashanti forests. Because mining required costly investments in time, labor, and security, the Europeans quickly shifted their focus toward the slave trade. Although slavery had existed since antiquity, the Atlantic slave trade generated one of the most significant movements of people in world history. It led to the forced migration of more than 10 million Africans to South America, the Caribbean islands, and North America. It ensured the success of several imperial conquests, and it transformed the demographic, cultural, and political landscape on four continents.

Originally driven by their quest to circumnavigate Africa and open a lucrative trade route with India, the Portuguese initiated a systematic exploration of the West African coastline. The architect of this system, Henry the Navigator, pioneered the use of military force and naval superiority to annex African islands and open up new trade routes, and he increased Portugal's southern frontier with every acquisition. In 1415 his ships captured Ceuta, a prosperous trade center located on the Mediterranean coast overlooking North African trade routes. Over the next four decades, Henry laid claim to the Madeira Islands, the Canary Islands, the Azores, and Cape Verde. After his death, other Portuguese explorers continued his pursuit of circumnavigation of Africa.

Diego Cão reached the Congo River in 1483 and sent several excursions up the river before returning to Lisbon. Two explorers completed the Portuguese mission at the end of the fifteenth century. Vasco da Gama, sailing from 1497 to 1499, and Bartholomeu Dias, from 1498 to 1499, sailed past the southern tip of Africa and eventually reached India. Since Muslims had already created a number of trade links between East Africa, Arabia, and India, Portuguese exploration furthered the integration of various regions into an emerging capitalist world system.

When the Portuguese shifted their trading from gold to slaves, the other European powers followed suit. The Netherlands, Spain, France, and England used their expanding naval technology to explore the Atlantic Ocean and ship millions of slaves across the ocean. A highly efficient and organized trade route quickly materialized. Since the Europeans were unwilling to venture beyond the walls of their coastal fortresses, merchants relied on African sources for slaves, supplying local kings and chiefs with the means to conduct profitable slave-raiding parties in the interior. In both the Congo and the Gold Coast region, many Africans became quite wealthy trading slaves.

In 1750 merchants paid King Tegbessou of Dahomey 250,000 pounds for 9,000 slaves, and his income exceeded the earnings of many in England's merchant and landowning class. After purchasing slaves, dealers sold them in the Americas to work in the mines or on plantations. Commodities such as coffee and sugar were exported back to Europe for home consumption. Merchants then sold alcohol, tobacco, textiles, and firearms to Africans in exchange for more slaves. This practice was abolished by the end of the nineteenth century, but not before more than 10 million Africans had been violently removed from their homeland. The Atlantic slave trade, however, joined port cities from the Gold Coast and Guinea in Africa with Rio de Janeiro, Hispaniola, Havana, Virginia, Charleston, and Liverpool, and constituted a pivotal step toward the rise of a unified global economy.

Magellan and Zheng He
The Portuguese explorer Ferdinand Magellan generated considerable interest in the Asian markets

when he led an expedition that sailed around the world from 1519 to 1522. Looking for a quick route to Asia and the Spice Islands, he secured financial backing from the king of Spain. Magellan sailed from Spain in 1519, canvassed the eastern coastline of South America, and visited Argentina. He ultimately traversed the narrow straits along the southern tip of the continent and ventured into the uncharted waters of the Pacific Ocean.

Magellan explored the islands of Guam and the Philippines but was killed in a skirmish on Mactan in 1521. Some of his crew managed to return to Spain in 1522, and one member subsequently published a journal of the expedition that dramatically enhanced the world's understanding of the major sea lanes that connected the continents.

China also opened up new avenues of trade and exploration in Southeast Asia during the fifteenth century. Under the direction of Chinese emperor Yongle, explorer Zheng He organized seven overseas trips from 1405 to 1433 and investigated economic opportunities in Korea, Vietnam, the Indian Ocean, and Egypt. His first voyage consisted of more than 28,000 men and 400 ships and represented the largest naval force assembled prior to World War I.

Zheng's armada carried porcelains, silks, lacquerware, and artifacts to Malacca, the vital port city in Indonesia. He purchased an Arab medical text on drug therapy and had it translated into Chinese. He introduced giraffes and mahogany wood into the mainland's economy, and his efforts helped spread Chinese ideas, customs, diet, calendars, scales and measures, and music throughout the global economy. Zheng He's discoveries, coupled with all the material gathered by the European explorers, provided cartographers and geographers with a credible store of knowledge concerning world geography.

Emerging Global Trade Networks

From 1400 to 1900, several regional economic systems facilitated the exchange of goods and services throughout a growing world system. Building on the triangular relationships produced by the slave trade, the Atlantic region helped spread new food-

stuffs around the globe. Plants and plantation crops provided societies with a plentiful supply of sweet potatoes, squash, beans, and maize. This system, often referred to as the Columbian exchange, also assisted development in other regions by supplying the global economy with an ample money supply in gold and silver. Europeans sent textiles and other manufactures to the Americas. In return, they received minerals from Mexico; sugar and molasses from the Caribbean; money, rum, and tobacco from North America; and foodstuffs from South America. Trade routes also closed the distance between the Pacific coastline in the Americas and the Pacific Rim.

Additional thriving trade routes existed in the African-West Asian region. Linking Europe and Africa with Arabia and India, this area experienced a considerable amount of trade over land and through the sea lanes in the Persian Gulf and Red Sea. Europeans received grains, timber, furs, iron, and hemp from Russia in exchange for wool textiles and silver. Central Asians secured stores of cotton textiles, silk, wheat, rice, and tobacco from India and sold silver, horses, camel, and sheep to the Indians. Ivory, blankets, paper, saltpeter, fruits, dates, incense, coffee, and wine were regularly exchanged among merchants situated along the trade route connecting India, Persia, the Ottoman Empire, and Europe.

Finally, a Russian-Asian-Chinese market provided Russia's ruling czars with arms, sugar, tobacco, and grain, and a sufficient supply of drugs, medicines, livestock, paper money, and silver moved eastward. Overall, this system linked the economies of three continents and guaranteed that a nation could acquire essential foodstuffs, resources, and money from a variety of sources.

Several profitable trade routes existed in the Indian Ocean sector. After Malacca emerged as a key trading port in the sixteenth century, this territory served as an international clearinghouse for the global economy. Indians sent tin, elephants, and wood into Burma and Siam. Rice, silk, and sugar were sold to Bengal. Pepper and other spices were shipped westward across the Arabian Sea, while Ceylon furnished India with vast quantities of jewels, cinnamon, pearls, and elephants. The booming interregional trade routes positioned along the Indian coastline ensured that many of the vast commodities produced in the world system could be obtained in India.

The final region of crucial trade routes was between Southeast Asia and China. While the extent of Asian overseas trade prior to the twentieth century is usually downplayed, an abundance of products flowed across the Bay of Bengal and the South China Sea. Japan procured silver, copper, iron, swords, and sulphur from Cantonese merchants,

and Japanese-finished textiles, dyes, tea, lead, and manufactures were in high demand on the mainland. The Chinese also purchased silk and ceramics from the Philippines in exchange for silver. Burma and Siam traded pepper, sappan wood, tin, lead, and saltpeter to China for satin, velvet, thread, and labor. As goods increasingly moved from the Malabar coast in India to the northern boundaries of Korea and Japan, the Pacific Rim played a prominent role in the global economy.

Robert D. Ubriaco, Jr.

ROAD TRANSPORTATION

Roads—the most common surfaces on which people and vehicles move—are a key part of human and economic geography. Transportation activities form part of a nation's economic product: They strengthen regional economy, influence land and natural resource use, facilitate communication and commerce, expand choices, support industry, aid agriculture, and increase human mobility. The need for roads closely correlates with the relative location of centers of population, commerce, industry, and other transportation.

History of Road Making

The great highway systems of modern civilization have their origin in the remote past. The earliest travel was by foot on paths and trails. Later, pack animals and crude sleds were used. The development of the wheel opened new options. As various ancient civilizations reached a higher level, many of them realized the importance of improved roads.

The most advanced highway system of the ancient world was that of the Romans. When Roman civilization was at its peak, a great system of military roads reached to the limits of the empire. The typical Roman road was bold in conception and con-

struction, built in a straight line when possible, with a deep multilayer foundation, perfect for wheeled vehicles.

After the decline of the Roman Empire, rural road building in Europe practically ceased, and roads fell into centuries of disrepair. Commerce traveled by water or on pack trains that could negotiate the badly maintained roads. Eventually, a commercial revival set in, and roads and wheeled vehicles increased.

Interest in the art of road building was revived in Europe in the late eighteenth century. P. Trésaguet, a noted French engineer, developed a new method of lightweight road building. The regime of French dictator Napoleon Bonaparte (1800–1814) encouraged road construction, chiefly for military purposes. At about the same time, two Scottish engineers, Thomas Telford and John McAdam, also developed road-building techniques.

Roads in the United States

Toward the end of the eighteenth century, public demand in the United States led to the improvement of some roads by private enterprise. These improvements generally took the form of toll roads, called "turnpikes" because a pike was rotated in each road to allow entry after the fee was paid, and generally were located in areas adjacent to larger cities. In the early nineteenth century, the federal government paid for an 800-mile-long macadam road from Cumberland, Maryland, to Vandalia, Illinois.

With the development of railroads, interest in road building began to wane. By 1900, however, demand for better roads came from farmers, who wanted to move their agricultural products to market more easily. The bicycle craze of the 1890s and the advent of motorized vehicles also added to the demand for more and better roads. Asphalt and concrete technology was well developed by then; now, the problem was financing. Roads had been primarily a local issue, but the growing demand led to greater state and federal involvement in funding.

The Federal-Aid Highway Act of 1956 was a milestone in the development of highway transportation in the United States; it marked the beginning of the largest peacetime public works program in the history of the world, creating a 41,000-mile National System of Interstate and Defense Highways, built to high standards. Later legislation expanded funding, improved planning, addressed environmental concerns, and provided for more balanced transportation. Other developed countries also developed highway programs but were more restrained in construction.

Roads and Development

Transportation presents a severe challenge for sustainable development. The number of motor vehicles at the end of the second decade of the twenty-first century—estimated at more than 1.2 billion worldwide—is growing almost everywhere at higher rates than either population or the gross domestic product. Overall road traffic grows even more quickly. The tiny nation of San Marino has nearly 1.2 cars per person. Americans own one car for every 1.88 residents. In Great Britain, there is one car for every 5.3 people.

Highways around the world have been built to help strengthen national unity. The Trans-Canada Highway, the world's longest national road, for ex-

---

HIGHWAY CLASSIFICATION

Modern roads can be classified by roadway design or traffic function. The basic type of roadway is the conventional, undivided two-way road. Divided highways have median strips or other physical barriers separating the lanes going in opposite directions.

Another quality of a roadway is its right-of-way control. The least expensive type of system controls most side access and some minor at-grade intersections; the more expensive type has side access fully controlled and no at-grade intersections. The amount of traffic determines the number of lanes. Two or three lanes in each direction is typical, but some roads in Los Angeles have five lanes, while some sections of the Trans-Canada Highway have only one lane. Some highways are paid for entirely from public funds; if users pay directly when they use the road, they are called tollways or turnpikes.

Roads are classified as expressway, arterial, collector, and local in urban areas, with a similar hierarchy in rural areas. The highest level—expressway—is intended for long-distance travel.

---

ample, extends east-west across the breadth of the country. Completed in the 1960s, it had the same goal as the Canadian Pacific Railroad a century before, to improve east-west commerce within Canada.

Sometimes, existing highways need to be upgraded; in less developed countries, this can simply mean paving a road for all-weather operation. An example of a late-1990s project of this nature was the Brazil-Venezuela Highway project, which had this description: Improve the Brazil-Venezuela highway link by completion of paving along the BR-174, which runs northward from Manaus in the Amazon, through Boa Vista and up to the frontier, so opening a route to the Caribbean. Besides the investment opportunities in building the road itself, the highway would result in investment opportunities in mining, tourism, telecommunications, soy and rice production, trade with Venezuela, manufacturing in the Manaus Free Trade Zone, ecotourism in the Amazon, and energy integration.

Growing road traffic has required increasingly significant national contributions to road construction. Beginning in the 1960s, the World Bank began to finance road construction in several countries. It required that projects be organized to the highest technical and economic standards, with private contracting and international competitive bidding rather than government workers. Still, there were questions as to whether these economic assessments had a road-sector bias and properly incorporated environmental costs. Sustainability was also a question—could the facilities be maintained once they were built?

In the 1990s, the World Bank financed a program to build an asphalt road network in Mozambique. Asphalt makes very smooth roads but is very maintenance-intensive, requiring expensive imported equipment and raw materials. By the end of the decade, the roads required resurfacing but the debt was still outstanding. Alternative materials would have given a rougher road, but it could have been built with local materials and labor.

The European Investment Bank has become a major player in the construction of highways linking Eastern and Western Europe to further European integration. Some of the fastest growth in the world in ownership of autos has been in Eastern Europe. There is a two-way feedback effect between highway construction and auto ownership.

Environment Consequences

Highways and highway vehicles have social, economic, and environmental consequences. Compromise is often necessary to balance transportation needs against these constraints. For example, in Israel, there has been a debate over construction of the Trans-Israel highway, a US$1.3 billon, six-lane highway stretching 180 miles (300 km.) from Galilee to the Negev.

Demand on resources for worldwide road infrastructure far exceeds available funds; governments increasingly are looking to external sources such as tolls. Private toll roads, common in the nineteenth century, are making a comeback. This has spread from the United States to Europe, where private and government-owned highway operators have begun to sell shares on the stock market. Private companies are not only operating and financing roads in Europe, they are also designing and building them. In Eastern Europe, where road construction languished under communism, private financing and toll collecting are seen as the means of supporting badly needed construction.

Industrial development in poor countries is adversely affected by limited transportation. Costs are high-unreliable delivery schedules make it necessary to maintain excessive inventories of raw materials and finished goods. Poor transport limits the radius of trade and makes it difficult for manufacturers to realize the economies of large-scale operations to compete internationally.

In more difficult terrain, roads become more expensive because of a need for cuts and fills, bridges, and tunnels. To save money, such roads often have steeper grades, sharper curves, and reduced width than might be desired. Severe weather changes also damage roads, further increasing maintenance costs.

Stephen B. Dobrow

RAILWAYS

Railroads were the first successful attempts by early industrial societies to develop integrated communication systems. Today, global societies are linked by Internet systems dependent upon communication satellites orbiting around Earth. The speed by which information and ideas can reach remote places breaks down isolation and aids in the developing of a world community. In the nineteenth century, railroads had a similar impact. Railroads were critical for the creation of an urban-industrial society: They linked regions and remote places together, were important contributors in developing nation-states, and revolutionized the way business was conducted through the creation of corporations. Although alternative forms of transportation exist, railroads remain important in the twenty-first century.

The Industrial Revolution and the Railroad

Development of the steam engine gave birth to the railroad. Late in the eighteenth century, James Watt perfected his steam engine in England. Water was superheated by a boiler and vaporized into steam, which was confined to a cylinder behind a piston. Pressure from expanding steam pushes the cylinder forward, causing it to do work if it is attached to wheels. Watt's engine was used in the manufacturing of textiles, thus beginning the Industrial Revolution whereby machine technology mass produced goods for mass consumption. Robert Fulton was the first innovator to commercially apply the steam engine to water transportation. His steamboat *Clermont* made its maiden voyage up the Hudson River in 1807.

Not until the 1820s was a steam engine used for land transportation. Rivers and lakes were natural features where no road needed to be built. Applying steam to land movement required some type of roadbed. In England, George Stephenson ran a locomotive over iron strips attached to wooden rails. Within a short time, England's forges were able to roll rails made completely of iron shaped like an inverted "U."

The amount of profit a manufacturer could make was determined partially by the cost of transportation. The lower the cost of moving cargo and people, the higher the profitability. Several alternatives existed before the emergence of railroads, although there were drawbacks compared to rail transportation. Toll roads were too slow. A loaded wagon pulled by four horses could average 15 miles (25 km.) a day. Canals were more efficient than early railroads, because barges pulled by mules moved faster over waterways. However, canals could not be built everywhere, especially over mountains.

The application of railroad technology, using steam as a power source, made it possible to overcome obstacles in moving goods and people over considerable distances and at profitable costs. Railroads transformed the way goods were purchased by reducing the costs for consumers, thus raising the living standards in industrial societies. Railroads transformed the human landscape by strengthening the link between farm and city, changed commercial cities into industrial centers, and started early forms of suburban growth well before automobiles arrived.

Financing Railroads

Constructing railroads was costly. Tunnels had to be blasted through mountains, and rivers had to be crossed by bridges. Early in the building of U.S. railroads, the nation's iron foundries could not meet the demands for rolled rails. Rails had to be imported from England until local forges developed more efficient technologies. Once a railway was completed, there was a constant need to maintain the right-of-way so that traffic flow would not be disrupted. Accidents were frequent, and it was an early practice to burn damaged cars because salvaging them was too expensive.

In some countries, railroads were built and operated by national governments. In the United States, railroads were privately owned; however, it was impossible for any single individual to finance and operate a rail system with miles of track. Businessmen raised money by selling stocks and bonds. Just as investors buy stocks in modern high-technology companies, investors purchased stocks and bonds in railroads.

Investing in railroads was good as long as they earned profits and returned money to their investors, but not all railroads made sufficient profits to reward their investors. Competition among railroads was heavy in the United States, and some railroads charged artificially low fares to attract as much business as they could. When ambitious investment schemes collapsed, railroads went bankrupt and were taken over by financiers.

Selling shares of common stock and bonds was made possible by creating corporations. Railroads were granted permission from state governments to organize a corporation. Every investor owned a portion of the railroad. Stockholders' interests were served by boards of directors, and all business transactions were opened for public inspection. One important factor of the corporation was that it relieved individuals from the responsibilities associated with accidents. The railroad, as a corporation, was held accountable, and any compensation for claims made against the company came out of corporate funds, not from individual pockets. This had an impact on the law profession, as law schools began specializing in legal matters relevant to railroads and interstate commerce.

The Success of Railroads

Railroads usually began by radiating outward from port cities where merchants engaged in transoceanic trade. A classic example, in the United States, is the country's first regional railroad—the Baltimore and Ohio. Construction commenced from Baltimore in 1828; by 1850, the railroad had crossed the Appalachian Mountains and was on the Ohio River at Wheeling, Virginia.

Once trunk lines were established, rail networks became more intensive as branch lines were built to link smaller cities and towns. Countries with extremely large continental dimensions developed interior articulating cities where railroads from all directions converged. Chicago and Atlanta are two such cities in the United States. Chicago was surrounded by three circular railroads (belts) whose only function was to interchange cars. Railroads from the Pacific Coast converged with lines from the Atlantic Coast as well as routes moving north from the Gulf Coast.

Mechanized farms and heavy industries developed within the network. Railroads made possible the extraction of fossil fuels and metallic ores, the necessary ingredients for industrial growth. Extension of railroads deep into Eastern Europe helped to generate massive waves of immigration into both North and South America, creating multicultural societies.

Building railroads in Africa and South Asia made it possible for Europe to increase its political control over native populations. The ultimate aim of the colonial railroad was to develop a colony's economy according to the needs of the mother country. Railroads were usually single-line routes transhipping commodities from interior centers to coastal ports for exportation. Nairobi, Kenya, began as a rail hub linking British interests in Uganda with Kenya's port city of Mombasa. Similar examples existed in Malaysia and Indonesia.

Railroads generated conflicts among colonial powers as nations attempted to acquire strategic resources. In 1904–1905 Russia and Japan fought a war in the Chinese province of Manchuria over railroad rights; Imperial Germany attempted to get around British interests in the Middle East by building a railroad linking Berlin with Baghdad to give Germany access to lucrative oil fields. India was a region of loosely connected provinces until British railroads helped establish unification. The resulting sense of national unity led to the termination of British rule in 1947 and independence for India and Pakistan.

In the United States, private railroads discontinued passenger service among cities early in the 1970s and the responsibility was assumed by the federal government (Amtrak). Most Americans riding trains do so as commuters traveling from the suburbs to jobs in the city. The U.S. has no true high-speed trains, aside from sections of Amtrak's Acela line in the Northeast Corridor, where it can reach 150 mph for only 34 miles of its 457-mile span. Passenger service remains popular in Japan and Europe. France, Germany, and Japan operate high-speed luxury trains with speeds averaging above 100 miles (160 km.) per hour.

Railroads are no longer the exclusive means of land transportation as they were early in the twentieth century. Although competition from motor vehicles and air freight provide alternate choices, railroads have remained important. France and England have direct rail linkage beneath the English Channel. In the United States, great railroad mergers and the application of computer technology have reduced operating costs while increasing profits. Transoceanic container traffic has been aided by railroads hauling trailers on flatcars. Railroads began the process of bringing regions within a nation together in the nineteenth century just as the computer and the World Wide Web began uniting nations throughout the world at the end of the twentieth century.

Sherman E. Silverman

Air Transportation

The movement of goods and people among places is an important field of geographic study. Transportation routes form part of an intricate global network through which commodities flow. Speed and cost determine the nature and volume of the materials transported, so air transportation has both advantages and disadvantages when compared with road, rail, or water transport.

Early Flying Machines
The transport of people and freight by air is less than a century old. Although hot-air balloons were used in the late eighteenth century for military purposes, aerial mapping, and even early photography, they were never commercially important as a means of transportation. In the late nineteenth century, the German count Ferdinand von Zeppelin began experimenting with dirigibles, which added self-propulsion to lighter-than-air craft. These aircraft were used for military purposes, such as the bombing of Paris in World War I. However, by the 1920s zeppelins had become a successful means of passenger transportation. They carried thousands of passengers on trips in Europe or across the Atlantic Ocean and also were used for exploration. Nevertheless, they had major problems and were soon superseded by flying machines heavier than air. The early term for such a machine was an aeroplane, which is still the word used for airplane in Great Britain.

Following pioneering advances with the internal combustion engine and in aerodynamic theory using gliders, the development of powered flight in a heavier-than-air machine was achieved by Wilbur and Orville Wright in December, 1903. From that time, the United States moved to the forefront of aviation, with Great Britain and Germany also making significant contributions to air transport. World War I saw the further development of aviation for military purposes, evidenced by the infamous bombing of Guernica.

Early Commercial Service
Two decades after the Wright brothers' brief flight, the world's first commercial air service began, covering the short distance from Tampa to St. Petersburg in Florida. The introduction of airmail service by the U.S. Post Office provided a new, regular source of income for commercial airlines in the United States, and from these beginnings arose the modern Boeing Company, United Airlines, and American Airlines. Europe, however, was the home of the world's first commercial airlines. These include the Deutsche Luftreederie in Germany, which connected Berlin, Leipzig, and Weimar in 1919; Farman in France, which flew from Paris to London; and KLM in the Netherlands (Amsterdam to London), followed by Qantas—the Queensland and Northern Territory Aerial Services, Limited—in Australia. The last two are the world's oldest still operating airlines.

Aircraft played a vital role in World War II, as a means of attacking enemy territory, defending territory, and transporting people and equipment. A humanitarian use of air power was the Berlin Airlift of 1948, when Western nations used airplanes to deliver food and medical supplies to the people of West Berlin, which the Soviet Union briefly blockaded on the ground.

Cargo and Passenger Service

The jet engine was developed and used for fighter aircraft during World War II by the Germans, the British, and the United States. Further research led to civil jet transport, and by the 1970s, jet planes accounted for most of the world's air transportation. Air travel in the early days was extremely expensive, but technological advances enabled longer flights with heavier loads, so commercial air travel became both faster and more economical.

Most air travel is made for business purposes. Of business trips between 750 and 1,500 miles (1,207 and 2,414 km.), air captures almost 85 percent, and of trips more than 1,500 miles (2,414 km.), 90 percent are made by air. The United States had 5,087 public airports in 2018, a slight decrease from the 5,145 public airports operating in 2014. Conversely, the number of private airports increased over this period from 13,863 to 14,549.

The biggest air cargo carriers in 2019 were Federal Express, which carried more than 15.71 billion freight tonne kilometres (FTK), Emirates Skycargo (12.27 billion FTK), and United Parcel Service (11.26 billion FTK).

The first commercial supersonic airliner, the British-French Concorde, which could fly at more than twice the speed of sound, began regular service in early 1976. However, the fleet was grounded after a Concorde crash in France in mid-2000. The first space shuttle flew in 1981. There have been 135 shuttle missions since then, ending with the successful landing of Space Shuttle Orbiter Atlantis on July 21, 2011. The shuttles have transported 600 people and 3 million pounds (1.36 million kilograms) of cargo into space.

Health Problems Transported by Air

The high speed of intercontinental air travel and the increasing numbers of air travelers have increased the risk of exotic diseases being carried into destination countries, thereby globalizing diseases previously restricted to certain parts of the world. Passengers traveling by air might be unaware that they are carrying infections or viruses. The worldwide spread of HIV/AIDS after the 1980s was accelerated by international air travel.

Disease vectors such as flies or mosquitoes can also make air journeys unnoticed inside airplanes. At some airports, both airplane interiors and passengers are subjected to spraying with insecticide upon arrival and before deplaning. The West Nile virus (West Nile encephalitis) was previously found only in Africa, Eastern Europe, and West Asia, but in the 1990s it appeared in the northeastern United States, transported there by birds, mosquitos, or people.

It was feared in the mid-1990s that the highly infectious and deadly Ebola virus, which originated in tropical Africa, might spread to Europe and the United States, by air passengers or through the importing of monkeys. The devastation of native bird communities on the island of Guam has been traced to the emergence there of a large population of brown tree snakes, whose ancestors are thought to have arrived as accidental stowaways on a military airplane in the late 1940s.

Ray Sumner

ENERGY AND ENGINEERING

ENERGY SOURCES

Energy is essential for powering the processes of modern industrial society: refining ores, manufacturing products, moving vehicles, heating buildings, and powering appliances. In 1999 energy costs were half a trillion dollars in the United States alone. All technological progress has been based on harnessing more energy and using it more effectively. Energy use has been shaped by geography and also has shaped economic and political geography.

Ancient to Modern Energy
Energy use in traditional tribal societies illustrates all aspects of energy use that apply in modern human societies. Early Stone Age peoples had only their own muscle power, fueled by meat and raw vegetable matter. Warmth for living came from tropical or subtropical climates. Then a new energy source, fire, came into use. It made cold climates livable. It enabled the cooking of roots, grains, and heavy animal bones, vastly increasing the edible food supply. Its heat also hardened wood tools, cured pottery, and eventually allowed metalworking.

Nearly as important as fire was the domestication of animals, which multiplied available muscle energy. Domestic animals carried and pulled heavy loads. Domesticated horses could move as fast as the game to be hunted or large animals to be herded.

Increased energy efficiency was as important as new energy sources in making tribal societies more successful. Cured animal hides and woven cloth were additional factors enabling people to move to cooler climates. Cooking fires also allowed drying meat into jerky to preserve it against times of limited supply. Fire-cured pottery helped protect food against pests and kept water close by. However, energy benefits had costs. Fire drives for hunting may have caused major animal extinctions. Periodic burning of areas for primitive agriculture caused erosion. Trees became scarce near the best campsites because they had been used for camp fires—the first fuel shortage.

Energy Fundamentals
Human use of energy revolves about four interrelated factors: energy sources, methods of harnessing the sources, means of transporting or storing energy, and methods of using energy. The potential energies and energy flows that might be harnessed are many times greater than present use.

The Sun is the primary source of most energy on Earth. Sunlight warms the planet. Plants use photosynthesis to transform water and carbon dioxide into the sugars that power their growth and indirectly power plant-eating and meat-eating animals. Many other energies come indirectly from the Sun. Remains of plants and animals become fossil fuels. Solar heat evaporates water, which then falls as rain, causing water flow in rivers. Regional differences in the amount of sunlight received and reflected cause temperature differences that generate winds, ocean currents, and temperature differences between different ocean layers. Food for muscle power of humans and animals is the most basic energy system.

Energy Sources

Biomass—wood or other vegetable matter that can be burned—is still the most important energy source in much of the world. Its basic use is to provide heat for cooking and warmth. Biomass fuels are often agricultural or forestry wastes. The advantage of biomass is that it is grown, so it can be replaced. However, it has several limitations. Its low energy content per unit volume and unit mass makes it unprofitable to ship, so its use is limited to the amount nearby. Collecting and processing biomass fuels costs energy, so the net energy is less. Biomass energy production may compete with food production, since both come from the soil. Finally, other fuels can be cheaper.

Greater concentration of biomass energy or more efficient use would enable it to better compete against other energy sources. For example, fermenting sugars into fuel alcohol is one means of concentrating energy, but energy losses in processing make it expensive.

Fossil fuels have more concentrated chemical energy than biomass. Underground heat and pressure compacts trees and swampy brush into the progressively more energy-concentrated peat, lignite coal, bituminous coal, and anthracite or black coal, which is mostly carbon. Industrializing regions turned to coal when they had exhausted their firewood. Like wood, coal could be stored and shoveled into the fire box as needed. Large deposits of coal are still available, but growth in the use of coal slowed by the mid-twentieth century because of two competing fossil fuels, petroleum and natural gas.

Petroleum includes gasoline, diesel fuel, and fuel oil. It forms from remains of one-celled plants and animals in the ocean that decompose from sugars into simpler hydrogen and carbon compounds (hydrocarbons). Petroleum yields more energy per unit than coal, and it is pumped rather than shoveled. These advantages mean that an oil-fired vehicle can be cheaper and have greater range than a coal-fired vehicle.

There are also hydrocarbon gases associated with petroleum and coal. The most common is the natural gas methane. Methane does not have the energy density of hydrocarbon liquids, but it burns cleanly and is a fuel of choice for end uses such as homes and businesses.

Petroleum and natural gas deposits are widely scattered throughout the world, but the greatest known deposits are in an area extending from Saudi Arabia north through the Caucasus Mountains. Deposits extend out to sea in areas such as the Persian Gulf, the North Sea, and the Gulf of Mexico. Other sources, such as oil tar sands and shale oil, are currently seen as a potentially important source of energy, but controversies surrounding the extraction, refining, and delivery processes make these energy sources a matter of significant debate and concern.

Heat engines transform the potential of chemical energies. James Watt's steam engine (1782) takes heat from burning wood or coal (external combustion), boils water to steam, and expands it through pistons to make mechanical motion. In the twentieth century, propeller-like steam turbines were developed to increase efficiency and decrease complexity. Auto and diesel engines burn fuel inside the engine (internal combustion), and the hot gases expand through pistons to make mechanical motion. Expanding them through a gas turbine is a jet engine. Heat engines can create energy from other sources, such as concentrated sunlight, nuclear fission, or nuclear fusion. The electrical generator transforms mechanical motion into electricity that can move by wire to uses far away. Such transportation (or wheeling) of electricity means that one power plant can serve many customers in different locations.

Flowing water and wind are two of the oldest sources of industrial power. The Industrial Revolution began with water power and wind power, but they could only be used in certain locations, and they were not as dependable as steam engines. In the early twentieth century, electricity made river power practical again. Large dams along river valleys with adequate water and steep enough slopes enabled areas like the Tennessee Valley to be industrial centers. In the 1970s wind power began to be used again, this time for generating electricity.

Solar energy can be tapped directly for heat or to make electricity. Although sunlight is free, it is not concentrated energy, so getting usable energy requires more equipment cost. Consequently, fossil-fueled heat is cheaper than solar heat, and power from the conventional utility grid has been much less expensive than solar-generated electricity. However, prices of solar equipment continue to drop as technologies improve.

Future Energy Sources
Possible future energy sources are nuclear fission, nuclear fusion, geothermal heat, and tides. Fission reactors contain a critical mass of radioactive heavy elements that sustains a chain reaction of atoms splitting (fissioning) into lighter elements—releasing heat to run a steam turbine. Tremendous amounts of fission energy are available, but reactor costs and safety issues have kept nuclear prices higher than that of coal.

Nuclear fusion involves the same reaction that powers the Sun: four hydrogen atoms fusing into one helium atom. However, duplicating the Sun's heat in a small area without damaging the surrounding reactor may be too expensive to allow profitable fusion reactors.

Geothermal power plants, tapping heat energy from within the earth, have operated since 1904, but widespread use depends on cheaper drilling to make them practical in more than highly volcanic areas. Tidal power is limited to the few bays that concentrate tidal energy.

Energy and Warfare
Much of ancient energy use revolved about herding animals and conducting warfare. Horse riders moved faster and hit harder than warriors on foot. The bow and arrow did not change appreciably for thousands of years. Herders on the plains rode horses and used the bow and arrow as part of tending their flocks, and the small amounts of metal needed for weapons was easily acquired. Consequently, the herders could invade and plunder much more advanced peoples. From Scythians to Parthians to Mongols, these people consistently destroyed the more advanced civilizations.

The geographical effect was that ancient civilizations generally developed only if they had physical barriers separating them from the flat plains of herding peoples. Egypt had deserts and seas. The Greeks and Romans lived on mountainous peninsulas, safe from easy attack. The Chinese built the Great Wall along their northern frontier to block invasions.

Nomadic riders dominated until the advent of an energy system of gunpowder and steel barrels began delivering lead bullets. With them, the Russians broke the power of the Tartars in Eurasia in the late fifteenth century, and various peoples from Europe conquered most of the world. Energy and industrial might became progressively more important in war with automatic weapons, high explosives, aircraft, rockets, and nuclear weapons.

By World War II, oil had become a reason for war and a crucial input for war. The Germans attempted to seize petroleum fields around Baku on the Caspian. Later in the war, major Allied attacks targeted oil fields in Romania and plants in Germany synthesizing liquid fuels. During the Arab-Israeli War of 1973, Arabs countered Western support of Israel with an oil boycott that rocked Western economies. In 1990 Iraq attempted to solve a border dispute with its oil-rich neighbor, Kuwait, by seizing all of Kuwait. An alliance, led by the United States, ejected the Iraqis.

Other wars occur over petroleum deposits that extend out to sea. European nations bordering on the North Sea negotiated a complete demarcation of economic rights throughout that body. Tensions between China and other Asian countries continue to mount over rights to the resources available in the South China Sea. Current estimates suggest that there may be 90 trillion cubic feet of natural gas and 11 billion barrels of oil in proved and probable reserves, with much more potentially undiscovered. The area is claimed by China, Vietnam, Malaysia, and the Philippines. Turkey and Greece have not resolved ownership division of Aegean waters that might have oil deposits.

Energy, Development, and Energy Efficiency
Ancient civilizations tended to grow and use locally available food and firewood. Soils and wood supplies often were depleted at the same time, which often coincided with declines in those civilizations.

The Industrial Revolution caused development to concentrate in new wooded areas where rivers suitable for power, iron ore, and coal were close together, for example, England, Silesia, and the Pittsburgh area. The iron ore of Alsace in France combined with nearby coal from the Ruhr in Germany fueled tremendous growth, not always peacefully.

By the late nineteenth century, the development of Birmingham, Alabama, demonstrated that railroads enabled a wider spread between coal deposits, iron ore deposits, and existing population centers. By the 1920s, the Soviet Union developed entirely new cities to connect with resources. By the 1970s, unit trains and ore-carrying ships transported coal from the thick coal beds in Montana and Wyoming to the United States' East Coast and to countries in Asia.

The mechanized transport of electrical distribution and distribution of natural gas in pipelines also changed settlement patterns. Trains and subway trains allowed cities to spread along rail corridors in the late nineteenth century and early twentieth century. By the 1940s, cars and trucks enabled cities such as Los Angeles and Phoenix to spread into suburbs. The trend continues with independent solar power that allows houses to be sited anywhere.

Advances in technology have allowed people to get more while using less energy. For example, early peoples stampeded herds of animals over cliffs for food, which was mostly wasted. Horseback hunting was vastly more efficient. Likewise, fireplaces in colonial North America were inefficient, sending most of their heat up the chimney. In the late eighteenth century, inventor and statesman Benjamin Franklin developed a metallic cylinder radiating heat in all directions, which saved firewood.

The ancient Greeks and others pioneered the use of passive solar energy and efficiency after they exhausted available firewood. They sited buildings to absorb as much low winter sun as possible and constructed overhanging roofs to shade buildings from the high summer sun. That siting was augmented by heavy masonry building materials that buffered the buildings from extremes of heat and cold. Later,

metal pipes and glass meant that solar energy could be used for water and space heating.

The first seven decades of the twentieth century saw major declines in energy prices, and cars and appliances became less efficient. That changed abruptly with the energy crises and high prices of the 1970s. Since then, countries such as Japan, with few local energy resources, have worked to increase efficiency so they will be less sensitive to energy shocks and be able to thrive with minimal energy inputs. This trend could lead eventually to economies functioning on only solar and biomass inputs.

Solid-state electronics, use of light emitting diode (LED) or compact fluorescent lamps (CFLs) rather than incandescent bulbs, and fuel cells, which convert fuel directly into electricity more efficiently than combustion engines, all could lead to less energy use. The speed of their adoption depends on the price of competing energies. According to the U.S. Energy Information Administration's (EIA) International Energy Outlook 2019 (IEO2019), the global supply of crude oil, other liquid hydrocarbons, and biofuels is expected to be adequate to meet world demand through 2050. However, many have noted that continuing to burn fossil fuels at our current rate is not sustainable, not because reserves will disappear, but because the damage to the climate would be unacceptable.

Energy and Environment

Energy affects the environment in three major ways. First, firewood gathering in underdeveloped countries contributes to deforestation and resulting erosion. Although more efficient stoves and small solar cookers have been designed, efficiency increases are competing against population increases.

Energy production also frequently causes toxic pollutant by-products. Sulfur dioxide (from sulfur impurities in coal and oil) and nitrogen oxides (from nitrogen being formed during combustion) damage lungs and corrode the surfaces of buildings. Lead additives in gasoline make internal combustion engines run more efficiently, but they cause low-grade lead poisoning. Spent radioactive fuel from nuclear fission reactors is so poisonous that it must be guarded for centuries.

Finally, carbon dioxide from the burning of fossil fuels may be accelerating the greenhouse effect, whereby atmospheric carbon dioxide slows the planetary loss of heat. If the effect is as strong as some research suggests, global temperatures may increase several degrees on average in the twenty-first century, with unknown effects on climate and sea level.

Roger V. Carlson

ALTERNATIVE ENERGIES

The energy that lights homes and powers industry is indispensable in modern societies. This energy usually comes from mechanical energy that is converted into electrical energy by means of generators—complex machines that harness basic energy captured when such sources as coal, oil, or wood are burned under controlled conditions. This energy, in turn, provides the thermal energy used for heating, cooling, and lighting and for powering automobiles, locomotives, steamships, and airplanes. Because such natural resources as coal, oil, and wood are being used up, it is vital that these nonrenewable sources of energy be replaced by sources that are renewable and abundant. It is also desirable that alternative sources of energy be developed in order to cut down on the pollution that results from the combustion of the hydrocarbons that make the nonrenewable fuels burn.

The Sun as an Energy Source

Energy is heat. The Sun provides the heat that makes Earth habitable. As today's commonly used fuel resources are used less, solar energy will be used increasingly to provide the power that societies need in order to function and flourish.

There are two forms of solar energy: passive and active. Humankind has long employed passive solar energy, which requires no special equipment. Ancient cave dwellers soon realized that if they inhabited caves that faced the Sun, those caves would be warmer than those that faced away from the Sun. They also observed that dark surfaces retained heat and that dark rocks heated by the Sun would radiate the heat they contained after the Sun had set. Modern builders often capitalize on this same knowledge by constructing structures that face south in the Northern Hemisphere and north in the Southern Hemisphere. The windows that face the Sun are often large and unobstructed by draperies and curtains. Sunlight beats through the glass and, in passive solar houses, usually heats a dark stone or brick floor that will emit heat during the hours when there is no sunlight. Just as an automobile parked in the sunlight will become hot and retain its heat, so do passive solar buildings become hot and retain their heat.

Active solar energy is derived by placing specially designed panels so that they face the Sun. These panels, called flat plate collectors, have a flat glass top beneath which is a panel, often made of copper with a black overlay of paint, that retains heat. These panels are constructed so that heat cannot escape from them easily. When water circulated through pipes in the panels becomes hot, it is either pumped into tanks where it can be stored or circulated through a central heating system.

Some active solar devices are quite complex and best suited to industrial use. Among these is the focusing collector, a saucer-shaped mirror that centers the Sun's rays on a small area that becomes extremely hot. A power plant at Odeillo in the French Pyrenees Mountains uses such a system to concentrate the Sun's rays on a concave mirror. The mirror directs its incredible heat to an enormous, confined body of water that the heat turns to steam, which is then used to generate electricity.

Another active solar device is the solar or photo-voltaic cell, which gathers heat from the Sun and turns it into energy directly. Such cells help to power spacecraft that cannot carry enough conventional fuel to sustain them through long missions in outer space.

Geothermal Heating

The earth's core is incredibly hot. Its heat extends far into the lower surfaces of the planet, at times causing eruptions in the form of geysers or volcanoes. Many places on Earth have springs that are warmed by heat from the earth's core.

In some countries, such as Iceland, warm springs are so abundant that people throughout the country bathe in them through the coldest of winters. In Iceland, geothermal energy is used to heat and light homes, making the use of fossil fuels unnecessary.

Hot areas exist beneath every acre of land on Earth. When such areas are near the surface, it is easy to use them to produce the energy that humans require. As dependence on fossil fuels decreases, means will increasingly be found of drawing on Earth's subterranean heat as a major source of energy.

Wind Power

Anyone who has watched a sailboat move effortlessly through the water has observed how the wind can be used as a source of kinetic energy—the kind of energy that involves motion—whose movement is transferred to objects that it touches. Wind power has been used throughout human history. In its more refined aspects, it has been employed to power windmills that cause turbines to rotate, providing generators with the power they require to produce electricity.

Windmills typically have from two to twenty blades made of wood or of heavy cloth such as canvas. Windmills are most effective when they are located in places where the wind regularly blows with considerable velocity. As their blades turn, they cause the shafts of turbines to rotate, thus powering generators. The electricity created is usually trans-mitted over metal cables for immediate use or for storage.

Modern vertical-axis wind turbines have two or three strips of curved metal that are attached at both ends to a vertical pole. They can operate efficiently even if they are not turned toward the wind. These windmills are a great improvement over the old horizontal axis windmills that have been in use for many years. From 2000 to 2015, cumulative wind capacity around the world increased from 17,000 megawatts to more than 430,000 megawatts. In 2015, China also surpassed the EU in the number of installed wind turbines and continues to lead installation efforts. Production of wind electricity in 2016 accounted for 16 percent of the electricity generated by renewables.

Oceans as Energy Sources

Seventy percent of the earth's surface is covered by oceans. Their tides, which rise and fall with predictable regularity twice a day, would offer a ready source of energy once it becomes economically feasible to harness them and store the electrical energy they can provide. The most promising spots to build facilities to create electrical energy from the tides are places where the tides are regularly quite dramatic, such as Nova Scotia's Bay of Fundy, where the difference between high and low tides averages about 55 feet (17 meters).

Some tidal power stations that currently exist were created by building dams across estuaries. The sluices of these dams are opened when the tide comes in and closed after the resulting reservoir fills. The water captured in the reservoir is held for several hours until the tide is low enough to create a considerable difference between the level of the wa-

---

OCEAN ENERGY

The oceans have tremendous untapped energy flows in currents and tremendous potential energy in the temperature differences between warmer tropical surface waters and colder deep waters, known as ocean thermal energy conversion. In both cases, the insurmountable cost has been in transporting energy to users on shore.

ter in the reservoir and that outside it. Then the sluice gates are opened and, as the water rushes out at a high rate of speed, it turns turbines that generate electricity.

The world's first large-scale tidal power plant was the Rance Tidal Power Station in France, which became operational in 1966. It was the largest tidal power station in terms of output until Sihwa Lake Tidal Power Station opened in South Korea in August 2011.

Future of Renewable Energy

As pollution becomes a huge problem throughout the world, the race to find nonpolluting sources of energy is accelerating rapidly. Scientists are working on unlocking the potential of the electricity generated by microbes as a fuel source, for example. New technologies are making renewable energy sources economically practical. As supplies of fossil fuels have diminished, pressure to become less dependent on them has grown worldwide. Alternative energy sources are the wave of the future.

R. Baird Shuman

ENGINEERING PROJECTS

Human beings attempt to overcome the physical landscape by building forms and structures on the earth. Most structures are small-scale, like houses, telephone poles, and schools. Other structures are great engineering works, such as hydroelectric projects, dams, canals, tunnels, bridges, and buildings.

Hydroelectric Projects

The potential for hydroelectricity generation is greatest in rapidly flowing rivers in mountainous or hilly terrain. The moving water turns turbines that, in turn, generate electricity. Hydroelectric power projects also can be built on escarpments and fall lines, where there is tremendous untapped energy in the falling water.

Most of the potential for hydroelectricity remains untapped. Only about one-sixth of the suitable rivers and falls are used for hydroelectric power. Certain areas of the world have used more of their potential than others. The percent of potential hydropower capacity that has not been developed is 71 percent in Europe, 75 percent in North America, 79 percent in South America, 95 percent in Africa, 95 percent in the Middle East, and 82 percent in Asia-Pacific. China, Brazil, Canada, and the United States currently produce the most hydroelectric power.

In Africa, only Zambia, Zimbabwe, and Ghana produce significant hydroelectricity. The region's total generating capacity needs to increase by 6 per cent per year to 2040 from the current total of 125 GW to keep pace with rising electricity demand. In Southeast Asia, countries continue to grapple with the need to build up their hydroelectric plants without causing harm to the rivers that are used to supply food, water, and transportation.

Dams

Dams serve several purposes. One purpose is the generation of hydroelectric power, as discussed above. Dams also provide flood control and irrigation. Rivers in their natural state tend to rise and fall with the seasons. This can cause serious problems for people living in downstream valleys. Flood-control dams also can be used to regulate the flow of water used for irrigation and other projects. A final reason to build dams is to reduce swampland, in order to control insects and the diseases they carry.

Famous dams are found in all regions of the world. In North America, two of the most notable dams are Hoover Dam, completed in 1936, on the Colorado River between Arizona and Nevada; and the Grand Coulee Dam, completed in 1942, on the Columbia River in Washington State.

In South America, the most famous dam is the Itaipu Dam, completed in 1983, on the Paraná River between Brazil and Paraguay. In Africa, the Aswan High Dam was completed in 1970, on the Nile River in Egypt, and the Kariba Dam was completed in 1958, on the Zambezi River between Zambia and Zimbabwe. In Asia, the Three Gorges Dam spans the Yangtze River by the town of Sandouping in Hubei province, China. The Three Gorges Dam has been the world's largest power station in terms of installed capacity (22,500 MW) since 2012.

Bridges

Bridges are built to span low-lying land between two high places. Most commonly, there is a river or other body of water in the way, but other features that might be spanned include ravines, deep valleys and trenches, and swamps. A related engineering

ENGINEERING WORKS AND ENVIRONMENTAL PROBLEMS

Although engineering allows humans to overcome natural obstacles, works of engineering often have unintended consequences. Many engineering projects have caused unanticipated environmental problems.

Dams, for instance, create large lakes behind them by trapping water that is released slowly. This water typically contains silt and other material that eventually would have formed soil downstream had the water been allowed to flow naturally. Instead, the silt builds up behind the dam, eventually diminishing the lake's usefulness. As an additional consequence, there is less silt available for soil-building downstream.

Canals also can cause environmental harm by diverting water from its natural course. The river from which water is diverted may dry up, negatively affecting fish, animals, and the people who live downstream.

The benefits of engineering works must be weighed against the damage they do to the environment. They may be worthwhile, but they are neither all good nor all bad: There are benefits and drawbacks in building any engineering project.

project is the causeway, in which land in a low-lying area is built up and a road is then constructed on it.

The longest bridge in the world is the Akashi Kaikyo in Japan near Osaka. It was built in 1998 and spans 6,529 feet (1,990 meters), connecting the island of Hōnshū to the small island of Awaji. The Storebælt Bridge in Denmark, also completed in 1998, spans 5,328 feet (1,624 meters), connecting the island of Sjaelland, on which Copenhagen is situated, with the rest of Denmark. Another bridge spanning more than 5,300 feet is the Osman Gazi Bridge in Turkey. The bridge was opened on 1 July 1, 2016, ad to become the longest bridge in Turkey and the fourth-longest suspension bridge in the world by the length of its central span. The length of the bridge is expected to be surpassed by the Çanakkale 1915 Bridge, which is currently under construction across the Dardanelles strait.

Other long bridges can be found across the Humber River in Hull, England; across the Chiang Jiang (Yangtze River) in China; in Hong Kong, Norway, Sweden, and Turkey and elsewhere in Japan.

The longest bridge in the United States is the Lake Pontchartrain Causeway, Louisiana, which spans 24 miles (38.5 km), the Verrazano-Narrows Bridge in New York City between Staten Island and Brooklyn was once the longest suspension bridge in the world. Completed in 1964, its main span measures 4,260 feet (1,298 meters).

Canals

Moving goods and people by water is generally cheaper and easier, if a bit slower, than moving them by land. Before the twentieth century, that cost savings overwhelmed the advantages of land travel—speed and versatility. Therefore, human beings have wanted to move things by water whenever possible. To do so, they had two choices: locate factories and people near water, such as rivers, lakes, and oceans, or bring water to where the factories and people are, by digging canals.

One of the most famous canals in the world is the Erie Canal, which runs from Albany to Buffalo in New York State. Built in 1825 and running a length of 363 miles (584 km.), the Erie Canal opened up the Great Lakes region of North America to development and led to the rise of New York City as one of the world's dominant cities.

Two other important canals in world history are the Panama Canal and the Suez Canal. The Panama Canal connects the Atlantic and Pacific Oceans over a length of 50.7 miles (81.6 km.) on the isthmus of Panama in Central America. Completed in 1914, the Panama Canal eliminated the long and dangerous sea journey around the tip of South America. The Suez Canal in Egypt, which runs for 100 miles (162 km.) and was completed in 1856, eliminates a similar journey around the Cape of Good Hope in South Africa.

The longest canal in the world is the Grand Canal in China, which was built in the seventh century and stretches a length of 1,085 miles (2,904 km.). It connects Tianjin, near Beijing in the north of China, with Nanjing on the Chang Jiang (Yangtze River) in Central China. The Karakum Canal runs across the Central Asian desert in Turkmenistan from the Amu Darya River westward to Ashkhabad. It was begun in the 1954, and completed in 1988 and is navigable over much of its 854-mile (1,375-km.) length. The Karakum Canal and carries 13 cubic kilometres (3.1 cu mi) of water annually from the Amu-Darya River across the Karakum Desert to irrigate the dry lands of Turkmenistan.

Many canals are found in Europe, particularly in England, France, Belgium, the Netherlands, and Germany, and in the United States and Canada, especially connecting the Great Lakes to each other and to the Ohio and Mississippi Rivers.

Tunnels

Tunnels connect two places separated by physical features that would make it extremely difficult, if not impossible, for them to be connected without cutting directly through them. Tunnels can be used in place of bridges over water bodies so that water traffic is not impeded by a bridge span. Tunnels of this type are often found in port cities, and cities with them include Montreal, Quebec; New York City; Hampton Roads, Virginia; Liverpool, England; or Rio de Janeiro, Brazil.

Tunnels are often used to go through mountains that might be too tall to climb over. Trains especially are sensitive to changes in slope, and train tunnels are found all over the world. Less common are automobile and truck tunnels, although these are also found in many places. Train and automotive tunnels through mountains are common in the Appalachian Mountains in Pennsylvania, the Rockies in the United States and Canada, Japan, and the Alps in Italy, France, Switzerland, and Austria.

The Chunnel

Arguably the most famous—and one of the most ambitious—tunnels in the world goes by the name Chunnel. Completed in 1994, it connects Dover, England, to Calais, France, and runs 31 miles (50 km.). "Chunnel" is short for the Channel Tunnel, named for the English Channel, the body of water that it goes under. It was built as a train tunnel, but cars and trucks can be carried through it on trains. In the year 2000 plans were underway to cut a second tunnel, to carry automobiles and trucks, that would run parallel to the first Chunnel.

The Seikan Tunnel in Japan, connects the large island of Hōnshū with the northern island of Hokkaidō. The Seikan Tunnel is nearly 2.4 miles (4 km.) longer than Europe's Chunnel; however, the undersea portion of the tunnel is not as long as that of the Chunnel.

Buildings

Historically, North America has been home to the tallest buildings in the world. Chicago has been called the birthplace of the skyscraper and was at one time home to the world's tallest building. In 1998, however, the two Petronas Towers (each 1,483 feet/452 meters tall) were completed in Kuala Lumpur, Malaysia, surpassing the height of the world's tallest building, Chicago's Sears Tower (1,450 feet/442 meters), which had been completed in 1974. In 2019, the tallest completed building in the world is the 2,717-foot (828-metre) tall Burj Khalifa in Dubai, the tallest building since 2008.

Of the twenty tallest buildings standing in the year 2019, China is home to ten (Shanghai Tower, Ping An Finance Center, Goldin Finance 117, Guangzhou CTF Finance Center, Tianjin CFT Finance Center, China Zun, Shanghai World Financial Center, International Commerce Center, Wuhan Greenland Center, Changsha); Malaysia (the Petronas towers) and the United States (One World Trade Center and Central Park Towers) boast two each; Vietnam has one (Landmark 81 in Ho Chi Minh City), as does Russia (Lakhta Center), Taiwan (Taipei 101), South Korea (Lotte World Trade Center), Saudi Aragia (Abraj Al-Bait Clock Tower in Mecca).

Timothy C. Pitts

INDUSTRY AND TRADE

MANUFACTURING

Manufacturing is the process by which value is added to materials by changing their physical form—shape, function, or composition. For example, an automobile is manufactured by piecing together thousands of different component parts, such as seats, bumpers, and tires. The component parts in unassembled form have little or no utility, but pieced together to produce a fully functional automobile, the resulting product has significant utility. The more utility something has, the greater its value. In other words, the value of the component parts increases when they are combined with the other parts to produce a useful product.

Employment in Manufacturing

On a global scale, 28 percent of the world's working population had jobs in the manufacturing sector in the third decade of the century. The rest worked in agriculture (28 percent) and services (49 percent). The importance of each of these sectors varies from country to country and from time period to time period. High-income countries have a higher percentage of their labor force employed in manufacturing than low-income countries do. For example, in the United States 19 percent of the labor force worked in manufacturing by 2019, whereas the African country of Tanzania had only 7 percent of its labor force employed in the manufacturing sector at that time.

At the end of the twentieth century, the vast majority of the U.S. labor force (74 percent) worked in services, a sector that includes jobs such as computer programmers, lawyers, and teachers. By the end of the second decade of the twenty-first century, the percentage has risen to slightly more than 79

percent. Only 1 percent worked in agriculture and mining. This employment structure is typical for a high-income country. In low-income countries, in contrast, the majority of the labor force have agricultural jobs. In Tanzania, for example, 66 percent of the labor force worked in agriculture, while services accounted for 27 percent of the jobs.

The importance of manufacturing as an employer changes over time. In 1950 manufacturing accounted for 38 percent of all jobs in the United States. The percentage of jobs accounted for by the manufacturing sector in high-income countries has decreased in the post-World War II period. The decreasing share of manufacturing jobs in high-income countries is partly attributable to the fact that many manufacturing companies have replaced people with machines on assembly lines. Because one machine can do the work of many people, manufacturing has become less labor-intensive (uses fewer people to perform a particular task) and more capital-intensive (uses machines to perform tasks formerly done by people). In the future, manufacturing in high-income countries is expected to become increasingly capital-intensive. It is not inconceivable that manufacturing's share of the U.S. labor force could fall below 10 percent over the course of the twenty-first century.

Geography of Manufacturing

Every country produces manufactured goods, but the vast bulk of manufacturing activity is concentrated geographically. Four countries—China, the United States, Japan, and Germany—produce almost 60 percent of the world's manufactured goods. The concentration of manufacturing activity in a small number of regions means that there are other regions where very little manufacturing occurs. Africa is a prime example of a region with little manufacturing.

Different countries tend to specialize in the production of different products. For example, 50 percent of the automobiles that were produced in that late 1990s were produced in three countries—Germany, Japan, and the United States. In the production of television sets, the top three countries were China, Japan, and South Korea, which together produced 48 percent of the world's television sets. It is important to note that these patterns change over time. For example, in 1960 the top three automobile-producing countries were Germany, the United Kingdom, and the United States, which together produced 76 percent of the world's automobiles.

Multinational Corporations
A multinational corporation is a corporation that is headquartered in one country but owns business facilities, for example, manufacturing plants, in other countries. Some examples of multinational corporations from the manufacturing sector include the automobile maker Ford, whose headquarters are the in the United States, the pharmaceutical company Bayer, whose headquarters are in Germany, and the candy manufacturer Nestlé, whose headquarters are in Switzerland. Since the end of World War II, multinational corporations have become increasingly important in the world economy. Most multinational corporations are headquartered in high-income countries, such as Japan, the United Kingdom, and the United States.

Companies open manufacturing plants in other countries for a variety of reasons. One of the most common reasons is that it allows them to circumvent barriers to trade that are imposed by foreign governments, especially tariffs and quotas. A tariff is an import tax that is imposed upon foreign-manufactured goods as they enter a country. A quota is a limitation imposed on the volume of a particular good that a particular country can export to another country. The net effect of tariffs and quotas is to increase the cost of imported goods for consumers.

Governments impose tariffs and quotas partly to raise revenue and partly to encourage consumers to purchase goods manufactured in their own country. Foreign manufacturers faced with tariffs and quotas often begin manufacturing their product in the country imposing the tariffs and quotas. As tariffs and quotas apply to imported goods only, producing in the country imposing the quotas or tariffs effectively makes these trade barriers obsolete.

Companies also open manufacturing plants in other countries because of differences in labor costs among countries. While most manufacturing takes place in high-income countries, some low-income countries have become increasingly attractive as production locations because their workers can be hired much more cheaply than in high-income countries. For example, in late 2019, the average manufacturing job in the United States paid more than US$22.50 per hour. By comparison, manufacturing employees in the Phillippines earned a few cents more than US$2.50 per hour.

This dramatic differences in labor costs have prompted some companies to close down their manufacturing plants in high-income countries and open up new plants in low-income countries. This has resulted in high-income countries purchasing more manufactured goods from low-income countries.

More than half the clothing imported into the United States came from Asian countries, for example, China, Taiwan, and South Korea, where labor costs were much lower than in the United States. Much of this clothing was made in factories where workers were paid by companies headquartered in the United States. For example, most of the Nike sports shoes that were sold in the United States were made in China, Indonesia, Vietnam, and Pakistan.

Transportation and Communications Technology
The ability of companies to have manufacturing plants in other countries stems from the fact that the world has a sophisticated and efficient transportation and communications system. An advanced

transportation and communications system makes it relatively easy and relatively cheap to transfer information and goods between geographically distant locations. Thus, Nike can manufacture soccer balls in Pakistan and transport them quickly and cheaply to customers in the United States.

The extent to which transportation and communications systems have improved during the last two centuries can be illustrated by a few simple examples. In 1800, when the stagecoach was the primary method of overland transportation, it took twenty hours to travel the ninety miles from Lansing, Michigan, to Detroit, Michigan. Today, with the automobile, the same journey takes approximately ninety minutes. In 1800 sailing ships traveling at an average speed of ten miles per hour were used to transport people and goods between geographically distant countries. In the year 2019 jet-engine aircraft could traverse the globe at speeds in excess of 600 miles per hour. Communications technology has also improved over time.

In 1930 a three-minute telephone call between New York and London, England, cost more than US$385 in 2018 dollars. In the year 2019 the same telephone call could be made for less than a dime.

In addition to modern telephones, there are fax machines, email, videoconferencing capabilities, and a host of other technologies that make communication with other parts of the world both inexpensive and swift.

Future Prospects

The global economy of the twenty-first century presents a wide variety of opportunities and challenges. Sophisticated communications and transportation networks provide increasing numbers of manufacturing companies with more choices as to where to locate their factories. However, high-income countries like the United States are increasingly in competition with other countries (both high- and low-income) to maintain existing and manufacturing investments and attract new ones. Persuading existing companies to keep their U.S. factories open and not move overseas has been a major challenge. Likewise, making the United States as an attractive place for foreign companies to locate their manufacturing plants is an equally challenging task.

Neil Reid

GLOBALIZATION OF MANUFACTURING AND TRADE

Why are most of the patents issued worldwide assigned to Asian corporations? How did a Taiwanese earthquake prevent millions of Americans from purchasing memory upgrades for their computers? Why have personal incomes in Beijing nearly doubled in less than a decade?

Answers to these questions can be found in the geography of globalization. Globalization is an economic, political, and social process characterized by the integration of the world's many systems of manufacturing and trade into a single and increasingly

seamless marketplace. The result: a new world geography.

This new geography is associated with the expansion of manufacturing and trade as capitalist principles replace old ideologies and state-controlled economies. With expanded free markets, the process of manufacturing and trading is constantly changing. Globalization delivers economic growth through improved manufacturing processes, newly developed goods, foreign investment in overseas manufacturing, and expanded employment.

The economies of developing countries are slowly transitioning from agricultural to industrial activities. Nevertheless, more than 65 percent of workers in these countries continue to work in agriculture. Meanwhile, developed countries, such as Australia and Germany, are experiencing high-technology service sector growth and reduced manufacturing employment. In the United States, nearly 30 percent of all workers were employed in manufacturing during the 1950s, but by 2019, less than 8.5 percent were.

In between these extremes, former state-controlled economies, like Romania, are adopting more efficient economic development strategies. Other nations and economic models, such as Indonesia and China, are pulled into the global marketplace by the growth and expansion of market economies. Despite the different economic paths of developing, transitioning, and developed nations, manufacturing and trade link all nations together and represent an economic convergence with important implications for political, business, and labor leaders—as well as all the world's citizens.

The geographies of manufacturing and trade can be examined as the distribution and location of economic activities in response to technological change and political and economic change.

Distribution and Location

Questions about where people live, work, and spend their money can be answered by reading product labels in any shopping mall, supermarket, or automobile dealership. They reveal the fact that manufacturing is a multistage process of component fabrication and final product assembly that can occur continents apart. For example, a shirt may be designed in New Jersey, assembled in Costa Rica from North Carolina fabric, and sold in British Columbia. To understand how goods produced in faraway locations are sold at neighborhood stores, geographers investigate the spatial, or geographic, distribution of natural resources, manufacturing plants, trading patterns, and consumption.

Historically, the geography of manufacturing and trade has been closely linked to the distribution of raw materials, workers, and buyers. In earlier times, this meant that manufacturing and trade were highly localized functions. In the eighteenth century, every North American town had cobblers or blacksmiths who produced goods from local resources for sale in local markets. By the start of the Industrial Revolution, improved transportation and manufacturing techniques had significantly enlarged the geography of manufacturing and trade. As distances increased, new manufacturing and trading centers developed. The location of these centers was contingent upon site and situation. Site and situation refer to a physical location, or site, relative to needed materials, transportation networks, and markets. For example, Pittsburgh, Pennsylvania, became the site of a major steel industry because it was near coal and iron resources. Pittsburgh also benefited from its historical role as a port town on a major river system that provided access to both western and eastern markets.

While relative location and transportation costs continue to be important factors, the geographic distribution of production and movement of goods across space is more complex than the simple calculus of site and situation. New global and local geographies of manufacturing and trade have been fueled by two major factors: technology and political change.

Technological Change

The old saying that time is money partially explains where goods are manufactured and traded. By compressing time and space, technology has enabled people, goods, and information to go farther more quickly. In the process, technology has reduced interaction costs, such as telecommunications. Just as steel enabled railroads to push farther westward, new technologies reduce the distance between places and people.

By increasing physical and virtual access to people, places, and things, technology has eliminated many barriers to global trade. However, improved telecommunications and transportation are only part of technology's contribution to globalization. If time is money, new efficient manufacturing processes also have reduced costs and facilitated globalization.

Armed with more efficient production processes, reliable telecommunications infrastructures, and transportation improvements, businesses can increase profits and remain competitive by seeking out lower-cost labor markets thousands of miles from consumers. As trade and manufacturing are increasingly spatially separate activities, the geographic distribution of manufacturing promotes an uneven distribution of income. The global distribution of manufacturing plants is closely related to industry-specific skill and wage requirements. For example, low-wage and low-skill jobs tend to concentrate in the developing regions of Asia, South America, and Africa. Alternately, high-technology and high-wage manufacturing activities concentrate in more developed regions.

In some cases, high wages and global competition force corporations to move their manufacturing plants to save costs and remain competitive. During the early 1990s, this byproduct of globalization was a major issue during the U.S. and Canadian debates to ratify the North American Free Trade Agreement (NAFTA). Focusing on primarily U.S. and Canadian companies that moved jobs to Mexico, the debate contributed to growing anxiety over job security as plants relocate to low-cost labor markets in South America and around the world.

As global competition increases, the geography of manufacturing and trade is increasingly global and rapidly changing. One company that has adapted to the shifting nature of global trade and manufacturing is Nike. Based in Beaverton, Oregon, Nike designs and develops new products at its Oregon world headquarters. However, Nike has internationalized much of its manufacturing capacity to compete in an aggressive athletic apparel industry. Over the last twenty-five years, Nike's strategy has meant shifts in production from high-wage U.S. locations to numerous low-wage labor markets around Pacific Rim.

Political and Economic Change: A New World Order

In order for companies such as Nike to successfully adapt to changing global dynamics, a stable international, or multilateral, trading system must be in place. In 1948 the General Agreement on Tariffs and Trade (GATT) was the first major step toward developing this stable global trading infrastructure. During that same period, the World Bank and International Monetary Fund were created to stabilize and standardize financial markets and practices. However, Cold War politics postponed complete economic integration for nearly half a century. Since the collapse of communism, globalization has accelerated as economies coalesce around the principles of free markets and capitalism. These important changes have become institutionalized through multilateral trade agreements and international trading organizations.

International trading organizations try to minimize or eliminate barriers to free and fair trade between nations. Trade barriers include tariffs (taxes levied on imported goods), product quotas, government subsidies to domestic industry, domestic content rules, and other regulations. Barriers prevent competitive access to domestic markets by artificially raising the prices of imported goods too high or preventing foreign firms from achieving economies of scale. In some cases, tariffs can also be used to promote fair trade by effectively leveling the playing field.

Because tariffs can be used both to promote fair trade and to unfairly protect markets, trading organizations are responsible for distinguishing between the two. For example, the Asian Pacific Economic Cooperation (APEC) forum has established guidelines to promote fair trade and attract foreign investment. APEC initiatives include a public Web-based database of member state tariff schedules and related links. Through programs such as the APEC information-sharing project, trading organizations are streamlining the international business process and promoting the overall stability of international markets.

The Future

As the globalization of manufacturing and trade continues, a new world geography is emerging. Unlike the Cold War's east-west geography and politics of ideology, an economic politics divides the developed and developing world along a north-south axis. While the types of conflicts associated with these new politics and the rules of engagement are unclear, it is evident that a new hierarchy of nations is emerging.

Globalization will raise the economic standard of living in most nations, but it has also widened the gap between richer and poorer countries. A small group of nations generates and controls most of the world's wealth. Conversely, the poorest countries account for roughly two-thirds of the world's population and less than 10 percent of its wealth.

This fundamental question of economic justice was a motive behind globalization's first major political clash. During the 1999 World Trade Organization (WTO) meetings in Seattle, Washington, approximately 50,000 environmentalists, labor unionists, and human and animal rights activists protested against numerous issues, including cultural intolerance, economic injustice, environmental degradation, political repression, and unfair labor practices they attribute to free trade. While the protesters managed to cancel the opening ceremonies, the United Nations secretary-general, Kofi Annan, expressed the general sentiment of most WTO member states. Agreeing that the protesters' concerns were important, Annan also asserted that the globalization of manufacturing and trade should not be used as a scapegoat for domestic failures to protect individual rights. More important, the secretary-general feared that those issues could

be little more than a pretext for a return to unilateral trade policies, or protectionism.

Like the Seattle protesters, supporters of multilateral trade advocate political and economic reforms. Proponents emphasize that open markets promote open societies. Free traders earnestly believe economic engagement encourages rogue nations to improve poor human rights, environmental, and labor records. It is argued that economic engagement raises the expectations of citizens, thereby promoting change.

Conclusion

Technological and political change have made global labor and consumer markets more accessible and established an economic world hierarchy. At the top, one-fifth of the world's population consumes the vast majority of produced goods and controls more than 80 percent of the wealth. At the bottom of this hierarchy, poor nations are industrializing but possess less than 10 percent of the world's wealth. In political, social, and cultural terms, this global economic reality defines the contours and cleavages of a changing world geography. Whether geographers calculate the economic and political costs of a widening gap between rich and poor or chart the flow of funds from Tokyo to Toronto, the globalization of manufacturing and trade will remain central to the study of geography well into the twenty-first century.

Jay D. Gatrell

MODERN WORLD TRADE PATTERNS

Trade, its routes, and its patterns are an integral part of modern society. Trade is primarily based on need. People trade the goods that they have, including money, to obtain the goods that they don't have. Some nations are very rich in agriculture or natural resources, while others are centers of industrial or technical activity. Because nations' needs change

only slowly, trade routes and trading patterns develop that last for long periods of time.

Types of Trade

The movement of goods can occur among neighboring countries, such as the United States and Mexico, or across the globe, as between Japan and

Italy. Some trade routes are well established with regularly scheduled service connecting points. Such service is called liner service. Liners may also serve intermediate points along a trade route to increase their revenue.

Some trade occurs only seasonally, such as the movement of fresh fruits from Chile to California. Some trade occurs only when certain goods are demanded, such as special orders of industrial goods. This type of service is provided by operators called tramps. They go where the business of trade takes them, rather than along fixed liner schedules and routes.

Many people think of international trade as being carried on great ships plying the oceans of the world. Such trade is important; however, a considerable amount of trade is carried by other modes of transportation. Ships and airplanes carry large volumes of freight over large distances, while trucks, trains, barges, and even animal transport are used to move goods over trade routes among neighboring or landlocked countries.

Trade Routes

Through much of human history, trade routes were limited. Shipping trade carried on sailing vessels, for example, was limited by the prevailing winds that powered the ships. Land routes were limited by the location of water, mountain ranges, and the slow development of roads through thick forests and difficult terrain. The mechanization of transportation eventually freed ships and other forms of transport to follow more direct trade routes. Also, the development of canals and transcontinental highway systems allowed trade routes to develop based solely upon economic requirements.

Other changes in trade routes have occurred with industrialization of transport systems. The world began to have a great need for coal. Trade routes ran to the countries in which coal was mined. Ships and trains delivered coal to the power industry worldwide. Later, trade shifted to locations where oil (petroleum) was drilled. Now, oil is delivered to those same powerplants and industrial sites around the world.

Noneconomic Factors

Some trade is not purely economic in nature. Political relationships among countries can play an important part in their trade relations. For example, many national governments try to protect their countries' automobile and electronics industries from outside competition by not allowing foreign goods to be imported easily. Governments control imports by assessing duties, or tariffs, on selected imports.

Some national governments use the concept of cabotage to protect their home transportation industries by requiring that certain percentages of imported and exported trade goods be carried by their own carriers. For example, the U.S. government might require that 50 percent of its trade use American ships, planes, or trucks. The government might also require that all American carriers employ only American citizens.

Nations also can exert pressure on their trading partners by limiting access to port or airport facilities. Stronger nations may force weaker nations into accepting unequal trade agreements. For example, the United States once had an agreement with Germany concerning air passenger service between the two countries. The agreement allowed United States carriers to carry 80 percent of the passengers, while German carriers were permitted to carry only 20 percent of the passengers.

Multilateral Trade

In situations in which pairs of trading nations do not have direct diplomatic contact with each other, they make their trade arrangements through other nations. Such trade is referred to as multilateral. Certain carriers cater to this type of trade. They operate their ships or planes in around-the-world service. They literally travel around the globe picking up and depositing cargo along the way for a variety of nations.

Trade Patterns

For many years, world populations were coast centered. This means that most of the people in the country lived close to the coast. This was due primarily to the availability of water transportation systems to move both goods and people. At this time, major railroad, highway and airline systems did not exist. As railway and highway systems pushed into the interiors of nations, the population followed, and goods were needed as well as produced in these areas. Thus, over the years many inland population centers have developed that require transportation systems to move goods into and away from this area.

In these cases, international trade to these inland centers required the use of a number of different modes of transportation. Each of the different modes required additional paperwork and time for repackaging and securing of the cargo. For example, cargo coming off ships from overseas was unloaded and placed in warehouse storage. At some later time, it was loaded onto trucks that carried it to railyards. There it would be unloaded, stored, and then loaded onto railcars. At the destination, the cargo would once again be shifted to trucks for the final delivery. During the course of the trip, the cargo would have been handled a number of times, with the possibility of damage or loss occurring each time.

Containerization

As more goods began to move in international trade, the systems for packaging and securing of cargo became more standardized. In the 1960s, shipments began to move in containers. These are highway truck trailers which have been removed from the chassis leaving only the box. Container packaging has become the standard for most cargos moving today in both domestic and international trade. With the advent of containerization of cargo in international trade, cargo movements could quickly move intermodally. Intermodal shipping involves the movement of cargo by using more than a single mode of transportation.

Land, water, and air carriers have attempted to make the intermodal movement of cargo in international trade as seamless as possible. They have not only standardized the box for carrying cargo, but they have also standardized the handling equipment, so that containers move quickly from one mode to another. Advances in communications and

THE WORLD TRADE ORGANIZATION AND GLOBAL TRADING

In 1998 domestic political pressures and an expected domestic surplus of rice prompted the Japanese government to unilaterally implement a 355-percent tariff on foreign rice, violating the United Nations' General Agreement on Tariffs and Trade (GATT). On April 1, 1999, Japan agreed to return to GATT import levels and imposed new over-quota tariffs. While domestic Japanese politics could have prompted a trade war with rice-exporting countries, the crisis demonstrates how multilateral trading initiatives promote stability. Without an agreement, rice exporters might not have gained access to Japanese markets. By returning to GATT minimum quotas and implementing over-quota taxes, the compromise addressed the interests of both domestic and foreign rice growers.

electronic banking allow the paperwork and payments also to be completed and transferred rapidly.

As the demands for products have grown and as the size of industrial plants has grown, the size of movements of raw materials and containerized cargo has also grown. Thus, the sizes of the ships and trains required to move these large volumes of cargo have also increased.

The development of VLCC's (very large crude carriers) has allowed shippers to move large volumes of oil products. The development of large bulk carriers has allowed for the carriage of large volumes of dry raw materials such as grains or iron ore. These large vessels take advantage of what is known as economies of scale. Goods can be moved more cheaply when large volumes of them are moved at the same time. This is because the doubling of the volume of cargo moved does not double the cost to build or operate the vessels in which it is carried. This savings reduces the cost to move large volumes of cargo.

Intermodal Transportation

Intermodal transportation has allowed cargo to move seamlessly across both international boundaries and through different modes of transportation. This seamless movement has changed ocean trade routes over recent years.

The development of the Pacific Rim nations created a demand for trade between East Asia and both the United States and Europe. This trade has usually taken the all-water routes between Asia and Europe. Ships moving from East Asia across the Pacific Ocean pass through the Panama Canal and cross the Atlantic Ocean to reach Western Europe. This journey is in excess of 10,000 miles (16,000 km.) and usually takes about thirty days for most ships to complete. The all-water route from Asia to New York is similar. The distance is almost as great as that to Europe and requires about twenty-one to twenty-four days to complete.

Intermodal transportation has given shippers alternatives to all-water routes. A great volume of Asian goods is now shipped to such western U.S. ports as Seattle, Oakland, and Los Angeles, from which these goods are carried by trains across the United States to New York. The overall lengths of these routes to New York are only about 7,400 miles (12,000 km.) and take between only fifteen and nineteen days to complete. Cargos continuing to Europe are put back on ships in New York and complete their journeys in an additional seven to ten days. Such intermodal shipping can save as much as a week in delivery time.

Airfreight

Another changing trend in trade patterns is the development of airfreight as an international competitor. Modern aircraft have improved dramatically both in their ability to lift large weights of cargo as well as their ability to carry cargos over long distances. Because of the speed at which aircraft travel in comparison to other modes of transportation, goods can be moved quickly over large distances. Thus, high-value cargos or very fragile cargos can move very quickly by aircraft.

The drawback to airfreight movement of cargo is that it is more expensive than other modes of travel. However, for businesses that need to move perishable commodities, such as flowers of the Netherlands, or expensive commodities, such as Paris fashions or Singapore-made computer chips, airfreight has become both economic and essential.

Robert J. Stewart

Political Geography

Forms of Government

Philosophers and political scientists have studied forms of government for many centuries. Ancient Greek philosophers such as Plato and Aristotle wrote about what they believed to be good and bad forms of government. According to Plato's famous work, *The Republic*, the best form of government was one ruled by philosopher-kings. Aristotle wrote that good governments, whether headed by one person (a kingship), a few people (an aristocracy), or many people (a polity), were those that ruled for the benefit of all. Those that were based on narrow, selfish interests were considered bad forms of government, whether ruled by an individual (a tyranny), a few people (an oligarchy), or many people (a democracy). Thus, democracy was not always considered a good form of government.

Constitutions and Political Institutions

All governments have certain things in common: institutions that carry out legislative, executive, and judicial functions. How these institutions are supposed to function is usually spelled out in a country's constitution, which is a guide to organizing a country's political system. Most, but not all, countries have written constitutions. Great Britain, for example, has an unwritten constitution based on documents such as the Magna Carta, the English Bill of Rights, and the Treaty of Rome, and on unwritten codes of behavior expected of politicians and members of the royal family.

The world's oldest written constitution still in use is that of the United States. All countries have written or unwritten constitutions, and most follow them most of the time. Some countries do not follow their constitutions—for example, the Soviet Union did not; other countries, for example France, change their constitutions frequently.

Constitutions usually first specify if the country is to be a monarchy or a republic. Few countries still have monarchies, and those that do usually grant the monarch only ceremonial powers and duties. Countries with monarchies at the beginning of the twenty-first century included Spain, Great Britain, Lesotho, Swaziland, Sweden, Saudi Arabia, and Jordan. Most countries that do not have monarchies are republics.

Constitutions also specify if power is to be concentrated in the hands of a strong national government, which is a unitary system; if it is to be divided between a national and various subnational governments such as states, provinces, or territories, which is a federal system; or if it is to be spread among various subnational governments that might delegate some power to a weak national government, which is a confederate system.

Examples of countries with unitary systems include Great Britain, France, and China; federal systems include the United States, Germany, Russia, Canada, India, and Brazil. There were no confederate systems by the third decade of the twenty-first century, although there are examples from history as well as confederations of various groups and nations. The United States under its eighteenth-century Articles of Confederation and the nineteenth-century Confederate States of America, made up of the rebelling Southern states, were confederate systems. Switzerland was a confederation for much of the nineteenth century. The concept of dividing power between the national and subnational governments is called the vertical axis of power.

MONARCHIES OF THE WORLD

Realm/Kingdom	Monarch	Type
Principality of Andorra	Co-Prince Emmanuel Macron; Co-Prince Archbishop Joan Enric Vives Sicília	Constitutional
Antigua and Barbuda	Queen Elizabeth II	Constitutional
Commonwealth of Australia	Queen Elizabeth II	Constitutional
Commonwealth of the Bahamas	Queen Elizabeth II	Constitutional
Barbados	Queen Elizabeth II	Constitutional
Belize	Queen Elizabeth II	Constitutional
Canada	Queen Elizabeth II	Constitutional
Grenada	Queen Elizabeth II	Constitutional
Jamaica	Queen Elizabeth II	Constitutional
New Zealand	Queen Elizabeth II	Constitutional
Independent State of Papua New Guinea	Queen Elizabeth II	Constitutional
Federation of Saint Kitts and Nevis	Queen Elizabeth II	Constitutional
Saint Lucia	Queen Elizabeth II	Constitutional
Saint Vincent and the Grenadines	Queen Elizabeth II	Constitutional
Solomon Islands	Queen Elizabeth II	Constitutional
Tuvalu	Queen Elizabeth II	Constitutional
United Kingdom of Great Britain and Northern Ireland	Queen Elizabeth II	Constitutional
Kingdom of Bahrain	King Hamad bin Isa	Mixed
Kingdom of Belgium	King Philippe	Constitutional
Kingdom of Bhutan	King Jigme Khesar Namgyel	Constitutional
Brunei Darussalam	Sultan Hassanal Bolkiah	Absolute
Kingdom of Cambodia	King Norodom Sihamoni	Constitutional
Kingdom of Denmark	Queen Margrethe II	Constitutional
Kingdom of Eswatini	King Mswati III	Absolute
Japan	Emperor Naruhito	Constitutional
Hashemite Kingdom of Jordan	King Abdullah II	Constitutional
State of Kuwait	Emir Sabah al-Ahmad	Constitutional
Kingdom of Lesotho	King Letsie III	Constitutional
Principality of Liechtenstein	Prince Regnant Hans-Adam II (Regent: The Hereditary Prince Alois)	Constitutional
Grand Duchy of Luxembourg	Grand Duke Henri	Constitutional
Malaysia	Yang di-Pertuan Agong Abdullah	Constitutional
Principality of Monaco	Sovereign Prince Albert II	Constitutional

MONARCHIES OF THE WORLD *(continued)*

Realm/Kingdom	Monarch	Type
Kingdom of Morocco	King Mohammed VI	Constitutional
Kingdom of the Netherlands	King Willem-Alexander	Constitutional
Kingdom of Norway	King Harald V	Constitutional
Sultanate of Oman	Sultan Haitham bin Tariq	Absolute
State of Qatar	Emir Tamim bin Hamad	Mixed
Kingdom of Saudi Arabia	King Salman bin Abdulaziz	Absolute theocracy
Kingdom of Spain	King Felipe VI	Constitutional
Kingdom of Sweden	King Carl XVI Gustaf	Constitutional
Kingdom of Thailand	King Vajiralongkorn	Constitutional
Kingdom of Tonga	King Tupou VI	Constitutional
United Arab Emirates	President Khalifa bin Zayed	Mixed
Vatican City State	Pope Francis	Absolute theocracy

Whether governments share power with subnational governments or not, there must be institutions to make laws, enforce laws, and interpret laws: the legislative, executive, and judicial branches of government. How these branches interact is what determines whether governments are parliamentary, presidential, or mixed parliamentary-presidential. In a presidential system, such as in the United States, the three branches—legislative, executive, and judicial—are separate, independent, and designed to check and balance each other according to a constitution. In a parliamentary system, the three branches are not entirely separate, and the legislative branch is much more powerful than the executive and judicial branches.

Great Britain is a good example of a parliamentary system. Some countries, such as France and Russia, have created a mixed parliamentary-presidential system, wherein the three branches are separate but are not designed to check and balance each other. In a mixed parliamentary-presidential system, the executive (led by a president) is the most powerful branch of government.

Looking at political systems in this way—how the legislative, executive, and judicial branches of government interact—is to examine the horizontal axis of power. All governments are unitary, federal, or confederate, and all are parliamentary, presidential, or mixed parliamentary-presidential. One can find examples of different combinations. Great Britain is unitary and parliamentary. Germany is federal and parliamentary. The United States is federal and presidential. France is unitary and mixed parliamentary-presidential. Russia is federal and mixed parliamentary-presidential. Furthermore, virtually all countries are either republics or monarchies.

Types of Government

Constitutions describe how the country's political institutions are supposed to interact and provide a guide to the relationship between the government and its citizens. Thus, while governments may have similar political institutions—for example, Germany and India are both federal, parliamentary republics—how the leaders treat their citizens can vary widely. However, governments may have political systems that function similarly although they have different forms of constitutions and institutions. For example, Great Britain, a unitary, parliamentary monarchy with an unwritten constitution, treats its citizens very similarly to the United States, which is a federal, presidential republic with a written constitution.

The three most common terms used to describe the relationships between those who govern and those who are governed are democratic, authoritarian, and totalitarian. Characteristics of democracies are free, fair, and meaningfully contested elections; majority rule and respect for minority rights and opinions; a willingness to hand power to the opposition after an election; the rule of law; and civil rights and liberties, including freedom of speech and press, freedom of association, and freedom to travel. The United States, Canada, Japan, and most European countries are democratic.

An authoritarian system is one that curtails some or all of the characteristics of a democratic regime. For example, authoritarian regimes might permit token electoral opposition by allowing other political parties to run in elections, but they do not allow the opposition to win those elections. If the opposition did win, the authoritarian regime would not hand over power. Authoritarian regimes do not respect the rule of law, the rights of minorities to dissent, or freedom of the press, speech, or association. Authoritarian governments use the police, courts, prisons, and the military to intimidate and threaten their citizens, thus preventing people from uniting to challenge the existing political rulers. Afghanistan, Cuba, Iran, Uzbekistan, Saudi Arabia, Chad, Syria, Libya, Sudan, Belarus, and China are examples of countries with authoritarian regimes.

Totalitarian regimes are similar to authoritarian regimes but are even more extreme. Under a totalitarian regime, there is no legal opposition, no freedom of speech, and no rule of law whatsoever. Totalitarian regimes attempt to control totally all members of the society to the point where everyone always must actively demonstrate their loyalty to and support for the regime. Nazi Germany under Adolf Hitler's rule (1933-1945) and the Soviet Union under Joseph Stalin's rule (1928-1953) are examples of totalitarian regimes. As of 2019, only Eritrea and North Korea are the still have governments classified as totalitarian dictatorships.

Forms of Government: Putting it All Together

In *The Republic*, Plato asserts that people have varied dispositions, and, therefore, there are various types of governments. In recent years, regimes have been created that some call mafiacracies (rule by criminal mafias), narcocracies (rule by narcotics gangs), gerontocracies (rule by very old people), theocracies (rule by religious leaders), and so forth. Such variations show the ingenuity of the human mind in devising forms of government.

Whatever labels that are given to a political system, there remain basic questions to be asked about that regime: Is it a monarchy or a republic? Is all power concentrated in the hands of a national government, or is power shared between a national government and the states or provinces? Are its institutions those of a parliamentary, presidential, or mixed parliamentary-presidential system? Is it democratic, authoritarian, or totalitarian? Finally, does it live up to its constitution, both in terms of how power is supposed to be distributed among institutions and in its relationship between the government and the people? To paraphrase Aristotle, how many rulers are there, and in whose interests do they rule?

Nathaniel Richmond

POLITICAL GEOGRAPHY

Students of politics have been aware that there is a significant relationship between physical and political geography since the time of ancient Greece.

The ancient Greek philosopher Plato argued that a *polis* (politically organized society) must be of limited geographical size and limited population or it would lack cohesion. The ideal *polis* would be only as geographically large as required to feed about 5,000 people, its maximum population.

Plato's illustrious pupil, Aristotle, agreed that stable states must be small. "One can build a wall around the Hellespont," the main territory of ancient Greece, he wrote in his treatise *Politics*, "but that will not make it a polis." Today human ideas differ about the maximum area of a successful state or nation-state, but the close influence of physical geography on political geography and their profound mutual effects on politics itself are not in question.

Geographical Influences on Politics

The physical shape and contours of states may be called their physical geography; the political shape and contours of states, starting with their basic structure as unified state, federation, or confederation, are primary features of their political geography. The idea of "political geography" also can refer to variations in a population's political attitudes and behavior that are influenced by geographical features. Thus, the combination of plentiful land and sparse population tend toward an independent spirit, especially where the economy is agriculturally based. This has historically been the case in the western United States; in the Pampas region of Argentina, where cattle are raised by independent-minded gauchos (cowboys); and on the Brazilian frontier, where government regulation is routinely resisted.

Likewise, where physical geography presents significant difficulties for inhabitants in earning a living or associating, as where there is rough terrain and poor soil or inhospitable climate, the populace is likely to exhibit a hardy, self-reliant character that strongly influences political preferences. Thus, physical geography helps to shape national character, including aspects of a nation's politics.

Furthermore, it is well known that where physical geography isolates one part of a country's population from the rest, political radicalism may take root. This tendency is found in coastal cities and remote regions, where labor union radicalism has often been pronounced. Populations in coastal loca-

tions with access to foreign trade often show a more liberal, tolerant, and outgoing spirit, as reflected in their political opinions. In ancient Greece, the coastal access enjoyed by Athens through a nearby port in the fifth century BCE had a strong influence on its liberal and democratic political order. In modern times, China's coastal cities, such as Tientsin, and North American cities such as San Francisco, show similar influences.

The Geographical Imperative

In many instances, political geography is shaped by what may be called the "geographical imperative." Physical geography in these instances demands, or at least strongly suggests, that political geography follow its course. The numerous valleys of mountainous Greece strongly influenced the emergence of the small, often fiercely independent, polis of ancient times. The formation and borders of Asian states such as Bhutan, Nepal, and Tibet have been strongly influenced by the Himalaya Mountains, and the Alps, which shape Switzerland.

As another example, physical geography demands that the land between the Pacific Ocean and the Andes Mountains along the western edge of South America be organized as a separate country—Chile. Island geography often plays a decisive role in its political geography. The qualified political unity of Great Britain can be directly traced to its insular status. Small islands often find themselves combined into larger units, such as the Hawaiian Islands.

The absence of the geographical imperative, however, leaves political geography an open question. For example, Indonesia comprises some 1,300 islands stretching 3,000 miles in bodies of water such as the Indian Ocean and the Celebes Sea. With so many islands, Indonesia lacks a geographical imperative to be a unified state. It also lacks the imperative of ethnic and cultural homogeneity and cohesion, a circumstance mirrored in its political life, since it has remained unified only through military force. As control by the military waned after the fall of the authoritarian General Suharto in 1998, conflicts among the nation's diverse peoples have threatened its breakup. No such threat, however, confronts Australia, an immense island continent where a European majority dominates a fragmented and primitive aboriginal minority. In Australia, the geographical imperative suggests a unity supported by the cultural unity of the majority.

As many examples show, the geographical imperative is not absolute. For example, mountainous Greece is politically united in the twenty-first century. Although long shielded geographically, Tibet lost its political independence after it was successfully invaded by China. The formerly independent Himalayan state Sikkim was taken over by India. Thus, political will trumps physical geography.

The frequency of exceptions to the geographical imperative illustrates that human freedom, while not unlimited, often plays a key role in shaping political geography. As one example, the Baltic Republics—Lithuania, Latvia, and Estonia—historically have been dominated, or largely swallowed up, by neighboring Russia. By the start of the twenty-first century, however, they had regained their independence through the political will to self-rule and the drive for cultural survival.

Strategically Significant Locations

Locations of great economic or military significance become focal points of political attention and, potentially, of military conflict. There are innumerable such places in the world, but several stand out as models of how important physical geography can be for political geography in the context of international politics.

One significant example is the Panama Canal, without which ships must sail around South America. The Suez Canal, which connects European and Asian shipping, is a similar waterway, saving passage around Africa. The canal's significance was reduced after 1956, however, when its blockage after the Arab-Israeli war of that year led to the building of supertankers too large to traverse it. Another example is Gibraltar, whose fortifications command the entrance to the Mediterranean Sea from the Atlantic Ocean. A final example is the Bosporus, the tiny entrance from the Black Sea to waters leading to the Mediterranean Sea. It is the only warm-water route to and from Eastern Russia and therefore is of great military and economic importance for regional and world power politics.

Charles F. Bahmueller

GEOPOLITICS

Geopolitics is a concept pertaining to the role of purely geographical features in the relations among states in international politics. Geopolitics is especially concerned with the geographical locations of the states in relationship to one another. Geopolitical relationships incorporate social, economic, political, and historical features of the states that interact with purely geographical elements to influence the strategic thinking and behavior of nations in the international sphere.

Coined in 1899 by the Swedish theorist Rudolf Kjellen, the term "geopolitics" combines the logic of the search for security and competition for dominance among states with geographical methodology. *Geopolitics* must not, however, be confused with *political geography*, which focuses on individual states' territorial sizes, boundaries, resources, internal political relations, and relations with other states.

Geopolitical is a term frequently used by military and political strategists, politicians and diplomats, political scientists, journalists, statesmen, and a variety of other government officials, such as policy planners and intelligence analysts.

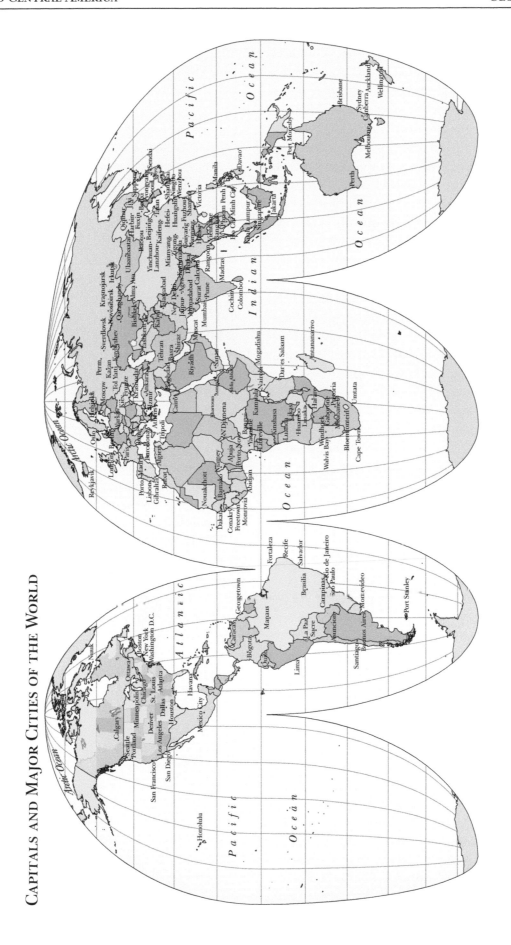

CAPITALS AND MAJOR CITIES OF THE WORLD

Power Struggles Among States

The idea of geopolitics arises in the course of what might be considered the universal struggle for power among the world's most powerful nations, which compete for political and military leadership. How one state can threaten another, for example, is often influenced by geographical factors in combination with technological, social, economic and other factors. The extent to which individual states can threaten each other depends in no small measure on purely geographical considerations.

By the close of twentieth century the Cold War that had dominated world security concerns was over. Nevertheless, the United States still worried about the danger of being attacked by nuclear missiles fired, not by the former Soviet Union, but by irresponsible, fanatical, or suicidal states. American political leaders and military planners were concerned with the geographical position of so-called "rogue states." or "states of concern." In 1994, North Korea, Cuba, Iran, Libya under Muammar Gaddafi, and Ba'athist Iraq were listed as states of concern. By 2019, a list of state sponsors of terrorism included Iran, North Korea, Sudan, and Syria.

Geographical factors play prominent roles in assessments of the different threats that those states presented to American interests. How far those states are located from American territory determines whether their missiles might pose a serious threat. A missile may be able to reach only the periphery of U.S. soil, or it might be able to carry only a small payload. Similar considerations determine the threat such states pose for U.S. forces stationed abroad, as well as for such important U.S. allies as Japan, Western Europe, or Israel. Such questions are thus said to constitute geopolitical, or geostrategic, considerations.

There are many examples of the influence of geopolitical factors on international relations among nations in the past. For example, the Bosporus, the narrow sea lane linking the Black Sea and the Mediterranean where Istanbul is situated, has long been considered of great strategic importance. In the nineteenth century, the Bosporus was the only direct route through which the Russian

A PEACEFULLY RESOLVED BORDER DISPUTE

The peaceful resolution of the border dispute between the Southern African states of Botswana and Namibia was hailed by observers of African politics. Instead of resorting to the armed warfare that so often has marked similar disputes on the continent, the two states chose a different course in 1996, when they found negotiations stalemated. They submitted their claims to the International Court of Justice in The Hague and agreed to accept the court's ruling. Late in 1999, by an eleven-to-four vote, the court ruled for Botswana, and Namibia kept its word to embrace the decision. At issue was a tiny island in the Chobe River on Botswana's northern border. An 1890 treaty between colonial rulers Great Britain and Germany had described the border at the disputed point vaguely, as the river's "main channel." The court took the course of the deepest channel to mark the agreed boundary, giving Botswana title to the 1.4-square-mile (3.5-sq. km.) territory.

navy could reach southern Europe and the Mediterranean Sea.

Because of Russia's nineteenth century history of expansionism and its integration into the pre-World War I European state system, with its networks of competing military alliances, the Bosporus took on added geopolitical meaning. It was the congested (and therefore vulnerable) space through which Russian naval power had to pass to reach the Mediterranean.

Historical Origins of Geopolitics

Although political geography was a well-established field by the late nineteenth century, geopolitics was just beginning to emerge as a field of study and political analysis at the end of the century. In 1896 the German theorist Friedrich Ratzel published his *Political Geography*, which put forward the idea of the state as territory occupied by a people bound together by an idea of the state. Ratzel's theory embraced Social Darwinist notions that justified the current boundaries of nations. Ratzel viewed the state as a biological organism in competition for land with other states. The ethical implication of his theory seemed to be that "might makes right."

That theme set the stage for later German geopolitical thought, especially the notion of the need for *Lebensraum* (living room)—space into which the people of a nation could expand. German dictator Adolf Hitler justified his attack on Russia during World War II partly upon his claim that the German people needed more *Lebensraum* to the east. To some modern geographers, the use of geopolitical theories to serve German fascism and to justify other instances of military aggression tarnished geopolitics itself as a field of study.

Historical Development of Geopolitics

Modern geopolitics has further origins in the work of the Scottish geographer Sir Halford John Mackinder. In 1904 he published a seminal article, "The Geographical Pivot of History," in which he argued that the world is made up of a Eurasian "heartland" and a secondary hinterland (the remainder of the world), which he called the "marginal crescent." According to his theory, international politics is the struggle to gain control of the heartland. Any state that managed that feat would dominate the world.

A major proposition of Mackinder's theory was that geographical factors are not merely causative factors, but coercive. He tried to describe the physical features of the world that he believed directed human actions. In his view, "Man and not nature initiates, but nature in large measure controls." Geopolitical factors were therefore to a great extent determinants of the behavior of states. If this were true, geopolitics as a science could have deep relevance and corresponding influence among governments.

After Mackinder's time, the concept of geopolitics had a double significance. On the one hand it was a purely descriptive theory of geographic causation in history. On the other hand, its purveyors also believed, as Mackinder argued in 1904, that geopolitics has "a practical value as setting into perspective some of the competing forces in current international politics." Mackinder sought to promote this field of study as a companion to British state-

craft, a tool to further Britain's national interest. By extension, geopolitical theory could assist any government in forming its political/military strategy.

As applied to the early twentieth-century world of international politics, however, Mackinder's theory had major weaknesses. Among his most glaring oversights were his failure to appreciate the rise of the United States, which attained considerable naval power after the turn of the century. Also, he failed to foresee the crucial strategic role that air power would play in warfare—and with it the immense change that air power could make in geopolitical considerations. Air power moves continents closer together, revolutionizing their geopolitical relationships.

One of Mackinder's chief critics was Nicolas John Spykman. Spykman argued that Mackinder had overvalued the potential economic, and therefore political, power of the Eurasian heartland, which could never reach its full potential because it could not overcome the obstacles to internal transportation. Moreover, the weaknesses of the remainder of the world—in effect, northern, western and southern Europe—could be overcome through forging alliances.

The dark side of geopolitical thought as handmaiden to political and military strategy became apparent in the Germany of the 1920s. At that time German theorists sought the resurrection of a German state broken by failure in World War I, the harsh terms of the Versailles Treaty that ended the war, and the hyperinflation that followed, wiping out the German middle class. In his 1925 article "Why Geopolitik?" Karl Haushofer urged the practical applications of *Geopolitik*. He urged that this form of analysis had not only "come to stay" but could also form important services for German political leaders, who should use all available tools "to carry on the fight for Germany's existence."

Haushofer ominously suggested that the "struggle" for German existence was becoming increasingly difficult because of the growth of the country's population. A people, he wrote, should study the living spaces of other nations so it could be prepared to "seize any possibility to recover lost ground." This discussion clearly implied that, from geopolitical necessity, Germany should seek additional territory to feed itself—a view carried into effect by Hitler in his quest for *Lebensraum* in attacking the Soviet Union, including its wheat-producing breadbasket, the Ukraine.

After World War II, a chastened Haushofer sought to soft-pedal both the direction and influence of his prewar writings. However, Hitler's morally heinous use of *Geopolitik* left geopolitical theorizing permanently tainted, in some eyes. Nevertheless, there is no necessary connection between geopolitics as a purely analytic description and geopolitics as the basis for a selfish search for power and advantage.

Geopolitics in the Twenty-first Century

Geopolitical considerations were unquestionably of profound relevance to the principal states of the post-World War II Cold War period. After the fall of the Berlin Wall in 1989, however, some theorists thought that the age of geopolitics had passed. In 1990 American strategic theorist Edward N. Luttwak, for example, argued that the importance of military power in international affairs had declined precipitously with the winding down of the Cold War. Military power had been overtaken in significance by economic prowess. Consequently, geopolitics had been eclipsed by what Luttwak called "geoeconomics," the waging of geopolitical struggle by economic means.

The view of Luttwak and various geographers of the declining significance of military power and geopolitical analysis, however, was soon proved to be overdrawn by events. As early as the first months of 1991, before the Soviet Union was officially dismantled, military power asserted itself as a key determinant on the international scene. Led by the United States, a far-flung alliance of nations participated in a war to remove Iraqi dictator Saddam Hussein's forces from neighboring Kuwait, which Iraq had illegally occupied. The decisive and successful use of military power in that war dramatically disproved assertions of its growing irrelevance.

Similarly, in the first three decades of the twenty-first century, military power retained its pre-

eminence in the dynamics of international politics, even as economic forces were seen to gather momentum. To states throughout Asia and the West (especially Western Europe and the United States), the relative military capability of potential adversaries, and therefore geopolitics, remained a vital feature of the international order. Central to this view of the world scene is the growing military rivalry of the United States and China in East Asia. As China modernizes and expands its nuclear and conventional forces, it may feel itself capable of challenging America's predominant military power and prestige in East Asia. This possibility heightens the use of geopolitical thinking, giving it currency in analyzing this emerging situation.

Geopolitics as Civilizational Clash

A sometimes controversial expression of geopolitical analysis has been offered by Samuel Huntington of Harvard University. In his *The Clash of Civilizations and the Remaking of World Order* (1996) Huntington constructs a theory to explain certain tendencies of international behavior. He divides the world into a number of cultural groupings, or "civilizations," and argues that the character of various international conflicts can best be explained as conflicts or clashes of civilizations. In his view, Western civilization differs from the civilization of Orthodox Christianity, with a variety of conflicts erupting between the two. An example is the attack by the North Atlantic Treaty Organization (NATO), the bastion of the West, on Serbia, which is part of the Orthodox East.

Huntington's other civilizations include Islamic, Jewish, Eastern Caribbean, Hindu, Sinic (Chinese), and Japanese. The clash between Israel and its neighbors, the struggle between Pakistan and India over Kashmir, the rivalries between the United States and China and between China and India, for example, can be viewed as civilizational conflicts. Huntington has stated, however, that his theory is not intended to explain all of the historical past, and he does not expect it to remain valid long into the future.

Charles F. Bahmueller

NATIONAL PARK SYSTEMS

The world's first national parks were established as a response to the exploitation of natural resources, disappearance of wildlife, and destruction of natural landscapes that took place during the late nineteenth century. Government efforts to preserve natural areas as parks began with the establishment of Yellowstone National Park in the United States in 1872 and were soon adopted in other countries, including Australia, Canada, and New Zealand.

While the preservation of nature continues to be an important benefit provided by national parks, worldwide increases in population and the pressures of urban living have raised public interest in setting aside places that provide opportunities for solitude and interaction with nature.

Because national parks have been established by nations with diverse cultural values, land resources, and management philosophies, there is no single definition of what constitutes a national park. In some countries, areas used principally for recreational purposes are designated as national parks; other countries emphasize preservation of outstanding scenic, geologic, or biological resources. The terminology used for national parks also varies among countries. For example, protected areas that are similar to national parks may be called reserves, preserves, or sanctuaries.

Diverse landscapes are protected within national parks, including swamps, river deltas, dune areas, mountains, prairies, tropical rain forests, temperate forests, arid lands, and marine environments. Individual parks within nations form networks that vary with respect to size, accessibility, function, and the type of natural landscapes preserved. Some national park areas are isolated and sparsely populated, such as Greenland National Park; others, such as Peak District National Park in Great Britain, contain numerous small towns and are easily accessible to urban populations.

The functions of national parks include the preservation of scenic landscapes, geological features, wilderness, and plants and animals within their natural habitats. National parks also serve as outdoor laboratories for education and scientific research and as reservoirs for genetic information. Many are components of the United Nations International Biosphere Reserve Program.

National parks also play important roles in preserving cultures, by protecting archaeological, cultural, and historical sites. The United Nations recognizes several national parks that possess important cultural attributes as World Heritage Sites. Tourism to national parks has become important to the economies of many developing nations, especially in Eastern and Southern Africa, India, Nepal, Ecuador, and Indonesia. Parks are sources of local employment and can stimulate improvements to transportation and other types of infrastructure while encouraging productive use of lands that are of marginal agricultural use.

The International Union for Conservation of Nature has developed a system for classifying the world's protected areas, with Category II areas designated as national parks. Using this definition, there are 3,044 national parks in the world, with a mean average size of 457 square miles (1,183 sq. km.) each. Together, they cover an area of about 1.5 million square miles (4 million sq. km.), accounting for about 2.7 percent of the total land area on Earth.

STEPHEN T. MATHER AND THE U.S. NATIONAL PARK SERVICE

In 1914 businessman and conservationist Stephen T. Mather wrote to Secretary of the Interior Franklin K. Lane about the poor condition of California's Yosemite and Sequoia National Parks. Lane wrote back, "if you don't like the way the national parks are being run, come on down to Washington and run them yourself." Mather accepted the challenge and became an assistant to Lane and later the first director of the U.S. National Park Service, from 1917 to 1929.

North America

In 1916 management of U.S. national parks and monuments was shifted from the U.S. Army to the newly established National Park Service (NPS). The system has since grown in size to protect sixty-one national parks, as well as other natural areas including national monuments, seashores, and preserves.

North America's second-largest system of national parks is Parks Canada, created in 1930. Among the best-known Canadian parks is Banff, established in southern Alberta in 1885. Preserved within this area are glacially carved valleys, evergreen forests, and turquoise lakes. Parks Canada has the goal of protecting representative examples of each of Canada's vegetation and physiographic regions.

Mexico began providing protection for natural areas in the late nineteenth century. Among its system of sixty-seven national parks is Dzibilchaltún, an important Mayan archaeological site on the Yucatán Peninsula. With fewer resources available for park management, the emphasis in Mexico remains the preservation of scenic beauty for public use.

South America

Two of South America's best-known national parks are located within Argentina's park system. Nahuel Huapi National Park preserves two rare deer species of the Andes, while Iguazú National Park, located on the border with Brazil, is home to tapir, ocelot, and jaguar.

Located on a plateau of the western slope of the Andes Mountains in Chile, Lauca National Park is one of the world's highest parks, with an average elevation of more than 14,000 feet (4,267 meters)-an altitude nearly as high as the tallest mountains in the continental United States. Huascarán, another mountain park located in western Peru, boasts twenty peaks that exceed 19,000 feet (5,791 meters) in elevation. The volcanic islands of Galapagos Islands National Park, managed by Ecuador, have been of interest to biologists since British naturalist Charles Darwin studied variation and adaptation in animal species there in 1835.

Australia and New Zealand

Established in 1886, Royal was Australia's first national park. Perhaps better known to tourists, Uluru National Park in Australia's Northern Territory protects two rock domes, Ayer's Rock and Mount Olga, that rise above the plains 15 miles (40 km.) apart.

Along with Australia and other former colonies of Great Britain, New Zealand was a leader in establishing early national parks. The first of these was Tongagiro, created in 1887 to protect sacred lands of the Maori people on the North Island. New Zealand's South Island features several national parks including Fiordland, created in 1904 to preserve high mountains, forests, rivers, waterfalls, and other spectacular features of glacial origin.

Africa

Game poaching continues to be a severe problem in Africa, where animals are slaughtered for ivory, meat, and hides. Many African national parks were established to protect large game. South Africa's national park system began in 1926, when the Sabie Game Preserve of the eastern Transvaal region became Kruger National Park. Among South Africa's greatest attractions to foreign visitors, Kruger is famous for its population of lions and elephants.

East Africa is also known for outstanding game sanctuaries, such as Serengeti National Park, created prior to Tanzania's independence from Great Britain. Another national park in Tanzania, Kilimanjaro, protects Africa's highest and best-known mountain. Other African countries with well-developed park systems include Kenya, the Democratic Republic of the Congo (formerly Zaire), and Zambia. Although there is now a network of national parks in Africa that protects a wide range of habitats in various regions, there remains a need to protect additional areas in the arid northern part of the continent that includes the Sahara Desert.

Europe

In comparison with the United States, the national park concept spread more slowly within Europe. In 1910 Germany set aside Luneburger Heide National Park near the Elbe River, and in 1913, Sweden established Sarek, Stora Sjöfallet, Peljekasje, and Abisko National Parks. Swiss National Park was founded in Switzerland in 1914, in the Lower Engadine region. Great Britain has several national parks, including Lake District, a favorite recreation destination for English poet William Wordsworth. Spain's Doñana National Park, located on its southwestern coast, preserves the largest dune area on the European continent.

Asia

The system of land tenure and rural economy in many Asian countries has made it difficult for national governments to set aside large areas free from human exploitation. Many national parks established by colonial powers prior to World War II were maintained or expanded by countries following independence. For example, Kaziranga National Park is a refuge for the largest heard of rhinocerous in India. Established in 1962, Thailand's Khao Yai National Park protects a sample of the country's wildlife, while Indonesia's Komodo Island National Park preserves the habitat for the large lizards known as Komodo dragons.

In Japan, high population density has made it difficult to limit human activities within large areas. Some Japanese national parks are principally recreation areas rather than wildlife sanctuaries and may contain cultural features such as Shinto shrines. One of the best known national parks in Japan is Fuji-Hakone-Izu, which contains world-famous Mount Fuji, a volcano with a nearly symmetrical shape.

The Future

National parks serve as relatively undisturbed enclaves that protect examples of the world's most outstanding natural and cultural resources. The movement to establish these areas is a relatively recent attempt to achieve an improved balance between human activities and the earth. In recent years, rising incomes and lower costs for international travel have improved the accessibility of national parks to a larger number of persons, meaning that park visitation is likely to continue to rise.

Thomas A. Wikle

Boundaries and Time Zones

International Boundaries

International boundaries are the marked or imaginary lines traversing natural terrain of land or water that mark off the territory of one politically organized society—a state or nation-state—from other states. In addition, states claim "air boundaries." While satellites circumnavigate the earth without nations' permission, airplanes and other air vessels that fly much lower must gain the permission of states over whose territory they travel.

The existence of international boundaries is a consequence of the "territoriality" that is a feature of modern human societies. All politically organized societies, except for nomadic tribes, claim to rule some exactly defined geographical territory. International boundaries provide the limits that define this territory.

International boundaries have ancient origins. For example, the oldest sections of the Great Wall of China date back to the Ch'in Dynasty of the second century BCE. The Roman Empire also maintained boundaries to its territories, such as Hadrian's Wall in the north of England, built by the Romans in 122 CE as a defensive barrier against marauders. In these and other ancient instances, however, there was little thought that borders must be exact.

The existence of precisely drawn boundaries among states is relatively recent. The modern state has existed for no more than a few hundred years. In addition, means to determine many boundaries have come into existence only in the nineteenth and twentieth centuries, with the invention of scientific methods and instruments, along with accompanying vocabulary, for determining exact boundaries. The most basic terms of this vocabulary begin with "latitude" and "longitude" and their subdivi-

sions into the "minutes" and "seconds" used in determining boundaries. In modern times, a new attitude toward states' territory was born, especially with the nineteenth century forms of nationalism, which tend to regard every acre of territory as sacred.

Types of Boundaries

There are several types of international boundaries. Some are geographical features, including rivers, lakes, oceans, and seas. Thus boundaries of the United States include the Great Lakes, which border Canada to the north; the Rio Grande, a river that forms part of the U.S. boundary with Mexico to the south; the Atlantic and Pacific Oceans, to the east and west, respectively; and the Gulf of Mexico, to the south. In Africa, Lake Victoria bounds parts of Tanzania, Uganda, and Kenya; and rivers, such as sections of the Congo and the Zambezi, form natural boundaries among many of the continent's states.

Other geographical features, such as mountains, often form international boundaries. The Pyrenees, for example, separate France and Spain and cradle the tiny state of Andorra. In South America, the Andes frequently serve as a boundary, such as between Argentina and Chile. The Himalayas in South Central Asia create a number of borders, such as between India, China, and Tibet and between Nepal, Butan, and their neighbors. When there are no clear geographical barriers between states, boundaries must be decided by mutual consent or the threat of force. In the 2016 presidential campaign, Donald Trump repeatedly called for a wall to be built between the United States and Mexico, claim-

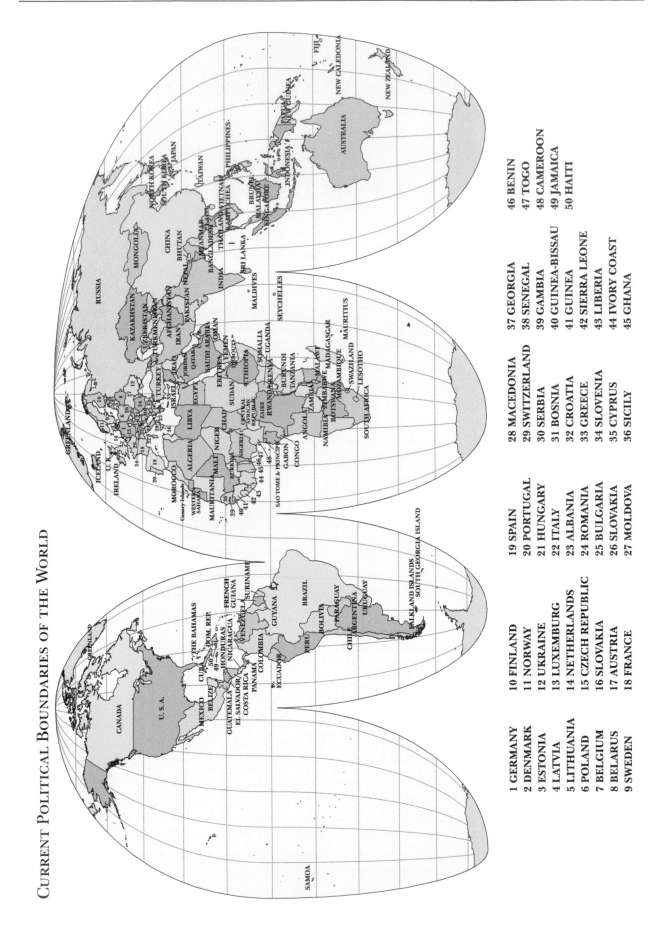

CURRENT POLITICAL BOUNDARIES OF THE WORLD

1 GERMANY
2 DENMARK
3 ESTONIA
4 LATVIA
5 LITHUANIA
6 POLAND
7 BELGIUM
8 BELARUS
9 SWEDEN

10 FINLAND
11 NORWAY
12 UKRAINE
13 LUXEMBURG
14 NETHERLANDS
15 CZECH REPUBLIC
16 SLOVAKIA
17 AUSTRIA
18 FRANCE

19 SPAIN
20 PORTUGAL
21 HUNGARY
22 ITALY
23 ALBANIA
24 ROMANIA
25 BULGARIA
26 SLOVAKIA
27 MOLDOVA

28 MACEDONIA
29 SWITZERLAND
30 SERBIA
31 BOSNIA
32 CROATIA
33 GREECE
34 SLOVENIA
35 CYPRUS
36 SICILY

37 GEORGIA
38 SENEGAL
39 GAMBIA
40 GUINEA-BISSAU
41 GUINEA
42 SIERRA LEONE
43 LIBERIA
44 IVORY COAST
45 GHANA

46 BENIN
47 TOGO
48 CAMEROON
49 JAMAICA
50 HAITI

ing that Mexico would pay for it. As of 2019, the wall has not been completed, however.

Creation and Change of International Boundaries

War and conquest often have been used to determine borders. Such wars, however, historically have created hostility among losers. Political pressures to recover lost lands build up among aggrieved losers, and such irredentist claims provide fuel for future wars. A classic example is the fate of the regions of Alsace and Lorraine between France and Germany. Although natural resources in the form of coal played a substantial role in the dispute over this area, national pride was also a potent element.

Whether boundaries are fixed through compelling geographical imperatives or in their absence, states typically sign treaties agreeing to their location. These may be treaties that conclude wars, or boundary commissions set up by those involved may draw up borders to which states give formal agreement. In 1846, for example, negotiators for Great Britain and the United States settled on the forty-ninth parallel as the boundary between the western United States and Canada, although in the United States, "Fifty-four [degrees latitude] Forty [minutes] or Fight" had been a popular motto in the presidential election campaign of 1844.

Sometimes no accepted borders exist because of chronic hostility between states. Thus, maps of the Kashmir region between India and Pakistan, claimed by both countries, show only a "line of control" or cease-fire line to divide the two warring states. Similarly, only a cease-fire line, drawn at the armistice of the Korean War of 1950-1953, divides North and South Korea; a mutually agreed-upon border remains unfixed.

In rare instances, no true boundary exists to mark where a state's territory begins and ends. Classic cases are found on the Arabian Peninsula, where the land borders of principalities, known as the Gulf Sheikdoms, are vague lines in the sand. Such circumstances usually create no difficulties where nothing is at stake, but when oil is discovered, states must come to agreement or risk coming to blows.

In other instances, negotiations and international arbitration have been effective for determining borders. Perhaps the most important principle for determining the borders of newly created states is found in the Latin phrase, *Uti possidetis iurus.* This principle is used when states become independent after having been colonies or constituent parts of a larger state that has broken up. The principle holds that states shall respect the borders in place when they were colonies. *Uti possidetis* was first extensively used in South America in the nineteenth century, when European colonial powers withdrew, leaving several newly born states to determine their own boundaries. The principle may be used as a basis for border agreements among the fifteen states of the former Soviet Union.

Besides war and negotiation, purchase has sometimes been a means of creating international boundaries. For example, in 1853 the United States purchased territory from Mexico in the southwest; in 1867, it purchased Alaska from Russia.

In rare cases, natural boundaries may change naturally or be changed deliberately by one side, incurring resentment among victims. An example occurred in 1997, when Vietnam complained that China had built an embankment on a border river embankment that caused the river to change its course; China countered that Vietnam had built a dam altering the river's course.

Other border difficulties among states include conflicts over water that flows from one country to another. In the 1990s, for example, Mexico complained of excessive U.S. use of Colorado River waters and demanded adjustment.

Border Disputes

Border disputes among states in the past two centuries have been numerous and lethal. In the twentieth century, numerous such controversies degenerated into violence. In Asia, India and Pakistan fought over Kashmir, beginning in 1947-1949 and recurring in 1965 and 1999. China has been involved in violent border disputes with India, especially in 1962; Vietnam in 1979; and Russia in 1969. In South America, border wars between Ecuador and Peru broke out in 1941, 1981, and 1995. This dispute was settled by negotiation in 1998. In Africa, among numerous recent armed conflicts, the bloody border conflict between Eritrea and Ethiopia in the 1990s was notable.

Other recent disputes have ended peacefully. Eritrea avoided violence with Yemen over several Red Sea islands by accepting arbitration by an international tribunal. In 1995 Saudi Arabia and the United Arab Emirates negotiated a peaceful agreement to their border dispute involving oil rights.

As of 2019, there are four ongoing border conflicts: Israeli-Syrian ceasefire line incidents during the Syrian Civil War, the War in Donbass, India-Pakistan military confrontation, and the 2019 Turkish offensive into north-eastern Syria, code-named by Turkey as Operation Peace Spring. Many unresolved boundary disputes might yet lead to conflicts. Among the most complex is the multinational dispute over the 600 tiny Spratley Islands in the South China Sea. Uninhabited but potentially valuable because of oil, the Spratleys are claimed by China, Brunei, Malaysia, Indonesia, the Philippines, Taiwan, and Vietnam.

Border Policies

Problems with international borders are not limited to territorial disputes. Policies regarding how borders should be operated—including the key questions of who and what should be allowed entrance and exit under what conditions—can be expected to continue as long as independent states exist. While the members of the European Union have agreed to allow free passage of people and goods among themselves, this policy does not extent to nonmembers.

The most important purpose of states is to protect the lives and property of their citizens. One of the principal purposes of international boundaries is to further this purpose. Most states insist on controlling their borders, although borders seem increasingly porous. Given the imperatives of control and the increasing difficulties of maintaining it, issues surrounding international borders are expected to continue indefinitely in the twenty-first century.

Charles F. Bahmueller

GLOBAL TIME AND TIME ZONES

Before the nineteenth century, people kept time by local reckoning of the position of the Sun; consequently, thousands of local times existed. In medieval Europe, "hours" varied in length, depending upon the seasons: Each hour was determined by the Roman Catholic Church. In the sixteenth century, Holy Roman emperor Charles V was the first secular ruler to decree hours to be of equal length. As the industrial and scientific revolutions swept Europe, North America, and other areas, some form of time standardization became necessary as communities and regions increasingly interacted. In 1780 Geneva, Switzerland, was the first locality known to employ a standard time, set by the town-hall clockkeeper, throughout the town and its immediate vicinity.

The growth and expansion of railroads, providing the first relatively fast movement of people and goods from city to city, underscored the need for a standard system in Great Britain. As early as 1828, Sir John Herschel, Astronomer Royal, called for a national standard time system based on instruments at the Royal Observatory at Greenwich. That practice began in 1852, when the British telegraph system had developed sufficiently for the Greenwich time signals to be sent instantly to any point in the country.

As railroads expanded through North America, they exposed a problem of local time variation similar to that in Great Britain but on a far larger scale, since the distances between the East and West Coasts were much greater than in Great Britain. In order for long-distance train schedules to work, different parts of the country had to coordinate their clocks. The first to suggest a standard time framework for the United States was Charles F. Dowd, president of Temple Grove Seminary for Women in Saratoga Springs, New York. Initially, Dowd proposed putting all U.S. railroads on a single standard time, based on the time in Washington, D.C. When he realized that the time in California would be behind such a standard by almost four hours, he produced a revised system, establishing four time zones in the United States. Dowd's plan, published in 1870, included the first known map of a time zone system for the country.

Not everyone was happy with the designation of Washington, D.C., as the administrative center of time in the United States. Northeastern railroad executives urged that New York, the commercial capital of the nation, be used instead: Many cities and towns in the region already had standardized to New York time out of practical necessity. Dowd proposed a compromise: to set the entire national time zone system in the United States using the Greenwich prime meridian, already in use in many parts of the world for maritime and scientific purposes. In 1873 the American Association of Railways (AAR) flatly rejected the proposal.

In the end, Dowd proved to be a visionary. In 1878 Sandford Fleming, chief engineer of the government of Canada, proposed a worldwide system of twenty-four time zones, each fifteen degrees of longitude in width, and each bisected by a meridian, beginning with the prime meridian of Greenwich. William F. Allen, general secretary of the AAR and armed with a deep knowledge of railroad practices and politics, took up the crusade and persuaded the railroads to agree to a system. At noon on Sunday, November 18, 1883, most of the more than six hundred U.S. railroad lines dropped the fifty-three arbitrary times they had been using and adopted Greenwich-indexed meridians that defined the times in each of four times zones: eastern,

central, mountain, and Pacific. Most major cities in the United States and Canada followed suit.

Time System for the World
Almost at the same time that American railroads adopted a standard time zone system, the State Department, authorized by the United States Congress, invited governments from around the world to assemble delegates in Washington, D.C., to adopt a global system. The International Meridian Conference assembled in the autumn of 1884, attended by representatives of twenty-five countries. Led by Great Britain and the United States, most favored adoption of Greenwich as the official prime meridian and Greenwich mean time as universal time.

There were other contenders: The French wanted the prime meridian to be set in Paris, and the Germans wanted it in Berlin; others proposed a mountaintop in the Azores or the tip of the Great Pyramid in Egypt. Greenwich won handily. The conference also agreed officially to start the universal day at midnight, rather than at noon or at sunrise, as practiced in many parts of the world. Each time zone in the world eventually came to have a local name, although technically, each goes by a letter in the alphabet in order eastward from Greenwich.

Once a global system was in place, there was a new issue: Many jurisdictions wanted to adjust their clocks for part of the year to account for differences in the number of hours of daylight between summer and winter months. In 1918 Congress decreed a system of daylight saving time for the United States but almost immediately abolished it, leaving state governments and communities to their local options. Daylight saving time, or a form of it, returned in the United States and many Allied nations during World War II. In the Uniform Time Act of 1966, Congress finally established a national system of daylight saving time, although with an option for states to abstain.

To the extent that it indicates how human communities want to manipulate time for social, political, or economic reasons, the issue of daylight saving time, rather than the establishment of a system of world time zones, is a better clue to the geographical issues involved in time administration. Both the history and the present format of the world time zone system show that the mathematically precise arrangement envisioned by many of the pioneers of time zones is not as important as things on the ground.

In the United States, the railroad time system adopted in 1883 drew the boundary between eastern time and central time more or less between the thirteen original states and the trans-Appalachian West: The entire Midwest, including Ohio, Indiana, and Michigan, fell in the central time zone. As the center of population migrated westward, train speeds increased, highways developed, and New York emerged as the center of mass media in the United States, the boundary between the eastern and central time zones marched steadily westward. In 1918 it ran down the middle of Ohio; by the 1960s, it was at the outskirts of Chicago.

One of the principal reasons for the popularity of Greenwich as the site of the prime meridian (zero degrees longitude), is that it places the international date line (180 degrees longitude)—where, in effect, time has to move forward to the next day rather than the next hour—far out in the Pacific Ocean where few people are affected by what otherwise would be an awkward arrangement. However, even this line is somewhat irregular, to avoid placing a small section of eastern Russia and some of the Aleutian Islands of the United States in different days.

Coordinated Universal Time Coordinated Universal Time (or UTC) is the primary time standard by which the world regulates clocks and time. It is within about 1 second of mean solar time at 0° longitude, and is not adjusted for daylight saving time. In some countries, the term Greenwich Mean Time is used. The co-ordination of time and frequency

transmissions around the world began on January 1, 1960. UTC was first officially adopted as CCIR Recommendation 374, Standard-Frequency and Time-Signal Emissions, in 1963, but the official abbreviation of UTC and the official English name of Coordinated Universal Time (along with the French equivalent) were not adopted until 1967. UTC uses a *slightly* different second called the *SI second*. That is based on *atomic clocks*. Atomic clocks are more regular than the slightly variable Earth's rotation period. Hence, the essential difference between GMT and UTC is that they use different definitions of exactly how long one second of time is.

By 1950 most nations had adopted the universal time zone system, although a few followed later: Saudi Arabia in 1962, Liberia in 1972. Despite adhering to the system in principle, many nations take considerable liberties with the zones, especially if their territory spans several. All of Western Europe, despite covering an area equivalent to two zones, remains on a single standard. The People's Republic of China, which stretches across five different time zones, arbitrarily sets the entire country officially on Beijing time, eight hours behind Greenwich. Iran, Afghanistan, India, and Myanmar, each of which straddle time zone boundaries, operate on half-hour compromise systems as their time standards (as does Newfoundland). As late as 1978, Guyana's standard time was three hours, forty-five minutes in advance of Greenwich.

It can be argued that adoption of a worldwide system of time zones in the late nineteenth century was one of the earliest manifestations of the emergence of a global economy and society, and has been a crucial factor in the unfolding of this process throughout the twentieth century and beyond.

Ronald W. Davis

GLOBAL EDUCATION

THEMES AND STANDARDS IN GEOGRAPHY EDUCATION

Many people believe that the study of geography consists of little more than knowing the locations of places. Indeed, in the past, whole generations of students grew up memorizing states, capitals, rivers, seas, mountains, and countries. Most students found that approach boring and irrelevant. During the 1990s, however, geography education in the United States underwent a remarkable transformation.

While it remains important to know the locations of places, geography educators know that place name recognition is just the beginning of geographic understanding. Geography classes now place greater emphasis on understanding the characteristics of and the connections between places. Three things have led to the renewal of geography education: the five themes of geography, the national geography standards, and the establishment of a network of geographic alliances.

The Five Themes of Geography

One of the first efforts to move geography education beyond simple memorization was the National Geographic Society's publication of five themes of geography in 1984: location, place, human-environment interactions, movement, and regions. Not intended to be a checklist or recipe for understanding the world, these themes merely provided a framework for teachers—many of whom did not have a background in the subject—to incorporate geography throughout a social studies curriculum. The five themes were promoted widely by the National Geographic Society and are still used by some teachers to organize their classes.

Location is about knowing where things are. Both the absolute location (where a place is on earth's surface) and relative location (the connections between places) are important. The concept of place involves the physical and human characteristics that distinguish one place from another. The theme of human/environment interaction recognizes that people have relationships within defined places and are influenced by their surroundings. For example, many different types of housing have been created as adaptations to the world's diverse climates. The theme of movement involves the flow of people, goods, and ideas around the world. Finally, regions are human creations to help organize and understand Earth, and geography studies how they form and change.

The National Geography Standards

Geography was one of six subjects identified by President George H. W. Bush and the governors of the U.S. states when they formulated the National Education Goals in 1989. While the goals themselves foundered amid the political debate that followed their adoption, one tangible result of the initiative was the creation of Geography for Life: The National Geography Standards. More than 1,000 teachers, professors, business people, and government officials were involved in the writing of Geography for Life. The project wassupported by four geography organizations: the American Geographical Society, the Association of American Geographers, the National Council for Geographic Education, and the National Geographic Society. The resulting book defines what every U.S. student

GEOGRAPHY STANDARDS

The geographically informed person knows and understands the following:
- how to use maps and other geographic representations, tools, and technologies to acquire, process, and report information from a spatial perspective;
- how to use mental maps to organize information about people, places, and environments in a spatial context;
- how to analyze the spatial organization of people, places, and environments on Earth's surface;
- the physical and human characteristics of places;
- that people create regions to interpret Earth's complexity;
- how culture and experience influence people's perceptions of places and regions;
- the physical processes that shape the patterns of Earth's surface;
- the characteristics and spatial distribution of ecosystems on Earth's surface;
- the characteristics, distribution, and migration of human populations on Earth's surface;
- the characteristics, distribution, and complexity of Earth's cultural mosaics;
- the patterns and networks of economic interdependence on Earth's surface;
- the processes, patterns, and functions of human settlement;
- how the forces of cooperation and conflict among people influence the division and control of Earth's surface;
- how human actions modify the physical environment;
- how physical systems affect human systems;
- the changes that occur in the meaning, use, distribution, and importance of resources;
- how to apply geography to interpret the past;
- how to apply geography to interpret the present and plan for the future.

Source: National Geography Standards Project. Geography for Life: National Geography Standards, Second Edition. Washington, D.C.: National Geographics Research and Exploration, 2012.

should know and be able to accomplish in geography.

Each of the eighteen standards is designed to develop students' geographic skills, including asking geographic questions; acquiring, organizing, and analyzing geographic information; and answering the questions. Each standard features explanations, examples, and specific requirements for students in grades four, eight, and twelve.

Geography Alliances and the Future of Geography Education

To publicize efforts in geography education, a network of geography alliances was established between 1986 and 1993. Today, each U.S. state has a geography alliance that links teachers and organizations such as the National Geographic Society and the National Council for Geographic Education to sponsor workshops, teacher training sessions, field experiences, and other ways of sharing the best in geographic teaching and learning.

A 2013 executive summary prepared by the National Geographic Society for the *Road Map for 21st Century Geography Education Project* continues to champion the goal of better geography education in K–12 schools. The Road Map Project represents the collaborative effort of four national organizations: the American Geographical Society (AGS), the Association of American Geographers (AAG), the National Council for Geographic Education (NCGE), and the National Geographic Society (NGS). The project partners share belief that geography education is essential for student success in all aspects of their adult lives—careers, civic lives, and personal decision making. It also is essential for the education of specialists who can help society addressing critical issues in the areas of social welfare, economic stability, environmental health, and international relations.

Eric J. Fournier

GLOBAL DATA

THE WORLD GAZETTEER OF OCEANS AND CONTINENTS

Places whose names are printed in SMALL CAPS *are subjects of their own entries in this gazetteer.*

Aden, Gulf of. Deep-water area between the RED and ARABIAN SEAS, bounded by Somalia, Africa, on the south and Yemen on the north. Water is warmer and saltier in the Gulf of Aden than in the Red and Arabian Seas, because little water enters from rain or land runoff.

Africa. Second-largest continent, connected to ASIA by the narrow isthmus of Suez. Bounded on the east by the INDIAN OCEAN and on the west by the ATLANTIC OCEAN. Countries of Africa are Algeria, Angola, Benin, Botswana, Burkina Faso, Burundi, Cameroon, Central African Republic, Chad, Congo, Côte d'Ivoire (Ivory Coast), the Democratic Republic of Congo, Egypt, Ethio-

pia, Gabon, Gambia, Ghana, Guinea, Kenya, Liberia, Libya, Madagascar, Malawi, Mali, Mauritania, Morocco, Mozambique, Namibia, Niger, Nigeria, Rio Muni (Mbini), Rwanda, Senegal, Sierra Leone, Somalia, South Africa, Sudan, Tanzania, Togo, Tunisia, Uganda, Western Sahara, Zambia, and Zimbabwe. Climate ranges from hot and rainy near the equator, to hot and dry in the huge Sahara Desert in the north and the Kalahari Desert in the south, to warm and fairly mild at the northern and southern extremes. Paleontological evidence indicates that humans originally evolved in Africa.

Agulhas Current. Warm, swift ocean current moving south along East AFRICA's coast. Part moves between AFRICA and MADAGASCAR to form the Mozambique Current. The warm water of the

OCEANS AND CONTINENTS

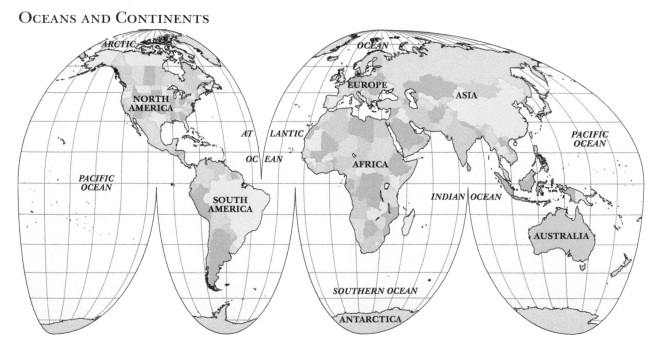

Agulhas Current increases the average temperatures in the eastern part of South Africa.

Agulhas Plateau. Relatively small ocean-bottom plateau that lies south of South AFRICA, at the area where the INDIAN and ATLANTIC OCEANS meet.

Aleutian Islands. Chain of volcanic islands that extends 1,100 miles (1,770 km.) from the tip of the Alaska Peninsula to the Kamchatka Peninsula in Russia and forms the boundary between the North PACIFIC OCEAN and the BERING SEA. The area is hazardous to navigation and has been called the "Home of Storms."

Aleutian Trench. Located on the northern margin of the PACIFIC OCEAN, stretching 3,666 miles (5,900 km.) from the western edge of the Aleutian Island chain to Prince William Sound, Alaska. Depth is 25,263 feet (7,700 meters).

American Highlands. Elevated region on the ANTARCTIC coast between Enderby Land and Wilkes Land, located far south of India. The Lambert and Fisher glaciers originate in the American Highlands and move down to feed the AMERY ICE SHELF.

Amery Ice Shelf. Year-round shelf of relatively flat ice in a bay of ANTARCTICA, located at approximately longitude 70 degrees east, between MAC. ROBERTSON LAND and the AMERICAN HIGHLANDS. The ice shelf is fed by the Lambert and Fisher glaciers.

Amundsen Sea. Portion of the southernmost PACIFIC OCEAN off the Wahlgreen Coast of ANTARCTICA, approximately longitude 100 to 120 degrees west. Named for the Norwegian explorer Roald Amundsen, who became the first person to reach the SOUTH POLE in 1911.

Antarctic Circle. Latitude of 66.3 degrees south. South of this line, the Sun does not set on the day of the summer solstice, about December 22 in the SOUTHERN HEMISPHERE, and does not rise on the day of the winter solstice, about June 21.

Antarctic Circumpolar Current. Eastward-flowing current that circles ANTARCTICA and extends from the surface to the deep ocean floor. The largest-volume current in the oceans. Extends northward to approximately 40 degrees south latitude and is driven by westerly winds.

Antarctic Convergence. Meeting place where cold Antarctic water sinks below the warmer sub-Antarctic water.

Antarctic Ocean. See SOUTHERN OCEAN.

Antarctica. Fifth-largest continent, located at the southernmost part of the world. There are two major regions; western Antarctica, which includes the mountainous Antarctic peninsula, and eastern Antarctica, which is mostly a low continental shield area. An ice cap up to 13,000 feet (4,000 meters) thick covers 95 percent of the continent's surface. Temperatures in the austral summer (December, January, and February) rarely rise above 0 degrees Fahrenheit (-18 degrees Celsius) except on the peninsula. By international treaty, the continent is not owned by any single country, and human access is largely regulated. There has never been a self-supporting human habitation on Antarctica.

Arabian Sea. Portion of the INDIAN OCEAN bounded by India on the east, Pakistan on the north, and Oman and Yemen of the Arabian Peninsula on the west.

Arctic Circle. Latitude of 66.3 degrees north. North of this line, the Sun does not set on the day of the summer solstice, about June 21 in the NORTHERN HEMISPHERE, and does not rise on the day of the winter solstice, about December 22.

Arctic Ocean. World's smallest ocean. It centers on the geographic NORTH POLE and connects to the PACIFIC OCEAN through the BERING SEA, and to the ATLANTIC OCEAN through the GREENLAND SEA. The Arctic Ocean is covered with ice up to 13 feet (4 meters) thick all year, except at its edges. Norwegian explorers on the ship *Fram* stayed locked in the icepack from 1893 to 1896, in order to study the movement of polar ice. They drifted in the ice a total of 1,028 miles (1,658 km.), from the Bering Sea to the Greenland Sea, proving that there was no land mass under the Arctic ice at the top of the world. Also known as Arctic Sea or Arctic Mediterranean Sea.

Argentine Basin. Basin on the floor of the western ATLANTIC OCEAN, off the coast of Argentina in SOUTH AMERICA. Among ocean basins, this one is unusually circular.

Ascension Island. Isolated volcanic island in the South ATLANTIC OCEAN, about midway between SOUTH AMERICA and AFRICA. One of the islands visited by British biologist Charles Darwin during his five-year voyage on the *Beagle*.

Asia. Largest continent; joins with EUROPE to form the great Eurasian landmass. Asia is bounded by the ARCTIC OCEAN on the north, the western PACIFIC OCEAN on the east, and the INDIAN OCEAN on the south. Its countries include Afghanistan, Bahrain, Bangladesh, Bhutan, Cambodia, China, India, Iran, Iraq, Irian Jaya, Israel, Japan, Jordan, Kalimantan, Kazakhstan, North and South Korea, Kyrgyzstan, Laos, Lebanon, Malaysia, Myanmar, Mongolia, Nepal, Oman, Pakistan, the Philippines, Russia, Sarawak, Saudi Arabia, Sri Lanka, Sumatra, Syria, Tajikistan, Thailand, Asian Turkey, Turkmenistan, United Arab Emirates, Uzbekistan, Vietnam, and Yemen. Climates include virtually all types on earth, from arctic to tropical, desert to rain forest. Asia has the highest (Mount Everest) and lowest (Dead Sea) surface points in the world. Nearly 60 percent of the world's people live in Asia.

Atlantic Ocean. Second-largest body of water in the world, covering more than 25 percent of Earth's surface. Bordered by NORTH and SOUTH AMERICA on the west, and EUROPE and East AFRICA on the east. The widest part (5,500 miles/8,800 km.) lies between West AFRICA and Mexico, along 20 degrees latitude. Scientists disagree on the north-south boundaries of the Atlantic; if one includes the ARCTIC OCEAN and the SOUTHERN OCEAN, the Atlantic Ocean extends about 13,300 miles (21,400 km.). The deepest spot (28,374 feet/8,648 meters) is found in the PUERTO RICO TRENCH. The Atlantic Ocean has been a major route for trade and communications, especially between North America and

Europe, for hundreds of years. This is because of its relatively narrow size and favorable currents, such as the GULF STREAM.

Australasia. Loosely defined term for the region, which, at the least, includes AUSTRALIA and New Zealand; at the most, it also includes other South Pacific Islands in the region.

Australia. Smallest continent, sometimes called the "island continent." Located between the INDIAN and PACIFIC OCEANS. It is the only continent occupied by a single nation, the Commonwealth of Australia. Australia is the flattest and driest continent; two-thirds is either desert or semiarid. Geologically, it is the oldest and most isolated continent. Unlike any other place on Earth, large mammals never evolved in Australia. Marsupials (pouched, warm-blooded animals) and unusual birds developed in their place.

Azores. Archipelago (group of islands) in the eastern ATLANTIC OCEAN lying about 994 miles (1,600 km.) west of Portugal. The islands are of volcanic origin and have been known, fought over, and used by the Europeans since before the fourteenth century. Spanish explorer Christopher Columbus stopped in the Azores to wait for favorable winds before his first trip across the ATLANTIC OCEAN.

Barents Sea. Partially enclosed section of the ARCTIC OCEAN, bounded by Russia and Norway on the south and the Russian island of Navaya Zemlaya on the east. The Barents Sea was important in World War II because Allied convoys had to cross it, through storms and submarine patrols, to deliver war supplies to Murmansk, the only ice-free port in western Russia. It was named for the Dutch explorer Willem Barents.

Bays. See under individual names.

Beaufort Sea. Area of the ARCTIC OCEAN located off the northern coast of Alaska and western Canada. It is usually frozen over and has no islands. Named for British admiral Sir Francis Beaufort, who devised the Beaufort Wind Scale as a means of classifying wind force at sea.

Bengal, Bay of. Northeast arm of the INDIAN OCEAN, bounded by India on the west and Myanmar on the east. The Ganges River empties into the Bay of Bengal. The great ports of Calcutta and Madras in India, and Rangoon in Myanmar lie in the bay, making it a busy and important area for shipping for centuries.

Benguela Current. Northward-flowing current along the western coast of Southern AFRICA. Normally, the Benguela Current carries cold, rich water that wells up from the ocean depths and supports a large fishing industry. A change in winds can reduce the oxygen supply and kill huge numbers of fish, similar to what may happen off the coast of Peru during El Niño weather conditions.

Bering Sea. Portion of the northernmost PACIFIC OCEAN that is bounded by the state of Alaska on the east, Russia and the Kamchatka Peninsula on the west, and the BERING STRAIT on the north. It is a valuable fishing ground, rich in shrimp, crabs, and fish. Whales, fur seals, sea otters, and walrus are also found there.

Bering Strait. Narrowest point of connection between the BERING SEA and the ARCTIC OCEAN, located between the easternmost point of Siberia on the west and Alaska on the east. The Bering Strait is 52 miles (84 km.) wide. During the Ice Age, when the sea level was lower, humans and animals were able to walk from the Asian continent across a land bridge—now known as Beringia—to the North American continent across the frozen strait, providing the first human access to the Americas.

Bikini Atoll. Small atoll in the Marshall Islands group in the western PACIFIC OCEAN. In the 1940s, the United States began testing nuclear bombs on Bikini and neighboring atolls. The U.S. Army removed the inhabitants of Bikini, and testing occurred from 1946 to 1958. The Bikini inhabitants were allowed to return in 1969, then removed again in 1978 when high levels of radioactivity were found to remain.

Black Sea. Large inland sea situated where southeastern EUROPE meets ASIA; connected to the MEDITERRANEAN SEA through Turkey's Bosporus strait. The sea covers an area of about 178,000 square miles (461,000 sq. km.), with a maximum depth of more than 7,250 feet (2,210 meters).

Brazil Current. Extension of part of the warm, westward-flowing South EQUATORIAL CURRENT, which turns south to the coast of Brazil. The Brazil Current has very salty water because of its long flow across the equator. It joins the WEST WIND DRIFT and moves eastward across the South ATLANTIC OCEAN as part of the SOUTH ATLANTIC GYRE.

California, Gulf of. Branch of the eastern PACIFIC OCEAN that separates Baja California from mainland Mexico. Warm, nutrient-rich water supports a variety of fish, oysters, and sponges. California gray whales migrate to the gulf to give birth and breed, January through March. Fisheries and tourism are important industries in the Gulf of California. Also known as the Sea of Cortés.

California Current. Cool water that flows southeast along the western coast of NORTH AMERICA from Washington State to Baja California. The eastern portion of the NORTH PACIFIC GYRE.

Canada Basin. Part of the ocean floor that lies north of northeastern Canada and Alaska. The BEAUFORT SEA lies above the Canada Basin.

Cape Horn. Southernmost tip of SOUTH AMERICA. It is the site of notoriously severe storms and is hazardous to shipping.

Cape Verde Plateau. ATLANTIC OCEAN plateau lying off the western bulge of the AFRICAN continent. The volcanic Cape Verde Islands lie on the plateau.

Caribbean Sea. Portion of the western ATLANTIC OCEAN bounded by CENTRAL and SOUTH AMERICA to the west and south, and the islands of the Antilles chain on the north and east. Mostly tropical in climate, the Caribbean Sea supports a large variety of plant and animal life. Its islands, including Puerto Rico, the Cayman Islands, and the Virgin Islands, are popular tourist sites.

Caspian Sea. World's largest inland sea. Located east of the Caucasus Mountains at EUROPE's southeasternmost extremity, it dominates the expanses of western Central ASIA. Its basin is 750 miles (1,200 kilometers) long, and its average width is 200 miles (320 kilometers). It covers 149,200 square miles (386,400 sq. km.).

Central America. Region generally understood to constitute the irregularly shaped neck of land linking North and South America, containing Belize, Guatemala, Honduras, El Salvador, Nicaragua, Costa Rica, and Panama.

Chukchi Sea. Portion of the Arctic Ocean, bounded by the Bering Strait on the south, Siberia on the southwest, and Alaska on the southeast. The Chukchi Sea is the area of exchange between waters and sea life of the Pacific and Arctic Oceans, and so is an area of interest to oceanographers and fishermen.

Clarion Fracture Zone. East-west-running fracture zone that begins off the west coast of Mexico and extends approximately 2.500 miles (4,023 km.) to the southwest.

Cocos Basin. Relatively small ocean basin located off the west coast of Sumatra in the northeast Indian Ocean.

Coral Sea. Area of the Pacific Ocean off the northeast coast of Australia, between Australia on the southwest, Papua New Guinea and the Solomon Islands on the northeast, and New Caledonia on the east. Site of a naval battle in 1942 that prevented the Japanese invasion of Australia.

Cortés, Sea of. See California, Gulf of.

Denmark Strait. Channel that separates Greenland and Iceland and connects the North Atlantic Ocean with the Arctic Ocean.

Dover, Strait of. Body of water between England and the European continent, separating the North Sea from the English Channel. It is 33 miles (53 km.) wide at its narrowest point. The tunnel between England and France (known as the "Chunnel") was cut into the rock under the Strait of Dover.

Drake Passage. Narrow part of the Southern Ocean that connects the Atlantic and Pacific Oceans between the southern tip of South America and the Antarctic peninsula. Named for sixteenth-century English navigator and explorer Sir Francis Drake, who discovered the passage when his ship was blown into it during a violent storm. Also called Drake Strait.

East China Sea. Area of the western Pacific Ocean bounded by China on the west, the Yellow Sea on the north, and Japan on the northeast. Large oil deposits were found under the East China Sea floor in the 1980s.

East Pacific Rise. Broad, nearly continuous undersea mountain range that extends from the southern end of Baja California southward, then curves east near Antarctica. It is formed along the southeast side of the Pacific Plate and is part of the Ring of Fire, a nearly continuous ring of volcanic and tectonic activity around the rim of the Pacific Ocean. Also called East Pacific Ridge.

East Siberian Sea. Portion of the Arctic Ocean bounded by the Chukchi Sea on the east, Siberia on the south, and the Laptev Sea on the west. Much of the East Siberian Sea is covered with ice year-round.

Eastern Hemisphere. The half of the earth containing Europe, Asia, and Africa; generally understood to fall between longitudes 20 degrees west and 160 degrees east.

El Niño. Conditions—also known as El Niño-Southern Oscillation (ENSO) events—that occur every two to ten years and cause weather and ocean temperature changes off the coast of Ecuador and Peru. Most of the time, the Peru Current causes cold, nutrient-rich water to well up off the coast of Ecuador and Peru. During ENSO years, the cold upwelling is replaced by warmer surface water that does not support plankton and fish. Fisheries decline and seabirds starve. Climatic changes of El Niño can bring floods to normally dry areas and drought to wet areas. Effects can extend across North and South America, and to the western Pacific Ocean. During the 1990s, the ENSO event fluctuated but did not go completely away, which caused tremendous damage to fisheries and agriculture, storms and droughts in North America, and numerous hurricanes.

Emperor Seamount Chain. Largest known example of submerged underwater volcanic ridges, located in the northern Pacific Ocean and extending southward from the Kamchatka Peninsula in Russia for about 2,500 miles (4,023 km.).

Enderby Land. Section of ANTARCTICA that lies between the INDIAN OCEAN and the South Polar Plateau, east of QUEEN MAUD LAND. Enderby Land lies between approximately longitude 45 and 60 degrees east.

English Channel. Strait water separating continental France from Great Britain. Runs for roughly 350 miles (560 km.), from the ATLANTIC OCEAN in the west to the Strait of Dover in the east.

Equatorial Current. Currents just north and south of the equator that flow from east to west. Equatorial currents are found in the PACIFIC and ATLANTIC OCEANS. The equatorial currents and the trade winds, which move in the same direction, greatly aid oceangoing traffic.

Eurasia. Term for the combined landmass of EUROPE and ASIA.

Europe. Sixth-largest continent, actually a large peninsula of the Eurasian landmass. Europe is densely populated and includes the countries of Albania, Andorra, Austria, Belarus, Belgium, Bulgaria, Bosnia-Herzegovina, Croatia, the Czech Republic, Denmark, Estonia, Finland, France, Germany, Greece, Hungary, Iceland, Ireland, Italy, Latvia, Lithuania, Macedonia, Malta, Moldova, Monaco, the Netherlands, Norway, Poland, Portugal, Romania, Slovakia, Spain, Switzerland, Turkey, and the United Kingdom (England, Northern Ireland, Scotland, and Wales). Climate ranges from near arctic in the north, to temperate and Mediterranean in the south.

Florida Current. Water moving northward along the east coast of Florida to Cape Hatteras, North Carolina, where it joins the GULF STREAM.

Fundy, Bay of. Large inlet on the North American Atlantic coast, northwest of Maine, separating New Brunswick and Nova Scotia in Canada. Renowned for having the largest tidal change in the world, more than 56 feet (17 meters).

Galápagos Islands. Located directly on the equator, 600 miles (965 km.) west of Ecuador. The islands are volcanic in origin and sit directly in the cold PERU CURRENT, which cools the islands and creates unusual microclimates and fogs. The extreme isolation of the islands allowed unique species to develop. Biologist Charles Darwin visited the Galápagos in the 1830s, and the unusual organisms he observed helped him to conceive the theory of evolution.

Galápagos Rift. Divergent plate boundary extending between the GALÁPAGOS ISLANDS and SOUTH AMERICA. The first hydrothermal vent community was discovered in 1977 in the Galápagos Rift. This unusual type of biological habitat is based on energy from bacteria that use heat and chemicals to make food, instead of sunlight.

Grand Banks. Portion of the northwest ATLANTIC OCEAN southeast of Nova Scotia and Newfoundland. The Grand Banks are extremely rich fishing grounds, although in the 1980s and 1990s catches of cod, flounder, and many other fish dropped dramatically due to overfishing and pollution.

Great Barrier Reef. Largest coral reef in the world, lying in the CORAL SEA off the east coast of AUSTRALIA. The reef system and its small islands stretch for more than 1,100 miles (1,750 km.) and is difficult to navigate through. The reefs are home to an incredible variety of tropical marine life, including large numbers of sharks.

Greenland. Largest island in the world that is not rated as a continent; lies between the northernmost part of the ATLANTIC OCEAN and the ARCTIC OCEAN, northeast of the North American continent. About 90 percent of Greenland is permanently covered with an ice sheet and glaciers. Residents engage in limited agriculture, growing potatoes, turnips, and cabbages. Most people live along the southwest coast, where the climate is warmed by the NORTH ATLANTIC CURRENT.

Greenland Sea. Body of water bounded by GREENLAND on the west, ICELAND on the north, and Spitsbergen on the east. It is often ice-covered.

Guinea, Gulf of. Arm of the North ATLANTIC OCEAN below the great bulge of West AFRICA.

Gulf Stream. Current of westward-moving warm water originating along the equator in the ATLANTIC OCEAN. The mass of water moves along the east coast of Florida as the Florida Current, then turns in a northeasterly direction off North Carolina to become the Gulf Stream. The Gulf Stream flows northeast past Newfoundland and the western edge of the British Isles. The warmer water of the Gulf Stream moderates the climate of northwestern Europe, causing temperatures in winter to be several degrees warmer than in areas of North America at the same latitudes. The Gulf Stream decreases the time required for ships to travel from North America to Europe. This was an important factor in trade and communication in American Colonial times and has continued to be significant.

Gulfs. See under individual names.

Hatteras Abyssal Plain. Part of the floor of the northwest Atlantic Ocean Basin, east of North Carolina. It rises to form shallow sandbars around Cape Hatteras, which are a notorious navigational hazard. In the seventeenth and eighteenth centuries, so many ships were lost in the area that Cape Hatteras became known as "The Graveyard of the Atlantic."

horse latitudes. Latitude belts between 30 and 35 degrees north and south latitude, where winds are usually light and variable and the climate mostly hot and dry.

Humboldt Current. See Peru Current

Iceland. Island country bounded by the Greenland Sea on the north, the Norwegian Sea on the east, and the Atlantic Ocean on the south and west. Total area of 39,768 square miles (103,000 sq. km.). The nearest land mass is Greenland, 200 miles (320 km.) to the northwest. Situated on top of the northern part of the Atlantic Mid-Oceanic Ridge, it is characterized by major volcanic activities, geothermal springs, and glaciers.

Idzu-Bonin Trench. Ocean trench in the western Pacific Ocean, about 6,082 miles (9,810 km.) long and 2,624 feet (800 meters) deep.

Indian Ocean. Third-largest of the world's oceans, bounded by the continents of Africa to the west, Asia to the north, Australia to the east, and Antarctica to the south. Most of the Indian Ocean lies below the equator. It has an approximate area of 33 million square miles (76 million sq. km.) and an average depth of about 13,120 feet (4,000 meters). Its deepest point is 24,442 feet (7,450 meters), in the Java Trench. The Indian Ocean was the first major ocean to be used as a trade route, particularly by the Egyptians. About 600 BCE, the Egyptian ruler Necho sent an expedition into the Indian Ocean, and the ship circumnavigated Africa, probably the first time this feat was accomplished. Warm winds blowing over the northern part of the Indian Ocean from May to September pick up huge amounts of moisture, which falls on India and Sri Lanka as monsoons. Fishing is important and mostly is done by small, family boats. About 40 percent of the world's offshore oil production comes from the Indian Ocean.

Indonesian Trench. See Java Trench.

Intracoastal Waterway. Series of bays, sounds, and channels, part natural and part human-made, that extends along the eastern coast of the United States from the Delaware River in New Jersey, south to the tip of Florida, then around the west coast of Florida. It extends around the Gulf Coast to the Rio Grande in Texas. It runs 2,455 miles (3,951 km.) and is an important, protected route for commercial and pleasure boat traffic.

Japan, Sea of. Marginal sea of the western Pacific Ocean that is bounded by Japan on the east and the Russian mainland on the west. Its surface area is approximately 377,600 square miles (978,000 sq. km.). It has an average depth of 5,750 feet (1,750 meters) and a maximum depth of 12,276 feet (3,742 meters).

Japan Trench. Ocean trench approximately 497 miles (800 km.) long, beginning at the eastern edge of the Japanese islands and stretching southward toward the Mariana Trench. Depth is 27,560 feet (8,400 meters).

Java Sea. Portion of the western Pacific Ocean between the islands of Java and Borneo. The sea has a total surface area of 167,000 square miles (433,000 sq. km.) and a comparatively shallow average depth of 151 feet (46 meters).

Java Trench. Ocean trench in the INDIAN OCEAN, 2,790 miles (4,500 km.) long and 24,443 feet (7,450 meters) deep. Also called the Indonesian Trench.

Kermadec Trench. Ocean trench approximately 930 miles (1,500 km.) long, located in the southwest PACIFIC OCEAN, beginning northeast of New Zealand. It has a depth of 32,800 feet (10,000 meters). Its northern end connects with the TONGA TRENCH.

Kurile Trench. Ocean trench approximately 1,367 miles (2,200 km.) long along the northeast rim of the PACIFIC OCEAN, beginning at the north end of the Japanese island chain and extending northeastward. Depth is 34,451 feet (10,500 meters).

Labrador Current. Cold current that begins in Baffin Bay between GREENLAND and northeastern Canada and flows southward. The Labrador Current sometimes carries icebergs into North Atlantic shipping channels; such an iceberg caused the famous sinking of the great passenger ship *Titanic* in 1912.

Laptev Sea. Marginal sea of the ARCTIC OCEAN off the coast of northern Siberia. The Taymyr Peninsula bounds it on the west and the New Siberian Islands on the east. Its area is about 276,000 square miles (714,000 sq. km.). Its average depth is 1,896 feet (578 meters), and the greatest depth is 9,774 feet (2,980 meters).

Lord Howe Rise. Elevation of the floor of the western PACIFIC OCEAN that lies between AUSTRALIA and New Guinea and under the TASMAN SEA.

Mac. Robertson Land. Land near the coast of ANTARCTICA, located between the INDIAN OCEAN and the south Polar Plateau, east of ENDERBY LAND. Mac.Robertson Land lies between approximately longitude 60 and 65 degrees east.

Macronesia. Loose grouping of islands in the ATLANTIC OCEAN that includes the Azores, Madeira, the Canary Islands and Cape Verde. The term is derived from Greek words meaning "large" and "island" and should not be confused with MICRONESIA, small islands in the central and North PACIFIC OCEAN.

Madagascar. Large island nation, officially called the Malagasy Republic, located in the INDIAN OCEAN about 200 miles from the southeast coast of AFRICA. Although geographically tied to the African continent, it has a culture more closely tied to those of France and Southeast Asia. Area is 226,657 square miles (587,042 sq. km.).

Magellan, Strait of. Waterway connecting the south ATLANTIC OCEAN with the South Pacific. Ships passing through the strait, north of Tierra del Fuego Island, avoid some of the world's roughest seas around CAPE HORN.

magnetic poles. The two points on the earth, one in the NORTHERN HEMISPHERE and one in the SOUTHERN HEMISPHERE, which are defined by the internal magnetism of the earth. Each point attracts one end of a compass needle and repels the opposite end.

Malacca, Strait of. Relatively narrow passage (200 miles/322 kilometers wide) bordered by Malaysia and Sumatra and linking the SOUTH CHINA SEA and the JAVA SEA. It is one of the most heavily traveled waterways in the world, with more than one thousand ships every week.

Mariana Trench. Lowest point on Earth's surface, with a maximum depth of 36,150 feet (11,022 meters) in the Challenger Deep. The Mariana Trench is located on the western margin of the PACIFIC OCEAN southeast of Japan, and is approximately 1,584 miles (2,550 km.) long.

Marie Byrd Land. Section of ANTARCTICA located at the base of the Antarctic peninsula and shaped like a large peninsula itself. It is bounded at its base by the ROSS ICE SHELF and the Ronne Ice Shelf.

Mediterranean Sea. Large sea that separates the continents of EUROPE, AFRICA, and ASIA. It takes its name from Latin words meaning "in the middle of land"—a reference to its nearly land-locked nature. Covers about 969,100 square miles (2.5 million sq. km.) and extends 2,200 miles (3,540 km.) from west to east and about 1,000 miles (1,600 km.) from north to south at its widest. Its greatest depth is 16,897 feet (5,150 meters).

Melanesia. One of three divisions of the Pacific Islands, along with MICRONESIA and POLYNESIA; located in the western Pacific. The name Melanesia, for "dark islands," was given to the area because of its inhabitants' dark skins

Mexico, Gulf of. Nearly enclosed arm of the western ATLANTIC OCEAN, bounded by the states of Florida, Alabama, Mississippi, Louisiana, and Texas, and Mexico and the Yucatan Peninsula. Cuba is located in the gap between the Yucatan Peninsula and Florida. Most ocean water enters through the Yucatan passage and exits the Gulf of Mexico around the tip of Florida, becoming the FLORIDA CURRENT. Fisheries, tourism, and oil production are important activities.

Micronesia. One of three divisions of the Pacific Islands, along with MELANESIA and POLYNESIA. Micronesia means "small islands." Micronesia's islands are mostly atolls and coral islands, but some are of volcanic origin. The more than 2,000 islands of Micronesia are located in the Pacific Ocean east of the Philippines, mostly north of the EQUATOR.

Mid-Atlantic Ridge. Steep-sided, underwater mountain range running down the middle of the ATLANTIC OCEAN. Formed by the divergent boundaries, or region where tectonic plates are separating.

Mozambique Current. See AGULHAS CURRENT.

New Britain Trench. Ocean trench in the southwest PACIFIC OCEAN, about 5,158 miles (8,320 km.) long and 2,460 feet (750 meters) deep.

New Hebrides Basin. Part of the CORAL SEA, located east of AUSTRALIA and west of the New Hebrides island chain. The basin contains volcanic islands, both old and recent.

New Hebrides Trench. Ocean trench in the southwest PACIFIC OCEAN, about 5,682 miles (9,165 km.) long and 3,936 feet (1,200 meters) deep.

North America. Third-largest continent, usually considered to contain all land and nearby islands in the WESTERN HEMISPHERE north of the Isthmus of Panama, which connects it to SOUTH AMERICA. The major mainland countries are Canada, the United States, Mexico, Guatemala, El Salvador, Honduras, Nicaragua, Costa Rica, and Panama. Island countries include the islands of the CARIBBEAN SEA and GREENLAND. Climate ranges from arctic to tropical.

North Atlantic Current. Continuation of the GULF STREAM, originating near the GRAND BANKS off Newfoundland. It curves eastward and divides into a northern branch, which flows into the NORWEGIAN SEA, a southern branch, which flows eastward, and a branch that forms the Canary Current and flows south along the coast of EUROPE.

North Atlantic Gyre. Large mass of water, located in the ATLANTIC OCEAN in the NORTHERN HEMISPHERE, that rotates clockwise. Warm water moves toward the pole and cold water moves toward the equator.

North Pacific Current. Eastward flow of water in the PACIFIC OCEAN in the NORTHERN HEMISPHERE. It originates as the Kuroshio Current and moves from Japan toward NORTH AMERICA.

North Pacific Gyre. Large mass of water, located in the PACIFIC OCEAN in the NORTHERN HEMISPHERE, that rotates clockwise. Warm water moves toward the pole and cold water moves toward the equator.

North Pole. Northern end of the earth's geographic axis, located at 90 degrees north latitude and longitude zero degrees. The North Pole itself is located on the Polar Abyssal Plain, about 14,000 feet (4,000 meters) deep in the ARCTIC OCEAN. U.S. explorer Robert Edwin is credited with being the first person to reach the North Pole, in 1909, although there is historical dispute over the claim. The North Pole is different from the North MAGNETIC POLE.

North Sea. Arm of the northeastern ATLANTIC OCEAN, bounded by Great Britain on the west and Norway, Denmark, and Germany on the east and south. The North Sea is one of the great fishing areas of the world and an important source of oil.

Northern Hemisphere. The half of the earth above the equator.

Norwegian Sea. Section of the North Atlantic Ocean. Norway borders it on the east and Iceland on the west. A submarine ridge linking

GREENLAND, ICELAND, the Faroe Islands, and northern Scotland separates the Norwegian Sea from the open ATLANTIC OCEAN. Cut by the ARCTIC CIRCLE, the sea is often associated with the ARCTIC OCEAN to the north. Reaches a maximum depth of about 13,020 feet (3,970 meters).

Oceania. Loosely applied term for the large island groups of the central and South Pacific; sometimes used to include AUSTRALIA and New Zealand.

Okhotsk, Sea of. Nearly enclosed area of the northwestern PACIFIC OCEAN bounded by Russia's Kamchatka Peninsula on the east and Siberia on the west. It is open to the Pacific Ocean on the south side only through Japan and the Kuril Islands, a string of islands belonging to Russia.

Pacific Ocean. Largest body of water in the world, covering more than one-third of Earth's surface—an area of about 70 million square miles (181 million sq. km.), more than the entire land area of the world. At its widest point, between Panama in CENTRAL AMERICA and the Philippines, it stretches 10,700 miles (17,200 km.). It runs 9,600 miles (15,450 km.) from the BERING STRAIT in the north to ANTARCTICA in the south. Bordered by NORTH and SOUTH AMERICA in the east, and ASIA and AUSTRALIA in the west. The average depth is about 12,900 feet (3,900 meters). It contains the deepest point on Earth (36,150 feet/11,022 meters), in the Challenger Deep of the MARIANA TRENCH, southwest of Japan. The Pacific Ocean bottom is more geologically varied than the INDIAN or ATLANTIC OCEANS; it has more volcanoes, ridges, trenches, seamounts, and islands. The vast size of the Pacific Ocean was a formidable barrier to travel, communications, and trade well into the nineteenth century. However, evidence shows that people crossed the Pacific Ocean in rafts or canoes as early as 3,000 BCE.

Pacific Rim. Modern term for the nations of ASIA and NORTH and SOUTH AMERICA that border, or are in, the PACIFIC OCEAN. Used mostly in discussions of economic growth.

Palau Trench. Ocean trench in the western PACIFIC OCEAN, about 250 miles (400 km.) long and 26,425 feet (8,054 meters) deep.

Palmer Land. Section of ANTARCTICA that occupies the base of the Antarctic peninsula.

Panama, Isthmus of. Narrow neck of land that joins CENTRAL and SOUTH AMERICA. In 1914 the Panama Canal was opened through the isthmus, creating a direct sea link between the PACIFIC OCEAN and the CARIBBEAN SEA. The canal stretches about 50 miles (80 km.) from Panama City on the Pacific to Colón on the Caribbean. More than 12,000 ships pass through the canal annually.

Persian Gulf. Large extension of the ARABIAN SEA that separates Iran from the Arabian Peninsula in the Middle East. It covers about 88,000 square miles (226,000 sq. km.) and is about 620 miles (1,000 km.) long and 125–185 miles (200–300 km.) wide.

Peru-Chile Trench. Ocean trench that runs along the eastern boundary of the PACIFIC OCEAN, off the western edge of SOUTH AMERICA. It is 3,666 miles (5,900 km.) long and 26,576 feet (8,100 meters) deep.

Peru Current. Cold, broad current that originates in the southernmost part of the SOUTH PACIFIC GYRE and flows up the west coast of SOUTH AMERICA. Off the coast of Peru, prevailing winds usually push the warmer surface water to the west. This causes the nutrient-rich, colder water of the Peru Current to well up to the surface, which provides excellent feeding for fish. At times, the upwelling ceases and biological, economic, and climatic catastrophe can result in EL NIÑO weather conditions. Also known as the Humboldt Current.

Philippine Trench. Ocean trench located on the western rim of the PACIFIC OCEAN, at the eastern margin of the PHILIPPINE ISLANDS. It is about 870 miles (1,400 km.) long and 34,451 feet (10,500 meters) deep.

Polynesia. One of three main divisions of the Pacific Islands, along with MELANESIA and MICRONESIA. The islands are spread through the central and South Pacific. Polynesia means "many islands." Mostly small, the islands are predominantly coral atolls, but some are of volcanic origin.

Puerto Rico Trench. Ocean trench in the western ATLANTIC OCEAN, about 27,500 feet (8,385 meters) deep and 963 miles (1,550 km.) long.

Queen Maud Land. Section of ANTARCTICA that lies between the ATLANTIC OCEAN and the south Polar Plateau, between approximately longitude 15 and 45 degrees east.

Red Sea. Narrow arm of water separating AFRICA from the ARABIAN PENINSULA. One of the saltiest bodies of ocean water on Earth, as a result of high evaporation and little freshwater input. It was used as a trade route for Mediterranean, Indian, and Chinese peoples for centuries before the Europeans discovered it in the fifteenth century. The Suez Canal was opened in 1869 between the MEDITERRANEAN SEA and the Red Sea, cutting the distance from the northern INDIAN OCEAN to northern EUROPE by about 5,590 miles (9,000 km.). This greatly increased the economic and military importance of the Red Sea.

Ring of Fire. Nearly continuous ring of volcanic and tectonic activity around the margins of the PACIFIC OCEAN.

Ross Ice Shelf. Thick layer of ice in the Ross SEA off the coast of ANTARCTICA. The relatively flat ice is attached to and nourished by a continental glacier.

Ross Sea. Bay in the SOUTHERN OCEAN off the coast of ANTARCTICA, located south of New Zealand. Named for English explorer James Clark Ross, the first person to break through the Antarctic ice pack in a ship, in 1841.

St. Peter and St. Paul. Cluster of rocks showing above the surface of the ATLANTIC OCEAN between Brazil and West AFRICA. Important landmarks in the days of slave ships.

Sargasso Sea. Warm, salty area of water located in the ATLANTIC OCEAN south and east of Bermuda, formed from water that circulates around the center of the NORTH ATLANTIC GYRE. Named for the seaweed, *Sargassum*, that floats on the surface in large amounts.

Scotia Sea. Area of the southernmost ATLANTIC OCEAN between the southern tip of SOUTH AMERICA and the ANTARCTIC peninsula. The area is known for severe storms.

Seas. See under individual names.

Siam, Gulf of. See THAILAND, GULF OF.

South America. Fourth-largest continent, usually considered to contain all land and nearby islands in the Western Hemisphere south of the Isthmus of Panama, which connects it to NORTH AMERICA. Countries are Argentina, Bolivia, Brazil, Chile, Colombia, Ecuador, French Guiana, Guyana, Paraguay, Peru, Suriname, Uruguay, and Venezuela. Climate ranges from tropical to cold, nearly sub-Antarctic.

South Atlantic Gyre. Large mass of water, located in the ATLANTIC OCEAN in the SOUTHERN HEMISPHERE, that rotates counterclockwise. Warm water moves toward the pole and cold water moves toward the equator.

South China Sea. Portion of the western PACIFIC OCEAN that lies along the east coast of China, Vietnam, and the southeastern part of the Gulf of Thailand. The eastern and southern edges are defined by the Philippine and Indonesian Islands.

South Equatorial Current. Part of the SOUTH ATLANTIC GYRE that is split in two by the eastern prominence of Brazil. One part moves along the northeastern coast of SOUTH AMERICA toward the CARIBBEAN SEA and the North ATLANTIC OCEAN; the other turns southward and forms the BRAZIL CURRENT.

South Pacific Gyre. Large mass of water, located in the PACIFIC OCEAN in the SOUTHERN HEMISPHERE, that rotates counterclockwise. Warm water moves toward the pole and cold water moves toward the equator.

South Pole. Southern end of the earth's geographic axis, located at 90 degrees south latitude and longitude zero degrees. The first person to reach the South Pole was Norwegian explorer Roald Amundsen, in 1911. The South Pole is different from the South MAGNETIC POLE.

Southeastern Pacific Plateau. Portion of the PACIFIC OCEAN floor closest to SOUTH AMERICA.

Southern Hemisphere. The half of the earth below the equator.

Southern Ocean. Not officially recognized as one of the major oceans, but a commonly used term for water surrounding ANTARCTICA and extending northward to 50 degrees south latitude. Also known as the Antarctic Ocean.

Straits. See under individual names.

Sunda Shelf. One of the largest continental shelves in the world, nearly 772,000 square miles (2 million sq. km.). Located in the JAVA SEA, SOUTH CHINA SEA, and Gulf of THAILAND. The area was above water in the Quaternary period, enabling large animals such as elephants and rhinoceros to migrate to Sumatra, Java, and Borneo.

Surtsey Island. Island formed by a volcanic explosion off the coast of ICELAND in 1963. It is valuable to scientists studying how island flora and fauna develop and is a popular tourist site.

Tashima Current. See TSUSHIMA CURRENT.

Tasman Sea. Area of the PACIFIC OCEAN off the southeast coast of AUSTRALIA, between Australia and Tasmania on the west and New Zealand on the east. First crossed by the Morioris people sometime before 1300 CE Also called the Tasmanian Sea.

Tasmanian Sea. See TASMAN SEA.

Thailand, Gulf of. Also known as the Gulf of Siam, inlet of the South China Sea, located between the Malay Archipelago and the Southeast Asian mainland. Bounded by Thailand, Cambodia, and Vietnam.

Tonga Trench. Ocean trench in the PACIFIC OCEAN, northeast of New Zealand. It stretches for 870 miles (1,400 km.), beginning at the northern end of the KERMADEC TRENCH. Depth is 32,810 feet (10,000 meters).

Tsushima Current. Warm current in the western PACIFIC OCEAN that flows out of the YELLOW SEA into the Sea of JAPAN in the spring and summer. Also called Tashima Current.

Walvis Ridge (Walfisch Ridge). Long, narrow undersea elevation near the southwestern coast of AFRICA, which extends about 1,900 miles (3,000 km.) in a southwesterly direction under the ATLANTIC OCEAN.

Weddell Sea. Bay in the SOUTHERN OCEAN bounded by the ANTARCTIC peninsula on the west and a northward bulge of ANTARCTICA on the east, stretching from approximately longitude 60 to 10 degrees west. One of the harshest environments on Earth; surface water temperatures stay near 32 degrees Fahrenheit (0 degrees Celsius) all year. The Weddell Sea was the site of much whaling and seal hunting in the nineteenth and twentieth centuries.

West Caroline Trench. See YAP TRENCH.

West Wind Drift. Surface portion of the ANTARCTIC CIRCUMPOLAR CURRENT, driven by westerly winds. Often extremely rough; seas as high as 98 feet (30 meters) have been reported.

Western Hemisphere. The half of the earth containing NORTH and SOUTH AMERICA; generally understood to fall between longitudes 160 degrees east and 20 degrees west.

Wilkes Land. Broad section near the coast of ANTARCTICA, which lies south of AUSTRALIA and east of the AMERICAN HIGHLANDS. Wilkes Land is the nearest landmass to the South MAGNETIC POLE.

Yap Trench. Ocean trench in the western PACIFIC OCEAN, about 435 miles (700 km.) long and 27,900 feet (8,527 meters) deep. Also called the West Caroline Trench.

Yellow Sea. Area of the PACIFIC OCEAN bounded by China on the north and west and Korea on the east. Named for the large amounts of yellow dust carried into it from central China by winds and by the Yangtze, Yalu, and Yellow Rivers. Parts of the sea often show a yellow color from the dust.

Kelly Howard

THE WORLD'S OCEANS AND SEAS

Name	Approximate Area		Average Depth	
	Sq. Miles	Sq. Km.	Feet	Meters
Pacific Ocean	64,000,000	165,760,000	13,215	4,028
Atlantic Ocean	31,815,000	82,400,000	12,880	3,926
Indian Ocean	25,300,000	65,526,700	13,002	3,963
Arctic Ocean	5,440,200	14,090,000	3,953	1,205
Mediterranean & Black Seas	1,145,100	2,965,800	4,688	1,429
Caribbean Sea	1,049,500	2,718,200	8,685	2,647
South China Sea	895,400	2,319,000	5,419	1,652
Bering Sea	884,900	2,291,900	5,075	1,547
Gulf of Mexico	615,000	1,592,800	4,874	1,486
Okhotsk Sea	613,800	1,589,700	2,749	838
East China Sea	482,300	1,249,200	617	188
Hudson Bay	475,800	1,232,300	420	128
Japan Sea	389,100	1,007,800	4,429	1,350
Andaman Sea	308,100	797,700	2,854	870
North Sea	222,100	575,200	308	94
Red Sea	169,100	438,000	1,611	491
Baltic Sea	163,000	422,200	180	55

MAJOR LAND AREAS OF THE WORLD

Area	Approximate Land Area		Percent of World Total
	Sq. Mi.	Sq. Km.	
World	57,308,738	148,429,000	100.0
Asia (including Middle East)	17,212,041	44,579,000	30.0
Africa	11,608,156	30,065,000	20.3
North America	9,365,290	24,256,000	16.3
Central America, South America, & Caribbean	6,879,952	17,819,000	8.9
Antarctica	5,100,021	13,209,000	8.9
Europe	3,837,082	9,938,000	6.7
Oceania, including Australia	2,967,966	7,687,000	5.2

MAJOR ISLANDS OF THE WORLD

| | | Area | |
		Sq. Mi.	Sq. Km.
Island	*Location*		
Greenland	North Atlantic Ocean	839,999	2,175,597
New Guinea	Western Pacific Ocean	316,615	820,033
Borneo	Western Pacific Ocean	286,914	743,107
Madagascar	Western Indian Ocean	226,657	587,042
Baffin	Canada, North Atlantic Ocean	183,810	476,068
Sumatra	Indonesia, northeast Indian Ocean	182,859	473,605
Hōnshū	Japan, western Pacific Ocean	88,925	230,316
Great Britain	North Atlantic Ocean	88,758	229,883
Ellesmere	Canada, Arctic Ocean	82,119	212,688
Victoria	Canada, Arctic Ocean	81,930	212,199
Sulawesi (Celebes)	Indonesia, western Pacific Ocean	72,986	189,034
South Island	New Zealand, South Pacific Ocean	58,093	150,461
Java	Indonesia, Indian Ocean	48,990	126,884
North Island	New Zealand, South Pacific Ocean	44,281	114,688
Cuba	Caribbean Sea	44,218	114,525
Newfoundland	Canada, North Atlantic Ocean	42,734	110,681
Luzon	Philippines, western Pacific Ocean	40,420	104,688
Iceland	North Atlantic Ocean	39,768	102,999
Mindanao	Philippines, western Pacific Ocean	36,537	94,631
Ireland	North Atlantic Ocean	32,597	84,426
Hokkaido	Japan, western Pacific Ocean	30,372	78,663
Hispaniola	Caribbean Sea	29,355	76,029
Tasmania	Australia, South Pacific Ocean	26,215	67,897
Sri Lanka	Indian Ocean	25,332	65,610
Sakhalin (Karafuto)	Russia, western Pacific Ocean	24,560	63,610
Banks	Canada, Arctic Ocean	23,230	60,166
Devon	Canada, Arctic Ocean	20,861	54,030
Tierra del Fuego	Southern tip of South America	18,605	48,187
Kyūshū	Japan, western Pacific Ocean	16,223	42,018
Melville	Canada, Arctic Ocean	16,141	41,805
Axel Heiberg	Canada, Arctic Ocean	15,779	40,868
Southampton	Hudson Bay, Canada	15,700	40,663

COUNTRIES OF THE WORLD

Country	Region	Population	Area Square Miles	Square Kilometers	Population Density Persons/ Sq. Mi.	Persons/ Sq. Km.
Afghanistan	Asia	31,575,018	249,347	645,807	127	49
Albania	Europe	2,862,427	11,082	28,703	259	100
Algeria	Africa	42,545,964	919,595	2,381,741	47	18
Andorra	Europe	76,177	179	464	425	164
Angola	Africa	29,250,009	481,354	1,246,700	60	23
Antigua and Barbuda	Caribbean	104,084	171	442	609	235
Argentina	South America	44,938,712	1,073,518	2,780,400	41	16
Armenia	Europe	2,962,100	11,484	29,743	259	100
Australia	Australia	25,576,880	2,969,907	7,692,024	9	3
Austria	Europe	8,877,036	32,386	83,879	275	106
Azerbaijan	Asia	10,027,874	33,436	86,600	300	116
Bahamas	Caribbean	386,870	5,382	13,940	73	28
Bahrain	Asia	1,543,300	300	778	5,136	1,983
Bangladesh	Asia	167,888,084	55,598	143,998	3,020	1,166
Barbados	Caribbean	287,025	166	430	1,730	668
Belarus	Europe	9,465,300	80,155	207,600	119	46
Belgium	Europe	11,515,793	11,787	30,528	976	377
Belize	Central America	398,050	8,867	22,965	44	17
Benin	Africa	11,733,059	43,484	112,622	269	104
Bhutan	Asia	821,592	14,824	38,394	55	21
Bolivia	South America	11,307,314	424,164	1,098,581	26	10
Bosnia and Herzegovina	Europe	3,511,372	19,772	51,209	179	69
Botswana	Africa	2,302,878	224,607	581,730	10.4	4
Brazil	South America	210,951,255	3,287,956	8,515,767	64	25
Brunei	Asia	421,300	2,226	5,765	189	73
Bulgaria	Europe	7,000,039	42,858	111,002	163	63
Burkina Faso	Africa	20,244,080	104,543	270,764	194	75
Burundi	Africa	10,953,317	10,740	27,816	1,020	394
Cambodia	Asia	16,289,270	69,898	181,035	233	90
Cameroon	Africa	24,348,251	179,943	466,050	135	52
Canada	North America	37,878,499	3,855,103	9,984,670	10	4
Cape Verde	Africa	550,483	1,557	4,033	352	136
Central African Republic	Africa	4,737,423	240,324	622,436	21	8
Chad	Africa	15,353,184	495,755	1,284,000	31	12
Chile	South America	17,373,831	291,930	756,096	60	23
China, People's Republic of	Asia	1,400,781,440	3,722,342	9,640,821	376	145
Colombia	South America	46,103,400	440,831	1,141,748	105	40
Comoros	Africa	873,724	719	1,861	1,215	469

Country	Region	Population	Area		Population Density	
			Square Miles	Square Kilometers	Persons/ Sq. Mi.	Persons/ Sq. Km.
Costa Rica	Central America	5,058,007	19,730	51,100	256	99
Côte d'Ivoire (Ivory Coast)	Africa	25,823,071	124,680	322,921	207	80
Croatia	Europe	4,087,843	21,831	56,542	186	72
Cuba	Caribbean	11,209,628	42,426	109,884	264	102
Cyprus	Europe	864,200	2,276	5,896	381	147
Czech Republic	Europe	10,681,161	30,451	78,867	350	135
Dem. Republic of the Congo	Africa	86,790,567	905,446	2,345,095	96	37
Denmark	Europe	5,814,461	16,640	43,098	350	135
Djibouti	Africa	1,078,373	8,880	23,000	122	47
Dominica	Caribbean	71,808	285	739	251	97
Dominican Republic	Caribbean	10,358,320	18,485	47,875	559	216
Ecuador	South America	17,398,588	106,889	276,841	163	63
Egypt	Africa	99,873,587	387,048	1,002,450	258	100
El Salvador	Central America	6,704,864	8,124	21,040	826	319
Equatorial Guinea	Africa	1,358,276	10,831	28,051	124	48
Eritrea	Africa	3,497,117	46,757	121,100	75	29
Estonia	Europe	1,324,820	17,505	45,339	75	29
Eswatini (Swaziland)	Africa	1,159,250	6,704	17,364	174	67
Ethiopia	Africa	107,534,882	410,678	1,063,652	262	101
Fed. States of Micronesia	Pacific Islands	105,300	271	701	388	150
Fiji	Pacific Islands	884,887	7,078	18,333	124	48
Finland	Europe	5,527,405	130,666	338,424	41	16
France	Europe	67,022,000	210,026	543,965	319	123
Gabon	Africa	2,067,561	103,347	267,667	21	8
Gambia	Africa	2,228,075	4,127	10,690	539	208
Georgia	Europe	3,729,600	26,911	69,700	140	54
Germany	Europe	83,073,100	137,903	357,168	603	233
Ghana	Africa	30,280,811	92,098	238,533	329	127
Greece	Europe	10,724,599	50,949	131,957	210	81
Grenada	Caribbean	108,825	133	344	818	316
Guatemala	Central America	17,679,735	42,042	108,889	420	162
Guinea	Africa	12,218,357	94,926	245,857	129	50
Guinea-Bissau	Africa	1,604,528	13,948	36,125	114	44
Guyana	South America	782,225	83,012	214,999	9.3	3.6
Haiti	Caribbean	11,263,077	10,450	27,065	1,077	416
Honduras	Central America	9,158,345	43,433	112,492	210	81
Hungary	Europe	9,764,000	35,919	93,029	272	105
Iceland	Europe	360,390	39,682	102,775	9.1	3.5
India	Asia	1,357,041,500	1,269,211	3,287,240	1,069	413
Indonesia	Asia	268,074,600	735,358	1,904,569	365	141
Iran	Asia	83,096,438	636,372	1,648,195	131	50

Country	Region	Population	Area Square Miles	Area Square Kilometers	Population Density Persons/ Sq. Mi.	Population Density Persons/ Sq. Km.
Iraq	Asia	39,309,783	169,235	438,317	233	90
Ireland	Europe	4,921,500	27,133	70,273	181	70
Israel	Asia	9,141,680	8,522	22,072	1,073	414
Italy	Europe	60,252,824	116,336	301,308	518	200
Jamaica	Caribbean	2,726,667	4,244	10,991	642	248
Japan	Asia	126,140,000	145,937	377,975	865	334
Jordan	Asia	10,587,132	34,495	89,342	307	119
Kazakhstan	Asia	18,592,700	1,052,090	2,724,900	18	7
Kenya	Africa	47,564,296	224,647	581,834	212	82
Kiribati	Pacific Islands	120,100	313	811	383	148
Kuwait	Asia	4,420,110	6,880	17,818	642	248
Kyrgyzstan	Asia	6,309,300	77,199	199,945	83	32
Laos	Asia	6,492,400	91,429	236,800	70	27
Latvia	Europe	1,910,400	24,928	64,562	78	30
Lebanon	Asia	6,855,713	4,036	10,452	1,740	672
Lesotho	Africa	2,263,010	11,720	30,355	194	75
Liberia	Africa	4,475,353	37,466	97,036	119	46
Libya	Africa	6,470,956	683,424	1,770,060	9.6	3.7
Liechtenstein	Europe	38,380	62	160	622	240
Lithuania	Europe	2,793,466	25,212	65,300	111	43
Luxembourg	Europe	613,894	998	2,586	614	237
Madagascar	Africa	25,680,342	226,658	587,041	114	44
Malawi	Africa	17,563,749	45,747	118,484	383	148
Malaysia	Asia	32,715,210	127,724	330,803	256	99
Maldives	Asia	378,114	115	298	3,287	1,269
Mali	Africa	19,107,706	482,077	1,248,574	39	15
Malta	Europe	493,559	122	315	3,911	1,510
Marshall Islands	Pacific Islands	55,500	70	181	795	307
Mauritania	Africa	3,984,233	397,955	1,030,700	10.4	4
Mauritius	Africa	1,265,577	788	2,040	1,606	620
Mexico	North America	126,577,691	759,516	1,967,138	166	64
Moldova	Europe	2,681,735	13,067	33,843	205	79
Monaco	Europe	38,300	0.78	2.02	49,106	18,960
Mongolia	Asia	3,000,000	603,902	1,564,100	4.9	1.9
Montenegro	Europe	622,182	5,333	13,812	117	45
Morocco	Africa	35,773,773	172,414	446,550	207	80
Mozambique	Africa	28,571,310	308,642	799,380	93	36
Myanmar (Burma)	Asia	54,339,766	261,228	676,577	207	80
Namibia	Africa	2,413,643	318,580	825,118	7.5	2.9
Nauru	Pacific Islands	11,000	8	21	1,357	524
Nepal	Asia	29,609,623	56,827	147,181	521	201

Country	Region	Population	Area		Population Density	
			Square Miles	Square Kilometers	Persons/ Sq. Mi.	Persons/ Sq. Km.
Netherlands	Europe	17,370,348	16,033	41,526	1,083	418
New Zealand	Pacific Islands	4,952,186	104,428	270,467	47	18
Nicaragua	Central America	6,393,824	46,884	121,428	137	53
Niger	Africa	21,466,863	458,075	1,186,408	47	18
Nigeria	Africa	200,962,000	356,669	923,768	565	218
North Korea	Asia	25,450,000	46,541	120,540	546	211
North Macedonia	Europe	2,077,132	9,928	25,713	210	81
Norway	Europe	5,328,212	125,013	323,782	41	16
Oman	Asia	4,183,841	119,499	309,500	36	14
Pakistan	Asia	218,198,000	310,403	803,940	703	271
Palau	Pacific Islands	17,900	171	444	104	40
Panama	Central America	4,158,783	28,640	74,177	145	56
Papua New Guinea	Pacific Islands	8,558,800	178,704	462,840	47	18
Paraguay	South America	7,052,983	157,048	406,752	44	17
Peru	South America	32,162,184	496,225	1,285,216	65	25
Philippines	Asia	108,785,760	115,831	300,000	939	363
Poland	Europe	38,386,000	120,728	312,685	319	123
Portugal	Europe	10,276,617	35,556	92,090	290	112
Qatar	Asia	2,740,479	4,468	11,571	614	237
Republic of the Congo	Africa	5,399,895	132,047	342,000	41	16
Romania	Europe	19,405,156	92,043	238,391	210	81
Russia[1]	Europe/Asia	146,877,088	6,612,093	17,125,242	23	9
Rwanda	Africa	12,374,397	10,169	26,338	1,217	470
St Kitts and Nevis	Caribbean	56,345	104	270	541	209
St Lucia	Caribbean	180,454	238	617	756	292
St Vincent and Grenadines	Caribbean	110,520	150	389	736	284
Samoa	Pacific Islands	199,052	1,093	2,831	181	70
San Marino	Europe	34,641	24	61	1,471	568
Sahrawi Arab Dem. Rep.[2]	Africa	567,421	97,344	252,120	6	2.3
São Tomé and Príncipe	Africa	201,784	386	1,001	523	202
Saudi Arabia	Asia	34,218,169	830,000	2,149,690	41	16
Senegal	Africa	16,209,125	75,955	196,722	212	82
Serbia	Europe	6,901,188	29,913	77,474	231	89
Seychelles	Africa	96,762	176	455	552	213
Sierra Leone	Africa	7,901,454	27,699	71,740	285	110
Singapore	Asia	5,638,700	279	722.5	20,212	7,804
Slovakia	Europe	5,450,421	18,933	49,036	287	111
Slovenia	Europe	2,084,301	7,827	20,273	267	103
Solomon Islands	Pacific Islands	682,500	10,954	28,370	62	24
Somalia	Africa	15,181,925	246,201	637,657	62	24
South Africa	Africa	58,775,022	471,359	1,220,813	124	48

Country	Region	Population	Area Square Miles	Area Square Kilometers	Population Density Persons/ Sq. Mi.	Population Density Persons/ Sq. Km.
South Korea	Asia	51,811,167	38,691	100,210	1,339	517
South Sudan	Africa	12,575,714	248,777	644,329	52	20
Spain	Europe	46,934,632	195,364	505,990	241	93
Sri Lanka	Asia	21,803,000	25,332	65,610	860	332
Sudan	Africa	40,782,742	710,251	1,839,542	57	22
Suriname	South America	568,301	63,251	163,820	9.1	3.5
Sweden	Europe	10,344,405	173,860	450,295	59	23
Switzerland	Europe	8,586,550	15,940	41,285	539	208
Syria	Asia	17,070,135	71,498	185,180	238	92
Taiwan	Asia	23,596,266	13,976	36,197	1,689	652
Tajikistan	Asia	9,127,000	55,251	143,100	166	64
Tanzania	Africa	55,890,747	364,900	945,087	153	59
Thailand	Asia	66,455,280	198,117	513,120	335	130
Timor-Leste	Asia	1,167,242	5,760	14,919	202	78
Togo	Africa	7,538,000	21,853	56,600	344	133
Tonga	Pacific Islands	100,651	278	720	362	140
Trinidad and Tobago	Caribbean	1,359,193	1,990	5,155	683	264
Tunisia	Africa	11,551,448	63,170	163,610	183	71
Turkey	Europe/Asia	82,003,882	302,535	783,562	271	105
Turkmenistan	Asia	5,851,466	189,657	491,210	31	12
Tuvalu	Pacific Islands	10,200	10	26	1,020	392
Uganda	Africa	40,006,700	93,263	241,551	429	166
Ukraine[3]	Europe	41,990,278	232,820	603,000	180	70
United Arab Emirates	Asia	9,770,529	32,278	83,600	303	117
United Kingdom	Europe	66,435,600	93,788	242,910	708	273
United States	North America	330,546,475	3,796,742	9,833,517	87	34
Uruguay	South America	3,518,553	68,037	176,215	52	20
Uzbekistan	Asia	32,653,900	172,742	447,400	189	73
Vanuatu	Pacific Islands	304,500	4,742	12,281	65	25
Vatican City	Europe	1,000	0.17	0.44	5,887	2,273
Venezuela	South America	32,219,521	353,841	916,445	91	35
Vietnam	Asia	96,208,984	127,882	331,212	751	290
Yemen	Asia	28,915,284	175,676	455,000	166	64
Zambia	Africa	16,405,229	290,585	752,612	57	22
Zimbabwe	Africa	15,159,624	150,872	390,757	101	39

Notes: (1) Including the population and area of Autonomous Republic of Crimea and City of Sevastopol, Ukraine's administrative areas on the Crimean Peninsula, which are claimed by Russia; (2) Administration is split between Morocco and the Sahrawi Arab Democratic Republic (Western Sahara), both of which claim the entire territory; (3) Excludes Crimea.
Source: U.S. Census Bureau, International Data Base

Past and Projected World Population Growth, 1950-2050

Year	Approximate World Population	Ten-Year Growth Rate (%)
1950	2,556,000,053	18.9
1960	3,039,451,023	22.0
1970	3,706,618,163	20.2
1980	4,453,831,714	18.5
1990	5,278,639,789	15.2
2000	6,082,966,429	12.6
2010	6,848,932,929	10.7
2020	7,584,821,144	8.7
2030	8,246,619,341	7.3
2040	8,850,045,889	5.6
2050	9,346,399,468	—

Note: The listed years are the baselines for the estimated ten-year growth rate figures; for example, the rate for 1950-1960 was 18.9%.
Source: U.S. Census Bureau, International Data Base

THE WORLD'S LARGEST COUNTRIES BY AREA

| | | | Area | |
Rank	Country	Region	Sq. Miles	Sq. Km.
1	Russia	Europe/Asia	6,612,093	17,125,242
2	Canada	North America	3,855,103	9,984,670
3	United States	North America	3,796,742	9,833,517
4	China	Asia	3,722,342	9,640,821
5	Brazil	South America	3,287,956	8,515,767
6	Australia	Australia	2,969,907	7,692,024
7	India	Asia	1,269,211	3,287,240
8	Argentina	South America	1,073,518	2,780,400
9	Kazakhstan	Asia	1,052,090	2,724,900
10	Algeria	Africa	919,595	2,381,741
11	Democratic Rep. of the Congo	Africa	905,446	2,345,095
12	Saudi Arabia	Asia	830,000	2,149,690
13	Mexico	North America	759,516	1,967,138
14	Indonesia	Asia	735,358	1,904,569
15	Sudan	Africa	710,251	1,839,542
16	Libya	Africa	683,424	1,770,060
17	Iran	Asia	636,372	1,648,195
18	Mongolia	Asia	603,902	1,564,100
19	Peru	South America	496,225	1,285,216
20	Chad	Africa	495,755	1,284,000
21	Mali	Africa	482,077	1,248,574
22	Angola	Africa	481,354	1,246,700
23	South Africa	Africa	471,359	1,220,813
24	Niger	Africa	458,075	1,186,408
25	Colombia	South America	440,831	1,141,748
26	Bolivia	South America	424,164	1,098,581
27	Ethiopia	Africa	410,678	1,063,652
28	Mauritania	Africa	397,955	1,030,700
29	Egypt	Africa	387,048	1,002,450
30	Tanzania	Africa	364,900	945,087

Source: U.S. Census Bureau, International Data Base

The World's Smallest Countries by Area

Rank	Country	Region	Area Sq. Miles	Sq. Km.
1	Vatican City*	Europe	0.17	0.44
2	Monaco*	Europe	0.78	2.02
3	Nauru	Pacific Islands	8	21
4	Tuvalu	Pacific Islands	10	26
5	San Marino*	Europe	24	61
6	Liechtenstein*	Europe	62	160
7	Marshall Islands	Pacific Islands	70	181
8	Saint Kitts and Nevis	Central America	104	270
9	Maldives	Asia	115	298
10	Malta	Europe	122	315
11	Grenada	Caribbean	133	344
12	Saint Vincent and the Grenadines	Central America	150	389
13	Barbados	Caribbean	166	430
14	Antigua and Barbuda	Caribbean	171	442
15	Palau	Pacific Islands	171	444
16	Seychelles	Africa	176	455
17	Andorra*	Europe	179	464
18	Saint Lucia	Central America	238	617
19	Federated States of Micronesia	Pacific Islands	271	701
20	Tonga	Pacific Islands	278	720

Note: Asterisks (*) denote countries on continents; all other countries are islands or island groups.
Source: U.S. Census Bureau, International Data Base

THE WORLD'S LARGEST COUNTRIES BY POPULATION

Rank	Country	Region	Population
1	China	Asia	1,401,028,280
2	India	Asia	1,357,746,150
3	United States	North America	329,229,067
4	Indonesia	Asia	268,074,600
5	Pakistan	Asia	218,385,000
6	Brazil	South America	211,032,216
7	Nigeria	Africa	200,962,000
8	Bangladesh	Asia	167,983,726
9	Russia	Europe/Asia	146,877,088
10	Mexico	North America	126,577,691
11	Japan	Asia	126,020,000
12	Philippines	Asia	108,210,625
13	Ethiopia	Africa	107,534,882
14	Egypt	Africa	99,930,038
15	Vietnam	Asia	96,208,984
16	Democratic Rep. of the Congo	Africa	86,790,567
17	Germany	Europe	83,149,300
18	Iran	Asia	83,142,818
19	Turkey	Europe/Asia	82,003,882
20	France	Europe	67,060,000
21	Thailand	Asia	66,461,867
22	United Kingdom	Europe	66,435,600
23	Italy	Europe	60,252,824
24	South Africa	Africa	58,775,022
25	Tanzania	Africa	55,890,747
26	Myanmar	Asia	54,339,766
27	South Korea	Asia	51,811,167
28	Kenya	Africa	47,564,296
29	Spain	Europe	46,934,632
30	Colombia	South America	46,127,200

Source: U.S. Census Bureau, International Data Base

THE WORLD'S SMALLEST COUNTRIES BY POPULATION

Rank	Country	Region	Population
1	Vatican City	Europe	1,000
2	Tuvalu	Pacific Islands	10,200
3	Nauru	Pacific Islands	11,000
4	Palau	Pacific Islands	17,900
5	San Marino	Europe	34,641
6	Monaco	Europe	38,300
7	Liechtenstein	Europe	38,380
8	Marshall Islands	Pacific Islands	55,500
9	Saint Kitts and Nevis	Central America	56,345
10	Dominica	Caribbean	71,808
11	Andorra	Europe	76,177
12	Seychelles	Africa	96,762
13	Tonga	Pacific Islands	100,651
14	Antigua and Barbuda	Caribbean	104,084
15	Federated States of Micronesia	Pacific Islands	105,300
16	Grenada	Caribbean	108,825
17	Saint Vincent and the Grenadines	Central America	110,520
18	Kiribati	Pacific Islands	120,100
19	Saint Lucia	Central America	180,454
20	Samoa	Pacific Islands	199,052
21	São Tomé and Príncipe	Africa	201,784
22	Barbados	Caribbean	287,025
23	Vanuatu	Pacific Islands	304,500
24	Iceland	Europe	360,390
25	Maldives	Asia	378,114
26	Bahamas	Caribbean	386,870
27	Belize	Central America	398,050
28	Brunei	Asia	421,300
29	Malta	Europe	493,559
30	Cape Verde	Africa	550,483

Source: U.S. Census Bureau, International Data Base.

THE WORLD'S MOST DENSELY POPULATED COUNTRIES

				Area		Persons Per Square	
Rank	Country	Region	Population	Sq. Miles	Sq. Km.	Mile	Km.
1	Monaco	Europe	38,300	0.78	2.02	49,106	18,960
2	Singapore	Asia	5,638,700	279	722.5	20,212	7,804
3	Vatican City	Europe	1,000	0.17	0.44	5,887	2,273
4	Bahrain	Asia	1,543,300	300	778	5,136	1,983
5	Malta	Europe	493,559	122	315	3,911	1,510
6	Maldives	Asia	378,114	115	298	3,287	1,269
7	Bangladesh	Asia	167,983,726	55,598	143,998	3,021	1,167
8	Lebanon	Asia	6,855,713	4,036	10,452	1,740	672
9	Barbados	Caribbean	287,025	166	430	1,730	668
10	Taiwan	Asia	23,596,266	13,976	36,197	1,689	652
11	Mauritius	Africa	1,265,577	788	2,040	1,606	620
12	San Marino	Europe	34,641	24	61	1,471	568
13	Nauru	Pacific Islands	11,000	8	21	1,344	519
14	South Korea	Asia	51,811,167	38,691	100,210	1,339	517
15	Rwanda	Africa	12,374,397	10,169	26,338	1,217	470
16	Comoros	Africa	873,724	719	1,861	1,215	469
17	Netherlands	Europe	17,426,881	16,033	41,526	1,087	420
18	Haiti	Caribbean	11,263,077	10,450	27,065	1,077	416
19	Israel	Asia	9,149,500	8,522	22,072	1,074	415
20	India	Asia	1,357,746,150	1,269,211	3,287,240	1,070	413
21	Burundi	Africa	10,953,317	10,740	27,816	1,020	394
22	Tuvalu	Pacific Islands	10,200	10	26	1,015	392
23	Belgium	Europe	11,515,793	11,849	30,689	974	376
24	Philippines	Asia	108,210,625	115,831	300,000	934	361
25	Japan	Asia	126,020,000	145,937	377,975	862	333
26	Sri Lanka	Asia	21,803,000	25,332	65,610	860	332
27	El Salvador	Central America	6,704,864	8,124	21,040	826	319
28	Grenada	Caribbean	108,825	133	344	818	316
29	Marshall Islands	Pacific Islands	55,500	70	181	795	307
30	Saint Lucia	Central America	180,454	238	617	756	292

Source: U.S. Census Bureau, International Data Base

THE WORLD'S LEAST DENSELY POPULATED COUNTRIES

Rank	Country	Region	Population	Area Sq. Miles	Sq. Km.	Persons Per Square Mile	Km.
1	Mongolia	Asia	3,000,000	603,902	1,564,100	4.9	1.9
2	Western Sahara	Africa	567,421	97,344	252,120	6	2.3
3	Namibia	Africa	2,413,643	318,580	825,118	7.5	2.9
4	Australia	Australia	25,594,366	2,969,907	7,692,024	9	3
5	Suriname	South America	568,301	63,251	163,820	9.1	3.5
6	Iceland	Europe	360,390	39,682	102,775	9.1	3.5
7	Guyana	South America	782,225	83,012	214,999	9.3	3.6
8	Libya	Africa	6,470,956	683,424	1,770,060	9.6	3.7
9	Canada	North America	37,898,384	3,855,103	9,984,670	10	4
10	Mauritania	Africa	3,984,233	397,955	1,030,700	10.4	4
11	Botswana	Africa	2,302,878	224,607	581,730	10.4	4
12	Kazakhstan	Asia	18,592,700	1,052,090	2,724,900	18	7
13	Central African Republic	Africa	4,737,423	240,324	622,436	21	8
14	Gabon	Africa	2,067,561	103,347	267,667	21	8
15	Russia	Europe/Asia	146,877,088	6,612,093	17,125,242	23	9
16	Bolivia	South America	11,307,314	424,164	1,098,581	26	10
17	Chad	Africa	15,353,184	495,755	1,284,000	31	12
18	Turkmenistan	Asia	5,851,466	189,657	491,210	31	12
19	Oman	Asia	4,183,841	119,499	309,500	36	14
20	Mali	Africa	19,107,706	482,077	1,248,574	39	15
21	Argentina	South America	44,938,712	1,073,518	2,780,400	41	16
22	Saudi Arabia	Asia	34,218,169	830,000	2,149,690	41	16
23	Republic of the Congo	Africa	5,399,895	132,047	342,000	41	16
24	Finland	Europe	5,527,405	130,666	338,424	41	16
25	Norway	Europe	5,328,212	125,013	323,782	41	16
26	Paraguay	South America	7,052,983	157,048	406,752	44	17
27	Belize	Central America	398,050	8,867	22,965	44	17
28	Algeria	Africa	42,545,964	919,595	2,381,741	47	18
29	Niger	Africa	21,466,863	458,075	1,186,408	47	18
30	Papua New Guinea	Pacific Islands	8,558,800	178,704	462,840	47	18

Source: U.S. Census Bureau, International Data Base

THE WORLD'S MOST POPULOUS CITIES

Rank	City	Country	Region	Population
1	Chongqing	China	Asia	30,484,300
2	Shanghai	China	Asia	24,256,800
3	Beijing	China	Asia	21,516,000
4	Chengdu	China	Asia	16,044,700
5	Karachi	Pakistan	Asia	14,910,352
6	Guangzhou	China	Asia	14,043,500
7	Istanbul	Turkey	Europe	14,025,000
8	Tokyo	Japan	Asia	13,839,910
9	Tianjin	China	Asia	12,784,000
10	Mumbai	India	Asia	12,478,447
11	São Paulo	Brazil	South America	12,252,023
12	Moscow	Russia	Europe/Asia	12,197,596
13	Kinshasa	Dem. Rep. of Congo	Africa	11,855,000
14	Baoding	China	Asia	11,194,372
15	Lahore	Pakistan	Asia	11,126,285
16	Wuhan	China	Asia	11,081,000
17	Delhi	India	Asia	11,034,555
18	Harbin	China	Asia	10,635,971
19	Suzhou	China	Asia	10,459,890
20	Cairo	Egypt	Africa	10,230,350
21	Seoul	South Korea	Asia	10,197,604
22	Jakarta	Indonesia	Asia	10,075,310
23	Lima	Peru	South America	9,174,855
24	Mexico City	Mexico	North America	9,041,395
25	Ho Chi Minh City	Vietnam	Asia	8,993,082
26	Dhaka	Bangladesh	Africa	8,906,039
27	London	United Kingdom	Europe	8,825,001
28	Bangkok	Thailand	Asia	8,750,600
29	Xi'an	China	Asia	8,705,600
30	New York	United States	North America	8,622,698
31	Bangalore	India	Asia	8,425,970
32	Shenzhen	China	Asia	8,378,900
33	Nanjing	China	Asia	8,230,000
34	Tehran	Iran	Asia	8,154,051
35	Rio de Janeiro	Brazil	South America	6,718,903
36	Shantou	China	Asia	5,391,028
37	Kolkata	India	Asia	4,486,679
38	Shijiazhuang	China	Asia	4,303,700
39	Los Angeles	United States	North America	3,884,307
40	Buenos Aires	Argentina	South America	3,054,300

MAJOR LAKES OF THE WORLD

		Surface Area		Maximum Depth	
Lake	Location	Sq. Mi.	Sq. Km.	Feet	Meters
Caspian Sea	Central Asia	152,239	394,299	3,104	946
Superior	North America	31,820	82,414	1,333	406
Victoria	East Africa	26,828	69,485	270	82
Huron	North America	23,010	59,596	750	229
Michigan	North America	22,400	58,016	923	281
Aral	Central Asia	13,000	33,800	223	68
Tanganyika	East Africa	12,700	32,893	4,708	1,435
Baikal	Russia	12,162	31,500	5,712	1,741
Great Bear	North America	12,000	31,080	270	82
Nyasa	East Africa	11,600	30,044	2,316	706
Great Slave	North America	11,170	28,930	2,015	614
Chad	West Africa	9,946	25,760	23	7
Erie	North America	9,930	25,719	210	64
Winnipeg	North America	9,094	23,553	204	62
Ontario	North America	7,520	19,477	778	237
Balkhash	Central Asia	7,115	18,428	87	27
Ladoga	Russia	7,000	18,130	738	225
Onega	Russia	3,819	9,891	361	110
Titicaca	South America	3,141	8,135	1,214	370
Nicaragua	Central America	3,089	8,001	230	70
Athabasca	North America	3,058	7,920	407	124
Rudolf	Kenya, East Africa	2,473	6,405	—	—
Reindeer	North America	2,444	6,330	—	—
Eyre	South Australia	2,400	6,216	varies	varies
Issyk-Kul	Central Asia	2,394	6,200	2,297	700
Urmia	Southwest Asia	2,317	6,001	49	15
Torrens	Australia	2,200	5,698	—	—
Vänern	Sweden	2,141	5,545	322	98
Winnipegosis	North America	2,086	5,403	59	18
Mobutu Sese Seko	East Africa	2,046	5,299	180	55
Nettilling	North America	1,950	5,051	—	—

Note: The sizes of some lakes vary with the seasons.

MAJOR RIVERS OF THE WORLD

River	Region	Source	Outflow	Approximate Length	
				Miles	Km.
Nile	N. Africa	Tributaries of Lake Victoria	Mediterranean Sea	4,180	6,690
Mississippi-Missouri-Red Rock	N. America	Montana	Gulf of Mexico	3,710	5,970
Yangtze Kiang	East Asia	Tibetan Plateau	China Sea	3,602	5,797
Ob	Russia	Altai Mountains	Gulf of Ob	3,459	5,567
Yellow (Huang He)	East Asia	Kunlun Mountains, west China	Gulf of Chihli	2,900	4,667
Yenisei	Russia	Tannu-Ola Mountains, western Tuva, Russia	Arctic Ocean	2,800	4,506
Paraná	S. America	Confluence of Paranaiba and Grande Rivers	Río de la Plata	2,795	4,498
Irtysh	Russia	Altai Mountains, Russia	Ob River	2,758	4,438
Congo	Africa	Confluence of Lualaba and Luapula Rivers, Congo	Atlantic Ocean	2,716	4,371
Heilong (Amur)	East Asia	Confluence of Shilka and Argun Rivers	Tatar Strait	2,704	4,352
Lena	Russia	Baikal Mountains, Russia	Arctic Ocean	2,652	4,268
Mackenzie	N. America	Head of Finlay River, British Columbia, Canada	Beaufort Sea	2,635	4,241
Niger	West Africa	Guinea	Gulf of Guinea	2,600	4,184
Mekong	Asia	Tibetan Plateau	South China Sea	2,500	4,023
Mississippi	N. America	Lake Itasca, Minnesota	Gulf of Mexico	2,348	3,779
Missouri	N. America	Confluence of Jefferson, Gallatin, and Madison Rivers, Montana	Mississippi River	2,315	3,726
Volga	Russia	Valdai Plateau, Russia	Caspian Sea	2,291	3,687
Madeira	S. America	Confluence of Beni and Maumoré Rivers, Bolivia-Brazil boundary	Amazon River	2,012	3,238
Purus	S. America	Peruvian Andes	Amazon River	1,993	3,207
São Francisco	S. America	S.W. Minas Gerais, Brazil	Atlantic Ocean	1,987	3,198

River	Region	Source	Outflow	Approximate Length	
				Miles	Km.
Yukon	N. America	Junction of Lewes and Pelly Rivers, Yukon Terr., Canada	Bering Sea	1,979	3,185
St. Lawrence	N. America	Lake Ontario	Gulf of St. Lawrence	1,900	3,058
Rio Grande	N. America	San Juan Mountains, Colorado	Gulf of Mexico	1,885	3,034
Brahmaputra	Asia	Himalayas	Ganges River	1,800	2,897
Indus	Asia	Himalayas	Arabian Sea	1,800	2,897
Danube	Europe	Black Forest, Germany	Black Sea	1,766	2,842
Euphrates	Asia	Confluence of Murat Nehri and Kara Su Rivers, Turkey	Shatt-al-Arab	1,739	2,799
Darling	Australia	Eastern Highlands, Australia	Murray River	1,702	2,739
Zambezi	Africa	Western Zambia	Mozambique Channel	1,700	2,736
Tocantins	S. America	Goiás, Brazil	Pará River	1,677	2,699
Murray	Australia	Australian Alps, New S. Wales	Indian Ocean	1,609	2,589
Nelson	N. America	Head of Bow River, western Alberta, Canada	Hudson Bay	1,600	2,575
Paraguay	S. America	Mato Grosso, Brazil	Paraná River	1,584	2,549
Ural	Russia	Southern Ural Mountains, Russia	Caspian Sea	1,574	2,533
Ganges	Asia	Himalayas	Bay of Bengal	1,557	2,506
Amu Darya (Oxus)	Asia	Nicholas Range, Pamir Mountains, Turkmenistan	Aral Sea	1,500	2,414
Japurá	S. America	Andes, Colombia	Amazon River	1,500	2,414
Salween	Asia	Tibet, south of Kunlun Mountains	Gulf of Martaban	1,500	2,414
Arkansas	N. America	Central Colorado	Mississippi River	1,459	2,348
Colorado	N. America	Grand County, Colorado	Gulf of California	1,450	2,333
Dnieper	Russia	Valdai Hills, Russia	Black Sea	1,419	2,284
Ohio-Allegheny	N. America	Potter County, Pennsylvania	Mississippi River	1,306	2,102
Irrawaddy	Asia	Confluence of Nmai and Mali rivers, northeast Burma	Bay of Bengal	1,300	2,092
Orange	Africa	Lesotho	Atlantic Ocean	1,300	2,092

| River | Region | Source | Outflow | Approximate Length | |
				Miles	Km.
Orinoco	S. America	Serra Parima Mountains, Venezuela	Atlantic Ocean	1,281	2,062
Pilcomayo	S. America	Andes Mountains, Bolivia	Paraguay River	1,242	1,999
Xi Jiang	East Asia	Eastern Yunnan Province, China	China Sea	1,236	1,989
Columbia	N. America	Columbia Lake, British Columbia, Canada	Pacific Ocean	1,232	1,983
Don	Russia	Tula, Russia	Sea of Azov	1,223	1,968
Sungari	East Asia	China-North Korea boundary	Amur River	1,215	1,955
Saskatchewan	N. America	Canadian Rocky Mountains	Lake Winnipeg	1,205	1,939
Peace	N. America	Stikine Mountains, British Columbia, Canada	Great Slave River	1,195	1,923
Tigris	Asia	Taurus Mountains, Turkey	Shatt-al-Arab	1,180	1,899

THE HIGHEST PEAKS IN EACH CONTINENT

| | | | Height | |
Continent	Mountain	Location	Feet	Meters
Asia	Everest	Tibet & Nepal	29,028	8,848
South America	Aconcagua	Argentina	22,834	6,960
North America	McKinley	Alaska	20,320	6,194
Africa	Kilimanjaro	Tanzania	19,340	5,895
Europe	Elbrus	Russia & Georgia	18,510	5,642
Antarctica	Vinson Massif	Ellsworth Mountains	16,066	4,897
Australia	Kosciusko	New South Wales	7,316	2,228

Note: The world's highest sixty-six mountains are all in Asia.

MAJOR DESERTS OF THE WORLD

| Desert | Location | Approximate Area | | Type |
		Sq. Miles	Sq. Km.	
Antarctic	Antarctica	5,400,000	14,002,200	polar
Sahara	North Africa	3,500,000	9,075,500	subtropical
Arabian	Southwest Asia	1,000,000	2,593,000	subtropical
Great Western (Gibson, Great Sandy, and Great Victoria)	Australia	520,000	1,348,360	subtropical
Gobi	East Asia	500,000	1,296,500	cold winter
Patagonian	Argentina, South America	260,000	674,180	cold winter
Kalahari	Southern Africa	220,000	570,460	subtropical
Great Basin	Western United States	190,000	492,670	cold winter
Thar	South Asia	175,000	453,775	subtropical
Chihuahuan	Mexico	175,000	453,775	subtropical
Karakum	Central Asia	135,000	350,055	cold winter
Colorado Plateau	Southwestern United States	130,000	337,090	cold winter
Sonoran	United States and Mexico	120,000	311,160	subtropical
Kyzylkum	Central Asia	115,000	298,195	cold winter
Taklimakan	China	105,000	272,265	cold winter
Iranian	Iran	100,000	259,300	cold winter
Simpson	Eastern Australia	56,000	145,208	subtropical
Mojave	Western United States	54,000	140,022	subtropical
Atacama	Chile, South America	54,000	140,022	cold coastal
Namib	Southern Africa	13,000	33,709	cold coastal
Arctic	Arctic Circle			polar

HIGHEST WATERFALLS OF THE WORLD

			Height	
Waterfall	*Location*	*Source*	*Feet*	*Meters*
Angel	Canaima National Park, Venezuela	Rio Caroni	3,212	979
Tugela	Natal National Park, South Africa	Tugela River	3,110	948
Utigord	Norway	glacier	2,625	800
Monge	Marstein, Norway	Mongebeck	2,540	774
Mutarazi	Nyanga National Park, Zimbabwe	Mutarazi River	2,499	762
Yosemite	Yosemite National Park, California, U.S.	Yosemite Creek	2,425	739
Espelands	Hardanger Fjord, Norway	Opo River	2,307	703
Lower Mar Valley	Eikesdal, Norway	Mardals Stream	2,151	655
Tyssestrengene	Odda, Norway	Tyssa River	2,123	647
Cuquenan	Kukenan Tepuy, Venezuela	Cuquenan River	2,000	610
Sutherland	Milford Sound, New Zealand	Arthur River	1,904	580
Kjell	Gudvanger, Norway	Gudvangen Glacier	1,841	561
Takkakaw	Yoho Natl Park, British Columbia, Canada	Takkakaw Creek	1,650	503
Ribbon	Yosemite National Park, California, U.S.	Ribbon Stream	1,612	491
Upper Mar Valley	near Eikesdal, Norway	Mardals Stream	1,536	468
Gavarnie	near Lourdes, France	Gave de Pau	1,388	423
Vettis	Jotunheimen, Norway	Utla River	1,215	370
Hunlen	British Columbia, Canada	Hunlen River	1,198	365
Tin Mine	Kosciusko National Park, Australia	Tin Mine Creek	1,182	360
Silver Strand	Yosemite National Park, California, U.S.	Silver Strand Creek	1,170	357
Basaseachic	Baranca del Cobre, Mexico	Piedra Volada Creek	1,120	311
Spray Stream	Lauterburnnental, Switzerland	Staubbach Brook	985	300
Fachoda	Tahiti, French Polynesia	Fautaua River	985	300
King Edward VIII	Guyana	Courantyne River	850	259
Wallaman	near Ingham, Australia	Wallaman Creek	844	257
Gersoppa	Western Ghats, India	Sharavati River	828	253
Kaieteur	Guyana	Rio Potaro	822	251
Montezuma	near Rosebery, Tasmania	Montezuma River	800	240
Wollomombi	near Armidale, Australia	Wollomombi River	722	2203

Source: Fifth Continent Australia Pty Limited

431

GLOSSARY

Places whose names are printed in SMALL CAPS *are subjects of their own entries in this glossary.*

Ablation. Loss of ice volume or mass by a GLACIER. Ablation includes melting of ice, SUBLIMATION, DEFLATION (removal by WIND), EVAPORATION, and CALVING. Ablation occurs in the lower portions of glaciers.

Abrasion. Wearing away of ROCKS in STREAMS by grinding, especially when rocks and SEDIMENT are carried along by stream water. The STREAMBED and VALLEY are carved out and eroded, and the rocks become rounded and smoothed by abrasion.

Absolute location. Position of any PLACE on the earth's surface. The absolute location can be given precisely in terms of DEGREES, MINUTES, and SECONDS of LATITUDE (0 to 90 degrees north or south) and of LONGITUDE (0 to 180 degrees east or west). The EQUATOR is 0 degrees latitude; the PRIME MERIDIAN, which runs through Greenwich in England, is 0 degrees longitude.

Abyss. Deepest part of the OCEAN. Modern TECHNOLOGY—especially sonar—has enabled accurate mapping of the ocean floors, showing that there are MOUNTAIN CHAINS, or RIDGES, in all the oceans, as well as deep CANYONS or TRENCHES closer to the edges of the oceans.

Acid rain. PRECIPITATION containing high levels of nitric or sulfuric acid; a major environmental problem in parts of North America, Europe, and Asia. Natural precipitation is slightly acidic (about 5.6 on the pH SCALE), because CARBON DIOXIDE—which occurs naturally in the ATMOSPHERE—is dissolved to form a weak carbonic acid.

Adiabatic. Change of TEMPERATURE within the ATMOSPHERE that is caused by compression or expansion without addition or loss of heat.

Advection. Horizontal movement of AIR from one PLACE to another in the ATMOSPHERE, associated with WINDS.

Advection fog. FOG that forms when a moist AIR mass moves over a colder surface. Commonly, warm moist air moves over a cool OCEAN CURRENT, so the air cools to SATURATION POINT and fog forms. This phenomenon, known as sea fog, occurs along subtropical west COASTS.

Aerosol. Substances held in SUSPENSION in the ATMOSPHERE, as solid particles or liquid droplets.

Aftershock. EARTHQUAKE that follows a larger earthquake and originates at or near the focus of the latter; many aftershocks may follow a major earthquake, decreasing in frequency and magnitude with time.

Agglomerate. Type of ROCK composed of volcanic fragments, usually of different sizes and rough or angular.

Aggradation. Accumulation of SEDIMENT in a STREAMBED. Aggradation often results from reduced flow in the channel during dry periods. It also occurs when the STREAM's load (BEDLOAD and SUSPENDED LOAD) is greater than the stream capacity. A BRAIDED STREAM pattern often results.

Air current. Air currents are caused by differential heating of the earth's surface, which causes heated air to rise. This causes WINDS at the surface as well as higher in the earth's ATMOSPHERE.

Air mass. Large body of air with distinctive homogeneous characteristics of TEMPERATURE, HUMIDITY, and stability. It forms when air remains stationary over a source REGION for a period of time, taking on the conditions of that region. An air mass can extend over a million square miles with a depth of more than a mile. Air masses are classified according to moisture content (*m* for maritime or *c* for continental) and temperature

433

(*A* for ARCTIC, *P* for polar, *T* for tropical, or *E* for equatorial). The air masses affecting North America are mP, cP, and mT. The interaction of AIR masses produces WEATHER. The line along which air masses meet is a FRONT.

Albedo. Measure of the reflective properties of a surface; the ratio of reflected ENERGY (INSOLATION) to the total incoming energy, expressed as a percentage. The albedo of Earth is 33 percent.

Alienation (land). Land alienation is the appropriation of land from its original owners by a more powerful force. In preindustrial societies, the ownership of agricultural land is of prime importance to subsistence farmers.

Alkali flat. Dry LAKEBED in an arid REGION, covered with a layer of SALTS. A well-known example is the Alkali Flat area of White Sands National Monument in New Mexico; it is the bed of a large lake that formed when the GLACIERS were melting. It is covered with a form of gypsum crystals called selenite. This material is blown off the surface into large SAND DUNES. Also called a salina. See also BITTER LAKE.

Allogenic sediment. SEDIMENT that originates outside the PLACE where it is finally deposited; SAND, SILT, and CLAY carried by a STREAM into a LAKE are examples.

Alluvial fan. Common LANDFORM at the mouth of a CANYON in arid REGIONS. Water flowing in a narrow canyon immediately slows as it leaves the canyon for the wider VALLEY floor, depositing the SEDIMENTS it was transporting. These spread out into a fan shape, usually with a BRAIDED STREAM pattern on its surface. When several alluvial fans grow side by side, they can merge into one continuous sloping surface between the HILLS and the valley. This is known by the Spanish word *bajada*, which means "slope."

Alluvial system. Any of various depositional systems, excluding DELTAS, that form from the activity of RIVERS and STREAMS. Much alluvial SEDIMENT is deposited when rivers top their BANKS and FLOOD the surrounding countryside. Buried alluvial sediments may be important water-bearing RESERVOIRS or may contain PETROLEUM.

Alluvium. Material deposited by running water. This includes not only fertile SOILS, but also CLAY, SILT, or SAND deposits resulting from FLUVIAL processes. FLOODPLAINS are covered in a thick layer of alluvium.

Altimeter. Instrument for measuring ALTITUDE, or height above the earth's surface, commonly used in airplanes. An altimeter is a type of ANEROID BAROMETER.

Altitudinal zonation. Existence of different ECOSYSTEMS at various ELEVATIONS above SEA LEVEL, due to TEMPERATURE and moisture differences. This is especially pronounced in Central America and South America. The hot and humid COASTAL PLAINS, where bananas and sugarcane thrive, is the *tierra caliente*. From about 2,500 to 6,000 feet (750–1,800 meters) is the *tierra templada*; crops grown here include coffee, wheat, and corn, and major cities are situated in this zone. From about 6,000 to 12,000 feet (1,800–3,600 meters) is the *tierra fria*; here only hardy crops such as potatoes and barley are grown, and large numbers of animals are kept. From about 12,000 to 15,000 feet (3,600 to 4,500 meters) lies the *tierra helada*, where hardy animals such as sheep and alpaca graze. Above 15,000 feet (4,500 meters) is the frozen *tierra nevada*; no permanent life is possible in the permanent SNOW and ICE FIELDS there.

Angle of repose. Maximum angle of steepness that a pile of loose materials such as SAND or ROCK can assume and remain stable; the angle varies with the size, shape, moisture, and angularity of the material.

Antecedent river. STREAM that was flowing before the land was uplifted and was able to erode at the pace of UPLIFT, thus creating a deep CANYON. Most deep canyons are attributed to antecedent rivers. In the Davisian CYCLE OF EROSION, this process was called REJUVENATION.

Anthropogeography. Branch of GEOGRAPHY founded in the late nineteenth century by German geographer Friedrich Ratzel. The field is closely related to human ECOLOGY—the study of humans, their DISTRIBUTION over the earth, and their interaction with their physical ENVIRON-MENT.

Anticline. Area where land has been UPFOLDED symmetrically. Its center contains stratigraphically older ROCKS. See also SYNCLINE.

Anticyclone. High-pressure system of rotating WINDS, descending and diverging, shown on a WEATHER chart by a series of closed ISOBARS, with a high in the center. In the NORTHERN HEMISPHERE, the rotation is CLOCKWISE; in the SOUTHERN HEMISPHERE, the rotation is COUNTERCLOCKWISE. An anticyclone brings warm weather.

Antidune. Undulatory upstream-moving bed form produced in free-surface flow of water over a SAND bed in a certain RANGE of high flow speeds and shallow flow depths.

Antipodes. TEMPERATE ZONE of the SOUTHERN HEMISPHERE. The term is now usually applied to the countries of Australia and New Zealand. The ancient Greeks had believed that if humans existed there, they must walk upside down. This idea was supported by the Christian Church in the Middle Ages.

Antitrade winds. WINDS in the upper ATMOSPHERE, or GEOSTROPHIC winds, that blow in the opposite direction to the TRADE WINDS. Antitrade winds blow toward the northeast in the NORTHERN HEMISPHERE and toward the southeast in the SOUTHERN HEMISPHERE.

Aperiodic. Irregularly occurring interval, such as found in most WEATHER CYCLES, rendering them virtually unpredictable.

Aphelion. Point in the earth's 365-DAY REVOLUTION when it is at its greatest distance from the SUN. This is caused by Earth's elliptical ORBIT around the Sun. The distance at aphelion is 94,555,000 miles (152,171,500 km.) and usually falls on July 4. The opposite of PERIHELION.

Aposelene. Earth's farthest point from the MOON.

Aquifer. Underground body of POROUS ROCK that contains water and allows water PERCOLATION through it. The largest aquifer in the United States is the Ogallala Aquifer, which extends south from South Dakota to Texas.

Arête. Serrated or saw-toothed ridge, produced in glaciated MOUNTAIN areas by CIRQUES eroding on either side of a RIDGE or mountain RANGE. From the French word for knife-edge.

Arroyo. Spanish word for a dry STREAMBED in an arid area. Called a WADI in Arabic and a WASH in English.

Artesian well. WELL from which GROUNDWATER flows without mechanical pumping, because the water comes from a CONFINED AQUIFER, and is therefore under pressure. The Great Artesian Basin of Australia has hundreds of artesian wells, called BORES, that provide drinking water for sheep and cattle. The name comes from the Artois REGION of France, where the phenomenon is common. A subartesian well is sunk into an UNCONFINED AQUIFER and requires a pump to raise water to the surface.

Asteroid belt. REGION between the ORBITS of Mars and Jupiter containing the majority of ASTEROIDS.

Asthenosphere. Part of the earth's UPPER MANTLE, beneath the LITHOSPHERE, in which PLATE movement takes place. Also known as the low-velocity zone.

Astrobleme. Remnant of a large IMPACT CRATER on Earth.

Astronomical unit (AU). Unit of measure used by astronomers that is equivalent to the average distance from the SUN to Earth (93 million miles/150 million km.).

Atmospheric pressure. Weight of the earth's ATMOSPHERE, equally distributed over earth's surface and pressing down as a result of GRAVITY. On average, the atmosphere has a force of 14.7 pounds per square inch (1 kilogram per centimeter) squared at SEA LEVEL, also expressed as 1013.2 millibars. Variations in atmospheric pressure, high or low, cause WINDS and WEATHER changes that affect CLIMATE. Pressure decreases rapidly with ALTITUDE or distance from the surface: Half of the total atmosphere is found below 18,000 feet (5,500 meters); more than 99 percent of the atmosphere is below 30 miles (50 km.) of the surface. Atmospheric pressure is measured with a BAROMETER.

Atoll. Ring-shaped growth of CORAL REEF, with a LA-GOON in the middle. Charles Darwin, who observed many Pacific atolls during his voyage on the *Beagle* in the nineteenth century, suggested that they were created from FRINGING REEFS around volcanic ISLANDS. As such islands sank beneath the water (or as SEA LEVELS rose), the coral continued growing upward. SAND resting atop an atoll enables plants to grow, and small human societies have arisen on some atolls. The world's largest atoll, Kwajalein in the Marshall Islands, measures about 40 by 18 miles (65 by 30 km.), but perhaps the most famous atoll is Bikini Atoll—the SITE of nuclear-bomb testing during the 1950s.

Aurora. Glowing and shimmering displays of colored lights in the upper ATMOSPHERE, caused by interaction of the SOLAR WIND and the charged particles of the IONOSPHERE. Auroras occur at high LATITUDES. Near the North Pole they are called aurora borealis or northern lights; near the South Pole, aurora australis or southern lights.

Austral. Referring to an object or occurrence that is located in the SOUTHERN HEMISPHERE or related to Australia.

Australopithecines. Erect-walking early human ancestors with a cranial capacity and body size within the RANGE of modern apes rather than of humans.

Avalanche. Mass of SNOW and ice falling suddenly down a MOUNTAIN slope, often taking with it earth, ROCKS, and trees.

Bank. Elevated area of land beneath the surface of the OCEAN. The term is also used for elevated ground lining a body of water.

Bar (climate). Measure of ATMOSPHERIC PRESSURE per unit surface area of 1 million dynes per square centimeter. Millibars (thousandths of a bar) are the MEASUREMENT used in the United States. Other countries use kilopascals (kPa); one kilopascal is ten millibars.

Bar (land). RIDGE or long deposit of SAND or gravel formed by DEPOSITION in a RIVER or at the COAST. Offshore bars and baymouth bars are common coastal features.

Barometer. Instrument used for measuring ATMO-SPHERIC PRESSURE. In the seventeenth century, Evangelista Torricelli devised the first barometer—a glass tube sealed at one end, filled with mercury, and upended into a bowl of mercury. He noticed how the height of the mercury column changed and realized this was a result of the pressure of air on the mercury in the bowl. Early MEASUREMENTS of atmospheric pressure were, therefore, expressed as centimeters of mercury, with average pressure at SEA LEVEL being 29.92 inches (760 millimeters). This cumbersome barometer was replaced with the ANEROID BAROME-TER—a sealed and partially evacuated box connected to a needle and dial, which shows changes in atmospheric pressure. See also ALTIMETER.

Barrier island. Long chain of SAND islands that forms offshore, close to the COAST. LAGOONS or shallower MARSHES separate the barrier islands from the mainland. Such LOCATIONS are hazardous for SETTLEMENTS because they are easily swept away in STORMS and HURRICANES.

Basalt. IGNEOUS EXTRUSIVE ROCK formed when LAVA cools; often black in color. Sometimes basalt occurs in tall hexagonal columns, such as the Giant's Causeway in Ireland, or the Devils Postpile at Mammoth, California.

Basement. Crystalline, usually PRECAMBRIAN, IG-NEOUS and METAMORPHIC ROCKS that occur beneath the SEDIMENTARY ROCK on the CONTI-NENTS.

Basin order. Approximate measure of the size of a STREAM BASIN, based on a numbering scheme applied to RIVER channels as they join together in their progress downstream.

Batholith. Large LANDFORM produced by IGNEOUS INTRUSION, composed of CRYSTALLINE ROCK, such as GRANITE; a large PLUTON with a surface area greater than 40 square miles (100 sq. km.). Most mountain RANGES have a batholith underneath.

Bathymetric contour. Line on a MAP of the OCEAN floor that connects points of equal depth.

Beaufort scale. SCALE that measures WIND force, expressed in numbers from 0 to 12. The original Beaufort scale was based on descriptions of the state of the SEA. It was adapted to land conditions, using descriptions of chimney smoke, leaves of trees, and similar factors. The scale was devised in the early nineteenth century by Sir Francis Beaufort, a British naval officer.

Belt. Geographical REGION that is distinctive in some way.

Bergeron process. PRECIPITATION formation in COLD CLOUDS whereby ice crystals grow at the expense of supercooled water droplets.

Bight. Wide or open BAY formed by a curve in the COASTLINE, such as the Great Australian Bight.

Biogenic sediment. SEDIMENT particles formed from skeletons or shells of microscopic plants and animals living in seawater.

Biostratigraphy. Identification and organization of STRATA based on their FOSSIL content and the use of fossils in stratigraphic correlation.

Bitter lake. Saline or BRACKISH LAKE in an arid area, which may dry up in the summer or in periods of DROUGHT. The water is not suitable for drinking. Another name for this feature is "salina." See also ALKALI FLAT.

Block lava. LAVA flows whose surfaces are composed of large, angular blocks; these blocks are generally larger than those of AA flows and have smooth, not jagged, faces.

Block mountain. MOUNTAIN or mountain RANGE with one side having a gentle slope to the crest, while the other slope, which is the exposed FAULT SCARP, is quite steep. It is formed when a large block of the earth's CRUST is thrust upward on one side only, while the opposite side remains in place. The Sierra Nevada in California are a good example of block mountains. Also known as fault-block mountain.

Blowhole. SEA CAVE or tunnel formed on some rocky, rugged COASTLINES. The pressure of the seawater rushing into the opening can force a jet of seawater to rise or spout through an opening in the roof of the cave. Blowholes are found in Scotland, Tasmania, and Mexico, and on the Hawaiian ISLANDS of Kauai and Maui.

Bluff. Steep slope that marks the farthest edge of a FLOODPLAIN.

Bog. Damp, spongy ground surface covered with decayed or decaying VEGETATION. Bogs usually are formed in cool CLIMATES through the in-filling, or silting up, of a LAKE. Moss and other plants grow outward toward the edge of the lake, which gradually becomes shallower, until the surface is completely covered. Bogs also can form on cold, damp MOUNTAIN surfaces. Many bogs are filled with PEAT.

Bore. Standing WAVE, or wall, of water created in a narrow ESTUARY when the strong incoming, or FLOOD, TIDE meets the RIVER water flowing outward; it moves upstream with the advancing tide, and downstream with the EBB TIDE. South America's Amazon River and Asia's Mekong River have large bores. In North America, the bore in the Bay of Fundy is visited by many tourists each year. Its St. Andrew's wharf is designed to handle changes in water level of as much as 53 feet (15 meters) in one DAY.

Boreal. Alluding to an item or event that is in the NORTHERN HEMISPHERE.

Bottom current. Deep-sea current that flows parallel to BATHYMETRIC CONTOURS.

Brackish water. Water with SALT content between that of SALT WATER and FRESH WATER; it is common in arid areas on the surface, in coastal MARSHES, and in salt-contaminated GROUNDWATER.

Braided stream. STREAM having a CHANNEL consisting of a maze of interconnected small channels within a broader STREAMBED. Braiding occurs when the stream's load exceeds its capacity, usually because of reduced flow.

Breaker. WAVE that becomes oversteepened as it approaches the SHORE, reaching a point at which it cannot maintain its vertical shape. It then breaks, and the water washes toward the shore.

Breakwater. Large structure, usually of ROCK, built offshore and parallel to the COAST, to absorb WAVE ENERGY and thus protect the SHORE. Between the breakwater and the shore is an area of calm water, often used as a boat anchorage or

HARBOR. A similar but smaller structure is a seawall.

Breeze. Gentle WIND with a speed of 4 to 31 miles (6 to 50 km.) per hour. On the BEAUFORT SCALE, the numbers 2 through 6 represent breezes of increasing strength.

Butte. Flat-topped HILL, smaller than a MESA, found in arid REGIONS.

Caldera. Large circular depression with steep sides, formed when a VOLCANO explodes, blowing away its top. The ERUPTION of Mount St. Helens produced a caldera. Crater Lake in Oregon is a caldera that has filled with water. From the Spanish word for kettle.

Calms of Cancer. Subtropical BELT of high pressure and light WINDS, located over the OCEAN near 25 DEGREES north LATITUDE. Also known as the HORSE LATITUDES.

Calms of Capricorn. Subtropical BELT of high pressure and light WINDS, located over the OCEAN near 25 DEGREES south LATITUDE.

Calving. Loss of glacial mass when GLACIERS reach the SEA and large blocks of ice break off, forming ICEBERGS.

Cancer, tropic of. PARALLEL of LATITUDE at 23.5 DEGREES north; this line is the latitude farthest north on the earth where the noon SUN is ever directly overhead. The REGION between it and the tropic of CAPRICORN is known as the TROPICS.

Capricorn, tropic of. Line of LATITUDE at 23.5 DEGREES south; this line is the latitude farthest south on the earth where the noon SUN is ever directly overhead. The REGION between it and the tropic of CANCER is known as the TROPICS.

Carbon dating. Method employed by physicists to determine the age of organic matter—such as a piece of wood or animal tissue—to determine the age of an archaeological or paleontological SITE. The method works on the principle that the amount of radioactive carbon in living matter diminishes at a steady, and measurable, rate after the matter dies. Technique is also known as carbon-14 dating, after the radioactive car-

bon-14 isotope it uses. Also known as radiocarbon dating.

Carrying capacity. Number of animals that a given area of land can support, without additional feed being necessary. Lush GRASSLAND may have a carrying capacity of twenty sheep per acre, while more arid, SEMIDESERT land may support only two sheep per acre. The term sometimes is used to refer to the number of humans who can be supported in a given area.

Catastrophism. Theory, popular in the eighteenth and nineteenth centuries, that explained the shape of LANDFORMS and CONTINENTS and the EXTINCTION of species as the results of intense or catastrophic events. The biblical FLOOD of Noah was one such event, which supposedly explained many extinctions. Catastrophism is linked closely to the belief that the earth is only about 6,000 years old, and therefore tremendous forces must have acted swiftly to create present LANDSCAPES. An alternative or contrasting theory is UNIFORMITARIANISM.

Catchment basin. Area of land receiving the PRECIPITATION that flows into a STREAM. Also called catchment or catchment area.

Central place theory. Theory that explains why some SETTLEMENTS remain small while others grow to be middle-sized TOWNS, and a few become large cities or METROPOLISES. The explanation is based on the provision of goods and services and how far people will travel to acquire these. The German geographer Walter Christaller developed this theory in the 1930s.

Centrality. Measure of the number of functions, or services, offered by any CITY in a hierarchy of cities within a COUNTRY or a REGION. See also CENTRAL PLACE THEORY.

Chain, mountain. Another term for mountain RANGE.

Chemical farming. Application of artificial FERTILIZERS to the SOIL and the use of chemical products such as insecticides, fungicides, and herbicides to ensure crop success. Chemical farming is practiced mainly in high-income countries, because the cost of the chemical products is high. Farmers in low-income economies rely more on natural organic fertilizers such as animal waste.

Chemical weathering. Chemical decomposition of solid ROCK by processes involving water that change its original materials into new chemical combinations.

Chlorofluorocarbons (CFCs). Manufactured compounds, not occurring in nature, consisting of chlorine, fluorine, and carbon. CFCs are stable and have heat-absorbing properties, so they have been used extensively for cooling in refrigeration and air-conditioning units. Previously, they were used as propellants for aerosol products. CFCs rise into the STRATOSPHERE where ULTRAVIOLET RADIATION causes them to react with OZONE, changing it to oxygen and exposing the earth to higher levels of ultraviolet (UV) radiation. Therefore, the manufacture and use of CFCs was banned in many countries. The commercial name for CFCs is Freon.

Chorology. Description or mapping of a REGION. Also known as chorography.

Chronometer. Highly accurate CLOCK or timekeeping device. The first accurate and effective chronometers were constructed in the mid-eighteenth century by John Harrison, who realized that accurate timekeeping was the secret to NAVIGATION at SEA.

Cinder cone. Small conical HILL produced by PYROCLASTIC materials from a VOLCANO. The material of the cone is loose SCORIA.

Circle of illumination. Line separating the sunlit part of the earth from the part in darkness. The circle of illumination moves around the earth once in every approximately 24 hours. At the VERNAL and autumnal EQUINOXES, the circle of illumination passes through the POLES.

Cirque. Circular BASIN at the head of an ALPINE GLACIER, shaped like an armchair. Many cirques can be seen in MOUNTAIN areas where glaciers have completely melted since the last ICE AGE.

City Beautiful movement. Planning and architectural movement that was at its height from around 1890 to the 1920s in the United States. It was believed that classical architecture, wide and carefully laid-out streets, parks, and urban monuments would reflect the higher values of the society and be a civilizing, even uplifting, experience for the citizens of such cities. Civic pride was fostered through remodeling or modernizing older URBAN AREAS. Chicago, Illinois, and Pasadena, California, are cities where the planners of the City Beautiful movement left their imprint.

Clastic. ROCK or sedimentary matter formed from fragments of older rocks.

Climatology. Study of Earth CLIMATES by analysis of long-term WEATHER patterns over a minimum of thirty years of statistical records. Climatologists—scientists who study climate—seek similarities to enable grouping into climatic REGIONS. Climate patterns are closely related to natural VEGETATION. Computer TECHNOLOGY has enabled investigation of phenomena such as the EL NIÑO effect and global climate change. The KÖPPEN CLIMATE CLASSIFICATION system is the most commonly used scheme for climate classification.

Climograph. Graph that plots TEMPERATURE and PRECIPITATION for a selected LOCATION. The most commonly used climographs plot monthly temperatures and monthly precipitation, as used in the KÖPPEN CLIMATE CLASSIFICATION. Also spelled "climagraph." The term climagram is rarely used.

Clinometer. Instrument used by surveyors to measure the ELEVATION of land or the inclination (slope) of the land surface.

Cloud seeding. Injection of CLOUD-nucleating particles into likely clouds to enhance PRECIPITATION.

Cloudburst. Heavy rain that falls suddenly.

Coal. One of the FOSSIL FUELS. Coal was formed from fossilized plant material, which was originally FOREST. It was then buried and compacted, which led to chemical changes. Most coal was formed during the CARBONIFEROUS PERIOD (286 million to 360 million years ago) when the earth's CLIMATE was wetter and warmer than at present.

Coastal plain. Large area of flat land near the OCEAN. Coastal plains can form in various ways, but FLUVIAL DEPOSITION is an important process. In the United States, the coastal plain extends from Texas to North Carolina.

Coastal wetlands. Shallow, wet, or flooded shelves that extend back from the freshwater-saltwater interface and may consist of MARSHES, BAYS, LAGOONS, tidal flats, or MANGROVE SWAMPS.

Cognitive map. Mental image that each person has of the world, which includes LOCATIONS and connections. These maps expand as children mature, from plans of their rooms, to their houses, to their neighborhoods. Adults know certain parts of the CITY and the streets connecting them.

Coke. Type of fuel produced by heating COAL.

Col. Lower section of a RIDGE, usually formed by the headward EROSION of two CIRQUE GLACIERS at an ARÊTE. Sometimes called a saddle.

Colonialism. Control of one COUNTRY over another STATE and its people. Many European countries have created colonial empires, including Great Britain, France, Spain, Portugal, the Netherlands, and Russia.

Columbian exchange. Interaction that occurred between the Americas and Europe after the voyages of Christopher Columbus. Food crops from the New World transformed the diet of many European countries.

Comet. Small body in the SOLAR SYSTEM, consisting of a solid head with a long gaseous tail. The elliptical ORBIT of a comet causes it to range from very close to the SUN to very far away. In ancient times, the appearance of a comet in the sky was thought to be an omen of great events or changes, such as war or the death of a king.

Comfort index. Number that expresses the combined effects of TEMPERATURE and HUMIDITY on human bodily comfort. The index number is obtained by measuring ambient conditions and comparing these to a chart.

Commodity chain. Network linking labor, production, delivery, and sale for any product. The chain begins with the production of the raw material, such as the extraction of MINERALS by miners, and extends to the acquisition of the finished product by a consumer.

Complex crater. IMPACT CRATER of large diameter and low depth-to-diameter ratio caused by the presence of a central UPLIFT or ring structure.

Composite cone. Cone or VOLCANO formed by volcanic explosions in which the LAVA is of different composition, sometimes fluid, sometimes PYROCLASTS such as cinders. The alternation of layers allows a concave shape for the cone. These are generally regarded as the world's most beautiful volcanoes. Composite volcanoes are sometimes called STRATOVOLCANOES.

Condensation nuclei. Microscopic particles that may have originated as DUST, soot, ASH from fires or VOLCANOES, or even SEA SALT; an essential part of CLOUD formation. When AIR rises and cools to the DEW POINT (saturation), the moisture droplets condense around the nuclei, leading to the creation of raindrops or snowflakes. A typical air mass might contain 10 billion condensation nuclei in a single cubic yard (1 cubic meter) of air.

Cone of depression. Cone-shaped depression produced in the WATER TABLE by pumping from a WELL.

Confined aquifer. AQUIFER that is completely filled with water and whose upper BOUNDARY is a CONFINING BED; it is also called an artesian aquifer.

Confining bed. Impermeable layer in the earth that inhibits vertical water movement.

Confluence. PLACE where two STREAMS or RIVERS flow together and join. The smaller of the two streams is called a TRIBUTARY.

Conglomerate. Type of SEDIMENTARY ROCK consisting of smaller rounded fragments naturally cemented together by another MINERAL. If the cemented fragments are jagged or angular, the rock is called breccia.

Conical projection. MAP PROJECTION that can be imagined as a cone of paper resting like a witch's hat on a globe with a light source at its center; the images of the CONTINENTS would be projected onto the paper. In reality, maps are constructed mathematically. A conic projection can show only part of one HEMISPHERE. This projection is suitable for constructing a MAP of the United States, as a good EQUAL-AREA representation can be achieved. Also called conic projection.

Consequent river. RIVER that flows across a LANDSCAPE because of GRAVITY. Its direction is determined by the original slope of the land. TRIBUTARY streams, which develop later as EROSION proceeds, are called subsequent streams.

Continental climate. CLIMATE experienced over the central REGIONS of large LANDMASSES; drier and subject to greater seasonal extremes of TEMPERATURE than at the CONTINENTAL MARGINS.

Continental rift zones. Continental rift zones are PLACES where the CONTINENTAL CRUST is stretched and thinned. Distinctive features include active VOLCANOES and long, straight VALLEY systems formed by normal FAULTS. Continental rifting in some cases has evolved into the breaking apart of a CONTINENT by SEAFLOOR SPREADING to form a new OCEAN.

Continental shelf. Shallow, gently sloping part of the seafloor adjacent to the mainland. The continental shelf is geologically part of the CONTINENT and is made of CONTINENTAL CRUST, whereas the OCEAN floor is OCEANIC CRUST. Although continental shelves vary greatly in width, on average they are about 45 miles (75 km.) wide and have slopes of 7 minutes (about one-tenth of a degree). The average depth of a continental shelf is about 200 feet (60 meters). The outer edge of the continental shelf is marked by a sharp change in angle where the CONTINENTAL SLOPE begins. Most continental shelves were exposed above current SEA LEVEL during the PLEISTOCENE EPOCH and have been submerged by rising sea levels over the past 18,000 years.

Continental shield. Area of a CONTINENT that contains the oldest ROCKS on Earth, called CRATONS. These are areas of granitic rocks, part of the CONTINENTAL CRUST, where there are ancient MOUNTAINS. The Canadian Shield in North America is an example.

Convectional rain. Type of PRECIPITATION caused when AIR over a warm surface is warmed and rises, leading to ADIABATIC cooling, CONDENSATION, and, if the air is moist enough, rain.

Convective overturn. Renewal of the bottom waters caused by the sinking of SURFACE WATERS that have become denser, usually because of decreased TEMPERATURE.

Convergence (climate). AIR flowing in toward a central point.

Convergence (physiography). Process that occurs during the second half of a SUPERCONTINENT CYCLE, whereby crustal PLATES collide and intervening OCEANS disappear as a result of plate SUBDUCTION.

Convergent plate boundary. Compressional PLATE BOUNDARY at which an oceanic PLATE is subducted or two continental plates collide.

Convergent plate margin. Area where the earth's LITHOSPHERE is returned to the MANTLE at a SUBDUCTION ZONE, forming volcanic "ISLAND ARCS" and associated HYDROTHERMAL activity.

Conveyor belt current. Large CYCLE of water movement that carries warm water from the north Pacific westward across the Indian Ocean, around Southern Africa, and into the Atlantic, where it warms the ATMOSPHERE, then returns at a deeper OCEAN level to rise and begin the process again.

Coordinated universal time (UTC). International basis of time, introduced to the world in 1964. The basis for UTC is a small number of ATOMIC CLOCKS. Leap seconds are occasionally added to UTC to keep it synchronized with universal time.

Core-mantle boundary. SEISMIC discontinuity 1,790 miles (2,890 km.) below the earth's surface that separates the MANTLE from the OUTER CORE.

Core region. Area, generally around a COUNTRY's CAPITAL CITY, that has a large, dense POPULATION and is the center of TRADE, financial services, and production. The rest of the country is referred to as the PERIPHERY. On a larger scale, the CONTINENT of Europe has a core region, which includes London, Paris, and Berlin; Iceland, Portugal, and Greece are peripheral LOCATIONS.

Coriolis effect. Apparent deflection of moving objects above the earth because of the earth's ROTA-

441

TION. The deflection is to the right in the NORTHERN HEMISPHERE and to the left in the SOUTHERN HEMISPHERE. The deflection is inversely proportional to the speed of the earth's rotation, being negligible at the EQUATOR but at its maximum near the POLES. The Coriolis effect is a major influence on the direction of surface WINDS. Sometimes called Coriolis force.

Corrasion. EROSION and lowering of a STREAMBED by FLUVIAL action, especially by ABRASION of the bedload (material transported by the STREAM) but also including SOLUTION by the water.

Cosmogony. Study of the origin and nature of the SOLAR SYSTEM.

Cotton Belt. Part of the United States extending from South Carolina through Georgia, Alabama, Mississippi, Tennessee, Louisiana, Arkansas, Texas, and Oklahoma, where cotton was grown on PLANTATIONS using slave labor before the Civil War. After that war, the South stagnated for almost a century. Racial SEGREGATION contributed to cultural isolation from the rest of the United States. Cotton is still produced in this REGION, but California has overtaken the Southern STATES as a cotton producer, and other agricultural products, such as soybeans and poultry, have become dominant crops in the old Cotton Belt. In-migration, due to the SUN BELT attraction, has led to rapid urban growth.

Counterurbanization. Out-migration of people from URBAN AREAS to smaller TOWNS or RURAL areas. As large modern cities are perceived to be overcrowded, stressful, polluted, and dángerous, many of their residents move to areas they regard as more favorable. Such moves are often related to individuals' retirements; however, younger workers and families are also part of counterurbanization.

Crater morphology. Structure or form of CRATERS and the related processes that developed them.

Craton. Large, geologically old, relatively stable CORE of a continental LITHOSPHERIC PLATE, sometimes termed a CONTINENTAL SHIELD.

Creep. Slow, gradual downslope movement of SOIL materials under gravitational stress. Creep tests are experiments conducted to assess the effects of time on ROCK properties, in which environmental conditions (surrounding pressure, TEMPERATURE) and the deforming stress are held constant.

Crestal plane. Plane or surface that goes through the highest points of all beds in a fold; it is coincident with the axial plane when the axial plane is vertical.

Cross-bedding. Layers of ROCK or SAND that lie at an angle to horizontal bedding or to the ground.

Crown land. Land belonging to a NATION'S MONARCHY.

Crude oil. Unrefined OIL, as it occurs naturally. Also called PETROLEUM.

Crustal movements. PLATE TECTONICS theorizes that Earth's CRUST is not a single rigid shell, but comprises a number of large pieces that are in motion, separating or colliding. There are two types of crust—the older continental and the much younger OCEANIC CRUST. When PLATES diverge, at SEAFLOOR SPREADING zones, new (oceanic) crust is created from the MAGMA that flows out at the MID-OCEAN RIDGES. When plates converge and collide, denser oceanic crust is SUBDUCTED under the lighter CONTINENTAL CRUST. The boundaries at the areas where plates slide laterally, neither diverging nor converging, are called TRANSFORM FAULTS. The San Andreas Fault represents the world's best-known transform BOUNDARY. As a result of crustal movements, the earth can be deformed in several ways. Where PLATE BOUNDARIES converge, compression can occur, leading to FOLDING and the creation of SYNCLINES and ANTICLINES. Other stresses of the crust can lead to fracture, or faulting, and accompanying EARTHQUAKES. LANDFORMS created in this way include HORSTS, GRABEN, and BLOCK MOUNTAINS.

Culture hearth. LOCATION in which a CULTURE has developed; a CORE REGION from which the culture later spread or diffused outward through a larger REGION. Mesopotamia, the Nile Valley, and the Peruvian ALTIPLANO are examples of culture hearths.

Curie point. TEMPERATURE at which a magnetic MINERAL locks in its magnetization. Also known as Curie temperature.

Cycle of erosion. Influential MODEL of LANDSCAPE change proposed by William Morris Davis near the end of the nineteenth century. The UPLIFT of a relatively flat surface, or PLAIN, in an area of moderate RAINFALL and TEMPERATURE, led to gradual EROSION of the initial surface in a sequence Davis categorized as Youth, Maturity, and Old Age. The final landscape was called PENEPLAIN. Davis also recognized the stage of REJUVENATION, when a new uplift could give new ENERGY to the cycle, leading to further downcutting and erosion. The model also was used to explain the sequence of LANDFORMS developed in REGIONS of ALPINE GLACIERS. The model has been criticized as misleading, since CRUSTAL MOVEMENT is continuous and more frequent than Davis perhaps envisaged, but remained useful as a description of TOPOGRAPHY. Also known as the Davisian cycle or geomorphic cycle.

Cyclonic rain. In the NORTHERN HEMISPHERE winter, two low-pressure systems or CYCLONES—the Aleutian Low and the Icelandic Low—develop over the OCEAN near 60 DEGREES north LATITUDE. The polar FRONT forms where the cold and relatively dry ARCTIC AIR meets the warmer, moist air carried by westerly WINDS. The warm air is forced upward, cools, and condenses. These cyclonic STORMS often move south, bringing winter PRECIPITATION to North America, especially to the STATES of Washington and Oregon.

Cylindrical projection. MAP PROJECTION that represents the earth's surface as a rectangle. It can be imagined as a cylinder of paper wrapped around a globe with a light source at its center; the images of the CONTINENTS would be projected onto the paper. In reality, MAPS are constructed mathematically. It is impossible to show the North Pole or South Pole on a cylindrical projection. Although the map is conformal, distortion of area is extreme beyond 50 DEGREES north and south LATITUDES. The Mercator projection, developed in the sixteenth century by the Flemish cartographer Gerardus Mercator, is the best-known cylindrical projection. It has been popular with seamen because the shortest route between two PORTS (the GREAT CIRCLE route) can be plotted as straight lines that show the COMPASS direction that should be followed. Use of this projection for other purposes, however, can lead to misunderstandings about size; for example, compare Greenland on a globe and on a Mercator map.

Datum level. Baseline or level from which other heights are measured, above or below. MEAN SEA LEVEL is the datum commonly used in surveying and in the construction of TOPOGRAPHIC MAPS.

Daylight saving time. System of seasonal adjustments in CLOCK settings designed to increase hours of evening sunlight during summer months. In the spring, clocks are set ahead one hour; in the fall, they are put back to standard time. In North America, these changes are made on the first Sunday in April and the last Sunday in October. The U.S. Congress standardized daylight saving time in 1966; however, parts of Arizona, Indiana, and Hawaii do not follow the system.

Débâcle. In a scientific context, this French word means the sudden breaking up of ice in a RIVER in the spring, which can lead to serious, sudden flooding.

Debris avalanche. Large mass of SOIL and ROCK that falls and then slides on a cushion of AIR downhill rapidly as a unit.

Debris flow. Flowing mass consisting of water and a high concentration of SEDIMENT with a wide RANGE of size, from fine muds to coarse gravels.

Declination, magnetic. Measure of the difference, in DEGREES, between the earth's NORTH MAGNETIC pole and the North Pole on a MAP; this difference changes slightly each year. The needle of a magnetic COMPASS points to the earth's geomagnetic pole, which is not exactly the same as the North Pole of the geographic GRID or the set of lines of LATITUDE and LONGITUDE. The geomagnetic poles, north and south, mark the ends

443

of the AXIS of the earth's MAGNETIC FIELD, but this field is not stationary. In fact, the geomagnetic poles have completely reversed hundreds of times throughout earth history. Lines of equal magnetic declination are called ISOGONIC LINES.

Declination of the Sun. LATITUDE of the SUBSOLAR POINT, the PLACE on the earth's surface where the SUN is directly overhead. In the course of a year, the declination of the Sun migrates from 23.5 DEGREES north LATITUDE, at the (northern) summer SOLSTICE, to 23.5 degrees south latitude, at the (northern) WINTER SOLSTICE. Hawaii is the only part of the United States that experiences the Sun directly overhead twice a year.

Deep-focus earthquakes. EARTHQUAKES occurring at depths ranging from 40 to 400 miles (70–700 km.) below the earth's surface. This RANGE of depths represents the zone from the base of the earth's CRUST to approximately one-quarter of the distance into Earth's MANTLE. Deep-focus earthquakes provide scientists information about the PLANET's interior structure, its composition, and SEISMICITY. Observation of deep-focus earthquakes has played a fundamental role in the discovery and understanding of PLATE TECTONICS.

Deep-ocean currents. Deep-ocean currents involve significant vertical and horizontal movements of seawater. They distribute oxygen- and nutrient-rich waters throughout the world's OCEANS, thereby enhancing biological productivity.

Defile. Narrow MOUNTAIN PASS or GORGE through which troops could march only in single file.

Deflation. EROSION by WIND, resulting in the removal of fine particles. The LANDFORM that typically results is a deflation hollow.

Deforestation. Removal or destruction of FORESTS. In the late twentieth century, there was widespread concern about tropical deforestation—destruction of the tropical RAIN FOREST—especially that of Brazil. Forest clearing in the TROPICS is uneconomic because of low SOIL fertility. Deforestation causes severe EROSION and environmental damage; it also destroys habitat, which leads to the EXTINCTION of both plant and animal species.

Degradation. Process of CRATER EROSION from all processes, including WIND and other meteorological mechanisms.

Degree (geography). Unit of LATITUDE or LONGITUDE in the geographic GRID, used to determine ABSOLUTE LOCATION. One degree of latitude is about 69 miles (111 km.) on the earth's surface. It is not exactly the same everywhere, because the earth is not a perfect sphere. One degree of longitude varies greatly in length, because the MERIDIANS converge at the POLES. At the EQUATOR, it is 69 miles (111 km.), but at the North or South Pole it is zero.

Degree (temperature). Unit of MEASUREMENT of TEMPERATURE, based on the CELSIUS SCALE, except in the United States, which uses the FAHRENHEIT SCALE. On the Celsius scale, one degree is one-hundredth of the difference between the freezing point of water and the boiling point of water.

Demographic measure. Statistical data relating to POPULATION.

Demographic transition. MODEL of POPULATION change that fits the experience of many European countries, showing changes in birth and death rates. In the first stage, in preindustrial countries, population size was stable because both BIRTH RATES and DEATH RATES were high. Agricultural reforms, together with the INDUSTRIAL REVOLUTION and subsequent medical advances, led to a rapid fall in the death rate, so that the second and third stages of the model were periods of rapid population growth, often called the POPULATION EXPLOSION. In the fourth stage of the model, birth rates fall markedly, leading again to stable population size.

Dendritic drainage. Most common pattern of STREAMS and their TRIBUTARIES, occurring in areas of uniform ROCK type and regular slope. A MAP, or aerial photograph, shows a pattern like the veins on a leaf—smaller streams join the main stream at an acute angle.

Denudation. General word for all LANDFORM processes that lead to a lowering of the LANDSCAPE, including WEATHERING, mass movement, EROSION, and transport.

Deposition. Laying down of SEDIMENTS that have been transported by water, WIND, or ice.

Deranged drainage. LANDSCAPE whose integrated drainage network has been destroyed by irregular glacial DEPOSITION, yielding numerous shallow LAKE BASINS.

Derivative maps. MAPS that are prepared or derived by combining information from several other maps.

Desalinization. Process of removing SALT and MINERALS from seawater or from saline water occurring in AQUIFERS beneath the land surface to render it fit for AGRICULTURE or other human use.

Desert climate. Low PRECIPITATION, low HUMIDITY, high daytime TEMPERATURES, and abundant sunlight are characteristics of desert climates. The hot DESERTS of the world generally are located on the western sides of CONTINENTS, at LATITUDES from fifteen to thirty DEGREES north or south of the EQUATOR. One definition, based on precipitation, defines deserts as areas that receive between 0 and 9 inches (0 to 250 millimeters) of precipitation per year. REGIONS receiving more precipitation are considered to have a SEMIDESERT climate, in which some AGRICULTURE is possible.

Desert pavement. Surface covered with smoothed PEBBLES and gravels, found in arid areas where DEFLATION (WIND EROSION) has removed the smaller particles. Called a "gibber plain" in Australia.

Desertification. Increase in DESERT areas worldwide, largely as a result of overgrazing or poor agricultural practices in semiarid and marginal CLIMATES. DEFORESTATION, DROUGHT, and POPULATION increase also contribute to desertification. The REGION of Africa just south of the Sahara Desert, known as the SAHEL, is the largest and most dramatic demonstration of desertification.

Detrital rock. SEDIMENTARY ROCK composed mainly of grains of silicate MINERALS as opposed to grains of calcite or CLAYS.

Devolution. Breaking up of a large COUNTRY into smaller independent political units is the final and most extreme form of devolution. The Soviet Union devolved from one single country into fifteen separate countries in 1991. At an intermediate level, devolution refers to the granting of political autonomy or self-government to a REGION, without a complete split. The reopening of the Scottish Parliament in 1999 and the Northern Ireland parliament in 2000 are examples of devolution; the Parliament of the United Kingdom had previously met only in London and made laws there for all parts of the country. Canada experienced devolution with the creation of the new territory of Nunavut, whose residents elect the members of their own legislative assembly.

Dew point. TEMPERATURE at which an AIR mass becomes saturated and can hold no more moisture. Further cooling leads to CONDENSATION. At ground level, this produces DEW.

Diagenesis. Conversion of unconsolidated SEDIMENT into consolidated ROCK after burial by the processes of compaction, cementation, recrystallization, and replacement.

Diaspora. Dispersion of a group of people from one CULTURE to a variety of other REGIONS or to other lands. A Greek word, used originally to refer to the Jewish diaspora. Jewish people now live in many countries, although they have Israel as a HOMELAND. Similar to this are the diasporas of the Irish and the overseas Chinese.

Diastrophism. Deformation of the earth's CRUST by faulting or FOLDING.

Diatom ooze. Deposit of soft mud on the OCEAN floor consisting of the shells of diatoms, which are microscopic single-celled creatures with SILICA-rich shells. Diatom ooze deposits are located in the southern Pacific around Antarctica and in the northern Pacific. Other PELAGIC, or deep-ocean, SEDIMENTS include CLAYS and calcareous ooze.

Dike (geology). LANDFORM created by IGNEOUS intrusion when MAGMA or molten material within the earth forces its way in a narrow band through overlying ROCK. The dike can be exposed at the surface through EROSION.

Dike (water). Earth wall or DAM built to prevent flooding; an EMBANKMENT or artificial LEVEE. Sometimes specifically associated with structures built in the Netherlands to prevent the entry of seawater. The land behind the dikes was reclaimed for AGRICULTURE; these new fields are called POLDERS.

Distance-decay function. Rate at which an activity diminishes with increasing distance. The effect that distance has as a deterrent on human activity is sometimes described as the FRICTION OF DISTANCE. It occurs because of the time and cost of overcoming distances between people and their desired activity. An example of the distance-decay function is the rate of visitors to a football stadium. The farther people have to travel, the less likely they are to make this journey.

Distributary. STREAM that takes waters away from the main CHANNEL of a RIVER. A DELTA usually comprises many distributaries. Also called distributary channel.

Diurnal range. Difference between the highest and lowest TEMPERATURES registered in one twenty-four-hour period.

Diurnal tide. Having only one high tide and one low tide each lunar DAY; TIDES on some parts of the Gulf of Mexico are diurnal.

Divergent boundary. BOUNDARY that results where two PLATES are moving apart from each other, as is the case along MID-OCEANIC RIDGES.

Divergent margin. Area where the earth's CRUST and LITHOSPHERE form by SEAFLOOR SPREADING.

Doline. Large SINKHOLE or circular depression formed in LIMESTONE areas through the CHEMICAL WEATHERING process of carbonation.

Dolomite. MINERAL consisting of calcium and magnesium carbonate compounds that often forms from PRECIPITATION from seawater; it is abundant in ancient ROCKS.

Downwelling. Sinking of OCEAN water.

Drainage basin. Area of the earth's surface that is drained by a STREAM. Drainage basins vary greatly in size, but each is separated from the next by RIDGES, or drainage DIVIDES. The CATCHMENT of the drainage basin is the WATERSHED.

Drift ice. ARCTIC or ANTARCTIC ice floating in the open SEA.

Drumlin. Low HILL, shaped like half an egg, formed by DEPOSITION by CONTINENTAL GLACIERS. A drumlin is composed of TILL, or mixed-size materials. The wider end faces upstream of the glacier's movement; the tapered end points in the direction of the ice movement. Drumlins usually occur in groups or swarms.

Dust devil. Whirling cloud of DUST and small debris, formed when a small patch of the earth's surface becomes heated, causing hot AIR to rise; cooler air then flows in and begins to spin. The resulting dust devil can grow to heights of 150 feet (50 meters) and reach speeds of 35 miles (60 km.) per hour.

Dust dome. Dome of AIR POLLUTION, composed of industrial gases and particles, covering every large CITY in the world. The pollution sometimes is carried downwind to outlying areas.

Earth pillar. Formation produced when a boulder or caprock prevents EROSION of the material directly beneath it, usually CLAY. The clay is easily eroded away by water during RAINFALL, except where the overlying ROCK protects it. The result is a tall, slender column, as high as 20 feet (6.5 meters) in exceptional cases.

Earth radiation. Portion of the electromagnetic spectrum, from about 4 to 80 microns, in which the earth emits about 99 percent of its RADIATION.

Earth tide. Slight deformation of Earth resulting from the same forces that cause OCEAN TIDES, those that are exerted by the MOON and the SUN.

Earthflow. Term applied to both the process and the LANDFORM characterized by fluid downslope movement of SOIL and ROCK over a discrete plane of failure; the landform has a HUMMOCKY surface and usually terminates in discrete lobes.

Earth's heat budget. Balance between the incoming SOLAR RADIATION and the outgoing terrestrial reradiation.

Eclipse, lunar. Obscuring of all or part of the light of the MOON by the shadow of the earth. A lunar eclipse occurs at the full moon up to three times a year. The surface of the Moon changes from gray to a reddish color, then back to gray. The sequence may last several hours.

Eclipse, solar. At least twice a year, the SUN, MOON, and Earth are aligned in one straight line. At that time, the Moon obscures all the light of the Sun along a narrow band of the earth's surface, causing a total eclipse; in REGIONS of Earth adjoining that area, there is a partial eclipse. A corona (halo of light) can be seen around the Sun at the total eclipse. Viewing a solar eclipse with naked eyes is extremely dangerous and can cause blindness.

Ecliptic, plane of. Imaginary plane that would touch all points in the earth's ORBIT as it moves around the SUN. The angle between the plane of the ecliptic and the earth's AXIS is 66.5 DEGREES.

Edge cities. Forms of suburban downtown in which there are nodal concentrations of office space and shopping facilities. Edge cities are located close to major freeways or highway intersections, on the outer edges of METROPOLITAN AREAS.

Effective temperature. TEMPERATURE of a PLANET based solely on the amount of SOLAR RADIATION that the planet's surface receives; the effective temperature of a planet does not include the GREENHOUSE temperature enhancement effect.

Ejecta. Material ejected from the CRATER made by a meteoric impact.

Ekman layer. REGION of the SEA, from the surface to about 100 meters down, in which the WIND directly affects water movement.

Eluviation. Removal of materials from the upper layers of a SOIL by water. Fine material may be removed by SUSPENSION in the water; other material is removed by SOLUTION. The removal by solution is called LEACHING. Eluviation from an upper layer leads to illuviation in a lower layer.

Enclave. Piece of territory completely surrounded by another COUNTRY. Two examples are Lesotho, which is surrounded by the Republic of South Africa, and the Nagorno-Karabakh REGION, populated by Armenians but surrounded by Azerbaijan. The term is also used for smaller regions, such as ethnic neighborhoods within larger cities. See also EXCLAVE.

Endemic species. Species confined to a restricted area in a restricted ENVIRONMENT.

Endogenic sediment. SEDIMENT produced within the water column of the body in which it is deposited; for example, calcite precipitated in a LAKE in summer.

Environmental degradation. Situation that occurs in slum areas and SQUATTER SETTLEMENTS because of poverty and inadequate INFRASTRUCTURE. Too-rapid human POPULATION growth can lead to the accumulation of human waste and garbage, the POLLUTION of GROUNDWATER, and DENUDATION of nearby FORESTS. As a result, LIFE EXPECTANCY in such degraded areas is lower than in the RURAL communities from which many of the settlers came. INFANT MORTALITY is particularly high. When people leave an area because of such environmental degradation, that is referred to as ecomigration.

Environmental determinism. Theory that the major influence on human behavior is the physical ENVIRONMENT. Some evidence suggests that TEMPERATURE, PRECIPITATION, sunlight, and TOPOGRAPHY influence human activities. Originally espoused by early German geographers, this theory has led to some extreme stances, however, by authors who have sought to explain the dominance of Europeans as a result of a cool temperate CLIMATE.

Eolian (aeolian). Relating to, or caused by, WIND. In Greek mythology, Aeolus was the ruler of the winds. EROSION, TRANSPORT, and DEPOSITION are common eolian processes that produce LANDFORMS in DESERT REGIONS.

Eolian deposits. Material transported by the WIND.

Eolian erosion. Mechanism of EROSION or CRATER DEGRADATION caused by WIND.

Eon. Largest subdivision of geologic time; the two main eons are the PRECAMBRIAN (c. 4.6 billion years ago to 544 million years ago) and the PHANEROZOIC (c. 544 million years ago to the present).

Ephemeral stream. Watercourse that has water for only a DAY or so.

Epicontinental sea. Shallow SEAS that are located on the CONTINENTAL SHELF, such as the North Sea or Hudson Bay. Also called an EPEIRIC SEA.

Epifauna. Organisms that live on the seafloor.

Epilimnion. Warmer surface layer of water that occurs in a LAKE during summer stratification; during spring, warmer water rises from great depths, and it heats up through the summer SEASON.

Equal-area projection. MAP PROJECTION that maintains the correct area of surfaces on7 a MAP, although shape distortion occurs. The property of such a map is called equivalence.

Erg. Sandy DESERT, sometimes called a SEA of SAND. Erg deserts account for less than 30 percent of the world's deserts. "Erg" is an Arabic word.

Eruption, volcanic. Emergence of MAGMA (molten material) at the earth's surface as LAVA. There are various types of volcanic eruptions, depending on the chemistry of the magma and its viscosity. Scientists refer to effusive and explosive eruptions. Low-viscosity magma generally produces effusive eruptions, where the lava emerges gently, as in Hawaii and Iceland, although explosive events can occur at those SITES as well. Gently sloping SHIELD VOLCANOES are formed by effusive eruptions; FLOODS, such as the Columbia Plateau, can also result. Explosive eruptions are generally associated with SUBDUCTION. Much gas, including steam, is associated with magma formed from OCEANIC CRUST, and the compressed gas helps propel the explosion. COMPOSITE CONES, such as Mount Saint Helens, are created by explosive eruptions.

Escarpment. Steep slope, often almost vertical, formed by faulting. Sometimes called a FAULT SCARP.

Esker. Deposit of coarse gravels that has a sinuous, winding shape. An esker is formed by a STREAM of MELTWATER that flowed through a tunnel it formed under a CONTINENTAL GLACIER. Now that the continental glaciers have melted, eskers can be found exposed at the surface in many PLACES in North America.

Estuarine zone. Area near the COASTLINE that consists of estuaries and coastal saltwater WETLANDS.

Etesian winds. WINDS that blow from the north over the Mediterranean during July and August.

Ethnocentrism. Belief that one's own ETHNIC GROUP and its CULTURE are superior to any other group.

Ethnography. Study of different CULTURES and human societies.

Eustacy. Any change in global SEA LEVEL resulting from a change in the absolute volume of available sea water. Also known as eustatic sea-level change.

Eustatic movement. Changes in SEA LEVEL.

Exclave. Territory that is part of one COUNTRY but separated from the main part of that country by another country. Alaska is an exclave of the United States; Kaliningrad is an exclave of Russia. See also ENCLAVE.

Exfoliation. When GRANITE rocks cooled and solidified, removal of the overlyingrock that was present reduced the pressure on the granite mass, allowing it to expand and causing sheets or layers of rock to break off. An exfoliation DOME, such as Half Dome in Yosemite National Park, is the resultant LANDFORM.

Exotic stream. RIVER that has its source in an area of high RAINFALL and then flows through an arid REGION or DESERT. The Nile River is the most famous exotic STREAM. In the United States, the Colorado River is a good example of an exotic stream.

Expansion-contraction cycles. Processes of wetting-drying, heating-cooling, or freezing-thawing, which affect SOIL particles differently according to their size.

Extrusive rock. Fine-grained, or glassy, ROCK which was formed from a MAGMA that cooled on the surface of the earth.

Fall line. Edge of an area of uplifted land, marked by WATERFALLS where STREAMS flow over the edge.

Fata morgana. Large mirage. Originally, the name given to a multiple mirage phenomenon often observed over the Straits of Messina and supposed to be the work of the fairy ("fata") Morgana. Another famous fata morgana is located in Antarctica.

Fathometer. Instrument that uses sound waves or sonar to determine the depth of water or the depth of an object below the water.

Fault drag. Bending of ROCKS adjacent to a FAULT.

Fault line. Line of breakage on the earth's surface. FAULTS may be quite short, but many are extremely long, even hundreds of miles. The origin of the faulting may lie at a considerable depth below the surface. Movement along the fault line generates EARTHQUAKES.

Fault plane. Angle of a FAULT. When fault blocks move on either side of a fault or fracture, the movement can be vertical, steeply inclined, or sometimes horizontal. In a NORMAL FAULT, the fault plane is steep to almost vertical. In a REVERSE FAULT, one block rides over the other, forming an overhanging FAULT SCARP. The angle of inclination of the fault plane from the horizontal is called the dip. The inclination of a fault plane is generally constant throughout the length of the fault, but there can be local variations in slope. In a STRIKE-SLIP FAULT the movement is horizontal, so no fault scarp is produced, although the FAULT LINE may be seen on the surface.

Fault scarp. FAULTS are produced through breaking or fracture of the surface ROCKS of the earth's CRUST as a result of stresses arising from tectonic movement. A NORMAL FAULT, one in which the earth movement is predominantly vertical, produces a steep fault scarp. A STRIKE-SLIP FAULT does not produce a fault scarp.

Feldspar. Family name for a group of common MINERALS found in such ROCKS as GRANITE and composed of silicates of aluminum together with potassium, sodium, and calcium. Feldspars are the most abundant group of minerals within the earth's CRUST. There are many varieties of feldspar, distinguished by variations in chemistry and crystal structure. Although feldspars have some economic uses, their principal importance lies in their role as rock-forming minerals.

Felsic rocks. IGNEOUS ROCKS rich in potassium, sodium, aluminum, and SILICA, including GRANITES and related rocks.

Fertility rate. DEMOGRAPHIC MEASURE of the average number of children per adult female in any given POPULATION. Religious beliefs, education, and other cultural considerations influence fertility rates.

Fetch. Distance along a large water surface over which a WIND of almost uniform direction and speed blows.

Feudalism. Social and economic system that prevailed in Europe before the INDUSTRIAL REVOLUTION. The land was owned and controlled by a minority comprising noblemen or lords; all other people were peasants or serfs, who worked as agricultural laborers on the lords' land. The peasants were not free to leave, or to do anything without their lord's permission. Other REGIONS such as China and Japan also had a feudal system in the past.

Firn. Intermediate stage between SNOW and glacial ice. Firn has a granular TEXTURE, due to compaction. Also called NÉVÉ.

Fission, nuclear. Splitting of an atomic nucleus into two lighter nuclei, resulting in the release of neutrons and some of the binding ENERGY that held the nucleus together.

Fissure. Fracture or crack in ROCK along which there is a distinct separation.

Flash flood. Sudden rush of water down a STREAM CHANNEL, usually in the DESERT after a short but intense STORM. Other causes, such as a DAM failure, could lead to a flash flood.

Flood control. Attempts by humans to prevent flooding of STREAMS. Humans have consistently settled on FLOODPLAINS and DELTAS because of the fertile SOIL for AGRICULTURE, and attempts at flood control date back thousands of years. In strictly agricultural societies such as ancient Egypt, people built VILLAGES above the FLOOD levels, but transport and industry made riverside LOCATIONS desirabl and engineers devised technological means to try to prevent flood damage. Artificial LEVEES, RESERVOIRS, and DAMS of ever-increasing size were built on

RIVERS, as well as bypass CHANNELS leading to artificial floodplains. In many modern dam construction projects, the production of HYDROELECTRIC POWER was more important than flood control. Despite modern TECHNOLOGY, floods cause the largest loss of human life of all natural disasters, especially in low-income countries such as Bangladesh.

Flood tide. Rising or incoming tide. Most parts of the world experience two flood TIDES in each 24-hour period.

Floodplain. Flat, low-lying land on either side of a STREAM, created by the DEPOSITION of ALLUVIUM from floods. Also called ALLUVIAL PLAIN.

Fluvial. Pertaining to running water; for example, fluvial processes are those in which running water is the dominant agent.

Fog deserts. Coastal DESERTS where FOG is an important source of moisture for plants, animals, and humans. The fog forms because of a cold OCEAN CURRENT close to the SHORE. The Namib Desert of southwestern Africa, the west COAST of California, and the Atacama Desert of Peru are coastal deserts.

Föhn wind. WIND warmed and dried by descent, usually on the LEE side of a MOUNTAIN. In North America, these winds are called the CHINOOK.

Fold mountains. ROCKS in the earth's CRUST can be bent by compression, producing folds. The Swiss Alps are an example of complex FOLDING, accompanied by faulting. Simple upward folds are ANTICLINES, downward folds are SYNCLINES; but subsequent EROSION can produce LANDSCAPES with synclinal MOUNTAINS.

Folding. Bending of ROCKS in the earth's CRUST, caused by compression. The rocks are deformed, sometimes pushed up to form mountain RANGES.

Foliation. TEXTURE or structure in which MINERAL grains are arranged in parallel planes.

Food web. Complex network of FOOD CHAINS. Food chains are interconnected, because many organisms feed on a variety of others, and in turn may be eaten by any of a number of predators.

Forced migration. MIGRATION that occurs when people are moved against their will. The Atlantic slave trade is an example of forced migration. People were shipped from Africa to countries in Europe, Asia, and the New World as forced immigrants. Within the United States, some NATIVE AMERICANS were forced by the federal government to migrate to new reservations.

Ford. Short shallow section of a RIVER, where a person can cross easily, usually by walking or riding a horse. To cross a STREAM in such a manner.

Formal region. Cultural REGION in which one trait, or group of traits, is uniform. LANGUAGE might be the basis of delineation of a formal cultural region. For example, the Francophone region of Canada constitutes a formal region based on one single trait. One might also identify a formal Mormon region centered on the STATE of Utah, combining RELIGION and LANDSCAPE as defining traits. Cultural geographers generally identify formal regions using a combination of traits.

Fossil fuel. Deposit rich in hydrocarbons, formed from organic materials compressed in ROCK layers—COAL, OIL, and NATURAL GAS.

Fossil record. Fossil record provides evidence that addresses fundamental questions about the origin and history of life on the earth: When life evolved; how new groups of organisms originated; how major groups of organisms are related. This record is neither complete nor without biases, but as scientists' understanding of the limits and potential of the fossil record grows, the interpretations drawn from it are strengthened.

Fossilization. Processes by which the remains of an organism become preserved in the ROCK record.

Fracture zones. Large, linear zones of the seafloor characterized by steep CLIFFS, irregular TOPOGRAPHY, and FAULTS; such zones commonly cross and displace oceanic RIDGES by faulting.

Free association. Relationship between sovereign NATIONS in which one nation—invariably the larger—has responsibility for the other nation's defense. The Cook Islands in the South Pacific have such a relationship with New Zealand.

Friction of distance. Distance is of prime importance in social, political, economic, and other relationships. Large distance has a negative effect on human activity. The time and cost of overcoming distance can be a deterrent to various activities. This has been called the friction of distance.

Frigid zone. Coldest of the three CLIMATE zones proposed by the ancient Greeks on the basis of their theories about the earth. There were two frigid zones, one around each POLE. The Greeks believed that human life was possible only in the TEMPERATE ZONE.

Fringing reef. Type of CORAL REEF formed at the SHORELINE, extending out from the land in shallow water. The top of the coral may be exposed at low TIDE.

Frontier Thesis. Thesis first advanced by the U.S. historian Frederick Jackson Turner, who declared that U.S. history and the U.S. character were shaped by the existence of empty, FRONTIER lands that led to exploration and westward expansion and DEVELOPMENT. The closing of the frontier occurred when transcontinental railroads linked the East and West Coasts and SETTLEMENTS spread across the United States. This thesis was used by later historians to explain the history of South Africa, Canada, and Australia. Critics of the Frontier Thesis point out that minorities and women were excluded from this view of history.

Frost wedging. Powerful form of PHYSICAL WEATHERING of ROCK, in which the expansion of water as it freezes in JOINTS or cracks shatters the rock into smaller pieces. Also known as frost shattering.

Fumarole. Crack in the earth's surface from which steam and other gases emerge. Fumaroles are found in volcanic areas and areas of GEOTHERMAL activity, such as Yellowstone National Park.

Fusion energy. Heat derived from the natural or human-induced union of atomic nuclei; in effect, the opposite of FISSION energy.

Gall's projection. MAP PROJECTION constructed by projecting the earth onto a cylinder that intersects the sphere at 45 DEGREES north and 45 degrees south LATITUDE. The resulting map has less distortion of area than the more familiar CYLINDRICAL PROJECTION of Mercator.

Gangue. Apparently worthless ROCK or earth in which valuable gems or MINERALS are found.

Garigue. VEGETATION cover of small shrubs found in Mediterranean areas. Similar to the larger *maquis.*

Genus (plural, genera). Group of closely related species; for example, *Homo* is the genus of humans, and it includes the species *Homo sapiens* (modern humans) and *Homo erectus* (Peking Man, Java Man).

Geochronology. Study of the time SCALE of the earth; it attempts to develop methods that allow the scientist to reconstruct the past by dating events such as the formation of ROCKS.

Geodesy. Branch of applied mathematics that determines the exact positions of points on the earth's surface, the size and shape of the earth, and the variations of terrestrial GRAVITY and MAGNETISM.

Geoid. Figure of the earth considered as a MEAN SEA LEVEL surface extended continuously through the CONTINENTS.

Geologic terrane. Crustal block with a distinct group of ROCKS and structures resulting from a particular geologic history; assemblages of TERRANES form the CONTINENTS.

Geological column. Order of ROCK layers formed during the course of the earth's history.

Geomagnetic elements. MEASUREMENTS that describe the direction and intensity of the earth's MAGNETIC FIELD.

Geomagnetism. External MAGNETIC FIELD generated by forces within the earth; this force attracts materials having similar properties, inducing them to line up (point) along field lines of force.

Geostationary orbit. ORBIT in which a SATELLITE appears to hover over one spot on the PLANET's EQUATOR; this procedure requires that the orbit be high enough that its period matches the planet's rotational period, and have no inclination relative to the equator; for Earth, the ALTITUDE is 22,260 miles (35,903 km.).

Geostrophic. Force that causes directional change because of the earth's ROTATION.

Geotherm. Curve on a TEMPERATURE-depth graph that describes how temperature changes in the subsurface.

Geothermal power. Power having its source in the earth's internal heat.

Glacial erratic. ROCK that has been moved from its original position and transported by becoming incorporated in the ice of a GLACIER. Deposited in a new LOCATION, the rock is noteworthy because its geology is completely different from that of the surrounding rocks. Glacial erratics provide information about the direction of glacial movement and strength of the flow. They can be as small as PEBBLES, but the most interesting erratics are large boulders. Erratics become smoothed and rounded by the transport and EROSION.

Glaciation. This term is used in two senses: first, in reference to the cyclic widespread growth and advance of ICE SHEETS over the polar and high- to mid-LATITUDE REGIONS of the CONTINENTS; second, in reference to the effect of a GLACIER on the TERRAIN it transverses as it advances and recedes.

Global Positioning System (GPS). Group of SATELLITES that ORBIT Earth every twenty-four hours, sending out signals that can be used to locate PLACES on Earth and in near-Earth orbits.

Global warming. Trend of Earth CLIMATES to grow increasingly warm as a result of the GREENHOUSE EFFECT. One of the most dramatic effects of global warming is the melting of the POLAR ICE CAPS and a consequent rise the level of the world's OCEANS.

Gondwanaland. Hypothesized ancient CONTINENT in the SOUTHERN HEMISPHERE that geologists theorize broke into at least two large segments; one segment became India and pushed northward to collide with the Eurasian LANDMASS, while the other, Africa, moved westward.

Graben. Roughly symmetrical crustal depression formed by the lowering of a crustal block between two NORMAL FAULTS that slope toward each other.

Granules. Small grains or pellets.

Gravimeter. Device that measures the attraction of GRAVITY.

Gravitational differentiation. Separation of MINERALS, elements, or both as a result of the influence of a gravitational field wherein heavy phases sink or light phases rise through a melt.

Great circle. Largest circle that goes around a sphere. On the earth, all lines of LONGITUDE are parts of great circles; however, the EQUATOR is the only line of LATITUDE that is a great circle.

Green mud. SOILS that develop under conditions of excess water, or waterlogged soils, can display colors of gray to blue to green, largely because of chemical reactions involving iron. Fine CLAY soils and muds in areas such as BOGS or ESTUARIES can be called green mud. This soil-forming process is called gleization.

Greenhouse effect. Trapping of the SUN's rays within the earth's ATMOSPHERE, with a consequence rise in TEMPERATURES that leads to GLOBAL WARMING.

Greenhouse gas. Atmospheric gas capable of absorbing electromagnetic radiation in the infrared part of the spectrum.

Greenwich mean time. Also known as universal time, the solar mean time on the MERIDIAN running through Greenwich, England—which is used as the basis for calculating time throughout most of the world.

Grid. Pattern of horizontal and vertical lines forming squares of uniform size.

Groundwater movement. Flow of water through the subsurface, known as groundwater movement, obeys set principles that allow hydrologists to predict flow directions and rates.

Groundwater recharge. Water that infiltrates from the surface of the earth downward through SOIL and ROCK pores to the WATER TABLE, causing its level to rise.

Growth pole. LOCATION where high-growth economic activity is deliberately encouraged and promoted. Governments often establish growth poles by creating industrial parks, open cities, special economic zones, new TOWNS, and other incentives. The plan is that the new industries will further stimulate economic growth in a cumulative trend. Automobile plants are a traditional form of growth industry but have been overtaken by high-tech industries and BIOTECHNOLOGY. In France, the term "technopole" is used for a high-tech growth pole. A related concept is SPREAD EFFECTS.

Guyot. Drowned volcanic ISLAND with a flat top caused by WAVE EROSION or coral growth. A type of SEAMOUNT.

Gyre. Large semiclosed circulation patterns of OCEAN CURRENTS in each of the major OCEAN BASINS that move in opposite directions in the Northern and Southern hemispheres.

Haff. Term used for various WETLANDS or LAGOONS located around the southern end of the Baltic Sea, from Latvia to Germany. Offshore BARS of SAND and shingle separate the haffs from the open SEA. One of the largest is the Stettiner Haff, which covers the BORDER REGION between Germany and Poland and is separated from the Baltic by the low-lying ISLAND of Usedom. The Kurisches Haff (in English, the Courtland Lagoon) is located on the Lithuanian border.

Harmonic tremor. Type of EARTHQUAKE activity in which the ground undergoes continuous shaking in response to subsurface movement of MAGMA.

Headland. Elevated land projecting into a body of water.

Headwaters. Source of a RIVER. Also called headstream.

Heat sink. Term applied to Antarctica, whose cold CLIMATE causes warm AIR masses flowing over it to chill quickly and lose ALTITUDE, affecting the entire world's WEATHER.

Heterosphere. Major realm of the ATMOSPHERE in which the gases hydrogen and helium become predominant.

High-frequency seismic waves. EARTHQUAKE WAVES that shake the ROCK through which they travel most rapidly.

Histogram. Bar graph in which vertical bars represent frequency and the horizontal axis represents categories. A POPULATION PYRAMID, or age-sex pyramid, is a histogram, as is a CLIMOGRAPH.

Historical inertia. Term used by economic geographers when heavy industries, such as steelmaking and large manufacture, that require huge capital investments in land and plant continue in operation for long periods, even after they become out of date, uncompetitive, or obsolete.

Hoar frost. Similar to DEW, except that moisture is deposited as ice crystals, not liquid dew, on surfaces such as grass or plant leaves. When moist AIR cools to saturation level at TEMPERATURES below the freezing point, CONDENSATION occurs directly as ice. Technically, hoar frost is not the same as frozen dew, but it is difficult to distinguish between the two.

Hogback. Steeply sloping homoclinal RIDGE, with a slope of 45 DEGREES or more. The angle of the slope is the same as the dip of the ROCK STRATA. These LANDFORMS develop in REGIONS where the underlying rocks, usually SEDIMENTARY, have been folded into anticlinal ridges and synclinal VALLEYS. Differential EROSION causes softer rock layers to wear away more rapidly than the harder layers of rock that form the hogback ridge. A similar feature with a gentler slope is called a CUESTA.

Homosphere. Lower part of the earth's ATMOSPHERE. In this area, 60 miles (100 km.) thick, the component gases are uniformly mixed together, largely through WINDS and turbulent AIR CURRENTS. Above the homosphere is the REGION of the atmosphere called the HETEROSPHERE. There, the individual gases separate out into layers on the basis of their molecular weight. The lighter gases, hydrogen and helium, are at the top of the heterosphere.

Hook. A long, narrow deposit of SAND and SILT that grows outward into the OCEAN from the land is called a SPIT or sandspit. A hook forms when currents or WAVES cause the deposited material to curve back toward the land. Cape Cod is the most famous spit and hook in the United States.

Horse latitudes. Parts of the OCEANS from about 30 to 35 DEGREES north or south of the EQUATOR. In these latitudes, AIR movement is usually light WINDS, or even complete calm, because there are semipermanent high-pressure cells called ANTI-CYCLONES, which are marked by dry subsiding air and fine clear WEATHER. The atmospheric circulation of an anticyclone is divergent and CLOCKWISE in the NORTHERN HEMISPHERE, so to the north of the horse latitudes are the westerly winds and to the south are the northeast TRADE WINDS. In the SOUTHERN HEMISPHERE, the circulation is reversed, producing the easterly winds and the southeast trade winds. It is believed that the name originated because when ships bringing immigrants to the Americas were becalmed for any length of time, horses were thrown overboard because they required too much FRESH WATER. Also called the CALMS OF CANCER.

Horst. FAULT block or piece of land that stands above the surrounding land. A horst usually has been uplifted by tectonic forces, but also could have originated by downward movement or lowering of the adjacent lands. Movement occurs along the parallel faults on either side of a horst. If the land is downthrown instead of uplifted, a VALLEY known as a GRABEN is formed. "Horst" comes from the German word for horse, because the flat-topped feature resembles a vaulting horse used in gymnastics.

Hot spot. PLACE on the earth's surface where heat and MAGMA rise from deep in the interior, perhaps from the lower MANTLE. Erupting VOLCANOES may be present, as in the formation of the Hawaiian Islands. More commonly, the heat from the rising magma causes GROUNDWATER to form HOT SPRINGS, GEYSERS, and other thermal and HYDROTHERMAL features. Yellowstone National Park is located on a hot spot. Also known as a MANTLE PLUME.

Hot spring. SPRING where hot water emerges at the earth's surface. The usual cause is that the GROUNDWATER is heated by MAGMA. A GEYSER is a special type of hot spring at which the water heats under pressure and that periodically spouts hot water and steam. Old Faithful is the best known of many geysers in Yellowstone National Park. In some countries, GEOTHERMAL EN-

ERGY from hot springs is used to generate electricity. Also called thermal spring.

Humus. Uppermost layer of a SOIL, containing decaying and decomposing organic matter such as leaves. This produces nutrients, leading to a fertile soil. Tropical soils are low in humus, because the rate of decay is so rapid. Soils of GRASSLANDS and DECIDUOUS FOREST develop thick layers of humus. In a SOIL PROFILE, the layer containing humus is the O Horizon.

Hydroelectric power. Electricity generated when falling water turns the blades of a turbine that converts the water's potential ENERGY to mechanical energy. Natural WATERFALLS can be used, but most hydroelectric power is generated by water from DAMS, because the flow of water from a dam can be controlled. Hydroelectric generation is a RENEWABLE, clean, cheap way to produce power, but dam construction inundates land, often displacing people, who lose their homes, VILLAGES, and farmland. Aquatic life is altered and disrupted also; for example, Pacific salmon cannot return upstream on the Columbia River to their spawning REGION. In a few coastal PLACES, TIDAL ENERGY is used to generate hydroelectricity; La Rance in France is the oldest successful tidal power plant.

Hydrography. Surveying of underwater features or those parts of the earth that are covered by water, especially OCEAN depths and OCEAN CURRENTS. Hydrographers make MAPS and CHARTS of the ocean floor and COASTLINES, which are used by mariners for NAVIGATION. For centuries, mariners used a leadline, a long rope with a lead weight at the bottom. The line was thrown overboard and the depth of water measured. The unit of MEASUREMENT was FATHOMS (6 feet/1.8 meters), which is one-thousandth of a NAUTICAL MILE. The invention of sonar (underwater echo sounding) has enabled mapping of large areas, and hydrographers currently use both television cameras and SATELLITE data.

Hydrologic cycle. Continuous circulation of the earth's HYDROSPHERE, or waters, through EVAPORATION, CONDENSATION, and PRECIPITATION.

Other parts of the hydrologic cycle include RUN-OFF, INFILTRATION, and TRANSPIRATION.

Hydrostatic pressure. Pressure imposed by the weight of an overlying column of water.

Hydrothermal vents. Areas on the OCEAN floor, typically along FAULT LINES or in the vicinity of undersea VOLCANOES, where water that has percolated into the ROCK reemerges much hotter than the surrounding water; such heated water carries various dissolved MINERALS, including metals and sulfides.

Hyetograph. Chart showing the DISTRIBUTION of RAINFALL over time. Typically, a hyetograph is constructed for a single STORM, showing the amount of total PRECIPITATION accumulating throughout the period. A hyetograph shows how rainfall intensity varies throughout the duration of a storm.

Hygrometer. Instrument for measuring the RELATIVE HUMIDITY of AIR, or the amount of water vapor in the ATMOSPHERE at any time.

Hypsometer. Instrument used for measuring ALTITUDE (height above SEA LEVEL), using boiling water that circulates around a THERMOMETER. Since ATMOSPHERIC PRESSURE falls with increased altitude, the boiling point of water is lower. The hypsometer relies on this difference in boiling point to calculate ELEVATION. A more common instrument for measuring altitude is the ALTIMETER.

Ice blink. Bright, usually yellowish-white glare or reflection on the underside of a CLOUD layer, produced by light reflected from an ice-covered surface such as pack ice. A similar phenomenon of reflection from a snow-covered surface is called snow blink.

Ice-cap climate. Earth's most severe CLIMATE, where the mean monthly TEMPERATURE is never above 32 DEGREES Fahrenheit (0 degrees Celsius). This climate is found in Greenland and Antarctica, which are high PLATEAUS, where KATABATIC WINDS blow strongly and frequently. At these high LATITUDES, INSOLATION (SOLAR ENERGY) is received for a short period in the summer months, but the high reflectivity of the ice and SNOW means that much is reflected back instead of being absorbed by the surface. No VEGETATION can grow, because the LANDSCAPE is permanently covered in ice and snow. Because AIR temperatures are so cold, PRECIPITATION is usually less than 5 inches (13 centimeters) annually. The POLES are REGIONS of stable, high-pressure air, where dry conditions prevail, but strong winds that blow the snow around are common. In the KÖPPEN CLIMATE CLASSIFICATION, the ice-cap climate is signified by the letters *EF.*

Ice sheet. Huge CONTINENTAL GLACIER. The only ice sheets remaining cover most of Antarctica and Greenland. At the peak of the last ICE AGE, around 18,000 years ago, ice covered as much as one-third of the earth's land surfaces. In the NORTHERN HEMISPHERE, there were two great ice sheets—the Laurentide ice sheet, covering North America, and the Scandinavian ice sheet, covering northwestern Europe and Scandinavia.

Ice shelf. Portion of an ICE SHEET extending into the OCEAN.

Ice storm. STORM characterized by a fall of freezing rain, with the formation of glaze on Earth objects.

Icefoot. Long, tapering extension of a GLACIER floating above the seawater where it enters the OCEAN. Eventually, it breaks away and forms an ICEBERG.

Igneous rock. ROCKS formed when molten material or MAGMA cools and crystallizes into solid rock. The type of rock varies with the composition of the magma and, more important, with the rate of cooling. Rocks that cool slowly, far beneath the earth's surface, are igneous INTRUSIVE ROCKS. These have large crystals and coarse grains. GRANITE is the most typical igneous intrusive rock. When cooling is more rapid, usually closer to or at the surface, finer-grained igneous EXTRUSIVE ROCKS such as rhyolite are formed. If the magma flows out to the surface as LAVA, it may cool quickly, forming a glassy rock called obsidian. If there is gas in the lava, rocks full of holes from bubbles of escaping gases form; PUMICE and BASALT are common igneous extrusive rocks.

Impact crater. Generally circular depression formed on the surface of a PLANET by the impact of a high-velocity projectile such as a METEORITE, ASTEROID, or COMET.

Impact volcanism. Process in which major impact events produce huge CRATERS along with MAGMA RESERVOIRS that subsequently produce volcanic activity. Such cratering is clearly visible on the MOON, Mars, Mercury, and probably Venus. It is assumed that Earth had similar craters, but EROSION has erased most of the evidence.

Import substitution. Economic process in which domestic producers manufacture or supply goods or services that were previously imported or purchased from overseas and foreign producers.

Index fossil. Remains of an ancient organism that are useful in establishing the age of ROCKS; index fossils are abundant and have a wide geographic DISTRIBUTION, a narrow stratigraphic RANGE, and a distinctive form.

Indian summer. Short period, usually not more than a week, of unusually warm WEATHER in late October or early November in the NORTHERN HEMISPHERE. Before the Indian summer, TEMPERATURES are cooler and there can be occurrences of FROST. Indian summer DAYS are marked by clear to hazy skies and calm to light WINDS, but nights are cool. The weather pattern is a high-pressure cell or ridge located for a few days over the East Coast of North America. The name originated in New England, referring to the practice of NATIVE AMERICANS gathering foods for winter storage over this brief spell. Similar weather in England is called an Old Wives' summer.

Infant mortality. DEMOGRAPHIC MEASURE calculated as the number of deaths in a year of infants, or children under one year of age, compared with the total number of live births in a COUNTRY for the same year. Low-income countries have high infant mortality rates, more than 100 infant deaths per thousand.

Infauna. Organisms that live in the seafloor.

Infiltration. Movement of water into and through the SOIL.

Initial advantage. In terms of economic DEVELOPMENT, not all LOCATIONS are suited for profitable investment. Some locations offer initial advantages, including an existing skilled labor pool, existing consumer markets, existing plants, and situational advantages. These advantages can also lead to clustering of a number of industries at a particular location and to further economic growth, which will provide the preconditions of initial advantage for further economic development.

Inlier. REGION of old ROCKS that is completely surrounded by younger rocks. These are often PLACES where ORES or MINERALS are found in commercial quantities.

Inner core. The innermost layer of the earth; the inner core is a solid ball with a radius of about 900 miles.

Inselberg. Exposed rocky HILL in a DESERT area, made of resistant ROCKS, rising steeply from the flat surrounding countryside. There are many inselbergs in Africa, but Uluru (Ayers Rock) in Australia is possibly the most famous inselberg. The word is German for "island mountain." A special type of inselberg is a bornhardt.

Insolation. ENERGY received by the earth from the SUN, which heats the earth's surface. The average insolation received at the top of the earth's ATMOSPHERE at an average distance from the Sun is called the SOLAR CONSTANT. Insolation is predominantly shortwave radiation, with wavelengths in the RANGE of 0.39 to 0.76 micrometers, which corresponds to the visible spectrum. Less than half of the incoming SOLAR ENERGY reaches the earth's surface-insolation is reflected back into space by CLOUDS; smaller amounts are reflected back by surfaces, absorbed, or scattered by the atmosphere. Insolation is not distributed evenly over the earth, because of Earth's curved surface. Where the rays are perpendicular, at the SUBSOLAR POINT, insolation is at the maximum. The word is a shortened form of incoming (or intercepted) SOLAR RADIATION.

Insular climate. Island climates are influenced by the fact that no PLACE is far from the SEA. Therefore, both the DIURNAL (daily) TEMPERATURE RANGE and the annual temperature range are small.

Insurgent state. STATE that arises when an uprising or guerrilla movement gains control of part of the territory of a COUNTRY, then establishes its own form of control or government. In effect, the insurgents create a state within a state. In Colombia, for example, the government and armed forces have been unable to control several REGIONS where insurgents have created their own domains. This is generally related to coca growing and the production of cocaine. Civilian farmers are unable to resist the drug-financed "armies."

Interfluve. Higher area between two STREAMS; the surface over which water flows into the stream. These surfaces are subject to RUNOFF and EROSION by RILL action and GULLYING. Over time, interfluves are lowered.

Interlocking spur. STREAM in a hilly or mountainous REGION that winds its way in a sinuous VALLEY between the different RIDGES, slowly eroding the ends of the spurs and straightening its course. The view of interlocking spurs looking upstream is a favorite of artists, as colors change with the receding distance of each interlocking spur.

Intermediate rock. IGNEOUS ROCK that is transitional between a basic and a silicic ROCK, having a SILICA content between 54 and 64 percent.

Internal migration. Movement of people within a COUNTRY, from one REGION to another. Internal MIGRATION in high-income economies is often urban-to-RURAL, such as the migration to the SUN BELT in the United States. In low-income economies, rural-to-URBAN migration is more common.

Intertillage. Mixed planting of different seeds and seedling crops within the same SWIDDEN or cleared patch of agricultural land. Potatoes, yams, corn, rice, and bananas might all be planted. The planting times are staggered throughout the year to increase the variety of crops or nutritional balance available to the subsistence farmer and his or her family.

Intrusive rock. IGNEOUS ROCK which was formed from a MAGMA that cooled below the surface of the earth; it is commonly coarse-grained.

Irredentism. Expansion of one COUNTRY into the territory of a nearby country, based on the residence of nationals in the neighboring country. Hitler used irredentist claims to invade Czechoslovakia, because small groups of German-speakers lived there in the Sudetenland. The term comes from Italian, referring to Italy's claims before World War I that all Italian-speaking territory should become part of Italy.

Isallobar. Imaginary line on a MAP or meteorological chart joining PLACES with an equal change in ATMOSPHERIC PRESSURE over a certain time, often three hours. Isallobars indicate a pressure tendency and are used in WEATHER FORECASTING.

Island arc. Chain of VOLCANOES next to an oceanic TRENCH in the OCEAN BASINS; an oceanic PLATE descends, or subducts, below another oceanic plate at ISLAND arcs.

Isobar. Imaginary line joining PLACES of equal ATMOSPHERIC PRESSURE. WEATHER MAPS show isobars encircling areas of high or low pressure. The spacing between isobars is related to the pressure gradient.

Isobath. Line on a MAP or CHART joining all PLACES where the water depth is the same; a kind of underwater CONTOUR LINE. This kind of map is a BATHYMETRIC CONTOUR.

Isoclinal folding. When the earth's CRUST is folded, the size and shape of the folds vary according to the force of compression and nature of the ROCKS. When the surface is compressed evenly so that the two sides of the fold are parallel, isoclinal folding results. When the sides or slopes of the fold are unequal or dissimilar in shape and angle, this can be an asymmetrical or overturned fold. See also ANTICLINE; SYNCLINE.

Isotherm. Line joining PLACES of equal TEMPERATURE. A world MAP with isotherms of average monthly temperature shows that over the OCEANS, temperature decreases uniformly from the EQUATOR to the POLES, and higher temperatures occur over the CONTINENTS in summer and lower temperatures in winter because of the unequal heating properties of land and water.

Isotropic surface. Hypothetical flat surface or PLAIN, with no variation in any physical attribute. An isotropic surface has uniform ELEVATION, SOIL type, CLIMATE, and VEGETATION. Economic geographic models study behavior on an isotropic surface before applying the results to the real world. For example, in an isotropic model, land value is highest at the CITY center and falls regularly with increasing distance from there. In the real world, land values are affected by elevation, water features, URBAN regulations, and other factors. The von Thuenen model of the Isolated State is based on a uniform plain or isotropic surface.

Isthmian links. Chains of ISLANDS between substantial LANDMASSES.

Isthmus. Narrow strip of land connecting two larger bodies of land. The Isthmus of Panama connects North and South America; the Isthmus of Suez connects Africa and Asia. Both of these have been cut by CANALS to shorten shipping routes.

Jet stream. WINDS that move from west to east in the upper ATMOSPHERE, 23,000 to 33,000 feet (7,000–10,000 meters) above the earth, at about 200 miles (300 km.) per hour. They are narrow bands, elliptical in cross section, traveling in irregular paths. Four jet streams of interest to earth scientists and meteorologists are the polar jet stream and the subtropical jet stream in the Northern and SOUTHERN HEMISPHERES. The polar jet stream is located at the TROPOPAUSE, the BOUNDARY between the TROPOSPHERE and the STRATOSPHERE, along the polar FRONT. There is a complex interaction between surface winds and jet streams. In winter the NORTHERN HEMISPHERE polar front can move as far south as Texas, bringing BLIZZARDS and extreme WEATHER conditions. In summer, the polar jet stream is located over Canada. The subtropical jet stream is located at the tropopause around 30 DEGREES north or south LATITUDE, but it also migrates north or south, depending on the SEASON.

At times, the polar and subtropical jet streams merge for a few DAYS. Aircraft take advantage of the jet stream, or avoid it, depending on the direction of their flight. Upper atmosphere winds are also known as GEOSTROPHIC winds.

Joint. Naturally occurring fine crack in a ROCK, formed by cooling or by other stresses. SEDIMENTARY ROCKS can split along bedding planes; other joints form at right angles to the STRATA, running vertically through the rocks. In IGNEOUS ROCKS such as GRANITE, the stresses of cooling and contraction cause three sets of joints, two vertical and one parallel to the surface, which leads to the formation of distinctive LANDFORMS such as TORS. BASALT often demonstrates columnar jointing, producing tall columns that are mostly hexagonal in section. The presence of joints in BEDROCK hastens WEATHERING, because water can penetrate into the joints. This is particularly obvious in LIMESTONE, where joints are rapidly enlarged by SOLUTION. FROST WEDGING is a type of PHYSICAL WEATHERING that can split large boulders through the expansion when water in a joint freezes to form ice. Compare with FAULTS, which occur through tectonic activity.

Jurassic. Second of the three PERIODS that make up

Kame. Small HILL of gravel or mixed-size deposits, SAND, and gravel. Kames are found in areas previously covered by CONTINENTAL GLACIERS or ICE SHEETS, near what was the outer edge of the ice. They may have formed by materials dropping out of the melting ice, or in a deltalike deposit by a STREAM of MELTWATER. These deposits of which kames are made are called drift. Small LAKES called KETTLES are often found nearby. A closely spaced group of kames is called a kame field.

Karst. LANDSCAPE of SINKHOLES, underground STREAMS and caverns, and associated features created by CHEMICAL WEATHERING, especially SOLUTION, in REGIONS where the BEDROCK is LIMESTONE. The name comes from a region in the southwest of what is now Slovenia, the Krs (Kras) Plateau, but the karst region extends south through the Dinaric Alps bordering the

Adriatic Sea, into Bosnia-Herzegovina and Montenegro. Where limestone is well jointed, RAINFALL penetrates the JOINTS and enters the GROUNDWATER, carrying the MINERALS, especially calcium, away in solution. Most of the famous CAVES and caverns of the world are found in karst areas. The Carlsbad Caverns in New Mexico are a good example. Kentucky, Tennessee, and Florida also have well-known areas of karst. In some tropical countries, a form called tower karst is found. Tall conical or steep-sided HILLS of limestone rise above the flat surrounding landscape. Around 15 percent of the earth's land surface is karst TOPOGRAPHY.

Katabatic wind. GRAVITY DRAINAGE WINDS similar to MOUNTAIN BREEZES but stronger in force and over a larger area than a single VALLEY. Cold AIR collects over an elevated REGION, and the dense cold air flows strongly downslope. The ICE-SHEETS of Antarctica and Greenland produce fierce katabatic winds, but they can occur in smaller regions. The BORA is a strong, cold, squally downslope wind on the Dalmatian COAST of Yugoslavia in winter.

Kettle. Small depression, often a small LAKE, produced as a result of continental GLACIATION. It is formed by an isolated block of ice remaining in the ground MORAINE after a GLACIER has retreated. Deposited material accumulates around the ice, and when it finally melts, a steep hole remains, which often fills with water. Walden Pond, made famous by writer Henry David Thoreau, is a glacial kettle.

Khamsin. Hot, dry, DUST-laden WIND that blows in the eastern Sahara, in Egypt, and in Saudi Arabia, bringing high TEMPERATURES for three or four DAYS. Winds can reach GALE force in intensity. The word Khamsin is Arabic for "fifty" and refers to the period between March and June when the khamsin can occur.

Knickpoint. Abrupt change in gradient of the bed of a RIVER or STREAM. It is marked by a WATERFALL, which over time is eroded by FLUVIAL action, restoring the smooth profile of the riverbed. The knickpoint acts as a TEMPORARY BASE LEVEL for the upper part of the stream.

Knickpoints can occur where a hard layer of ROCK is slower to erode than the rocks downstream, for example at Niagara Falls. Other knickpoints and waterfalls can develop as a result of tectonic forces. UPLIFT leads to new EROSION by a stream, creating a knickpoint that gradually moves upstream. The bed of a tributary GLACIER is often considerably higher than the VALLEY of the main glacier, so that after the glaciers have melted, a waterfall emerges over this knickpoint from the smaller hanging valley to join the main stream. Yosemite National Park has several such waterfalls.

Köppen climate classification. Commonly used scheme of CLIMATE classification that uses statistics of average monthly TEMPERATURE, average monthly PRECIPITATION, and total annual precipitation. The system was devised by Wladimir Köppen early in the twentieth century.

La Niña. WEATHER phenomenon that is the opposite part of EL NIÑO. When the SURFACE WATER in the eastern Pacific Ocean is cooler than average, the southeast TRADE WINDS blow strongly, bringing heavy rains to countries of the western Pacific. Scientists refer to the whole RANGE of TEMPERATURE, pressure, WIND, and SEA LEVEL changes as the SOUTHERN OSCILLATION (ENSO). The term "El Niño" gained wide currency in the U.S. media after a strong ENSO warm event in 1997-1998. A weak ENSO cold event, or La Niña, followed it in 1998. Means "the little girl" in Spanish. Alternative terms are "El Viejo" and "anti-El Niño."

Laccolith. LANDFORM of INTRUSIVE volcanism formed when viscous MAGMA is forced between overlying sedimentary STRATA, causing the surface to bulge upward in a domelike shape.

Lahar. Type of mass movement in which a MUD-FLOW occurs because of a volcanic explosion or ERUPTION. The usual cause is that the heat from the LAVA or other pyroclastic material melts ice and SNOW at the VOLCANO's SUMMIT, causing a hot mudflow that can move downslope with great speed. The eruption of Mount Saint Helens in 1985 was accompanied by a lahar.

Lake basin. Enclosed depression on the surface of the land in which SURFACE WATERS collect; BASINS are created primarily by glacial activity and tectonic movement.

Lakebed. Floor of a LAKE.

Land bridge. Piece of land connecting two CONTINENTS, which permits the MIGRATION of humans, animals, or plants from one area to another. Many former land bridges are now under water, because of the rise in SEA LEVEL after the last ICE AGE. The Bering Strait connecting Asia and North America was an important land bridge for the latter continent.

Land hemisphere. Because the DISTRIBUTION of land and water surfaces on Earth is quite asymmetrical on either side of the EQUATOR, the NORTHERN HEMISPHERE might well be called the land hemisphere. For many centuries, Europeans refused to believe that there was not an equal area of land in the SOUTHERN HEMISPHERE. Explorers such as James Cook were dispatched to seek such a "Great South Land."

Landmass. Large area of land—an ISLAND or a CONTINENT.

Landsat. Space-exploration project begun in 1972 to MAP the earth continuously with SATELLITE imaging. The satellites have collected data about the earth: its AGRICULTURE, FORESTS, flat lands, MINERALS, waters, and ENVIRONMENT. These were the first satellites to aid in Earth sciences, helping to produce the best maps available and assisting farmers around the world to improve their crop yields.

Language family. Group of related LANGUAGES believed to have originated from a common prehistoric language. English belongs in the Indo-European language family, which includes the languages spoken by half of the world's peoples.

Lapilli. Small ROCK fragments that are ejected during volcanic ERUPTIONS. A lapillus ranges from about the size of a pea to not larger than a walnut. Some lapilli form by accretion of VOLCANIC ASH around moisture droplets, in a manner similar to hailstone formation. Lapilli sometimes form into a textured rock called lapillistone.

Laterite. Bright red CLAY SOIL, rich in iron oxide, that forms in tropical CLIMATES, where both TEMPERATURE and PRECIPITATION are high year-round, as ROCKS weather. It can be used in brick making and is a source of iron. When the soil is rich in aluminum, it is called BAUXITE. When laterite or bauxite forms a hard layer at the surface, it is called duricrust. Australia and sub-Saharan Africa have large areas of duricrust, some of which is thought to have formed under previous conditions during the TRIASSIC period.

Laurasia. Hypothetical SUPERCONTINENT made up of approximately the present CONTINENTS of the NORTHERN HEMISPHERE.

Lava tube. Cavern structure formed by the draining out of liquid LAVA in a pahoehoe flow.

Layered plains. Smooth, flat REGIONS believed to be composed of materials other than sulfur compounds.

Leaching. Removal of nutrients from the upper horizon or layer of a SOIL, especially in the humid TROPICS, because of heavy RAINFALL. The remaining soil is often bright red in color because iron is left behind. Despite their bright color, tropical soils are infertile.

Leeward. Rear or protected side of a MOUNTAIN or RANGE is the leeward side. Compare to WINDWARD.

Legend. Explanation of the different colors and symbols used on a MAP. For example, a map of the world might use different colors for high-income, middle-income, and low-income economies. A historical map might use different colors for countries that were once colonies of Britain, France, or Spain.

Light year. Distance traveled by light in one year; widely used for measuring stellar distances, it is equal to roughly 6 trillion miles (9.5 million km.).

Lignite. Low-grade COAL, often called brown coal. It is mined and used extensively in eastern Germany, Slovakia, and the Moscow Basin.

Liquefaction. Loss in cohesiveness of water-saturated SOIL as a result of ground shaking caused by an EARTHQUAKE.

Lithification. Process whereby loose material is transformed into solid ROCK by compaction or cementation.

Lithology. Description of ROCKS, such as rock type, MINERAL makeup, and fluid in rock pores.

Lithosphere. Solid outermost layer of the earth. It varies in thickness from a few miles to more than 120 miles (200 km.). It is broken into pieces known as TECTONIC PLATES, some of which are extremely large, while others are quite small. The upper layer of the lithosphere is the CRUST, which may be CONTINENTAL CRUST or OCEANIC CRUST. Below the crust is a layer called the ASTHENOSPHERE, which is weaker and plastic, enabling the motion of tectonic plates.

Littoral. Adjacent to or related to a SEA.

Llanos. Grassy REGION in the Orinoco Basin of Venezuela and part of Colombia. SAVANNA VEGETATION gradually gives way to scrub at the outer edges of the *llanos.* The area is relatively undeveloped.

Loam. SOIL TEXTURE classification, indicating a soil that is approximately equal parts of SAND, SILT, and CLAY. Farmers generally consider a sandy loam to be the best soil texture because of its water-retaining qualities and the ease with which it can be cultivated.

Local sea-level change. Change in SEA LEVEL only in one area of the world, usually by land rising or sinking in that specific area.

Lode deposit. Primary deposit, generally a VEIN, formed by the filling of a FISSURE with MINERALS precipitated from a HYDROTHERMAL solution.

Loess. EOLIAN, or wind-blown, deposit of fine, silt-sized, light-colored material. Loess covers about 10 percent of the earth's land surface. The loess PLATEAU of China is good agricultural land, although susceptible to EROSION. Loess has the property of being able to form vertical CLIFFS or BLUFFS, and many people have built dwellings in the steep cliffs above the Huang He (Yellow) River. In the United States, loess deposits are found in the VALLEYS of the Platte, Missouri, Mississippi, and Ohio Rivers, and on the Columbia Plateau. A German word, meaning loose or unconsolidated, which comes from loess deposits along the Rhine River.

Longitudinal bar. Midchannel accumulation of SAND and gravel with its long end oriented roughly parallel to the RIVER flow.

Longshore current. Current in the OCEAN close to the SHORE, in the surf zone, produced by WAVES approaching the COAST at an angle. Also called a LITTORAL current. The longshore current combined with wave action can move large amounts of SAND and other BEACH materials down the coast, a process called LONGSHORE DRIFT.

Longshore drift. The movement of SEDIMENT parallel to the BEACH by a LONGSHORE CURRENT.

Maar. Explosion vent at the earth's surface where a volcanic cone has not formed. A small ring of pyroclastic materials surrounds the maar. Often a LAKE occupies the small CRATER of a maar. A larger form is called a TUFF RING.

Macroburst. Updrafts and downdrafts within a CUMULONIMBUS CLOUD or THUNDERSTORM can cause severe TURBULENCE. A DOWNBURST within a thunderstorm when windspeeds are greater than 130 miles (210 km.) per hour and over areas of 2.5 square miles (5 sq. km.) or more is called a macroburst. See also MICROBURST.

Magnetic poles. Locations on the earth's surface where the earth's MAGNETIC FIELD is perpendicular to the surface. The magnetic poles do not correspond exactly to the geographic North Pole and South Pole, or earth's AXIS; the difference is called magnetic variation or DECLINATION.

Magnetic reversal. Change in the earth's MAGNETIC FIELD from the North Pole to the South MAGNETIC POLE.

Magnetic storm. Rapid changes in the earth's MAGNETIC FIELD as a result of the bombardment of the earth by electrically charged particles from the SUN.

Magnetosphere. REGION surrounding a PLANET where the planet's own MAGNETIC FIELD predominates over magnetic influences from the SUN or other planets.

461

Mantle convection. Thermally driven flow in the earth's MANTLE thought to be the driving force of PLATE TECTONICS.

Mantle plume. Rising jet of hot MANTLE material that produces tremendous volumes of basaltic LAVA. See also HOT SPOT.

Map projection. Mathematical formula used to transform the curved surface of the earth onto a flat plane or sheet of paper. Projections are divided into three classes: CYLINDRICAL, CONICAL, and AZIMUTHAL.

Marchland. FRONTIER area where boundaries are poorly defined or absent. The marches themselves were a type of BOUNDARY REGION. Marchlands have changed hands frequently throughout history. The name is related to the fact that armies marched across them.

Mass balance. Summation of the net gain and loss of ice and SNOW mass on a GLACIER in a year.

Mass extinction. Die-off of a large percentage of species in a short time.

Mass wasting. Downslope movement of Earth materials under the direct influence of GRAVITY.

Massif. French term used in geology to describe very large, usually IGNEOUS INTRUSIVE bodies.

Meandering river. RIVER confined essentially to a single CHANNEL that transports much of its SEDIMENT load as fine-grained material in SUSPENSION.

Mechanical weathering. Another name for PHYSICAL WEATHERING, or the breaking down of ROCK into smaller pieces.

Mechanization. Replacement of human labor with machines. Mechanization occurred in AGRICULTURE as tractors, reapers, picking machinery, and similar technological inventions took the place of human farm labor. Mechanization in industry was part of the INDUSTRIAL REVOLUTION, as spinning and weaving machines were introduced into the textile industry.

Medical geography. Branch of geography specializing in the study of health and disease, with a particular emphasis on the areal spread or DIFFUSION of disease. The spatial perspective of geography can lead to new medical insights. Geographers working with medical researchers in Africa have made great contributions to understanding the role of disease on that CONTINENT. John Snow's studies of the origin and spread of cholera in London in 1854 mark the beginnings of medical geography.

Megalopolis. Conurbation formed when large cities coalesce physically into one huge built-up area. Originally coined by the French geographer Jean Gottman in the early 1960s for the northeastern part of the United States, from Boston to Washington, D.C.

Mesa. Flat-topped HILL with steep sides. EROSION removes the surrounding materials, while the mesa is protected by a cap of harder, more resistant ROCK. Usually found in arid REGIONS. A larger LANDFORM of this type is a PLATEAU; a smaller feature is a BUTTE. The Colorado Plateau and Grand Canyon in particular are rich in these landforms. From the Spanish word for table.

Mesosphere. Atmospheric layer above the STRATOSPHERE where TEMPERATURE drops rapidly.

Mestizo. Person of mixed European and Amerindian ancestry, especially in countries of LATIN AMERICA.

Metamorphic rock. Any ROCK whose mineralogy, MINERAL chemistry, or TEXTURE has been altered by heat, pressure, or changes in composition; metamorphic rocks may have IGNEOUS, SEDIMENTARY, or other, older metamorphic rocks as their precursors.

Metamorphic zone. Areas of ROCK affected by the same limited RANGE of TEMPERATURE and pressure conditions, commonly identified by the presence of a key individual MINERAL or group of minerals.

Meteor. METEOROID that enters the ATMOSPHERE of a PLANET and is destroyed through frictional heating as it comes in contact with the various gases present in the atmosphere.

Meteorite. Fragment of an ASTEROID that survives passage through the ATMOSPHERE and strikes the surface of the earth.

Meteoroid. Small planetary body that enters Earth's ATMOSPHERE because its path intersects the earth's ORBIT. Friction caused by the earth's atmosphere on the meteoroid creates a glowing METEOR, or "shooting star." This is a common phenomenon, and most meteors burn away completely. Those that are large enough to reach the ground are called METEORITES.

Microburst. Brief but intense downward WIND, lasting not more than fifteen minutes over an area of 0.6 to 0.9 square mile (1.5–8 sq. km.). Usually associated with THUNDERSTORMS, but are quite unpredictable. The sudden change in wind direction associated with a microburst can create wind shear that causes airplanes to crash, especially if it occurs during takeoff or landing. See also MACROBURST.

Microclimate. CLIMATE of a small area, at or within a few yards of the earth's surface. In this REGION, variations of TEMPERATURE, PRECIPITATION, and moisture can have a pronounced effect on the bioclimate, influencing the growth or well-being of plants and animals, including humans. DEW or FROST, RAIN SHADOW effects, wind-tunneling between tall buildings, and similar phenomena are studied by microclimatologists. Horticulturists know the variations in aspect that affect INSOLATION and temperature, so that certain plants grow best on south-facing walls, for example. The growing of grapes for wine production is a major industry where microclimatology is essential. The study of microclimatology was pioneered by the German meteorologist Rudolf Geiger.

Microcontinent. Independent LITHOSPHERIC PLATE that is smaller than a CONTINENT but possesses continental-type CRUST. Examples include Cuba and Japan.

Microstates. Tiny countries. In 2000, seventeen independent countries each had an area of less than 200 square miles (520 sq. km.). The smallest microstate is Vatican City, with an area of 0.2 square miles (0.5 sq. km.). Most of the world's microstates are island NATIONS, including Nauru, Tuvalu, Marshall Islands, Saint Kitts and Nevis, Seychelles, Maldives, Malta, Grenada, Saint Vincent and the Grenadines, Barbados, Antigua and Barbuda, and Palau.

Mineral species. Mineralogic division in which all the varieties in any one species have the same basic physical and chemical properties.

Monadnock. Isolated HILL far from a STREAM, composed of resistant BEDROCK. Monadnocks are found in humid temperate REGIONS. A similar LANDFORM in an arid region is an INSELBERG.

Monogenetic. Pertaining to a volcanic ERUPTION in which a single vent is used only once.

Moraine. Materials transported by a GLACIER, and often later deposited as a RIDGE of unsorted ROCKS and smaller material. Lateral moraine is found at the side of the glacier; medial moraine occurs when two glaciers join. Other types of moraine include ABLATION moraine, ground moraine, and push, RECESSIONAL, and TERMINAL MORAINE.

Mountain belts. Products of PLATE TECTONICS, produced by the CONVERGENCE of crustal PLATES. Topographic MOUNTAINS are only the surficial expression of processes that profoundly deform and modify the CRUST. Long after the mountains themselves have been worn away, their former existence is recognizable from the structures that mountain building forms within the ROCKS of the crust.

Nappe. Huge sheet of ROCK that was the upper part of an overthrust fold, and which has broken and traveled far from its original position due to the tremendous forces. The Swiss Alps have nappes in many LOCATIONS.

Narrows. STRAIT joining two bodies of water.

Nation-state. Political entity comprising a COUNTRY whose people are a national group occupying the area. The concept originated in eighteenth century France; in practice, such cultural homogeneity is rare today, even in France.

Natural increase, rate of. DEMOGRAPHIC MEASURE of POPULATION growth: the difference between births and deaths per year, expressed as a percentage of the POPULATION. The rate of natural increase for the United States in 2000 was 0.6 percent. In countries where the population is decreasing, the DEATH RATE is greater than the BIRTH RATE.

Natural selection. Main process of biological evolution; the production of the largest number of offspring by individuals with traits that are best adapted to their ENVIRONMENTS.

Nautical mile. Standard MEASUREMENT at SEA, equalling 6,076.12 feet (1.85 km.). The mile used for land measurements is called a statute mile and measures 5,280 feet (1.6 km.).

Neap tide. TIDE with the minimum RANGE, or when the level of the high tide is at its lowest.

Near-polar orbit. Earth ORBIT that lies in a plane that passes close to both the north and south POLES.

Nekton. PELAGIC organisms that can swim freely, without having to rely on OCEAN CURRENTS or WINDS. Nekton includes shrimp; crabs; oysters; MARINE reptiles such as turtles, crocodiles, and snakes; and even sharks; porpoises; and whales.

Net migration. Net balance of a COUNTRY or REGION's IMMIGRATION and EMIGRATION.

Nomadism. Lifestyle in which pastoral people move with grazing animals along a defined route, ensuring adequate pasturage and water for their flocks or herds. This lifestyle has decreased greatly as countries discourage INTERNATIONAL MIGRATION. A more restricted form of nomadism is TRANSHUMANCE.

North geographic pole. Northernmost REGION of the earth, located at the northern point of the PLANET's AXIS of ROTATION.

North magnetic pole. Small, nonstationary area in the Arctic Circle toward which a COMPASS needle points from any LOCATION on the earth.

Notch. Erosional feature found at the base of a SEA CLIFF as a result of undercutting by WAVE EROSION, bioabrasion from MARINE organisms, and dissolution of ROCK by GROUNDWATER seepage. Also known as a nip.

Nuclear energy. ENERGY produced from a naturally occurring isotope of uranium. In the process of nuclear FISSION, the unstable uranium isotope absorbs a neutron and splits to form tin and molybdenum. This releases more neurons, so a chain reaction proceeds, releasing vast amounts of heat energy. Nuclear energy was seen in the 1950s as the energy of the future, but safety fears and the problem of disposal of radioactive nuclear waste have led to public condemnation of nuclear power plants.

Nuée ardente. Hot cloud of ROCK fragments, ASH, and gases that suddenly and explosively erupt from some VOLCANOES and flow rapidly down their slopes.

Nunatak. Isolated MOUNTAIN PEAK or RIDGE that projects through a continental ICE SHEET. Found in Greenland and Antarctica.

Obduction. Tectonic collisional process, opposite in effect to SUBDUCTION, in which heavier OCEANIC CRUST is thrust up over lighter CONTINENTAL CRUST.

Oblate sphere. Flattened shape of the earth that is the result of ROTATION.

Occultation. ECLIPSE of any astronomical object other than the SUN or the MOON caused by the Moon or any PLANET, SATELLITE, or ASTEROID.

Ocean basins. Large worldwide depressions that form the ultimate RESERVOIR for the earth's water supply.

Ocean circulation. Worldwide movement of water in the SEA.

Ocean current. Predictable circulation of water in the OCEAN, caused by a combination of WIND friction, Earth's ROTATION, and differences in TEMPERATURE and density of the waters. The five great oceanic circulations, known as GYRES, are in the North Pacific, North Atlantic, South Pacific, South Atlantic, and Indian Oceans. Because of the CORIOLIS EFFECT, the direction of circulation is CLOCKWISE in the NORTHERN HEMISPHERE and COUNTERCLOCKWISE in the SOUTHERN HEMISPHERE, except in the Indian Ocean, where the direction changes annually with the pattern of winds associated with the Asian MONSOON. Currents flowing toward the EQUATOR are cold currents; those flowing away from the equator are warm currents. An important current is the warm Gulf Stream, which flows north from the Gulf of Mexico along the East Coast of the United States; it crosses the North Atlantic, where it is called the North Atlantic Drift, and brings warmer conditions to the western parts of Europe. The West Coast of the United States is affected by the cool, south-flowing California Current. The cool Humboldt, or Peru, Current, which flows north along the South American coast, is an important indicator of whether there will be an EL NIÑO event. Deep currents, below 300 feet (100 meters), are extremely complicated and difficult to study.

Oceanic crust. Portion of the earth's CRUST under its OCEAN BASINS.

Oceanic island. ISLANDS arising from seafloor volcanic ERUPTIONS, rather than from continental shelves. The Hawaiian Islands are the best-known examples of oceanic islands.

Off-planet. Pertaining to REGIONS off the earth in orbital or planetary space.

Ore deposit. Natural accumulation of MINERAL matter from which the owner expects to extract a metal at a profit.

Orogeny. MOUNTAIN-building episode, or event, that extends over a period usually measured in tens of millions of years; also termed a revolution.

Orographic precipitation. Phenomenon caused when an AIR mass meets a topographic barrier, such as a mountain RANGE, and is forced to rise; the air cools to saturation, and orographic precipitation falls on the WINDWARD side as rain or SNOW. The lee side is a RAIN SHADOW. This effect is noticeable on the West Coast of the United States, which has RAIN FOREST on the windward side of the MOUNTAINS and DESERTS on the lee.

Orography. Study of MOUNTAINS that incorporates assessment of how they influence and are affected by WEATHER and other variables.

Oscillatory flow. Flow of fluid with a regular back-and-forth pattern of motion.

Overland flow. Flow of water over the land surface caused by direct PRECIPITATION.

Oxbow lake. LAKE created when floodwaters make a new, shorter CHANNEL and abandon the loop of a MEANDER. Over time, water in the oxbow lake evaporates, leaving a dry, curving, low-lying area known as a meander scar. Oxbow lakes are common on FLOODPLAINS. Another name for this feature is a cut-off.

Ozone hole. Decrease in the abundance of ANTARCTIC OZONE as sunlight returns to the POLE in early springtime

Ozone layer. Narrow band of the STRATOSPHERE situated near 18 miles (30 km.) above the earth's surface, where molecules of OZONE are concentrated. The average concentration is only one in 4 million, but this thin layer protects the earth by absorbing much of the ultraviolet light from the SUN and reradiating it as longer-wavelength radiation. Scientists were disturbed to discover that the ozonosphere was being destroyed by photochemical reaction with CHLOROFLUOROCARBONS (CFCs). The OZONE HOLES over the South and North Poles negatively affect several animal species, as well as humans; skin cancer risk is increasing rapidly as a consequence of depletion of the ozone layer. Stratospheric ozone should not be confused with ozone at lower levels, which is a result of PHOTOCHEMICAL SMOG. Also called the ozonosphere.

P wave. Fastest elastic wave generated by an EARTHQUAKE or artificial ENERGY source; basically an acoustic or shock wave that compresses and stretches solid material in its path.

Pangaea. Name used by Alfred Wegener for the SUPERCONTINENT that broke apart to create the present CONTINENTS.

Parasitic cone. Small volcanic cone that appears on the flank of a larger VOLCANO, or perhaps inside a CALDERA.

Particulate matter. Mixture of small particles that adversely affect human health. The particles may come from smoke and DUST and are in their highest concentrations in large URBAN AREAS, where they contribute to the "DUST DOME." Increased occurrences of illnesses such as asthma and bronchitis, especially in children, are related to high concentrations of particulate matter.

Pastoralism. Type of AGRICULTURE involving the raising of grazing animals, such as cattle, goats, and sheep. Pastoral nomads migrate with their domesticated animals in order to ensure sufficient grass and water for the animals.

Paternoster lakes. Small circular LAKES joined by a STREAM. These lakes are the result of glacial EROSION. The name comes from the resemblance to rosary beads and the accompanying prayer (the Our Father).

Pedestal crater. A CRATER that has assumed the shape of a pedestal as a result of unique shaping processes caused by WIND.

Pedology. Scientific study of SOILS.

Pelagic. Relating to life-forms that live on or in open SEAS, rather than waters close to land.

Peneplain. In the geomorphic CYCLE, or cycle of LANDFORM development, described by W. M. Davis, the final stage of EROSION led to the creation of an extensive land surface with low RELIEF. Davis named this a peneplain, meaning "almost a plain." It is now known that tectonic forces are so frequent that there would be insufficient time for such a cycle to complete all stages required to complete this landform.

Percolation. Downward movement of part of the water that falls on the surface of the earth, through the upper layers of PERMEABLE SOIL and ROCKS under the influence of GRAVITY. Eventually, it accumulates in the zone of SATURATION as GROUNDWATER.

Perforated state. STATE whose territory completely surrounds another state. The classic example of a perforated state is South Africa, within which lies the COUNTRY of Lesotho. Technically, Italy is perforated by the MICROSTATES of San Marino and Vatican City.

Perihelion. Point in Earth's REVOLUTION when it is closest to the SUN (usually on January 3). At perihelion, the distance between the earth and the Sun is 91,500,000 miles (147,255,000 km.). The opposite of APHELION.

Periodicity. The recurrence of related phenomena at regular intervals.

Permafrost. Permanently frozen SUBSOIL. The condition occurs in perennially cold areas such as the ARCTIC. No trees can grow because their roots cannot penetrate the permafrost. The upper portion of the frozen SOIL can thaw briefly in the summer, allowing many smaller plants to thrive in the long daylight. Permafrost occurs in about 25 percent of the earth's land surface, and the condition even hampers construction in REGIONS such as Siberia and ARCTIC Canada.

Perturb. To change the path of an orbiting body by a gravitational force.

Petrochemical. Chemical substance obtained from NATURAL GAS or PETROLEUM.

Petrography. Description and systematic classification of ROCKS.

Photochemical smog. Mixture of gases produced by the interaction of sunlight on the gases emanating from automobile exhausts. The gases include OZONE, nitrogen dioxide, carbon monoxide, and peroxyacetyl nitrates. Many large cities suffer from poor AIR quality because of photochemical smog. Severe health problems arise from continued exposure to photochemical smog.

Photometry. Technique of measuring the brightness of astronomical objects, usually with a photoelectric cell.

Phylogeny. Study of the evolutionary relationships among organisms.

Phylum. Major grouping of organisms, distinguished on the basis of basic body plan, grade of anatomical complexity, and pattern of growth or development.

Physiography. The PHYSICAL GEOGRAPHY of a PLACE—the LANDFORMS, water features, CLIMATE, SOILS, and VEGETATION.

Piedmont glacier. GLACIER formed when several ALPINE GLACIERS join together into a spreading glacier at the base of a MOUNTAIN or RANGE. The Malaspina glacier in Alaska is a good example of a piedmont glacier.

Place. In geographic terms, space that is endowed with physical and human meaning. Geographers study the relationship between people, places, and ENVIRONMENTS. The five themes that geographers use to examine the world are LOCATION, place, human/environment interaction, movement, and REGIONS.

Placer. Accumulation of valuable MINERALS formed when grains of the minerals are physically deposited along with other, nonvaluable mineral grains.

Planetary wind system. Global atmospheric circulation pattern, as in the BELT of prevailing westerly WINDS.

Plantation. Form of AGRICULTURE in which a large area of agricultural land is devoted to the production of a single cash crop, for export. Many plantation crops are tropical, such as bananas, sugarcane, and rubber. Coffee and tea plantations require cooler CLIMATES. Formerly, slave labor was used on most plantations, and the owners were Europeans.

Plate boundary. REGION in which the earth's crustal PLATES meet, as a converging (SUBDUCTION ZONE), diverging (MID-OCEAN RIDGE), TRANSFORM FAULT, or collisional interaction.

Plate tectonics. Theory proposed by German scientist Alfred Wegener in 1910. Based on extensive study of ancient geology, STRATIGRAPHY, and CLIMATE, Wegener concluded that the CONTINENTS were formerly one single enormous LANDMASS, which he named PANGAEA. Over the past 250 million years, Pangaea broke apart, first into LAURASIA and GONDWANALAND, and subsequently into the present continents. Earth scientists now believe that the earth's CRUST is composed of a series of thin, rigid PLATES that are in motion, sometimes diverging, sometimes colliding.

Plinian eruption. Rapid ejection of large volumes of VOLCANIC ASH that is often accompanied by the collapse of the upper part of the VOLCANO. Named either for Pliny the Elder, a Roman naturalist who died while observing the ERUPTION of Mount Vesuvius in 79 CE, or for Pliny the Younger, his nephew, who chronicled the eruption.

Plucking. Term used to describe the way glacial ice can erode large pieces of ROCK as it makes its way downslope. The ice penetrates JOINTS, other openings on the floor, or perhaps the side wall, and freezes around the block of stone, tearing it away and carrying it along, as part of the glacial MORAINE. The rocks contribute greatly to glacial ABRASION, causing deep grooves or STRIATIONS in some places. The jagged torn surface left behind is subject to further plucking. ALPINE GLACIERS can erode steep VALLEYS called glacial TROUGHS.

Plutonic. IGNEOUS ROCKS made of MINERAL grains visible to the naked eye. These igneous rocks have cooled relatively slowly. GRANITE is a good example of a plutonic rock.

Pluvial period. Episode of time during which rains were abundant, especially during the last ICE AGE, from a few million to about 10,000 years ago.

Polar stratospheric clouds. CLOUDS of ice crystals formed at extremely low TEMPERATURES in the polar STRATOSPHERE.

Polder. Lands reclaimed from the SEA by constructing DIKES to hold back the sea and then pumping out the water retained between the dikes and the land. Before AGRICULTURE is possible, the SOIL must be specially treated to remove the SALT. Some polders are used for recreational land; cities also have been built on polders. The largest polders are in the Netherlands, where the northern part, known as the Low Netherlands, covers almost half of the total area of this COUNTRY.

Polygenetic. Pertaining to volcanism from several physically distinct vents or repeated ERUPTIONS from a single vent punctuated by long periods of quiescence.

Polygonal ground. Distinctive geological formation caused by the repetitive freezing and thawing of PERMAFROST.

Possibilism. Concept that arose among French geographers who rejected the concept of ENVIRONMENTAL DETERMINISM, instead asserting that the relationship between human beings and the ENVIRONMENT is interactive.

Potable water. FRESH WATER that is being used for domestic consumption.

Potholes. Circular depressions formed in the bed of a RIVER when the STREAM flows over BEDROCK. The scouring of PEBBLES as a result of water TURBULENCE wears away the sides of the depression, deepening it vertically and producing a smooth, rounded pothole. (In modern parlance, the term is also applied to holes in public roads.)

Primary minerals. MINERALS formed when MAGMA crystallizes.

Primary wave. Compressional type of EARTHQUAKE wave, which can travel in any medium and is the fastest wave.

Primate city. CITY that is at least twice as large as the next-largest city in that COUNTRY. The "law of the primate city" was developed by U.S. geographer Mark Jefferson, to analyze the phenomenon of countries where one huge city dominates the political, economic, and cultural life of that country. The size and dominance of a primate city is a PULL FACTOR and ensures its continuing dominance.

Principal parallels. The most important lines of LATITUDE. PARALLELS are imaginary lines, parallel to the EQUATOR. The principal parallels are the equator at zero DEGREES, the tropic of CANCER at 23.5 degrees North, the tropic of CAPRICORN at 23.5 degrees south, the Arctic Circle at 66.5 degrees north, and the Antarctic Circle at 66.5 degrees south.

Protectorate. COUNTRY that is a political DEPENDENCY of another NATION; similar to a COLONY, but usually having a less restrictive relationship with its overseeing power.

Proterozoic eon. Interval between 2.5 billion and 544 million years ago. During this PERIOD in the GEOLOGIC RECORD, processes presently active on Earth first appeared, notably the first clear evidence for PLATE TECTONICS. ROCKS of the Proterozoic eon also document changes in conditions on Earth, particularly an apparent increase in atmospheric oxygen.

Pull factors. Forces that attract immigrants to a new COUNTRY or LOCATION as permanent settlers. They include economic opportunities, educational facilities, land ownership, gold rushes, CLIMATE conditions, democracy, and similar factors of attraction.

Push factors. Forces that encourage people to migrate permanently from their HOMELANDS to settle in a new destination. They include war, persecution for religious or political reasons, hunger, and similar negative factors.

Pyroclasts. Materials that are ejected from a VOLCANO into the AIR. Pyroclastic materials return to Earth at greater or lesser distances, depending on their size and the height to which they are thrown by the explosion of the volcano. The largest pyroclasts are volcanic bombs. Smaller pieces are volcanic blocks and scoria. These generally fall back onto the volcano and roll down the sides. Even smaller pyroclasts are LAPILLI, cinders, and VOLCANIC ASH. The finest pyroclastic materials may be carried by WINDS for great distances, even completely around the earth, as was the case with DUST from the Krakatoa explosion in 1883 and the early 1990s explosions of Mount Pinatubo in the Philippines.

Qanat. Method used in arid REGIONS to bring GROUNDWATER from mountainous regions to lower and flatter agricultural land. A qanat is a long tunnel or series of tunnels, perhaps more than a mile long. The word *qanat* is Arabic, but the first qanats are thought to have been constructed in Farsi-speaking Persia more than 2,000 years ago. Qanats are still used there, as well as in Afghanistan and Morocco.

Quaternary sector. Economic activity that involves the collection and processing of information. The rapid spread of computers and the Internet caused a major increase in the importance of employment in the quaternary sector.

Radar imaging. Technique of transmitting radar toward an object and then receiving the reflected radiation so that time-of-flight MEASUREMENTS provide information about surface TOPOGRAPHY of the object under study.

Radial drainage. The pattern of STREAM courses often reveals the underlying geology or structure of a REGION. In a radial drainage pattern, streams radiate outward from a center, like spokes on a wheel, because they flow down the slopes of a VOLCANO.

Radioactive minerals. MINERALS combining uranium, thorium, and radium with other elements. Useful for nuclear TECHNOLOGY, these minerals furnish the basic isotopes necessary not only for nuclear reactors but also for advanced medical treatments, metallurgical analysis, and chemicophysical research.

Rain gauge. Instrument for measuring RAINFALL, usually consisting of a cylindrical container open to the sky.

Rain shadow. Area of low PRECIPITATION located on the LEEWARD side of a topographic barrier such as a mountain RANGE. Moisture-laden WINDS are forced to rise, so they cool ADIABATICALLY, leading to CONDENSATION and precipitation on the WINDWARD side of the barrier. When the AIR descends on the other side of the MOUNTAIN, it is dry and relatively warm. The area to the east of the Rocky Mountains is in a rain shadow.

Range, mountain. Linear series of MOUNTAINS close together, formed in an OROGENY, or mountain-building episode. Tall mountain ranges such as the Rocky Mountains are geologically much younger than older mountain ranges such as the Appalachians.

Rapids. Stretches of RIVERS where the water flow is swift and turbulent because of a steep and rocky CHANNEL. The turbulent conditions are called WHITE WATER. If the change in ELEVATION is greater, as for small WATERFALLS, they are called CATARACTS.

Recessional moraine. Type of TERMINAL MORAINE that marks a position of shrinkage or wasting or a GLACIER. Continued forward flow of ice is maintained so that the debris that forms the moraine continues to accumulate. Recessional moraines occur behind the terminal moraine.

Recumbent fold. Overturned fold in which the upper part of the fold is almost horizontal, lying on top of the nearest adjacent surface.

Reef (geology). VEIN of ORE, for example, a reef of gold.

Reef (marine). Underwater ridge made up of sand, rocks, or coral that rises near to the water's surface.

Refraction of waves. Bending of waves, which can occur in all kinds of waves. When OCEAN WAVES approach a COAST, they start to break as they approach the SHORE because the depth decreases.

The wave speed is retarded and the WAVE CREST seems to bend as the wavelength decreases. If waves are approaching a coast at an oblique angle, the crest line bends near the shore until it is almost parallel. If waves are approaching a BAY, the crests are refracted to fit the curve of the bay.

Regression. Retreat of the SEA from the land; it allows land EROSION to occur on material formerly below the sea surface.

Relative humidity. Measure of the HUMIDITY, or amount of moisture, in the ATMOSPHERE at any time and place compared with the total amount of moisture that same AIR could theoretically hold at that TEMPERATURE. Relative humidity is a ratio that is expressed as a percentage. When the air is saturated, the relative humidity reaches 100 percent and rain occurs. When there is little moisture in the air, the relative humidity is low, perhaps 20 percent. Relative humidity varies inversely with temperature, because warm air can hold more moisture than cooler air. Therefore, when temperatures fall overnight, the air often becomes saturated and DEW appears on grass and other surfaces. The human COMFORT INDEX is related to the relative humidity. Hot temperatures are more bearable when relative humidity is low. Media announcers frequently use the term "humidity" when they mean relative humidity.

Replacement rate. The rate at which females must reproduce to maintain the size of the POPULATION. It corresponds to a FERTILITY RATE of 2.1.

Reservoir rock. Geologic ROCK layer in which OIL and gas often accumulate; often SANDSTONE or LIMESTONE.

Retrograde orbit. ORBIT of a SATELLITE around a PLANET that is in the opposite sense (direction) in which the planet rotates.

Retrograde rotation. ROTATION of a PLANET in a direction opposite to that of its REVOLUTION.

Reverse fault. Feature produced by compression of the earth's CRUST, leading to crustal shortening. The UPTHROWN BLOCK overhangs the downthrown block, producing a FAULT SCARP where the overhang is prone to LANDSLIDES. When the movement is mostly horizontal, along a low angle FAULT, an overthrust fault is formed. This is commonly associated with extreme FOLDING.

Reverse polarity. Orientation of the earth's MAGNETIC FIELD so that a COMPASS needle points to the SOUTHERN HEMISPHERE.

Ria coast. Ria is a long narrow ESTUARY or RIVER MOUTH. COASTS where there are many rias show the effects of SUBMERGENCE of the land, with the SEA now occupying former RIVER VALLEYS. Generally, there are MOUNTAINS running at an angle to the coast, with river valleys between each RANGE, so that the ria coast is a succession of estuaries and promontories. The submergence can result from a rising SEA LEVEL, which is common since the melting of the PLEISTOCENE GLACIERS, or it can be the result of SUBSIDENCE of the land. There is often a great TIDAL RANGE in rias, and in some, a tidal BORE occurs with each TIDE. The eastern coast of the United States, from New York to South Carolina, is a ria coast. The southwest coast of Ireland is another. The name comes from Spain, where rias occur in the south.

Richter scale. SCALE used to measure the magnitude of EARTHQUAKES; named after U.S. physicist Charles Richter, who, together with Beno Gutenberg, developed the scale in 1935. The scale is a quantitative measure that replaced the older MERCALLI SCALE, which was a descriptive scale. Numbers range from zero to nine, although there is no upper limit. Each whole number increase represents an order of magnitude, or an increase by a factor of ten. The actual MEASUREMENT was logarithm to base 10 of the maximum SEISMIC WAVE amplitude (in thousandths of a millimeter) recorded on a standard SEISMOGRAPH at a distance of 60 miles (100 km.) from the earthquake EPICENTER.

Rift valley. Long, low REGION of the earth's surface; a VALLEY or TROUGH with FAULTS on either side. Unlike valleys produced by EROSION, rift valleys are produced by tectonic forces that have caused the faults or fractures to develop in the ROCKS of Earth's CRUST. TENSION can lead to the block of land between two faults dropping in ELEVATION compared to the surrounding blocks, thus forming the rift valley. A small LANDFORM produced in this way is called a GRABEN. A rift valley is a much larger feature. In Africa, the Great Rift Valley is partially occupied by Lake Malawi and Lake Tanganyika, as well as by the Red Sea.

Ring dike. Volcanic LANDFORM created when MAGMA is intruded into a series of concentric FAULTS. Later EROSION of the surrounding material may reveal the ring dike as a vertical feature of thick BASALT rising above the surroundings.

Ring of Fire. Zone of volcanic activity and associated EARTHQUAKES that marks the edges of various TECTONIC PLATES around the Pacific Ocean, especially those where SUBDUCTION is occurring.

Riparian. Term meaning related to the BANKS of a STREAM or RIVER. Riparian VEGETATION is generally trees, because of the availability of moisture. RIPARIAN RIGHTS allow owners of land adjacent to a river to use water from the river.

River terraces. LANDFORMS created when a RIVER first produces a FLOODPLAIN, by DEPOSITION of ALLUVIUM over a wide area, and then begins downcutting into that alluvium toward a lower BASE LEVEL. The renewed EROSION is generally because of a fall in SEA LEVEL, but can result from tectonic UPLIFT or a change in CLIMATE pattern due to increased PRECIPITATION. On either side of the river, there is a step up from the new VALLEY to the former alluvium-covered floodplain surface, which is now one of a pair of river terraces. This process may occur more than once, creating as many as three sets of terraces. These are called depositional terraces, because the terrace is cut into river deposits. Erosional terraces, in contrast, are formed by lateral migration of a river, from one part of the valley to another, as the river creates a floodplain. These terraces are cut into BEDROCK, with only a thin layer of alluvium from the point BAR deposits, and they do not occur in matching pairs.

River valleys. VALLEYS in which STREAMS flow are produced by those streams through long-term EROSION and DEPOSITION. The LANDFORMS produced by FLUVIAL action are quite diverse, ranging from spectacular CANYONS to wide, gently sloping valleys. The patterns formed by stream networks are complex and generally reflect the BEDROCK geology and TERRAIN characteristics.

Rock avalanche. Extreme case of a rockfall. It occurs when a large mass of ROCK moves rapidly down a steeply sloping surface, taking everything that lies in its path. It can be started by an EARTHQUAKE, rock-blasting operations, or vibrations from thunder or artillery fire.

Rock cycle. Cycle by which ROCKS are formed and reformed, changing from one type to another over long PERIODS of geologic time. IGNEOUS ROCKS are formed by cooling from molten MAGMA. Once exposed at the surface, they are subject to WEATHERING and EROSION. The products of erosion are compacted and cemented to form SEDIMENTARY ROCKS. The heat and pressure accompanying a volcanic intrusion causes adjacent rocks to be altered into METAMORPHIC ROCKS.

Rock slide. Event that occurs when water lubricates an unconsolidated mass of weathered ROCK on a steep slope, causing rapid downslope movement. In a RIVER VALLEY where there are steep SCREE slopes being constantly carried away by a swiftly flowing STREAM, the undercutting at the base can lead to constant rockslides of the surface layer of rock. A large rockslide is a ROCK AVALANCHE.

S waves. Type of SEISMIC disturbance of the earth when an EARTHQUAKE occurs. In an S wave, particles move about at right angles to the direction in which the wave is traveling. S waves cannot pass through the earth's CORE, which is why scientists believe the INNER CORE is liquid. Also called transverse wave, shear wave, or secondary wave.

Sahel. Southern edge of the Sahara Desert; a great stretch of semiarid land extending from the Atlantic Ocean in Senegal and Mauritania through Mali, Burkina Faso, Nigeria, Niger, Chad, and Sudan. Northern Ethiopia, Eritrea, Djibouti, and Somalia usually are included also. This transition zone between the hot DESERT and the tropical SAVANNA has low summer RAINFALL of less than 8 inches (200 millimeters) and a natural VEGETATION of low grasses with some small shrubs. The REGION traditionally has been used for PASTORALISM, raising goats, camels, and occasionally sheep. Since a prolonged DROUGHT in the 1970s, DESERTIFICATION, SOIL EROSION, and FAMINE have plagued the Sahel. The narrow band between the northern Sahara and the Mediterranean North African COAST is also called Sahel. "Sahel" is the Arabic word for edge.

Saline lake. LAKE with elevated levels of dissolved solids, primarily resulting from evaporative concentration of SALTS; saline lakes lack an outlet to the SEA. Well-known examples include Utah's Great Salt Lake, California's Mono Lake and Salton Sea, and the Dead Sea in the Middle East.

Salinization. Accumulation of SALT in SOIL. When IRRIGATION is used to grow crops in semiarid to arid REGIONS, salinization is frequently a problem. Because EVAPORATION is high, water is drawn upward through the soil, depositing dissolved salts at or near the surface. Over years, salinization can build up until the soil is no longer suitable for AGRICULTURE. The solution is to maintain a plentiful flow of water while ensuring that the water flows through the soil and is drained away.

Salt domes. Formations created when deeply buried salt layers are forced upwards. SALT under pressure is a plastic material, one that can flow or move slowly upward, because it is lighter than surrounding SEDIMENTARY ROCKS. The salt forms into a plug more than a half mile (1 km.) wide and as much as 5 miles (8 km.) deep, which passes through overlying sedimentary rock layers, pushing them up into a dome shape as it passes. Some salt domes emerge at the earth's surface; others are close to the surface and are easy to mine for ROCK SALT. OIL and NATURAL GAS often accumulate against the walls of a salt dome. Salt domes are numerous around the COAST of the Gulf of Mexico, in the North Sea REGION, and in Iran and Iraq, all of which are major oil-producing regions.

Sand dunes. Accumulations of SAND in the shape of mounds or RIDGES. They occur on some COASTS and in arid REGIONS. Coastal dunes are formed when the prevailing WINDS blow strongly on-shore, piling up sand into dunes, which may become stabilized when grasses grow on them. DESERT sand dunes are a product of DEFLATION, or wind EROSION removing fine materials to leave a DESERT PAVEMENT in one region and sand deposits in another. Sand dunes are classified by their shape into barchans, or crescent-shaped dunes; seifs or LONGITUDINAL DUNES; TRANSVERSE DUNES; star dunes; and sand drifts or sand sheets.

Sapping. Natural process of EROSION at the bases of HILL slopes or CLIFFS whereby support is removed by undercutting, thereby allowing overlying layers to collapse; SPRING SAPPING is the facilitation of this process by concentrated GROUNDWATER flow, generally at the heads of VALLEYS.

Saturation, zone of. Underground REGION below the zone of AERATION, where all pore space is filled with water. This water is called GROUNDWATER; the upper surface of the zone of saturation is the WATER TABLE.

Scale. Relationship between a distance on a MAP or diagram and the same distance on the earth. Scale can be represented in three ways. A linear, or graphic, scale uses a straight line, marked off in equally spaced intervals, to show how much of the map represents a mile or a kilometer. A representative fraction (RF) gives this scale as a ratio. A verbal scale uses words to explain the relationship between map size and actual size. For example, the RF 1:63,360 is the same as saying "one inch to the mile."

Scarp. Short version of the word "ESCARPMENT," a short steep slope, as at the edge of a PLATEAU. EARTHQUAKES lead to the formation of FAULT SCARPS.

Schist. METAMORPHIC ROCK that can be split easily into layers. Schist is commonly produced from the action of heat and pressure on SHALE or SLATE. The rock looks flaky in appearance. Mica-schists are shiny because of the development of visible mica. Other schists include talc-schist, which contains a large amount of talc,

and hornblende-schist, which develops from basaltic rocks.

Scree. Broken, loose ROCK material at the base of a slope or CLIFF. It is often the result of FROST WEDGING of BEDROCK cliffs, causing rockfall. Another name for scree is TALUS.

Sedimentary rocks. ROCKS formed from SEDIMENTS that are compressed and cemented together in a process called LITHIFICATION. Sedimentary rocks cover two-thirds of the earth's land surface but are only a small proportion of the earth's CRUST. SANDSTONE is a common sedimentary rock. Sedimentary rocks form STRATA, or layers, and sometimes contain FOSSILS.

Seif dunes. Long, narrow RIDGES of SAND, built up by WINDS blowing at different times of year from two different directions. Seif dunes occur in parallel lines of sand over large areas, running for hundreds of miles in the Sahara, Iran, and central Australia. Another name for seif dunes is LONGITUDINAL DUNES. The Arabic word means sword.

Seismic activity. Movements within the earth's CRUST that often cause various other geological phenomena to occur; the activity is measured by SEISMOGRAPHS.

Seismology. The scientific study of EARTHQUAKES. It is a branch of GEOPHYSICS. The study of SEISMIC WAVES has provided a great deal of knowledge about the composition of the earth's interior.

Shadow zone. When an EARTHQUAKE occurs at one LOCATION, its waves travel through the earth and are detected by SEISMOGRAPHS around the world. Every earthquake has a shadow zone, a band where neither P nor S WAVES from the earthquake will be detected. This shadow zone leads scientists to draw conclusions about the size, density, and composition of the earth's CORE.

Shale oil. SEDIMENTARY ROCK containing sufficient amounts of hydrocarbons that can be extracted by slow distillation to yield OIL.

Shallow-focus earthquakes. EARTHQUAKES having a focus less than 35 miles (60 km.) below the surface.

Shantytown. URBAN SQUATTER SETTLEMENT, usually housing poor newcomers.

Shield. Large part of the earth's CONTINENTAL CRUST, comprising very old ROCKS that have been eroded to REGIONS of low RELIEF. Each CONTINENT has a shield area. In North America, the Canadian Shield extends from north of the Great Lakes to the Arctic Ocean. Sometimes known as a CONTINENTAL SHIELD.

Shield volcano. VOLCANO created when the LAVA is quite viscous or fluid and highly basaltic. Such lava spreads out in a thin sheet of great radius but comparatively low height. As flows continue to build up the volcano, a low DOME shape is created. The greatest shield volcanoes on Earth are the ISLANDS of Hawaii, which rise to a height of almost 30,000 feet (10,000 meters) above SEA LEVEL.

Shock city. CITY that typifies disturbing changes in social and cultural conditions or in economic conditions. In the nineteenth century, the shock city of the United States was Chicago.

Sierra. Spanish word for a mountain RANGE with a serrated crest. In California, the Sierra Nevada is an important range, containing Mount Whitney, the highest PEAK in the continental United States.

Sill. Feature formed by INTRUSIVE volcanic activity. When LAVA is forced between two layers of ROCK, it can form a narrow horizontal layer of BASALT, parallel with the adjacent beds. Although it resembles a windowsill in its flatness, a sill may be hundreds of miles long and can range in thickness from a few centimeters to considerable thickness.

Siltation. Build-up of SILT and SAND in creeks and waterways as a result of SOIL EROSION, clogging water courses and creating DELTAS at RIVER MOUTHS. Siltation often results from DEFORESTATION or removal of tree cover. Such ENVIRONMENTAL DEGRADATION causes loss of agricultural productivity, worsening of water supply, and other problems.

Sima. Abbreviation for SILICA and *ma*gnesium. These are the two principal constituents of heavy ROCKS such as BASALT, which forms much of the OCEAN floor. Lighter, more abundant rock is SIAL.

Sinkhole. Circular depression in the ground surface, caused by WEATHERING of LIMESTONE, mainly through the effects of SOLUTION on JOINTS in the ROCK. If a STREAM flows above ground and then disappears down a sinkhole, the feature is called a swallow hole. In everyday language, many events that cause the surface to collapse are called sinkholes, even though they are rarely in limestone and rarely caused by weathering.

Sinking stream. STREAM or RIVER that loses part or all of its water to pathways dissolved underground in the BEDROCK.

Situation. Relationship between a PLACE, such as a TOWN or CITY, and its RELATIVE LOCATION within a REGION. A situation on the COAST is desirable in terms of overseas TRADE.

Slip-face. LEEWARD side of a SAND DUNE. As the WIND piles up sand on the WINDWARD side, it then slips down the rear or slip-face. The angle of the slip-face is gentler than the angle of the windward slope.

Slump. Type of LANDSLIDE in which the material moves downslope with a rotational motion, along a curved slip surface.

Snout. Terminal end of a GLACIER.

Snow line. The height or ELEVATION at which snow remains throughout the year, without melting away. Near the EQUATOR, the snow line is more than 15,000 feet (almost 5,000 meters); at higher LATITUDES, the snow line is correspondingly lower, reaching SEA LEVEL at the POLES. The actual snow line varies with the time of year, retreating in summer and coming lower in winter.

Soil horizon. SOIL consists of a series of layers called horizons. The uppermost layer, the O horizon, contains organic materials such as decayed leaves that have been changed into HUMUS. Beneath this is the A horizon, the TOPSOIL, where farmers plow and plant seeds. The B HORIZON often contains MINERALS that have been washed downwards from the A horizon, such as calcium, iron, and aluminum. The A and B horizons together comprise a solum, or true soil. The C horizon is weathered BEDROCK, which contains pieces of the original ROCK from which the soil formed. Another name for the C horizon is REGOLITH. Beneath this is the R horizon, or bedrock.

Soil moisture. Water contained in the unsaturated zone above the WATER TABLE.

Soil profile. Vertical section of a SOIL, extending through its horizon into the unweathered parent material.

Soil stabilization. Engineering measures designed to minimize the opportunity and/or ability of EXPANSIVE SOILS to shrink and swell.

Solar energy. One of the forms of ALTERNATIVE or RENEWABLE ENERGY. In the late 1990s, the world's largest solar power generating plant was located at Kramer Junction, California. There, solar energy heats huge OIL-filled containers with a parabolic shape, which produces steam to drive generating turbines. An alternative is the production of energy through photovoltaic cells, a TECHNOLOGY that was first developed for space exploration. Many individual homes, especially in isolated areas, use this technology.

Solar system. SUN and all the bodies that ORBIT it, including the PLANETS and their SATELLITES, plus numerous COMETS, ASTEROIDS, and METEOROIDS.

Solar wind. Gases from the SUN's ATMOSPHERE, expanding at high speeds as streams of charged particles.

Solifluction. Word meaning flowing SOIL. In some REGIONS of PERMAFROST, where the ground is permanently frozen, the uppermost layer thaws during the summer, creating a saturated layer of soil and REGOLITH above the hard layer of frozen ground. On slopes, the material can flow slowly downhill, creating a wavy appearance along the hillslope.

Solution. Form of CHEMICAL WEATHERING in which MINERALS in a ROCK are dissolved in water. Most substances are soluble, but the combination of water with CARBON DIOXIDE from the ATMOSPHERE means that RAINFALL is slightly acidic, so

that the chemical reaction is often a combination of solution and carbonation.

Sound. Long expanse of the SEA, close to the COAST, such as a large ESTUARY. It can also be the expanse of sea between the mainland and an ISLAND.

Source rock. ROCK unit or bed that contains sufficient organic carbon and has the proper thermal history to generate OIL or gas.

Spatial diffusion. Notion that things spread through space and over time. An understanding of geographic change depends on this concept. Spatial diffusion can occur in various ways. Geographers distinguish between expansion diffusion, relocation diffusion, and hierarchical diffusion.

Spheroidal weathering. Form of ROCK WEATHERING in which layers of rock break off parallel to the surface, producing a rounded shape. It results from a combination of physical and CHEMICAL WEATHERING. Spheroidal weathering is especially common in GRANITE, leading to the creation of TORS and similar rounded features. Onion-skin weathering is a term sometimes used, especially when this is seen on small rocks.

Spring tide. TIDE of maximum RANGE, occurring when lunar and solar tides reinforce each other, a few DAYS after the full and new MOONS.

Squall line. Line of vigorous THUNDERSTORMS created by a cold downdraft that spreads out ahead of a fast-moving COLD FRONT.

Stacks. Pieces of ROCK surrounded by SEA water, which were once part of the mainland. WAVE EROSION has caused them to be isolated. Also called sea stacks.

Stalactite. Long, tapering piece of calcium carbonate hanging from the roof of a LIMESTONE CAVE or cavern. Stalactites are formed as water containing the MINERAL in solution drips downward. The water evaporates, depositing the dissolved minerals.

Stalagmite. Column of calcium carbonate growing upward from the floor of a LIMESTONE CAVE or cavern.

Steppe. Huge REGION of GRASSLANDS in the midlatitudes of Eurasia, extending from central

Europe to northeast China. The region is not uniform in ELEVATION; most of it is rolling PLAINS, but some mountain RANGES also occur. These have not been a barrier to the migratory lifestyle of the herders who have occupied the steppe for many centuries. The Asian steppe is colder than the European steppe, because of greater elevation and greater continentality. The best-known rulers from the steppe were the Mongols, whose empire flourished in the thirteenth and fourteenth centuries. Geographers speak of a steppe CLIMATE, a semiarid climate where the EVAPORATION rate is double that of PRECIPITATION. South of the steppe are great DESERTS; to the north are midlatitude mixed FORESTS. In terms of climate and VEGETATION, the steppe is like the short-grass PRAIRIE vegetation west of the Mississippi River. Also called steppes.

Storm surge. General rise above normal water level, resulting from a HURRICANE or other severe coastal STORM.

Strait. Relatively narrow body of water, part of an OCEAN or SEA, separating two pieces of land. The world's busiest SEAWAY is the Johore Strait between the Malay Peninsula and the island of Sumatra.

Strata. Layers of SEDIMENT deposited at different times, and therefore of different composition and TEXTURE. When the sediments are laid down, strata are horizontal, but subsequent tectonic processes can lead to tilting, FOLDING, or faulting. Not all SEDIMENTARY ROCKS are stratified. Singular form of the word is stratum.

Stratified drift. Material deposited by glacial MELTWATERS; the water separates the material according to size, creating layers.

Stratigraphy. Study of sedimentary STRATA, which includes the concept of time, possible correlation of the ROCK units, and characteristics of the rocks themselves.

Stratovolcano. Type of VOLCANO in which the ERUPTIONS are of different types and produce different LAVAS. Sometimes an eruption ejects cinder and ASH; at other times, viscous lava flows down the sides. The materials flow, settle, and fall to produce a beautiful symmetrical LANDFORM with a broad circular base and concave slopes tapering upward to a small circular CRATER. Mount Rainier, Mount Saint Helens, and Mount Fuji are stratovolcanoes. Also known as a COMPOSITE CONE.

Streambed. Channel through which a STREAM flows. Dry streambeds are variously known as ARROYOS, DONGAS, WASHES, and WADIS.

Strike. Term used when earth scientists study tilted or inclined beds of SEDIMENTARY ROCK. The strike of the inclined bed is the direction of a horizontal line along a bedding plane. The strike is at right angles to the dip of the rocks.

Strike-slip fault. In a strike-slip fault, the surface on either side of the fault moves in a horizontal plane. There is no vertical displacement to form a FAULT SCARP, as there is with other types of faults. The San Andreas Fault is a strike-slip fault. Also called a transcurrent fault.

Subduction zone. CONVERGENT PLATE BOUNDARY where an oceanic PLATE is being thrust below another plate.

Sublimation. Process by which water changes directly from solid (ice) to vapor, or vapor to solid, without passing through a liquid stage.

Subsolar point. Point on the earth's surface where the SUN is directly overhead, making the Sun's rays perpendicular to the surface. The subsolar point receives maximum INSOLATION, compared with other PLACES, where the Sun's rays are oblique.

Sunspots. REGIONS of intense magnetic disturbances that appear as dark spots on the solar surface; they occur approximately every eleven years.

Supercontinent. Vast LANDMASS of the remote geologic past formed by the collision and amalgamation of crustal PLATES. Hypothesized supercontinents include PANGAEA, GONDWANALAND, and LAURASIA.

Supersaturation. State in which the AIR'S RELATIVE HUMIDITY exceeds 100 percent, the condition necessary for vapor to begin transformation to a liquid state.

Supratidal. Referring to the SHORE area marginal to shallow OCEANS that are just above high-tide level.

Swamp. WETLAND where trees grow in wet to water-logged conditions. Swamps are common close to the RIVER on FLOODPLAINS, as well as in some coastal areas.

Swidden. Area of land that has been cleared for SUBSISTENCE AGRICULTURE by a farmer using the technique of slash-and-burn. A variety of crops is planted, partly to reduce the risk of crop failure. Yields are low from a swidden because SOIL fertility is low and only human labor is used for CLEARING, planting, and harvesting. See also INTERTILLAGE.

Symbolic landscapes. LANDSCAPES centered on buildings or structures that are so visually emblematic that they represent an entire CITY.

Syncline. Downfold or TROUGH shape that is formed through compression of ROCKS. An upfold is an ANTICLINE.

Tableland. Large area of land with a mostly flat surface, surrounded by steeply sloping sides, or ESCARPMENTS. A small PLATEAU.

Taiga. Russian name for the vast BOREAL FORESTS that cover Siberia. The marshy ground supports a tree VEGETATION in which the trees are CONIFEROUS, comprising mostly pine, fir, and larch.

Talus. Broken and jagged pieces of ROCK, produced by WEATHERING of steep slopes, that fall to the base of the slope and accumulate as a talus cone. In high MOUNTAINS, a ROCK GLACIER may form in the talus. See also SCREE.

Tarn. Small circular LAKE, formed in a CIRQUE, which was previously occupied by a GLACIER.

Tectonism. The formation of MOUNTAINS because of the deformation of the CRUST of the earth on a large scale.

Temporary base level. STREAMS or RIVERS erode their beds down toward a BASE LEVEL—in most cases, SEA LEVEL. A section of hard ROCK may slow EROSION and act as a temporary, or local, base level. Erosion slows upstream of the temporary base level. A DAM is an artificially constructed temporary base level.

Tension. Type of stress that produces a stretching and thinning or pulling apart of the earth's CRUST. If the surface breaks, a NORMAL FAULT is created, with one side of the surface higher than the other.

Tephra. General term for volcanic materials that are ejected from a vent during an ERUPTION and transported through the AIR, including ASH (volcanic), BLOCKS (volcanic), cinders, LAPILLI, SCORIA, and PUMICE.

Terminal moraine. RIDGE of unsorted debris deposited by a GLACIER. When a glacier erodes it moves downslope, carrying ROCK debris and creating a ground MORAINE of material of various sizes, ranging from big angular blocks or boulders down to fine CLAY. At the terminus of the glacier, where the ice is melting, the ground moraine is deposited, building the ridge of unsorted debris called a terminal moraine.

Terrain. Physical features of a REGION, as in a description of rugged terrain. It should not be confused with TERRANE.

Terrane. Piece of CONTINENTAL CRUST that has broken off from one PLATE and subsequently been joined to a different plate. The terrane has quite different composition and structure from the adjacent continental materials. Alaska is composed mostly of terranes that have accreted, or joined, the North American plate.

Terrestrial planet. Any of the solid, rocky-surfaced bodies of the inner SOLAR SYSTEM, including the PLANETS Mercury, Venus, Earth, and Mars and Earth's SATELLITE, the MOON.

Terrigenous. Originating from the WEATHERING and EROSION of MOUNTAINS and other land formations.

Texture. One of the properties of SOILS. The three textures are SAND, SILT, and CLAY. Texture is measured by shaking the dried soil through a series of sieves with mesh of reducing diameters. A mixture of sand, silt, and clay gives a LOAM soil.

Thermal equator. Imaginary line connecting all PLACES on Earth with the highest mean daily TEMPERATURE. The thermal equator moves south of the EQUATOR in the SOUTHERN HEMISPHERE summer, especially over the CONTINENTS of South America, Africa, and Australia. In the northern summer, the thermal equator moves far into Asia, northern Africa, and North America.

Thermal pollution. Disruption of the ECOSYSTEM caused when hot water is discharged, usually as a thermal PLUME, into a relatively cooler body of water. The TEMPERATURE change affects the aquatic ecosystem, even if the water is chemically pure. Nuclear power-generating plants use large volumes of water in the process and are important sources of thermal pollution.

Thermocline. Depth interval at which the TEMPERATURE of OCEAN water changes abruptly, separating warm SURFACE WATER from cold, deep water.

Thermodynamics. Area of science that deals with the transformation of ENERGY and the laws that govern these changes; equilibrium thermodynamics is especially concerned with the reversible conversion of heat into other forms of energy.

Thermopause. Outer limit of the earth's ATMOSPHERE.

Thermosphere. Atmospheric zone beyond the MESOSPHERE in which TEMPERATURE rises rapidly with increasing distance from the earth's surface.

Thrust belt. Linear BELT of ROCKS that have been deformed by THRUST FAULTS.

Thrust fault. FAULT formed when extreme compression of the earth's CRUST pushes the surface into folds so closely spaced that they overturn and the ROCK then fractures along a fault.

Tidal force. Gravitational force whose strength and direction vary over a body and thus act to deform the body.

Tidal range. Difference in height between high TIDE and low tide at a given point.

Tidal wave. Incorrect name for a TSUNAMI.

Till. Mass of unsorted and unstratified SEDIMENTS deposited by a GLACIER. Boulders and smaller rounded ROCKS are mixed with CLAY-sized materials.

Timberline. Another term for tree line, the BOUNDARY of tree growth on MOUNTAIN slopes. Above the timberline, TEMPERATURES are too cold for tree growth.

Tombolo. Strip of SAND or other SEDIMENT that connects an ISLAND or SEA stack to the mainland. Mont-Saint-Michel is linked to the French mainland by a tombolo.

Topography. Description of the natural LANDSCAPE, including LANDFORMS, RIVERS and other waters, and VEGETATION cover.

Topological space. Space defined in terms of the connectivity between LOCATIONS in that space. The nature and frequency of the connections are measured, while distance between locations is not considered an important factor. An example of topological space is a transport network diagram, such as a bus route or a MAP of an underground rail system. Networks are most concerned with flows, and therefore with connectivity.

Toponyms. PLACE names. Sometimes, names of features and SETTLEMENTS reveal a good deal about the history of a REGION. For example, the many names starting with "San" or "Santa" in the Southwest of the United States recall the fact that Spain once controlled that area. The scientific study of place names is toponymics.

Tor. Rocky outcrop of blocks of ROCK, or corestones, exposed and rounded by WEATHERING. Tors frequently form in GRANITE, where three series of JOINTS often developed as the rock originally cooled when it was formed.

Transform faults. FAULTS that occur along DIVERGENT PLATE boundaries, or SEAFLOOR SPREADING zones. The faults run perpendicular to the spreading center, sometimes for hundreds of miles, some for more than five hundred miles. The motion along a transform fault is lateral or STRIKE-SLIP.

Transgression. Flooding of a large land area by the SEA, either by a regional downwarping of continental surface or by a global rise in SEA LEVEL.

Transmigration. Policy of the government of Indonesia to encourage people to move from the densely overcrowded ISLAND of Java to the sparsely populated other islands.

Transverse bar. Flat-topped body of SAND or gravel oriented transverse to the RIVER flow.

Trophic level. Different types of food relations that are found within an ECOSYSTEM. Organisms that derive food and ENERGY through PHOTOSYNTHESIS are called autotrophs (self-feeders) or producers. Organisms that rely on producers as their source of energy are called heterotrophs (feeders on others) or consumers. A third trophic level is represented by the organisms known as decomposers, which recycle organic waste.

Tropical cyclone. STORM that forms over tropical OCEANS and is characterized by extreme amounts of rain, a central area of calm AIR, and spinning WINDS that attain speeds of up to 180 miles (300 km.) per hour.

Tropical depression. STORM with WIND speeds up to 38 miles (64 km.) per hour.

Tropopause. BOUNDARY layer between the TROPOSPHERE and the STRATOSPHERE.

Troposphere. Lowest and densest of Earth's atmospheric layers, marked by considerable TURBULENCE and a decrease in TEMPERATURE with increasing ALTITUDE.

Tsunami. SEISMIC SEA WAVE caused by a disturbance of the OCEAN floor, usually an EARTHQUAKE, although undersea LANDSLIDES or volcanic ERUPTIONS can also trigger tsunami.

Tufa. LIMESTONE or calcium carbonate deposit formed by PRECIPITATION from an alkaline LAKE. Mono Lake is famous for the dramatic tufa towers exposed by the lowering of the level of lake water. Also known as TRAVERTINE.

Tumescence. Local swelling of the ground that commonly occurs when MAGMA rises toward the surface.

Tunnel vent. Central tube in a volcanic structure through which material from the earth's interior travels.

U-shaped valley. Steep-sided VALLEY carved out by a GLACIER. Also called a glacial TROUGH.

Ubac slope. Shady side of a MOUNTAIN, where local or microclimatic conditions permit lower TIMBERLINES and lower SNOW LINES than occur on a sunny side.

Ultimate base level. Level to which a STREAM can erode its bed. For most RIVERS, this is SEA LEVEL. For streams that flow into a LAKE, the ultimate base level is the level of the lakebed.

Unconfined aquifer. AQUIFER whose upper BOUNDARY is the WATER TABLE; it is also called a water table aquifer.

Underfit stream. STREAM that appears to be too small to have eroded the VALLEY in which it flows. A RIVER flowing in a glaciated valley is a good example of underfit.

Uniformitarianism. Theory introduced in the early nineteenth century to explain geologic processes. It used to be believed that the earth was only a few thousand years old, so the creation of LANDFORMS would have been rapid, even catastrophic. This theory, called CATASTROPHISM, explained most landforms as the result of the Great Flood of the Bible, when Noah, his family, and animals survived the deluge. Uniformitarian- ism, in contrast, stated that the processes in operation today are slow, so the earth must be immensely older than a mere few thousand years.

Universal time (UT). See GREENWICH MEAN TIME.

Universal Transverse Mercator. Projection in which the earth is divided into sixty zones, each SIX DEGREES of LONGITUDE wide. In a traditional Mercator projection, the earth is seen as a sphere with a cylinder wrapped around the EQUATOR. UTM can be visualized as a series of six-degree side strips running transverse, or north-south.

Unstable air. Condition that occurs when the AIR above rising air is unusually cool so that the rising air is warmer and accelerates upward.

Upthrown block. When EARTHQUAKE motion produces a FAULT, the block of land on one side is displaced vertically relative to the other. The higher is the upthrown block; the lower is the downthrown block.

Upwelling. OCEAN phenomenon in which warm SURFACE WATERS are pushed away from the

COAST and are replaced by cold waters that carry more nutrients up from depth.

Urban heat island. Cities experience a different MICROCLIMATE from surrounding REGIONS. The CITY TEMPERATURE is typically higher by a few DEGREES, both DAY and night, because of factors such as surfaces with higher heat absorption, decreased WIND strength, human heat-producing activities such as power generation, and the layer of AIR POLLUTION (DUST DOME).

Vadose zone. The part of the SOIL also known as the zone of AERATION, located above the WATER TABLE, where space between particles contains AIR.

Valley train. Fan-shaped deposit of glacial MORAINE that has been moved down-valley and redeposited by MELTWATER from the GLACIER.

Van Allen radiation belts. Bands of highly energetic, charged particles trapped in Earth's MAGNETIC FIELD. The particles that make up the inner BELT are energetic protons, while the outer belt consists mainly of electrons and is subject to DAY-night variations.

Varnish, desert. Shiny black coating often found over the surface of ROCKS in arid REGIONS. This is a form of OXIDATION or CHEMICAL WEATHERING, in which a coating of manganese oxides has formed over the exposed surface of the rock.

Varve. Pair of contrasting layers of SEDIMENT deposited over one year's time; the summer layer is light, and the winter layer is dark.

Ventifacts. PEBBLES on which one or more sides have been smoothed and faceted by ABRASION as the WIND has blown SAND particles.

Volcanic island arc. Curving or linear group of volcanic ISLANDS associated with a SUBDUCTION ZONE.

Volcanic rock. Type of IGNEOUS ROCK that is erupted at the surface of the earth; volcanic rocks are usually composed of larger crystals inside a fine-grained matrix of very small crystals and glass.

Volcanic tremor. Continuous vibration of long duration, detected only at active VOLCANOES.

Volcanology. Scientific study of VOLCANOES.

Voluntary migration. Movement of people who decide freely to move their place of permanent residence. It results from PULL FACTORS at the chosen destination, together with PUSH FACTORS in the home situation.

Warm temperate glacier. GLACIER that is at the melting TEMPERATURE throughout.

Water power. Generally means the generation of electricity using the ENERGY of falling water. Usually a DAM is constructed on a RIVER to provide the necessary height difference. The potential energy of the falling water is converted by a water turbine into mechanical energy. This is used to power a generator, which produces electricity. Also called HYDROELECTRIC POWER. Another form of water power is tidal power, which uses the force of the incoming and outgoing TIDE as its source of energy.

Water table. The depth below the surface where the zone of AERATION meets the zone of SATURATION. Above the water table, there may be some SOIL MOISTURE, but most of the pore space is filled with air. Below the water table, pore space of the ROCKS is occupied by water that has percolated down through the overlying earth material. This water is called GROUNDWATER. In practice, the water table is rarely as flat as a table, but curved, being far below the surface in some PLACES and even intersecting the surface in others. When GROUNDWATER emerges at the surface, because it intersects the water table, this is called a SPRING. The depth of the water table varies from SEASON to season, and with pumping of water from an AQUIFER.

Watershed. The whole surface area of land from which RAINFALL flows downslope into a STREAM. The watershed comprises the STREAMBED or CHANNEL, together with the VALLEY sides, extending up to the crest or INTERFLUVE, which separates that watershed from its neighbor. Each watershed is separated from the next by the drainage DIVIDE. Also called a DRAINAGE BASIN.

Waterspout. TORNADO that forms over water, or a tornado formed over land which then moves over water. The typical FUNNEL CLOUD, which reaches down from a CUMULONIMBUS CLOUD, is a narrow rotating STORM, with WIND speeds reaching hundreds of miles per hour.

Wave crest. Top of a WAVE.

Wave-cut platform. As SEA CLIFFS are eroded and worn back by WAVE attack, a wave-cut platform is created at the base of the cliffs. ABRASION by ROCK debris from the cliffs scours the platform further, as waves wash to and fro and TIDES ebb and flow. The upper part of the wave-cut platform is exposed at high tide. These areas contain rockpools, which are rich in interesting MARINE life-forms. Offshore beyond the platform, a wave-built TERRACE is formed by DEPOSITION.

Wave height. Vertical distance between one WAVE CREST and the adjacent WAVE TROUGH.

Wave length. Distance between two successive WAVE CRESTS or two successive WAVE TROUGHS.

Wave trough. The low part of a WAVE, between two WAVE CRESTS.

Weather analogue. Approach to WEATHER FORECASTING that uses the WEATHER behavior of the past to predict what a current weather pattern will do in the future.

Weather forecasting. Attempt to predict WEATHER patterns by analysis of current and past data.

Wilson cycle. Creation and destruction of an OCEAN BASIN through the process of SEAFLOOR SPREADING and SUBDUCTION of existing ocean basins.

Wind gap. Abandoned WATER GAP. The Appalachian Mountains contain both wind gaps and water gaps.

Windbreak. Barrier constructed at right angles to the prevailing WIND direction to prevent damage to crops or to shelter buildings. Generally, a row of trees or shrubs is planted to form a windbreak. The feature is also called a shelter belt.

Windchill. MEASUREMENT of apparent TEMPERATURE that quantifies the effects of ambient WIND and temperature on the rate of cooling of the human body.

World Aeronautical Chart. International project undertaken to map the entire world, begun during World War II.

World city. CITY in which an extremely large part of the world's economic, political, and cultural activity occurs. In the year 2018, the top ten world cities were London, New York City, Tokyo, Paris, Singapore, Amsterdam, Seoul, Berlin, Hong Kong, and Sydney.

Xenolith. Smaller piece of ROCK that has become embedded in an IGNEOUS ROCK during its formation. It is a piece of older rock that was incorporated into the fluid MAGMA.

Xeric. Description of SOILS in REGIONS with a MEDITERRANEAN CLIMATE, with moist cool winters and long, warm, dry summers. Since summer is the time when most plants grow, the lack of SOIL MOISTURE is a limiting factor on plant growth in a xeric ENVIRONMENT.

Xerophytic plants. Plants adapted to arid conditions with low PRECIPITATION. Adaptations include storage of moisture in tissue, as with cactus plants; long taproots reaching down to the WATER TABLE, as with DESERT shrubs; or tiny leaves that restrict TRANSPIRATION.

Yardangs. Small LANDFORMS produced by WIND EROSION. They are a series of sharp RIDGES, aligned in the direction of the wind.

Yazoo stream. TRIBUTARY that flows parallel to the main STREAM across the FLOODPLAIN for a considerable distance before joining that stream. This occurs because the main stream has built up NATURAL LEVEES through flooding, and because RELIEF is low on the floodplain. The yazoo stream flows in a low-lying wet area called backswamps. Named after the Yazoo River, a tributary of the Mississippi.

Zero population growth. Phenomenon that occurs when the number of deaths plus EMIGRATION is matched by the number of births plus IMMIGRATION. Some European countries have reached zero population growth.

BIBLIOGRAPHY

THE NATURE OF GEOGRAPHY

Adams, Simon, Anita Ganeri, and Ann Kay. *Geography of the World*. London: DK, 2010. Print.

Harley, J. B., and David Woodward, eds. *The History of Cartography: Cartography in the Traditional Islamic and South Asian Societies*. Vol. 2, book 1. Chicago: University of Chicago Press, 1992. Offers a critical look at maps, mapping, and mapmakers in the Islamic world and South Asia.

———, eds. *The History of Cartography: Cartography in the Traditional East and Southeast Asian Societies*. Vol. 2, book 2. Chicago: University of Chicago Press, 1994. Similar in thrust and breadth to volume 2, book 1.

Marshall, Tim, and John Scarlett. *Prisoners of Geography: Ten Maps That Tell You Everything You Need to Know About Global Politics*. London : Elliott and Thompson Limited, 2016. Print

Nijman, Jan. *Geography: Realms, Regions, and Concepts*. Hoboken, NJ : Wiley, 2020. Print.

Snow, Peter, Simon Mumford, and Peter Frances. *History of the World Map by Map*. New York: DK Smithsonian, 2018.

Woodward, David, et al., eds. *The History of Cartography: Cartography in the Traditional African, American, Arctic, Australian, and Pacific Societies*. Vol. 2, book 3. Chicago: University of Chicago Press, 1998. Investigates the roles that maps have played in the wayfinding, politics, and religions of diverse societies such as those in the Andes, the Trobriand Islanders of Papua-New Guinea, the Luba of central Africa, and the Mixtecs of Central America.

PHYSICAL GEOGRAPHY

Christopherson, Robert W, and Ginger H. Birkeland. *Elemental Geosystems*. Hoboken, NJ : Wiley, 2016. Print.

Lutgens, Frederick K., and Edward J. Tarbuck. *Foundations of Earth Science*. Upper Saddle River, N.J.: Prentice-Hall, 2017. Undergraduate text for an introductory course in earth science, consisting of seven units covering basic principles in geology, oceanography, meteorology, and astronomy, for those with little background in science.

McKnight, Tom. *Physical Geography: A Landscape Appreciation*. 12th ed. New York: Prentice Hall, 2017. Now-classic college textbook that has become popular because of its illustrations, clarity, and wit. Comes with a CD-ROM that takes readers on virtual-reality field trips.

Robinson, Andrew. *Earth Shock: Climate Complexity and the Force of Nature*. New York: W. W. Norton, 1993. Describes, illustrates, and analyzes the forces of nature responsible for earthquakes, volcanoes, hurricanes, floods, glaciers, deserts, and drought. Also recounts how humans have perceived their relationship with these phenomena throughout history.

Weigel, Marlene. *UxL Encyclopedia of Biomes*. Farmington Hills, Mich.: Gale Group, 1999. This three-volume set should meet the needs of seventh grade classes for research. Covers all biomes such as the forest, grasslands, and desert. Each biome includes sections on development of that particular biome, type, and climate, geography, and plant and animal life.

Woodward, Susan L. *Biomes of Earth*. Westport, CT: Greenwood Press, 2003. Print.

HUMAN GEOGRAPHY

Blum, Richard C, and Thomas C. Hayes. *An Accident of Geography: Compassion, Innovation, and the Fight against Poverty*. Austin, TX: Greenleaf Book Group, 2016. Print.

Dartnell, Lewis. *Origins: How the Earth Made Us*. New York: Hachette Book Group, 2019. Print.

Glantz, Michael H. *Currents of Change: El Niño's Impact on Climate and Society*. New York: Cambridge University Press, 1996. Aids readers in understanding the complexities of the earth's weather pattern, how it relates to El Niño, and the impact upon people around the globe.

Morland, Paul. *The Human Tide: How Population Shaped the Modern World*. New York: PublicAffairs, 2019. Print.

Novaresio, Paolo. *The Explorers: From the Ancient World to the Present*. New York: Stewart, Tabori and Chang, 1996. Describes amazing journeys and exhilarating discoveries from the earliest days of seafaring to the first landing on the moon and beyond.

Rosin, Christopher J, Paul Stock, and Hugh Campbell. *Food Systems Failure: The Global Food Crisis and the Future of Agriculture*. New York: Routledge, 2014. Print.

ECONOMIC GEOGRAPHY

Diamond, Jared M. *Guns, Germs, and Steel: The Fates of Human Societies*. New York: Norton, 2011. Print.

Esping-Andersen, Gosta. *Social Foundations of Postindustrial Economies*. New York: Cambridge University Press, 1999. Examines such topics as social risks and welfare states, the structural bases of postindustrial employment, and recasting welfare regimes for a postindustrial era.

Michaelides, Efstathios E. S.*Alternative Energy Sources*. Berlin: Springer Berlin, 2014. Print. This book offers a clear view of the role each form of alternative energy may play in supplying energy needs in the near future. It details the most common renewable energy sources as well as examines nuclear energy by fission and fusion energy.

Robertson, Noel, and Kenneth Blaxter. *From Dearth to Plenty: The Modern Revolution in Food Production*. New York: Cambridge University Press, 1995. Tells a story of scientific discovery and its exploitation for technological advance in agriculture. It encapsulates the history of an important period, 1936-86, when government policy sought to aid the competitiveness of the agricultural industry through fiscal measures and by encouraging scientific and technical innovation.

REGIONAL GEOGRAPHY

Biger, Gideon, ed. *The Encyclopedia of International Boundaries*. New York: Facts on File, 1995. Entries for approximately 200 countries are arranged alphabetically, each beginning with introductory information describing demographics, political structure, and political and cultural history. The boundaries of each state are then described with details of the geographical setting, historical background, and present political situation, including unresolved claims and disputes.

Leinen, Jo, Andreas Bummel, and Ray Cunningham. *A World Parliament: Governance and Democracy in the 21st Century*. Berlin Democracy Without Borders, 2018. Print.

Pitts, Jennifer. *Boundaries of the International: Law and Empire*. Cambridge, Mass: Harvard University Press, 2018. Print.

SOUTH AMERICA AND CENTRAL AMERICA

PHYSICAL GEOGRAPHY

Blouet, Brian W, and Olwyn M. Blouet. *Latin America and the Caribbean: a Systematic and Regional Survey*, 7th Ed. John Wiley & Sons, 2015. Print.

Georges, D. V. *South America*. Danbury, Conn.: Children's Press, 1986. Discusses characteristics of various sections of South America such as the Andes, the Amazon rain forest, and the pampas.

Matthews, Down, and Kevin Schaefer. *Beneath the Canopy: Wildlife of the Latin American Rain Forest*. San Francisco: Chronicle Books, 2007. Kevin Schaefer's photographs offer a rare, up-close look at the beautiful and elusive creatures that make their home in this natural paradise—from its leafy shadows to the forest canopy. Captions and text by nature writer Matthews give further insight into the lives of these amazing animals.

HUMAN GEOGRAPHY

Early, Edwin, et al., eds. *The History Atlas of South America*. Foster City, Calif.: IDG Books Worldwide, 1998. Describes South America's history, which is a rich tapestry of complex ancient civilizations, colonial clashes, and modern growth, economic challenges, and cultural vibrancy.

Kelly, Philip. *Checkerboards and Shatterbelts: The Geopolitics of South America*. Austin: University of Texas Press, 1997.

Uses the geographical concepts of "checkerboards" and "shatterbelts" to characterize much of South America's geopolitics and to explain why the continent has never been unified or dominated by a single nation.

Levine, Robert M., and John J. Crocitti, eds. *The Brazil Reader: History, Culture, Politics*. Durham, N.C.: Duke University Press, 2019. Selections range from early colonization to the present day, with sections on imperial and republican Brazil, the days of slavery, the Vargas years, and the more recent return to democracy.

Levinson, David, ed. *The Encyclopedia of World Culture, Vol. 7: South America*. Indianapolis, Ind.: Macmillan, 1994. Addresses the diverse cultures of South America south of Panama, with an emphasis on the American Indian cultures, although the African-American culture and the European and Asian immigrant cultures are also covered. Linguistics, historical and cultural relations, economy, kinship, marriage, sociopolitical organizations, and religious beliefs are among the topics discussed for each culture.

Webster, Donovan. "Orinoco River" *National Geographic* 193, no. 4 (April, 1998): 2-31. Examination of the Orinoco River Basin in the Amazonian portion of Venezuela. The article focuses on the fauna, flora, and the Yanomani, Yekwana, and Piaroa tribes and "tropical cowboys."

ECONOMIC GEOGRAPHY

Biondi-Morra, Brizio. *Hungry Dreams: The Failure of Food Policy in Revolutionary Nicaragua*. Ithaca, N.Y.: Cornell University Press, 1993. Examines how food policy was formulated in Nicaragua and the effects on foreign exchange, food prices, and the relationship to wages and credit.

Cupples, Julie, Marcela Palomino-Schalscha, and Manuel Prieto. *The Routledge Handbook of Latin American Development*. New York: Routledge, 2019. Print.

Folch, Christine. *Hydropolitics: The Itaipu Dam, Sovereignty, and the Engineering of Modern South America. Princeton, NJ: Princeton University Press, 2019. Print.*

Wilken, Gene C. *Good Farmers: Traditional Agriculture Resource Management in Mexico and Central America*. Berkeley, Calif.: University of California Press, 1987. Focusing on the farming practices of Mexico and Central America, this book examines in detail the effectiveness of sophisticated traditional methods of soil, water, climate, slope, and space management that rely primarily on human and animal power.

REGIONAL GEOGRAPHY

Frank, Zephyr L, Frederico Freitas, and Jacob Blanc. *Big Water: The Making of the Borderlands between Brazil, Argentina, and Paraguay*. Tucson: The University of Arizona Press, 2018. Internet resource.

Gheerbrandt, Alain. *The Amazon: Past, Present and Future*. New York: Harry N. Abrams, 1992. Presents the past, present, and uncertain future of the Amazon rain forest and its inhabitants. It includes spectacular illustrations and a section of historical documents.

Kent, Robert B. *Latin America: Regions and People*. New York: The Guilford Press, 2016. Print.

McClain, Michael E, Jeffrey E. Richey, and Reynaldo L. Victoria. *The Biogeochemistry of the Amzon Basin*. Oxford: Oxford University Press, 2001. Print.

Pulsipher, Lydia Mihelic. *World Regional Geography*. W H Freeman, 2019. Chacterizing global issues through the daily lives of individuals helps readers to better grasp circumstances affecting regions around the world such as environment; gender and population; urbanization; globalization and development; and power and politics.

INDEX

I

J

N

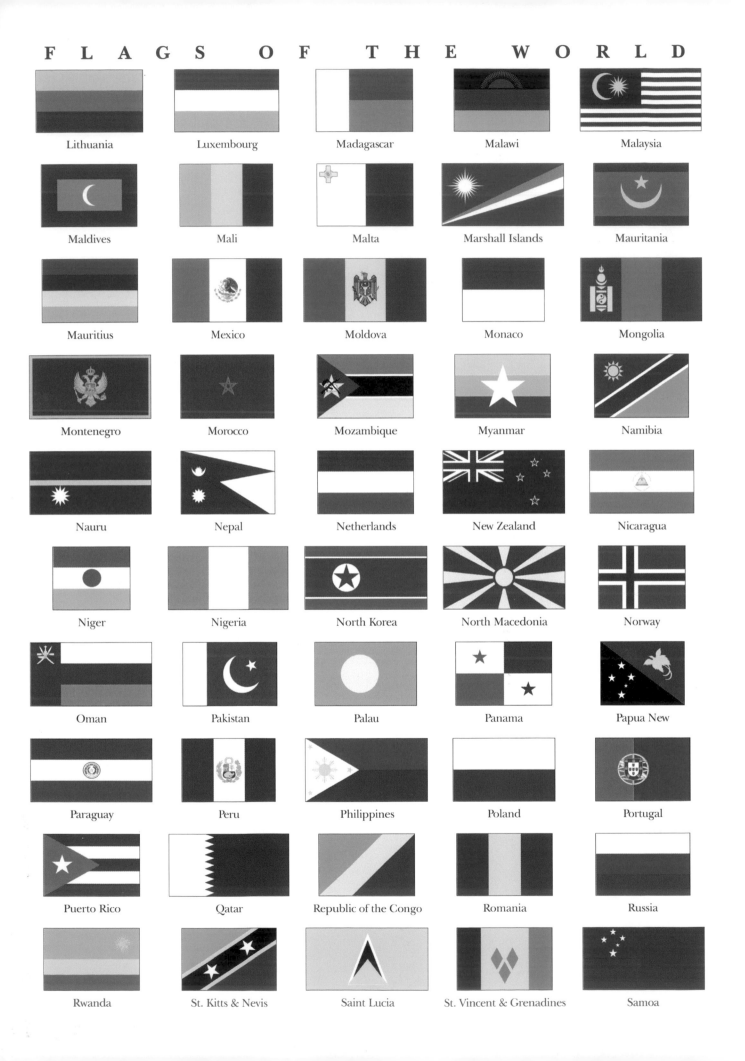

FLAGS OF THE WORLD

Lithuania	Luxembourg	Madagascar	Malawi	Malaysia
Maldives	Mali	Malta	Marshall Islands	Mauritania
Mauritius	Mexico	Moldova	Monaco	Mongolia
Montenegro	Morocco	Mozambique	Myanmar	Namibia
Nauru	Nepal	Netherlands	New Zealand	Nicaragua
Niger	Nigeria	North Korea	North Macedonia	Norway
Oman	Pakistan	Palau	Panama	Papua New
Paraguay	Peru	Philippines	Poland	Portugal
Puerto Rico	Qatar	Republic of the Congo	Romania	Russia
Rwanda	St. Kitts & Nevis	Saint Lucia	St. Vincent & Grenadines	Samoa